# HOLT
# PRE-ALGEBRA

Eugene D. Nichols
Bonnie H. Litwiller
Paul A. Kennedy

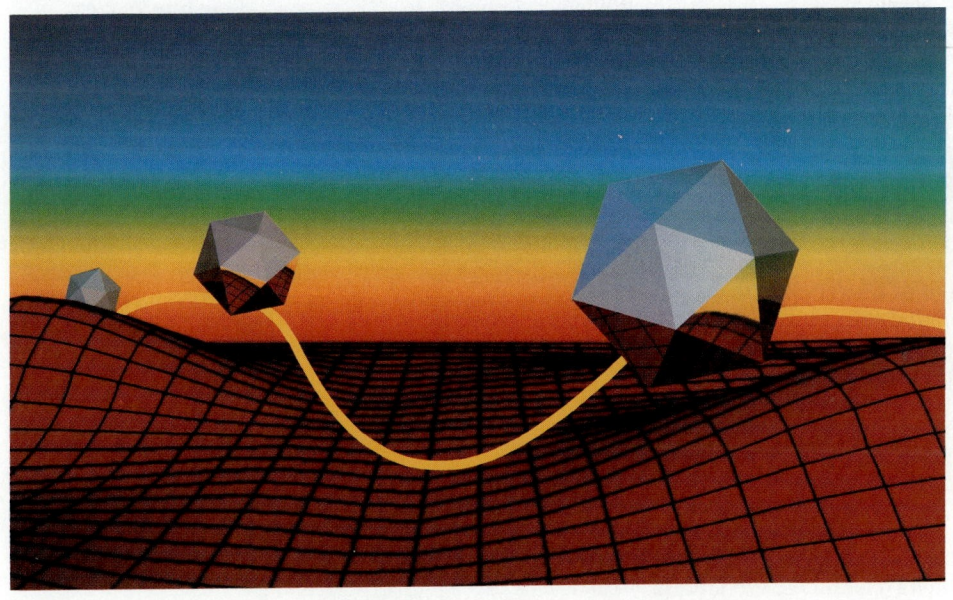

Austin • Orlando • San Diego • Chicago • Dallas • Toronto

## AUTHORS

**Eugene D. Nichols**
Distinguished Professor of
Mathematics Education
Florida State University
Tallahassee, Florida

**Bonnie H. Litwiller**
Professor of Mathematics
University of Northern Iowa
Cedar Falls, Iowa

**Paul A. Kennedy**
Professor of Mathematics
Southwest Texas State
University
San Marcos, Texas

## REVIEWERS

**Edna F. Bazik**
Mathematics Coordinator
Hinsdale Community Consolidated
School District 181
Hinsdale, Illinois

**Carmen Castillo**
Mathematics Teacher
United High School
Laredo, Texas

**Marilane G. Dusterhoff**
Mathematics Supervisor
Round Rock ISD
Round Rock, Texas

**William J. Geppert, Jr.**
Mathematics Supervisor
Delaware State Board of
Education
Dover, Delaware

**Carmen Saucedo**
Mathematics Teacher
Lamar High School
Houston, Texas

**Lee Stiff**
Associate Professor
Mathematics and Science
Education Department
North Carolina University
Raleigh, North Carolina

**Christine T. West**
Mathematics Supervisor
Charlestown County
School District
Charlestown, South Carolina

**Jim Zwick**
Chairman, Department of
Mathematics
Guion Creek Middle School
MSD of Pike County
Township Schools
Indianapolis, Indiana

The contributions of **Brother Neal Golden, S.C.,** who wrote the *Computer Explorations*, are gratefully acknowledged.

Copyright © 1992 by Holt, Rinehart and Winston, Inc.

All rights reserved. No part of this publication may be reproduced or transmitted in any form or by any means, electronic or mechanical, including photocopy, recording, or any information storage and retrieval system, without permission in writing from the publisher.

Requests for permission to make copies of any part of the work should be mailed to: Permissions Department, Holt, Rinehart and Winston, Inc., 8th Floor, Orlando, Florida 32887.

Printed in the United States of America

4 5 6 7     040     9 8 7 6 5                    ISBN 0-03-047068-4

# TO THE STUDENT

Dear Student,

Welcome to a new school year and to a new math course—*Pre-Algebra*. Beginnings are exciting, and fun, and perhaps a bit scary too. Here are a few ideas to help you learn and enjoy this new adventure.

Get **involved!** Think about math. Try solving problems on your own. Don't be afraid to make mistakes and try again. Math is not a spectator sport.

**Read** and **ask** questions. Read all the information in the textbook lessons. Ask yourself questions such as these: "How does this work? Do I already know something about this? How is this related to what I already know? How can I use what I know to solve this problem? Is there a pattern here?

**Start a math journal** after your first math class of the year. Write down your ideas, thoughts, and plans about how to solve math problems. At the end of the year, compare your notes from each unit. You will be surprised and pleased at how much your ability to use the power of math has increased.

**Enjoy** math! Look at the pictures in your textbook and think about how they are related to math. Learn to use a calculator and a computer to explore ideas and to do work for you.

Good luck! We know that you will have a successful year.

Your friends at Holt, Rinehart and Winston
Publishers of Pre-Algebra

# UNIT 1 INTRODUCTION TO ALGEBRA    xii

## CHAPTER 1  INTEGERS

- **1-1** The Integers .................................... 2
- **1-2** Comparing and Ordering Integers .................... 7
- **1-3** Problem Solving Exploration: Addition Model ........ 11
- **1-4** Adding Integers .................................... 13
- **1-5** Problem Solving Exploration: Subtraction Model ..... 21
- **1-6** Subtracting Integers ............................... 23
- **1-7** Problem Solving Strategies: Methods of Computation . 26
- **1-8** Multiplying Integers ............................... 28
- **1-9** Dividing Integers .................................. 32

**Special Features**
Computer Exploration ............. 19
Mathematical Footnote ............ 20
Math Team Problems .......... 10, 31

**Review and Testing**
Midchapter Review ................ 20
Chapter Summary ................. 36
Chapter Review ................... 37
Chapter Test ..................... 38

## CHAPTER 2  VARIABLES AND EXPRESSIONS

- **2-1** Evaluating Expressions ............................ 39
- **2-2** Problem Solving Strategies: Using a Step-by-Step Process ........................................ 44
- **2-3** Order of Operations ............................... 46
- **2-4** Problem Solving Exploration: Making Generalizations . 51
- **2-5** Basic Properties and Mental Computation .......... 54
- **2-6** Using the Distributive Property .................... 59
- **2-7** Formulas: Perimeter, Area, Average ............... 64
- **2-8** Area Formulas: Parallelograms, Triangles, Trapezoids . 69

**Special Features**
Computer Exploration ............. 75
Mathematical Footnotes ....... 50, 53
Math Team Problems ...... 43, 63, 74

**Review and Testing**
Midchapter Review ................ 53
Chapter Summary ................. 76
Chapter Review ................... 77
Chapter Test ..................... 78
Cumulative Review ............... 79

| CHAPTER 3 | EQUATIONS AND PROBLEM SOLVING |

| 3-1 | Equations and Inequalities | 80 |
| 3-2 | Solving Equations | 85 |
| 3-3 | Problem Solving Strategies: Choosing Strategies | 89 |
| 3-4 | Solving Addition and Subtraction Equations | 91 |
| 3-5 | Solving Multiplication and Division Equations | 95 |
| 3-6 | Inverse Operations | 101 |
| 3-7 | Solving Multi-Step Equations | 105 |
| 3-8 | Translating Word Expressions | 110 |
| 3-9 | Problem Solving: Writing an Equation | 114 |

**Special Features**
Mathematical Footnotes ...... **88, 100**
Math Team Problems .... **99, 104, 109**

**Review and Testing**
Midchapter Review ............... **100**
Chapter Summary ............... **119**
Chapter Review ................. **120**
Chapter Test .................... **121**
Cumulative Review ............. **122**

# UNIT 2  RATIONAL NUMBERS                    124

| CHAPTER 4 | NUMBER THEORY |

| 4-1 | Factors and Multiples | 126 |
| 4-2 | Tests for Divisibility | 131 |
| 4-3 | Problem Solving Exploration: Factors and Patterns | 137 |
| 4-4 | Prime Numbers | 139 |
| 4-5 | Prime Factorization | 143 |
| 4-6 | Problem Solving Strategies: Simpler Problems | 148 |
| 4-7 | Exponents | 150 |
| 4-8 | LCM and GCF | 154 |

**Special Features**
Computer Exploration ............ **147**
Mathematical Footnotes ...... **142, 153**
Math Team Problems ........ **136, 158**

**Review and Testing**
Midchapter Review ............... **142**
Chapter Summary ............... **159**
Chapter Review ................. **159**
Chapter Test .................... **161**
Cumulative Review ............. **162**

### CHAPTER 5 — FRACTIONS

| | | |
|---|---|---|
| 5-1 | Equivalent Fractions | 163 |
| 5-2 | Problem Solving Exploration: Multiplication | 169 |
| 5-3 | Multiplying Fractions | 171 |
| 5-4 | Multiplying Mixed Numbers | 175 |
| 5-5 | Problem Solving Strategies: Organizing Information | 180 |
| 5-6 | Using Reciprocals to Solve Equations | 183 |
| 5-7 | Dividing Fractions and Mixed Numbers | 189 |
| 5-8 | Fractions with Like Denominators | 194 |
| 5-9 | Fractions with Unlike Denominators | 200 |
| 5-10 | Subtracting Mixed Numbers | 206 |

**Special Features**
Computer Exploration............ **168**
Mathematical Footnotes .... **170, 182**
Math Team Problem............ **193**

**Review and Testing**
Midchapter Review............. **182**
Chapter Summary.............. **210**
Chapter Review................ **211**
Chapter Test .................. **212**
Cumulative Review ............ **213**

## UNIT 3  USING RATIOS — 214

### CHAPTER 6 — PROBABILITY

| | | |
|---|---|---|
| 6-1 | Ratio and Proportion | 216 |
| 6-2 | Problem Solving: Using Proportions | 221 |
| 6-3 | Ratio and Measurement | 226 |
| 6-4 | Problem Solving Strategies: Collecting Data | 231 |
| 6-5 | Problem Solving Exploration: Recording Chances | 234 |
| 6-6 | Probability | 236 |
| 6-7 | Independent and Dependent Events | 241 |
| 6-8 | Making Choices | 247 |
| 6-9 | Permutations | 252 |

**Special Features**
Mathematical Footnotes ...................... **233, 235**
Math Team Problems ......................... **225, 246**

**Review and Testing**
Midchapter Review .... **233**   Chapter Test ......... **260**
Chapter Summary ..... **258**   Cumulative Review .... **261**
Chapter Review ....... **258**

### CHAPTER 7   DECIMALS

| | | |
|---|---|---|
| 7-1 | Rational Expressions | 262 |
| 7-2 | Rational Numbers | 267 |
| 7-3 | Decimals and Fractions | 271 |
| 7-4 | Problem Solving Strategies: Drawing a Diagram | 275 |
| 7-5 | Repeating Decimals | 277 |
| 7-6 | Estimating Sums and Differences | 282 |
| 7-7 | Problem Solving Exploration: Powers of Ten | 287 |
| 7-8 | Scientific Notation | 289 |
| 7-9 | Problem Solving Exploration: The Metric System | 295 |
| 7-10 | Estimating Products and Quotients of Decimals | 297 |

**Special Features**
Mathematical Footnotes .... 281, 294
Math Team Problems .. 266, 288, 300

**Review and Testing**
Midchapter Review .............. 281
Chapter Summary ............... 301
Chapter Review ................. 302
Chapter Test ................... 303
Cumulative Review .............. 304

### CHAPTER 8   PERCENT

| | | |
|---|---|---|
| 8-1 | The Meaning of Percent | 305 |
| 8-2 | Decimals and Percents | 309 |
| 8-3 | Estimating the Percent of a Number | 313 |
| 8-4 | Problem Solving Strategies: Deciding on Estimates | 317 |
| 8-5 | Finding the Percent of a Number | 319 |
| 8-6 | Interest | 324 |
| 8-7 | Discount | 330 |
| 8-8 | Solving Percent Equations and Proportions | 334 |
| 8-9 | Percent Increase and Decrease | 338 |

**Special Features**
Computer Exploration ............ 329
Mathematical Footnote ........... 323

**Review and Testing**
Midchapter Review .............. 323
Chapter Summary ............... 342
Chapter Review ................. 343
Chapter Test ................... 344
Cumulative Review .............. 345

# UNIT 4  USING GRAPHS  346

## CHAPTER 9  ANALYZING DATA

| | | |
|---|---|---|
| 9-1 | Misleading Graphs | 348 |
| 9-2 | Using Data From Graphs and Tables | 351 |
| 9-3 | Organizing and Presenting Data | 355 |
| 9-4 | Measures of Central Tendency | 359 |
| 9-5 | Problem Solving Strategies: Analyzing Sample Data | 364 |
| 9-6 | Stem and Leaf Plots | 366 |
| 9-7 | Box and Whisker Plots | 370 |

### Special Features
Computer Exploration ............ **358**
Mathematical Footnote ........... **363**
Math Team Problem .............. **354**

### Review and Testing
Midchapter Review ............... **363**
Chapter Summary ................ **373**
Chapter Review .................. **373**
Chapter Test .................... **375**
Cumulative Review ............... **376**

## CHAPTER 10  THE NUMBER LINE

| | | |
|---|---|---|
| 10-1 | The Set of Real Numbers | 377 |
| 10-2 | The Addition Property of Inequality | 382 |
| 10-3 | The Multiplication Property of Inequality | 386 |
| 10-4 | Problem Solving Strategies: Using Generalizations | 390 |
| 10-5 | Solving Inequalities | 393 |
| 10-6 | Conjunctions | 399 |
| 10-7 | Disjunctions | 403 |

### Special Features
Computer Exploration............. **398**
Mathematical Footnotes .... **389, 392**
Math Team Problems....... **385, 397**

### Review and Testing
Midchapter Review................ **392**
Chapter Summary................ **406**
Chapter Review.................. **407**
Chapter Test .................... **408**
Cumulative Review ............... **409**

### CHAPTER 11   THE COORDINATE PLANE

| | | |
|---|---|---|
| 11-1 | Coordinate Graphs | 410 |
| 11-2 | Graphing Linear Equations | 414 |
| 11-3 | The Standard Form of a Linear Equation | 418 |
| 11-4 | The Slope of a Line | 422 |
| 11-5 | Problem Solving Strategies: Revising the Solution | 428 |
| 11-6 | Graphing Equations and Inequalities | 430 |
| 11-7 | Problem Solving: Using Two Variables | 435 |
| 11-8 | Translations | 440 |

**Special Features**
Mathematical Footnotes .... **427, 439**
Math Team Problems.. **417, 421, 434**

**Review and Testing**
Midchapter Review ............ **427**
Chapter Summary ............. **444**
Chapter Review .............. **445**
Chapter Test ................ **446**
Cumulative Review ........... **447**

## UNIT 5   USING REAL NUMBERS     448

### CHAPTER 12   SQUARE ROOTS AND RIGHT TRIANGLES

| | | |
|---|---|---|
| 12-1 | Problem Solving Exploration: Square Roots | 450 |
| 12-2 | Using Square Roots | 452 |
| 12-3 | The Pythagorean Theorem | 457 |
| 12-4 | Problem Solving Strategies: Formulating Questions | 463 |
| 12-5 | Similar Triangles | 466 |
| 12-6 | The Tangent Ratio | 474 |

**Special Features**
Computer Exploration............ **473**
Mathematical Footnotes **456, 465, 480**
Math Team Problems....... **472, 480**

**Review and Testing**
Midchapter Review ............ **465**
Chapter Summary ............. **481**
Chapter Review .............. **481**
Chapter Test ................ **483**
Cumulative Review ........... **484**

### CHAPTER 13 POLYNOMIALS

| | | |
|---|---|---|
| 13-1 | Adding Polynomials | 485 |
| 13-2 | Subtracting Polynomials | 491 |
| 13-3 | Problem Solving Exploration: A Multiplication Model | 494 |
| 13-4 | Multiplying Binomials | 496 |
| 13-5 | Problem Solving Strategies: Too Much/Too Little Data | 501 |
| 13-6 | Using the Distributive Property | 504 |
| 13-7 | Special Products | 508 |
| 13-8 | Common Factors | 512 |
| 13-9 | Factoring a Trinomial | 516 |

**Special Features**
Mathematical Footnote ............ 503
Math Team Problems ....... 500, 515

**Review and Testing**
Midchapter Review ................ 503
Chapter Summary ................. 520
Chapter Review ................... 520
Chapter Test ..................... 522
Cumulative Review ................ 523

## UNIT 6 USING EQUATIONS 524

### CHAPTER 14 EQUATIONS IN GEOMETRY

| | | |
|---|---|---|
| 14-1 | Angles and Angle Measures | 526 |
| 14-2 | Parallel and Perpendicular Lines | 532 |
| 14-3 | Problem Solving Strategies: Extending Patterns | 537 |
| 14-4 | Triangles | 539 |
| 14-5 | Polygons | 546 |
| 14-6 | Circumference and Area | 551 |
| 14-7 | Circle Graphs | 558 |

**Special Features**
Computer Exploration............. 557
Mathematical Footnotes .... 545, 556
Math Team Problem............. 544

**Review and Testing**
Midchapter Review ................ 545
Chapter Summary ................. 562
Chapter Review ................... 563
Chapter Test ..................... 564
Cumulative Review ................ 565

### CHAPTER 15  VOLUME AND SURFACE AREA

- **15-1** Problem Solving Exploration: Surface Area and Volume .................................. **566**
- **15-2** Volume of a Rectangular Prism ................. **568**
- **15-3** Surface Area of a Rectangular Prism ............ **572**
- **15-4** Volume of a Cylinder ......................... **576**
- **15-5** Volume of a Pyramid ......................... **582**
- **15-6** Volume of a Cone ........................... **586**
- **15-7** Problem Solving Strategies: Making a Model ...... **590**
- **15-8** Volume and Area of a Sphere .................. **592**

**Special Features**
Mathematical Footnotes .... **581, 589**
Math Team Problems ....... **575, 585**

**Review and Testing**
Midchapter Review .............. **581**
Chapter Summary ............... **596**
Chapter Review ................ **596**
Chapter Test .................. **598**
Cumulative Review ............. **599**

*Table of Squares and Square Roots* .................... **600**
*Table of Trigonometric Ratios* ....................... **601**
*Glossary* ......................................... **602**
*Selected Answers* .................................. **605**
*Index* ............................................ **624**

# UNIT 1

## PROJECT 1
How much electricity does your family use in a month? Work with a team to find how much electricity is used by all the appliances in your home and the cost of using them. Are there ways to reduce your family's monthly costs for electricity?

## PROJECT 2
You want to open a checking account. After a telephone survey, you learn that banks have up to a half-dozen plans each. You read the banks' brochures, hoping to compare "apples with apples." But you find that each bank has special account names and special ways of describing the accounts. For this project, your team will compare the checking accounts of three banks, item for item. You will determine what type of account would be suitable for a person of your age and for people in other situations also.

# Introduction to Algebra

## CONTENTS

### 1 Integers
The Integers
Comparing and Ordering Integers
Problem Solving Exploration:
   Using an Addition Model
Adding Integers
Problem Solving Exploration:
   Using a Subtraction Model
Subtracting Integers
Problem Solving Strategies:
   Methods of Computation
Multiplying Integers
Dividing Integers

### 2 Variables and Expressions
Evaluating Expressions
Problem Solving Strategies:
   Using a Step-by-Step Process
Order of Operations
Problem Solving Exploration:
   Making Generalizations
Basic Properties and
   Mental Computation
Using the Distributive Property
Formulas: Perimeter, Area, Average
Area Formulas: Parallelograms,
   Triangles, and Trapezoids

### 3 Equations and Problem Solving
Equations and Inequalities
Solving Equations
Problem Solving Strategies:
   Choosing Strategies
Addition and Subtraction Equations
Multiplication and Division
   Equations
Inverse Operations
Solving Multi-Step Equations
Translating Word Expressions to
   Algebraic Expressions
Problem Solving: Writing Equations

### PROJECT 3
The dog has long been man's best friend. Recently, however, cats have replaced dogs as the favorite household pet. Birds such as the parakeet and the parrot have also become popular.

You have permission to get one of these four pets. Work with a team to find out as much as you can about the needs and care of each of the four pets. Which pet will you choose?

# CHAPTER 1 | INTEGERS

| Heights of Mountains | |
|---|---|
| Everest (Nepal) | 29,029 ft |
| McKinley (Alaska) | 20,320 ft |
| Rainier (Washington) | 14,410 ft |

| Greatest Depths of Oceans | |
|---|---|
| Pacific | 36,200 ft |
| Atlantic | 30,246 ft |
| Indian | 24,442 ft |

## The Integers

**R**aul and Mara are completing their geography project about the world's mountains and oceans. They find most of the data they need in an almanac.

To make the data easier to use, Raul and Mara decide to display the data in a bar graph. First they round each height and depth to the nearest thousand. Then they construct the bar graph below.

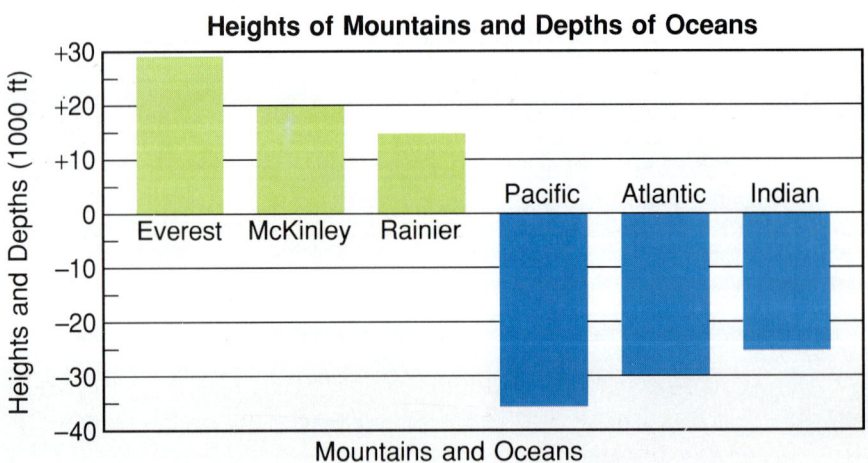

Raul and Mara choose zero to represent sea level. They consider the distance above sea level to be positive, and the distance below sea level to be negative.

Height of Mount Everest: +29,000   Read as "positive 29,000."
Depth of the Atlantic:    −30,000   Read as "negative 30,000."

When you use whole numbers and their negatives to represent amounts, you are using **integers**. Zero is also an integer, but zero is neither positive nor negative.

These are integers.

−20, 0, +4, −15, +250

These are **not** integers.

$2.5, -3\frac{1}{2}, \frac{2}{3}, -\frac{5}{4}, \pi$

Numbers such as +29 and −30 show direction as well as amount. For numbers that show direction, choose zero as the starting point. Choose one direction as positive. Then the other direction is negative.

**Positive Direction**

| 5 feet to the right | +5 |
| 11 points ahead | +11 |
| 3 seconds after | +3 |

**Negative Direction**

| 5 feet to the left | −5 |
| 25 points behind | −25 |
| 10 seconds before | −10 |

**EXAMPLE 1**   Write an integer to represent 115 meters above sea level.

**Solution**   Choose a positive number to represent the distance above sea level. Then **+115 m** represents 115 meters above sea level.

**Try This**   **Use an integer to represent each amount.**
1. A loss of 12 yards
2. 15 degrees below zero

You can use a horizontal number line to show integers. Negative integers are usually placed to the left of zero. Positive integers are usually placed to the right of zero.

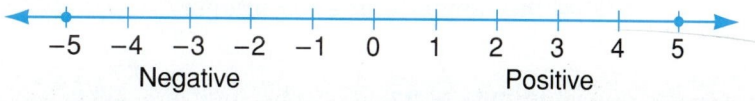

The whole numbers such as 0, 1, 2, 3, ... are sometimes called **non-negative** integers. Positive integers are usually written without the plus sign.

1-1   The Integers   **3**

Notice that both 5 and −5 are 5 units from zero on the number line below. The distance between an integer and zero is called the **absolute value** of the integer. The symbol | | indicates the absolute value of a number.

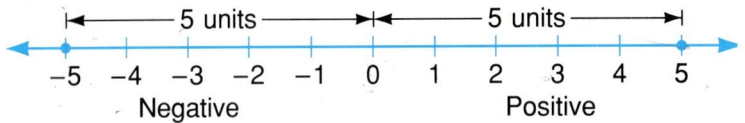

|5| = 5   Read: "The absolute value of five is five."

|−5| = 5   Read: "The absolute value of negative five is five."

**EXAMPLE 2**   Find the absolute value of each integer.
  a. −18         b. 0          c. 7

**Solutions**   a. |−18| = **18**   b. |0| = **0**   c. |7| = **7**

**Try This**   Find the absolute value of each integer.
  3. −11        4. 5          5. −15        6. −3

Two different integers with the same absolute value are called **opposites**. Thus, 3 is the opposite of −3 and −3 is the opposite of 3.

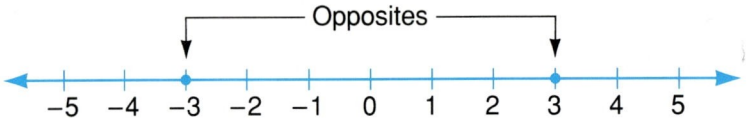

You can read −3 as the **opposite of** 3 or as **negative** 3. The opposite of zero is zero. Zero is the only integer that is its own opposite. The opposite of any negative integer is always positive.

−(3) = −3   Read: "The opposite of three is negative three."

−(−3) = 3   Read: "The opposite of negative three is three."
            or as
            "The opposite of the opposite of three is three."

**EXAMPLE 3**   Find the opposite of each integer.
  a. −25        b. 12          c. −6

**Solutions**   a. −(−25) = **25**   b. −(12) = **−12**   c. −(−6) = **6**

**Try This**   Find the opposite of each integer.
  7. 21         8. −15         9. 0          10. −100

# EXERCISES

**Objectives:** To determine the absolute value of an integer
To determine the opposite of an integer

## Check Your Understanding

**Skill Check** Write the letter of the correct answer.

1. Which of these is **not** an integer?
   a. −4   b. 0
   c. $\frac{2}{3}$   d. 9

2. Which represents the distance between 0 and −20 on the number line?
   a. 20   b. −20
   c. −|20|   d. −|−20|

3. Which cannot be represented by −8?
   a. A fall of 8° in temperature
   b. −(8)
   c. −|−8|
   d. |−8|

4. Which statement is **not** correct?
   a. Opposites have the same absolute value.
   b. Zero is its own opposite.
   c. The absolute value of an integer is always a positive number.
   d. −5 is a non-negative integer.

**Complete each statement.**

5. −8 is read as  ?  .
6. If +12 represents 12 meters to the right,  ?  represents 12 meters to the left.
7. The non-negative integers 0, 1, 2, 3, ... are sometimes called  ?  numbers.
8. The opposite of the opposite of 12 is  ?  .

## Practice and Apply

**Which of the following are integers? Answer *Yes* or *No*.**

9. −4    10. 0    11. −3.5    12. $\frac{1}{2}$    13. +7

**Write an integer to represent each amount.**

14. 20 degrees above zero
15. A withdrawal of $45
16. The depth of 500 meters
17. A stock price increase of 3
18. A gain of 15 yards
19. A weight loss of 9 pounds
20. A growth of 2 centimeters
21. A temperature drop of 6 degrees
22. A loss of $20
23. 10 years ago

**Write the absolute value of each integer.**

24. 14    25. 6    26. −4    27. −7    28. 8
29. −3    30. 0    31. −1    32. 1    33. 23

1-1  The Integers  **5**

Write the opposite of each integer.

**34.** −9    **35.** 5    **36.** +13    **37.** −11    **38.** 0
**39.** −7   **40.** −21   **41.** −36    **42.** 1     **43.** −1

**44.** This table shows the population of 4 cities in 1980 and 1990. Find the increase or decrease in population for each city. Then use a positive or negative integer to record the change in the table. Finally, draw a bar graph to show the changes in population.

|  | 1980 | 1990 | Change |
|---|---|---|---|
| Dallas | 904,078 | 990,957 | ? |
| Kansas City | 448,159 | 427,799 | ? |
| Oklahoma City | 403,213 | 441,154 | ? |
| San Francisco | 678,974 | 711,407 | ? |

## Connect and Extend

Use the number line to answer Exercises 45–50.

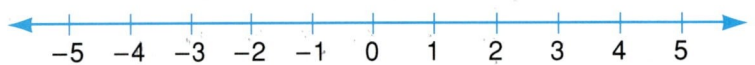

You start each time at −5 on the number line. Where would you be after each move?

**45.** 8 steps to the right    **46.** 4 steps to the left    **47.** 7 steps to the left

Which integer is farther from 2?

**48.** −10 or 7    **49.** −8 or 2    **50.** 0 or −4

Replace the  ?  to make the statement true.

**51.** |?| = 6    **52.** |?| = 0    **53.** |?| = −3

Give several examples to show why each sentence is true.

**54.** The opposite of the opposite of any integer is the integer itself.
**55.** If two integers are opposites, they have the same absolute value.
**56.** If you press the +/− key on your calculator twice after an integer is in the display window, the result is the original integer.

## Maintain Your Skills

Find each answer.

**57.** 768 − 296    **58.** 760 ÷ 21    **59.** $48.35 + $26.88
**60.** 75 · 18     **61.** $96.24 − $37.58  **62.** 8 · $24.03

**63.** A memory chip in a computer can find one bit of information in one-millionth of a second. Write this time as a decimal.

**64.** More than 71,400,000 homes in the United States have color TV sets. Round this number to the nearest million.

# LESSON 1-2
# Comparing and Ordering Integers

On a number line, the integers are written in **ascending order** as you move from left to right. As you move from right to left, the integers are written in **descending order.**

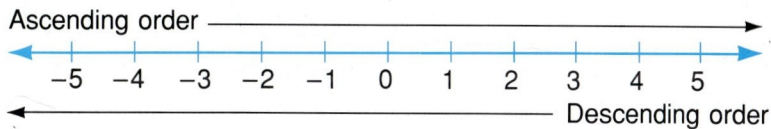

When you compare two integers on a number line, the integer on the right is always greater than the one on the left.

Which distance is greater, 3 feet or 5 feet?

Since 5 is to the right of 3, 5 is greater than 3.
5 feet is greater than 3 feet.

Which score is greater, −4 points or −2 points?

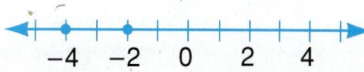

Since −2 is to the right of −4, −2 is greater than −4.
−2 points is greater than −4 points.

When you compare two integers on a number line, the number to the left is always less than the one on the right.

Which is less, −3 or 3?

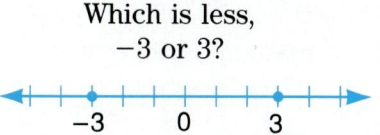

Since −3 is to the left of 3, −3 is less than 3.

Which is less, −5 or −2?

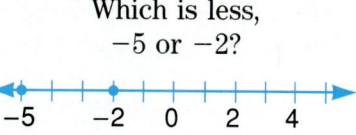

Since −5 is to the left of −2, −5 is less than −2.

These symbols are used to compare numbers.

| In Words | Using Symbols |
|---|---|
| Zero is less than five. | $0 < 5$ |
| Zero is equal to zero. | $0 = 0$ |
| Zero is greater than negative five. | $0 > -5$ |

1-2 Comparing and Ordering Integers

**EXAMPLE 1**  Write an inequality sign (< or >) to compare the integers.
  a. −20 and −16    b. −13 and 5

**Solutions**  a. **Think:** Since −20 is to the left of −16, **−20 < −16.**

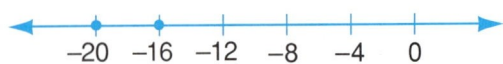

Or, since −16 is to the right of −20, −16 > −20.

b. **Think:** Since −13 is to the left of 5, −13 < 5, or **5 > −13.**

**Try This**  Write an inequality sign to compare the integers.
  **1.** −7 and −9    **2.** 0 and −8    **3.** −6 and 8

You can also use a number line to write integers in ascending or descending order. Integers *increase from least to greatest* as you move from *left to right* on the number line. Integers *decrease from greatest to least* as you move from *right to left* on the number line.

**EXAMPLE 2**  a. List 4, −7, and −2 in ascending order.
  b. List 4, −7, and −2 in descending order.

**Solutions**  Locate 4, −7, and −2 on the number line.

  a. For increasing order, list the integers from left to right.
     **Ascending order: −7, −2, 4**

  b. For decreasing order, list the integers from right to left.
     **Descending order: 4, −2, −7**

**Try This**  **List the integers in ascending order.**
  **4.** 0, −6, −2    **5.** −5, 3, −8    **6.** −1, −3, 0

**List the integers in descending order.**
  **7.** 6, −9, −1    **8.** −7, −3, −5    **9.** 0, 1, −1

# EXERCISES

**Objectives:** To compare integers
To write integers in ascending or descending order

## Check Your Understanding

**Skill Check** Write the letter of the correct answer.

**1.** Which statement is **not** correct?
  **a.** $-6$ is to the right of $-4$ on the number line.
  **b.** $-4$ is to the right of $-6$ on the number line.
  **c.** $-4 > -6$
  **d.** $-6$ is less than $-4$.

**2.** Which statement is correct?
  **a.** No integer equals itself.
  **b.** $-8$ points is less than $-9$ points.
  **c.** $-122 < -121$ is the same as $-121 > -122$.
  **d.** Integers to the left of 0 on the number line are greater than integers to the right of 0.

**3.** Which integers are written in ascending order?
  **a.** $1, -4, -6, -1$
  **b.** $1, -1, -4, -6$
  **c.** $-6, -4, -1, 1$
  **d.** $-6, -4, 1, -1$

**4.** Which integers are written in descending order?
  **a.** $0, -4, -3, -1$
  **b.** $9, -9, -10, -15$
  **c.** $2, -4, 8, -16$
  **d.** $-8, -7, -6, -5$

Replace each ? with < or > to make a true sentence.

**5.** $-5$ ? $-1$      **6.** $21$ ? $-15$      **7.** $-2$ ? $-8$      **8.** $-9$ ? $-3$

## Practice and Apply

Write an inequality sign (< or >) to compare the two integers.

**9.** $-3$ and $-4$     **10.** 6 and 5     **11.** 8 and 6     **12.** 0 and $-2$
**13.** $-8$ and $-10$     **14.** $-5$ and 5     **15.** 2 and 7     **16.** $-4$ and $-6$
**17.** $-3$ and $-7$     **18.** 5 and $-3$     **19.** 9 and 0     **20.** 4 and $-10$

First, locate the integers on the number line. Then list the integers in ascending order.

**21.** $-13, 8, -4$     **22.** $-2, -4, -6$     **23.** $-11, 5, -2$     **24.** $6, -4, -8, 4$

First, locate the integers on the number line. Then list the integers in descending order.

**25.** $-12, -2, -16$     **26.** $-3, -9, 0$     **27.** $-2, 1, -10$     **28.** $14, -15, -13, -10$

**29.** Look up the words *ascending* and *descending* in a dictionary. Explain how the definitions help you to understand ascending order and descending order of integers.

For Exercises 30–33, refer to the temperature chart at the right.

**30.** List the temperatures in order from coldest to warmest.

**31.** Which states have lowest recorded temperatures that are less than −40°F?

**32.** How many states have lowest recorded temperatures that are greater than −25°F?

**33.** Write an inequality that compares Alaska's and Hawaii's temperatures.

| Alaska | −80°F |
| Florida | −2°F |
| Georgia | −17°F |
| Hawaii | 12°F |
| Illinois | −35°F |
| Iowa | −47°F |
| Maryland | −40°F |
| Vermont | −50°F |

## Connect and Extend

**Rewrite each sentence using mathematical symbols.**

**34.** Negative 3 is less than ten.

**35.** The opposite of 6 is less than 5.

**36.** The opposite of negative two is greater than the opposite of the absolute value of two.

**Write each mathematical sentence in words.**

**37.** $-17 < 28$  **38.** $5 > -2$  **39.** $|-7| > |-3|$  **40.** $-|-3| < 0$

## Maintain Your Skills

**Find the absolute value of each integer. (Pages 2–6)**

**41.** $-9$   **42.** $13$   **43.** $0$   **44.** $-1$   **45.** $+2$

**Find the opposite of each integer. (Pages 2–6)**

**46.** $-3$   **47.** $-16$   **48.** $4$   **49.** $-5$   **50.** $|0|$

## ★ Math Team Problems

The graph at the right shows the motion of a car that starts from a stop, increases in speed, travels at a constant speed, then decreases in speed, and comes to a stop.

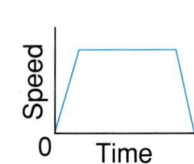

**For each of these graphs, write a description of the car's motion.**

**51.** **52.** 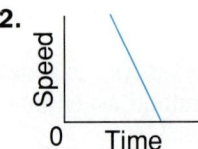 **53.**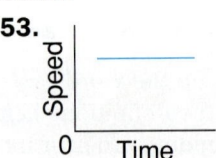

# LESSON 1-3
# Problem Solving Exploration Using an Addition Model

**T**hese are the materials you will need for this lesson.

1. Several ⊕ counters and several ⊖ counters.
2. A sheet of paper with a large square drawn on it

The value of a square is determined by the number and type of counters in the square.

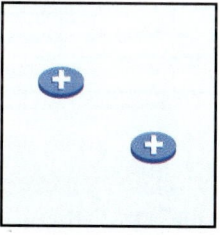
Value: +2

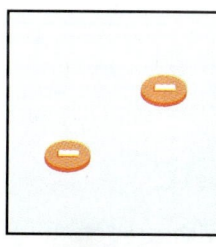
Value: −2

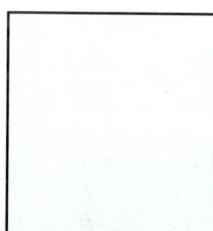
Value: 0

A ⊕ counter and a ⊖ counter have opposite signs. A pair of counters with opposite signs form a **neutral pair,** ⊕—⊖. A neutral pair does not change the value of the square.

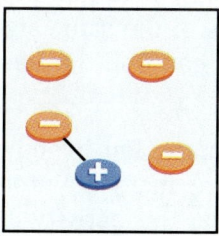
Value: −3

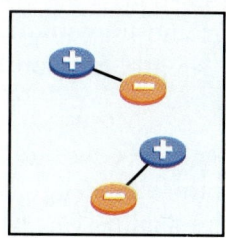
Value: 0

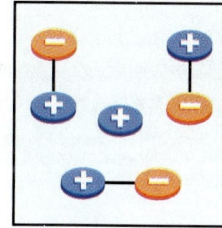
Value: +1

Here is an example of how to find the value of a square.

**A.** Put 3 ⊕ counters in the square.

**B.** Add 5 ⊖ counters to the square.

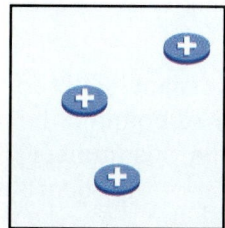

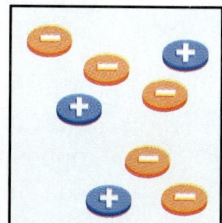

1-3 Problem Solving Exploration  **11**

C. Remove any ⊕—⊖ pairs.  D. Final Value: −2

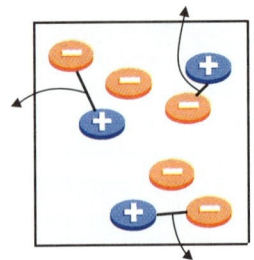
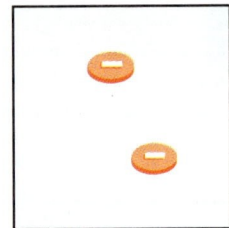

E. Record your activity like this.

| Put in | Add | Remove ⊕—⊖ pairs. | Record the sum. |
|---|---|---|---|
| +3 | −5 | ⊕—⊖, ⊕—⊖, ⊕—⊖ | +3 + (−5) = −2 |

# EXPLORE

A. Start with no counters in the square. Put in some counters, all having the same sign. Next add as many counters of the same sign as you wish. Record the sum. Repeat this activity three times using only ⊕ counters and three times using only ⊖ counters.

1. A classmate tells you how many counters there were at the start and how many counters of the same sign were added to the square. Explain how you can find the final value of the square.

B. Start with no counters in the square. Put in some counters, all having the same sign. Next add as many counters of the opposite sign as you wish. Repeat this activity 3 times using more ⊕ counters than ⊖ counters, and 3 times using more ⊖ counters than ⊕ counters.

2. A classmate tells you how many counters there were at the start and that the number of counters she added having the opposite sign was more than this number. Explain how you can find whether the value of the square is positive or negative.

3. A classmate tells you how many counters there were at the start and that the number of counters he added having the opposite sign was less than this number. Explain how you can find whether the value of the square is positive or negative.

## LESSON 1-4

# Adding Integers

**L**ana Sims records her daily stock transactions in a table.

What is her net profit after the first two days?
After the first three days?
At the end of one week?

| Profit in Dollars | |
|---|---|
| Mon. | −2 |
| Tues. | −3 |
| Wed. | 5 |
| Thurs. | 4 |
| Fri. | −2 |

To find Lana's profit after the first two days, you add two negative integers.

$$-2 + (-3) = \underline{\ ?\ }$$

You can use counters to model this sum.

Put in 2  counters.

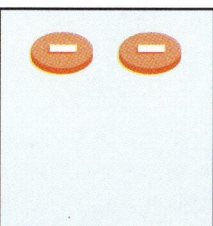

Add 3 ⊖ counters.

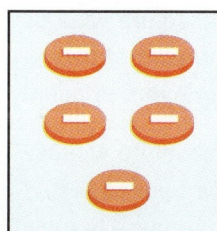

Record the sum.

$$-2 + (-3) = -5$$

The model shows that the sum is **−5**.

1-4  Adding Integers  **13**

You can also draw a number line to model $-2 + (-3)$.

Start at $-2$.
Move 3 units to the left.
Stop at $-5$.

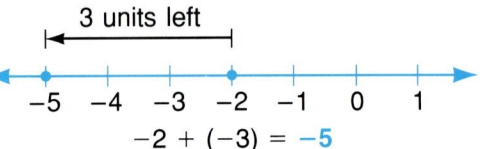

$-2 + (-3) = -5$

**EXAMPLE 1** Use a number line to find each sum.
    **a.** $4 + 2$      **b.** $-3 + (-4)$

**Solutions**
**a.** Start at 4.
Move 2 units to the right.
Stop at 6.

$4 + 2 = 6$

**b.** Start at $-3$.
Move 4 units to the left.
Stop at $-7$.

$-3 + (-4) = -7$

**Try This** Draw a number line to find each sum.
  **1.** $8 + 4$      **2.** $-8 + (-4)$      **3.** $(-1) + (-6)$

Example 1 suggests a rule for adding integers with like signs.

> **Adding Integers with Like Signs**
> 1. Find the sum of their absolute values.
> 2. The sign of the sum is the same as the sign of the integers.

**EXAMPLE 2** Find each sum:   **a.** $5 + 3$      **b.** $-2 + (-7)$

**Solutions**
**a.** Add the absolute values.
Think: The sum of two positive integers is positive.

$|5| + |3| = 8$

$5 + 3 = 8$

**b.** Add the absolute values.
Think: The sum of two negative integers is negative.

$|-2| + |-7| = 9$

$-2 + (-7) = -9$

**Try This** Find each sum:   **4.** $-7 + (-4)$      **5.** $(-6) + (-4)$

After the first two days, Lana has a profit of $−5. To find the profit after 3 days, find the sum of −5 and 5.

What do you think will happen if you use counters to model −5 + 5?

Since −5 and 5 are opposites, they form a **neutral pair.** The sum of two opposites, or a neutral pair, is 0. So Lana's profit after 3 days is $0. This suggests a special property of the set of integers.

> **Property of Opposites**
> The sum of two opposites is zero.

To find Lana's profit for the week, you add the profit after 4 days, which is 4, and the profit for Friday, −2. You can use counters to model 4 + (−2).

Put in 4 ⊕ counters.

Add 2 ⊖ counters.

Remove ⊕—⊖ pairs. Final value: **2**

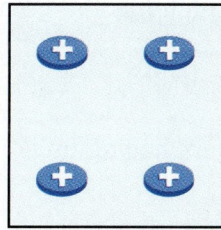 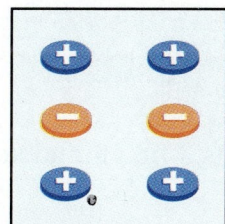 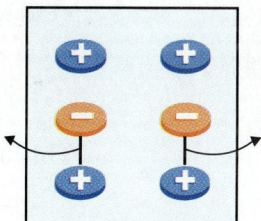

You have shown that 4 + (−2) = **2.**

You can also draw a number line to model 4 + (−2).

Start at 4.
Move 2 units to the left.
Stop at 2.

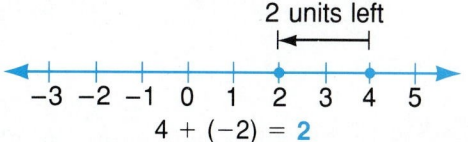

So Lana made a profit of **$2** for the week.

**EXAMPLE 3** Draw a number line to find each sum.
a. −4 + 3       b. 6 + (−2)       c. 2 + (−7)

**Solutions**  a. Start at −4.
Move 3 units to the right.
Stop at −1.

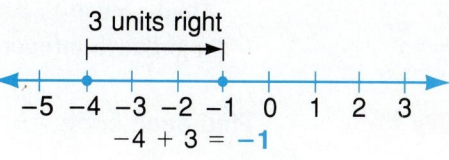

1-4  Adding Integers  **15**

b. Start at 6.
Move 2 units to the left.
Stop at 4.

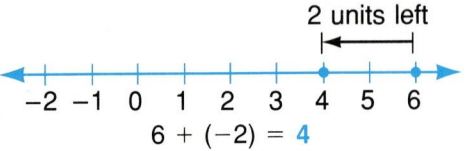

$6 + (-2) = 4$

c. Start at 2.
Move 7 units to the left.
Stop at −5.

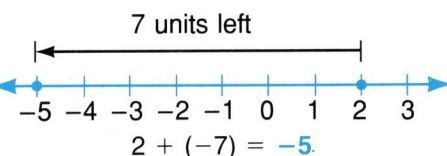

$2 + (-7) = -5$

**Try This**

Draw a number line to find each sum.

6. $-2 + 6$   7. $3 + (-5)$   8. $3 + (-3)$

Study the results of each addition problem in Example 3.

- **Example 3a:** Does $-4 + 3 = -1$? Which has the greater absolute value, $-4$ or $3$?
  Does $-4 + 3 = -(|-4| - |3|)$?

- **Example 3b:** Does $6 + (-2) = 4$? Which has the greater absolute value, $6$ or $(-2)$?
  Does $6 + (-2) = |6| - |-2|$?

This suggests a rule for adding integers having unlike signs.

> **Adding Integers with Unlike Signs**
> 1. Find the difference of the absolute values.
> 2. Write the sum with the same sign as the integer having the greater absolute value.

**EXAMPLE 4** Find each sum:   a. $7 + (-3)$   b. $2 + (-9)$

**Solutions**

a. Find the difference of the absolute values.
Think: Since $|7| > |-3|$, the sum will be a positive integer.

$|7| - |-3| = 7 - 3$

$7 + (-3) = 4$

b. Find the difference of the absolute values.
Think: Since $|-9| > |2|$, the sum will be a negative integer.

$|-9| - |2| = 7$

$2 + (-9) = -7$

**Try This**   Find each sum:   9. $3 + (-8)$   10. $-6 + 4$

# EXERCISES

**Objective:** To add integers

## Check Your Understanding

**Skill Check** Write the letter of the correct answer.

1. Find the sum: $-9 + (-5)$
   a. 14   b. 0   c. 4   d. $-14$

2. Find the sum: $-8 + 3$
   a. $-11$   b. $-5$   c. 5   d. 11

3. Find the sum: $-7 + 18$
   a. $-11$   b. 11   c. 25   d. $-25$

4. Find the sum: $-6 + (-5) + 6$
   a. 0   b. 5   c. $-5$   d. $-11$

Draw a picture of counters to show each sum.

5. $3 + (-8)$   6. $-2 + (-4)$   7. $-5 + 7$   8. $0 + 6$

Draw a number line to show each sum.

9. $-6 + (-3)$   10. $4 + (-5)$   11. $-2 + 2$   12. $6 + (-3)$

## Practice and Apply

Write an addition sentence to describe each change on the number line.

13.    14.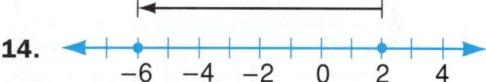

Write P if the sum is positive, Z if it is zero, and N if it is negative.

15. $-15 + (-3)$   16. $10 + (-3)$   17. $27 + 14$   18. $-19 + 14$
19. $-12 + -43$   20. $-29 + 29$   21. $-47 + 40$   22. $58 + (-45)$
23. $68 + (-25)$   24. $-72 + (-38)$   25. $96 + 74$   26. $-88 + (-56)$

Find each sum.

27. $-7 + (-3)$   28. $3 + (-5)$   29. $7 + (-7)$   30. $-4 + (-8)$
31. $-2 + 7$   32. $-9 + (-8)$   33. $14 + (-3)$   34. $-11 + (-5)$
35. $-8 + 4$   36. $16 + (-7)$   37. $-14 + 11$   38. $26 + (-12)$
39. $15 + (-20)$   40. $-26 + 13$   41. $30 + (-20)$   42. $-20 + 30$

Write the letter of the correct answer.

43. Paul gained 13 pounds in January. He lost 8 pounds in February. Which expression can you use to find the change in Paul's weight?
    a. $13 + 8$   b. $13 + (-8)$   c. $-13 + (-8)$

44. Rosa scored 10 points for the first answer. She lost 6 points on her second answer. Which expression can you use to find the net change in Rosa's score?
    a. $10 + (-6)$   b. $10 + 6$   c. $-10 + (-6)$

1-4 Adding Integers

**45.** In this lesson, you found Lana's net profit by adding the profit two days at a time.

$-2 + (-3) = -5$   $-5 + 5 = 0$   $0 + 4 = 4$   $4 + (-2) = 2$

Can you think of a shortcut method to find this sum mentally?
$$-2 + (-3) + 5 + 4 + (-2) = \underline{\ ?\ }$$

## Connect and Extend

**True or False? If a statement is false, rewrite it to make a true statement.**

**46.** The sum of two positive integers is always a positive integer.

**47.** The sum of two negative integers is always a negative integer.

**48.** The sum of a positive integer and a negative integer is always a negative integer.

This chart shows the total yards gained (+) or lost (−) in four successive plays in the Chiefs football game. Calculate the final position of the ball after each series of plays. The Chiefs are starting on their own given yard line in each case.

|  | Starting Yard Line | Yardage | Ending Yard Line |
|---|---|---|---|
| **49.** | 5 | gained 5, lost 3, gained 7, lost 2 | ? |
| **50.** | 12 | +3, −5, +12, −10 | ? |
| **51.** | 20 | −6, +7, −13, +9 | ? |

The sum of the eight consecutive integers from −3 to 4 is this.
$$-3 + (-2) + (-1) + 0 + 1 + 2 + 3 + 4 = 4$$

**52.** Refer to the pattern shown above to find the sum of the integers from −10 to 11.

**53.** Explain a shortcut for finding this sum.
$$-100 + (-99) + (-98) + \ldots + 98 + 99 + 100 + 101 = \underline{\ ?\ }$$

## Maintain Your Skills

**Write the answer only if it is over 500.**

**54.** $25 \cdot 20$   **55.** $40 \cdot 35$   **56.** $16 \cdot 31$   **57.** $17 \cdot 31$

**Write the opposite of each integer. (Pages 2–6)**

**58.** $-30$   **59.** $14$   **60.** $-16$   **61.** $0$   **62.** $-12$   **63.** $+25$

# COMPUTER EXPLORATION

## Adding Integers

**Objective:** To explore addition of integers using a computer

1. **a.** With the Holt *Pre-Algebra* disk in drive A (or drive 1), run the program, *Adding Integers*, as follows.
   **IBM:** Type **a:a** and press **ENTER**.
   **Apple:** Turn on the computer. When the menu appears on the screen, press **A**.
   **b.** Read the opening messages of the program. Press **RETURN** to continue.

2. Read the opening screen of the program. Then press **RETURN** (or **ENTER**).
   **a.** The computer shows a number line and the problem, $5 + (-4)$. Guess the answer. Then press any key to continue.
   **b.** The computer marks the point 5 on the number line. Press any key to continue.
   **c.** The computer counts 4 units to the left and draws an arrow. Press any key to continue.
   **d.** The sum is the coordinate of the point where the arrow stops. Read this point. Does your guess agree with the coordinate of the point shown on the graph? Press any key to continue.
   **e.** Type the answer to $5 + (-4)$ and press **RETURN.**
   **f.** The computer marks the coordinate of the point you just typed in. Then it asks if you want to do another problem. Press **Y** for yes.

3. Repeat step 2 for the problem, $3 + (-3)$.

4. Continue running the program until you have completed all the problems presented. The computer will tell you how many you have answered correctly.

5. Remove the disk and turn off the computer and the monitor. Return the disk to its storage place.

Computer Exploration **19**

**Write an integer to represent each situation. (Pages 2–6)**
1. Twenty degrees below zero
2. A deposit of $40
3. Seven years ago

**Find the absolute value of each integer. (Pages 2–6)**
4. $-6$
5. $3$
6. $17$
7. $-8$
8. $0$
9. $+4$

**Write the opposite of each integer. (Pages 2–6)**
10. $21$
11. $-14$
12. $0$
13. $7$
14. $-37$

**Write an inequality sign ($<$ or $>$) to compare the two integers. (Pages 7–10)**
15. $6$ and $-6$
16. $-4$ and $-8$
17. $-9$ and $-3$
18. $-9$ and $4$

**Write in descending order. (Pages 7–10)**
19. $-1, 4, -8$
20. $-7, -3, -9$
21. $6, -5, 0$

**Write P if the sum is positive; write N if the sum is negative. (Pages 13–18)**
22. $-8 + 6$
23. $-4 + (-3)$
24. $18 + (-20)$
25. $-9 + 6$

**Find each sum. (Pages 13–18)**
26. $15 + (-4)$
27. $-7 + (-9)$
28. $-8 + (-2)$
29. $12 + (-18)$

30. Write three numbers that are **not** integers.

# Mathematical Footnote

On April 6, 1909, Admiral Robert E. Peary, Matthew A. Henson, and four Eskimos became the first people to reach the North Pole. Peary calculated the exact spot and asked Henson, who had been with him on seven Arctic expeditions, to place the American flag at the Pole.

Henson has been recognized as the one indispensable person whose expertise in organizing, planning, carrying out, and overseeing the various stages of the expedition was responsible for its success. Henson received many honors for his part in the expedition of 1908–1909, and wrote a book about his experiences.

# LESSON 1-5
# Problem Solving Exploration Using a Subtraction Model

You can use ⊕ and ⊖ counters to model subtraction.

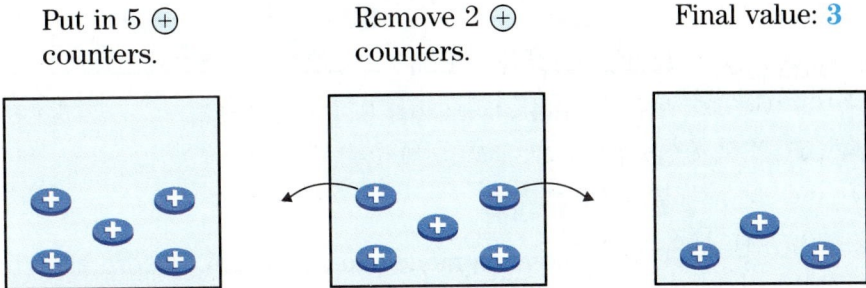

Put in 5 ⊕ counters.   Remove 2 ⊕ counters.   Final value: **3**

The model shows that $5 - 2 =$ **3**.

What happens when the number you subtract from 5 is greater than 5? For example, how can you use counters to model $5 - 7$?

How will you take 7 counters from a square if you started with 5 ⊕ counters only?

How can you add two ⊕ counters without changing the value of the square?

What happens to the value of the square when you put in pairs of neutral counters?

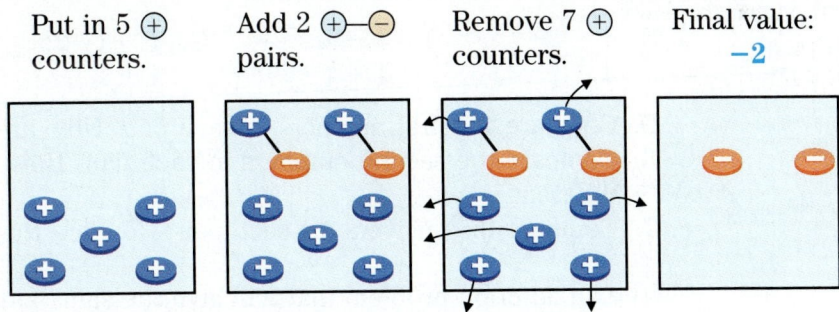

Put in 5 ⊕ counters.   Add 2 ⊕—⊖ pairs.   Remove 7 ⊕ counters.   Final value: **−2**

This model shows that $5 - 7 =$ **−2**.

What happens when you subtract a negative number from 5? For example, how can you use counters to model $5 - (-2)$?

How can you remove 2 ⊖ counters that are not already in the square?

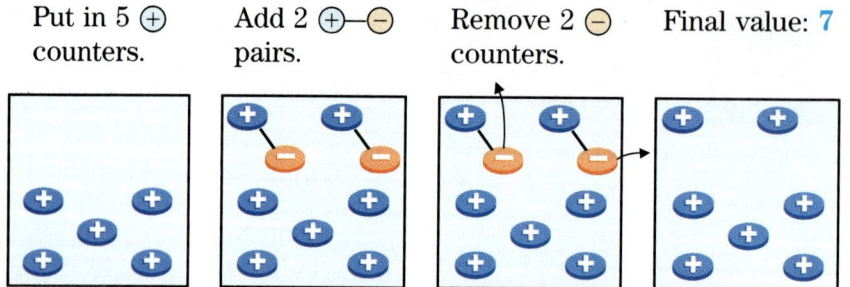

Put in 5 ⊕ counters.  Add 2 ⊕—⊖ pairs.  Remove 2 ⊖ counters.  Final value: 7

This model shows that $5 - (-2) = 7$.

# EXPLORE

Use counters to model each problem. Record each final value.

**A.** $-5 - 3$   **B.** $-5 - 7$   **C.** $-5 - (-2)$   **D.** $-5 - (-7)$

1. Explain what happens if you start with $-5$ and subtract a positive number having an absolute value less than $|-5|$.
2. Explain what happens if you start with $-5$ and subtract a positive number having an absolute value greater than $|-5|$.
3. Explain what happens if you start with $-5$ and subtract a negative number having an absolute value less than $|-5|$.
4. Explain what happens if you start with $-5$ and subtract a negative number having an absolute value greater than $|-5|$.

Use counters to model each pair of subtraction and addition problems. Record the final values.

**E.** $4 - 2$   **F.** $4 - 6$   **G.** $4 - (-2)$   **H.** $4 - (-6)$
$4 + (-2)$   $4 + (-6)$   $4 + 2$   $4 + 6$

5. Compare the first numbers in each pair. How are they alike?
6. Compare the second numbers in each pair. How are they different?
7. Compare the answers for each pair. How are they alike?

Write an addition problem that will give the same answer as each of these subtraction problems. Check by using counters to model each pair of addition and subtraction problems.

**8.** $3 - (-2)$   **9.** $2 - (-5)$   **10.** $-4 - (-3)$   **11.** $-1 - (-4)$

12. Think of a shortcut for subtracting an integer by using addition. Test your shortcut by modeling several problems. If your shortcut doesn't work, change it and test again.

# LESSON 1-6

## Subtracting Integers

To show how subtraction and addition are related, use counters to model the problems.

**Subtraction:** 2 − 3

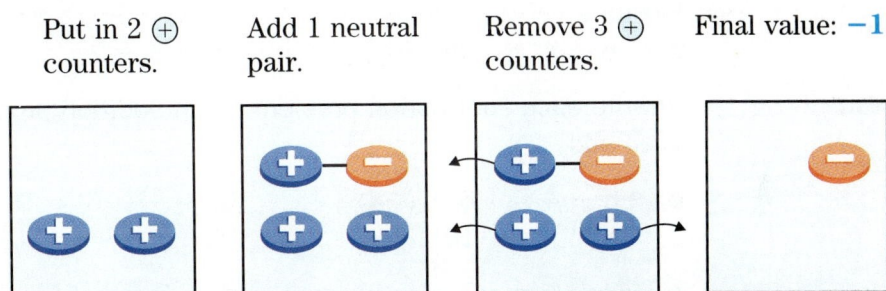

Put in 2 ⊕ counters.   Add 1 neutral pair.   Remove 3 ⊕ counters.   Final value: −1

This model shows that 2 − 3 = −1.

**Addition:** 2 + (−3)

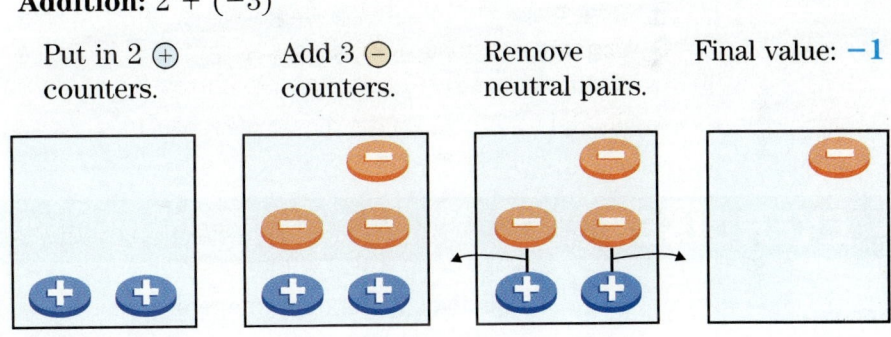

Put in 2 ⊕ counters.   Add 3 ⊖ counters.   Remove neutral pairs.   Final value: −1

This model shows that 2 + (−3) = −1.

The models show that subtracting 3 from 2 is the same as adding the opposite of 3 to 2. That is,

$$2 - 3 = 2 + (-3).$$

This suggests that you can write any subtraction problem as an addition problem.

| Subtraction | | Addition |
|---|---|---|
| 4 − 7 | is the same as | 4 + (−7) |
| −2 − 4 | is the same as | −2 + (−4) |
| 5 − (−3) | is the same as | 5 + 3 |
| −6 − (−8) | is the same as | −6 + 8 |

The subtraction rule generalizes these examples.

> **Subtracting Integers**
> To subtract an integer, add its opposite.

**EXAMPLE**  Evaluate.
  **a.** $-6 - 4$   **b.** $9 - (-6)$   **c.** $-3 - 2$   **d.** $-3 - (-5)$

**Solutions**  Rewrite each subtraction problem as an addition problem. Then add.

  **a.** $-6 - 4 = -6 + (-4)$
  $\phantom{-6 - 4\ } = -10$

  **b.** $9 - (-6) = 9 + 6$
  $\phantom{9 - (-6)\ } = 15$

  **c.** $-3 - 2 = -3 + (-2)$
  $\phantom{-3 - 2\ } = -5$

  **d.** $-3 - (-5) = -3 + 5$
  $\phantom{-3 - (-5)\ } = 2$

**Try This**  Evaluate.
  **1.** $-3 - 4$     **2.** $-8 - (-6)$    **3.** $4 - (-3)$
  **4.** $-2 - 5$    **5.** $8 - (-5)$     **6.** $-5 - (-9)$

# EXERCISES

**Objective:** To subtract integers

## Check Your Understanding

**Skill Check**  Write the letter of the correct answer.

**1.** Which expression means the same as $7 - (-9)$?
  **a.** $7 + (-9)$    **b.** $7 - 9$
  **c.** $7 + 9$       **d.** $9 - 7$

**2.** Which will give a positive integer for the answer?
  **a.** $1 - 6$       **b.** $1 + (-6)$
  **c.** $-6 - 1$      **d.** $1 - (-6)$

**3.** Which will give a negative integer for the answer?
  **a.** $-5 - (-10)$   **b.** $0 - (-5)$
  **c.** $-5 - (-2)$    **d.** $10 - (-5)$

**4.** Which expression has the same value as $10 - 21$?
  **a.** $10 + 21$     **b.** $21 - 10$
  **c.** $10 - (-21)$  **d.** $-21 + 10$

Tell whether the difference is positive, P, or negative, N.
  **5.** $-100 - (30)$    **6.** $45 - (-90)$    **7.** $140 - 60$    **8.** $60 - 140$

## Practice and Apply

**Rewrite each subtraction problem as an addition problem.**

**9.** $7 - (-2)$  **10.** $14 - 8$  **11.** $-9 - (-16)$  **12.** $-7 - (-3)$

**Subtract.**

**13.** $-5 - 6$  **14.** $-3 - (-7)$  **15.** $-9 - (-2)$  **16.** $7 - 16$
**17.** $8 - (-3)$  **18.** $-2 - (-8)$  **19.** $-7 - 7$  **20.** $5 - (-9)$
**21.** $-7 - (-11)$  **22.** $-15 - 9$  **23.** $6 - (-8)$  **24.** $-10 - 8$
**25.** $-19 - (-18)$  **26.** $-15 - 6$  **27.** $27 - 35$  **28.** $17 - (-10)$

**Find the difference only if it is greater than 9.**

**29.** $-18 - (-28)$  **30.** $-30 - (-30)$  **31.** $47 - 27$  **32.** $-59 - (-79)$

**33.** Lupe owed her cousin $3. As a birthday present, her cousin gave her a note that said "Your debt is paid." Represent this as the sum of two integers.

**34.** Alan's parents pay him for mowing the lawn by giving him $10 and canceling a $3 debt. Represent the amount Alan earned as the difference of two integers.

## Connect and Extend

To enter a negative number into your calculator, use the change-of-sign key, [+/-].

For example, to enter $-7.84$, press 7.84 [+/-]

To use a calculator to subtract $-7.84 - (-2.99)$, press these keys, 7.84 [+/-] − 2.99 [+/-] =   The display will show $-4.85$.

**Write the problem that each calculator sequence represents.**

**35.** 10.18 [+/-] [−] 3.9 [=]          **36.** 9.9 [+/-] [−] 10.58 [+/-] [=]

**Replace each ? with the integer that makes each statement true.**

**37.** $2 - ? = -6$  **38.** $-7 - ? = 12$  **39.** $? - (-3) = -20$
**40.** $-9 - ? = 10$  **41.** $-20 - ? = 0$  **42.** $? - (-18) = 12$

## Maintain Your Skills

**Find each answer.**

**43.** $276.55 + $128.19  **44.** $786 - 295$  **45.** $27 \cdot 35$  **46.** $405 \div 9$

**Write an inequality sign (< or >) to make a true statement. (Pages 7–9).**

**47.** $-9 \; ? \; -4$  **48.** $-2 \; ? \; 6$  **49.** $5 \; ? \; 3$  **50.** $-2 \; ? \; -8$.

**Add. (Pages 13–18)**

**51.** $6 + (-8)$  **52.** $-10 + (-10)$  **53.** $-16 + (-3)$  **54.** $24 + (-30)$

## LESSON 1-7
# Problem Solving Strategies Choosing a Method of Computation

**W**hen formulating a plan to solve a problem, it is helpful to think about ways that make problem-solving easier and more efficient.

What ways do you know to solve problems?
Here is a list of some ways to use the tools of problem-solving.

- Use a calculator
- Use mental math
- Use counters or tiles
- Use a model
- Use paper and pencil
- Use a computer

How do you decide on which tools to use?

Think about how the answers to this question are related to a particular problem you are trying to solve.

Do you need an exact answer, or will an estimate be close enough?
Can you see how to arrive at the answer by using mental math?
Are there calculations that must be repeated many times?
Does the problem involve large numbers?
Can you make a model to represent the problem?

## EXERCISES

**Objective:** To apply strategies in problem solving

**For Exercises 1–7, suggest two different ways to solve each problem. Explain your choices. Do NOT solve the problem.**

1. Find how many faces a cube has.
2. Write 25 numbers that are divisible by 19.
3. Is this sum closer to 300 or to 400?
   $93 + 86 + 7 + 14 + 100$
4. Find the product.
   $73 \cdot 73 \cdot 73 \cdot 73 \cdot 73$
5. What is the sum in years of the ages of the students in your math class?
6. The cost of a 10-pound bag of potatoes is $2.06. You can also buy the potatoes for 23¢ per pound. Which is the cheaper way to buy 10 pounds of potatoes?
7. Pins in a bowling alley are set up as shown in the figure at the right. How can you move three pins to form a triangle exactly like the original triangle, but facing in the opposite direction?

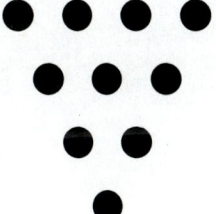

8. Work with a partner to choose one of the problems in Exercises 1–7. Solve the problem in two different ways. Compare your method of solution with other teams that chose the same problem. Discuss whether there is a "best" way to solve the problem. Think about accuracy, speed, and ease of solution.

1-7  Problem Solving Strategies  **27**

*Amish Breakdown* by Teresa Ueltschey

## LESSON 1-8
# Multiplying Integers

**Y**ou already know that the product of two positive numbers is a positive number. For example,

$$56 \times 43 = 2408 \qquad 1.5 \times 42.8 = 64.2$$

What is the product of a positive integer and a negative integer? To find out, study the pattern below.

## EXPLORE

Look for the pattern:
$$3 \cdot 3 = 9$$
$$3 \cdot 2 = 6$$
$$3 \cdot 1 = 3$$
$$3 \cdot 0 = 0$$
$$3 \cdot (-1) = \underline{\ ?\ }$$

1. What happens to the product (9, 6, 3, ...) as the second factor (3, 2, 1, ...) in the multiplication decreases by 1?
2. To continue this pattern, what must $3 \cdot (-1)$ equal?
3. Copy and complete these patterns.

   **a.** $5 \cdot 2 = 10$  
   $5 \cdot 1 = 5$  
   $5 \cdot 0 = 0$  
   $5 \cdot (-1) = \underline{\ ?\ }$  
   $5 \cdot (-2) = \underline{\ ?\ }$

   **b.** $2 \cdot 3 = 6$  
   $1 \cdot 3 = 3$  
   $0 \cdot 3 = 0$  
   $(-1) \cdot 3 = \underline{\ ?\ }$  
   $(-2) \cdot 3 = \underline{\ ?\ }$

4. When two integers with different signs are multiplied, will the product be a positive or a negative integer? How do you know?

Now that you know that the product of a negative and a positive integer is a negative integer, you can use this fact and patterns to find the product of two negative integers.

Copy and complete each pattern. Notice what happens to the product as one of the two factors decreases by 1 each time.

a. $-3 \cdot 2 = -6$
$-3 \cdot 1 = -3$
$-3 \cdot 0 = 0$
$-3 \cdot (-1) = 3$
$-3 \cdot (-2) = 6$

b. $2 \cdot (-6) = -12$
$1 \cdot (-6) = -6$
$0 \cdot (-6) = 0$
$-1 \cdot (-6) = 6$
$-2 \cdot (-6) = 12$

5. When two negative integers are multiplied, will the product be a positive or a negative integer? How do you know?

The patterns you have explained suggest these rules for multiplying integers.

> **Multiplying Integers with Like Signs**
> 1. Find the product of the absolute values of the integers.
> 2. Write the product as a positive integer.
>
> **Multiplying Integers with Unlike Signs**
> 1. Find the product of the absolute values of the integers.
> 2. Write the product as a negative integer.

**EXAMPLE**  Find each product.
a. $-2 \cdot (-5)$     b. $8 \cdot (-1)$     c. $-10 \cdot 7$

**Solutions**
a. **Think:** Since the integers have the same sign, the product is positive.
$-2 \cdot (-5) = 10$

b. **Think:** Since the integers have unlike signs, the product is negative.
$8 \cdot (-1) = -8$

c. **Think:** Since the integers have unlike signs, the product is negative.
$-10 \cdot 7 = -70$

**Try This**  Find each product.
1. $-4 \cdot (-5)$    2. $6 \cdot (-8)$    3. $-3 \cdot 7$    4. $8 \cdot 9$

# EXERCISES

**Objective:** To multiply integers

## Check Your Understanding

**Skill Check**  Write the letter of the correct answer.

1. Find the product: $-9 \cdot 8$
   a. $-1$    b. $72$    c. $-72$    d. $1$

2. Find the product: $-15 \cdot (-4)$
   a. $60$    b. $-60$    c. $-19$    d. $-11$

3. Which statement is **not** correct?
   a. $-9 \cdot (-4) = -12 \cdot (-3)$
   b. $-6 \cdot 6 = -6 \cdot (-6)$
   c. $2 \cdot (-18) = -18 \cdot 2$
   d. $36 \cdot (-1) = -6 \cdot 6$

4. Which statement is correct?
   a. $-4 \cdot (-1) < -1 \cdot (-4)$
   b. $2 \cdot (-3) = -2 \cdot (-3)$
   c. $8 \cdot (-5) < 8 \cdot (-7)$
   d. $5 \cdot (-4) > 5 \cdot (-5)$

**True or False? When a statement is false, tell why.**
5. The integers $-9$ and $-12$ have unlike signs.
6. The product of two integers with like signs has the same sign as the integers.
7. The product of two integers with unlike signs is negative.
8. The product of a negative integer and zero is always a negative integer.

## Practice and Apply

**Copy and complete each pattern.**

9. $7 \cdot 2 = 14$
   $7 \cdot 1 = \underline{?}$
   $7 \cdot 0 = \underline{?}$
   $7 \cdot (-1) = -7$

10. $-9 \cdot 2 = \underline{?}$
    $-9 \cdot 1 = \underline{?}$
    $-9 \cdot 0 = \underline{?}$
    $-9 \cdot (-1) = \underline{?}$

11. $-5 \cdot 2 = -10$
    $\underline{?} \cdot \underline{?} = -5$
    $\underline{?} \cdot \underline{?} = 0$
    $\underline{?} \cdot \underline{?} = 5$

**Write P if the product is positive. Write N if it is negative.**

12. $-8 \cdot 34$    13. $-9 \cdot 26$    14. $-7 \cdot (-18)$    15. $28 \cdot 4$
16. $-5 \cdot (-53)$    17. $42 \cdot (-6)$    18. $65 \cdot 8$    19. $-3 \cdot 116$

**Multiply.**

20. $-3 \cdot (-4)$    21. $-8 \cdot 7$    22. $6 \cdot 5$    23. $4 \cdot (-8)$
24. $-9 \cdot 7$    25. $-9 \cdot (-5)$    26. $-8 \cdot (-1)$    27. $-4 \cdot (-3)$
28. $-4 \cdot 9$    29. $-6 \cdot 4$    30. $7 \cdot (-8)$    31. $-1 \cdot 6$
32. $-9 \cdot 0$    33. $-7 \cdot (-3)$    34. $2 \cdot 9$    35. $4 \cdot (-5)$
36. $-10 \cdot (-10)$    37. $-9 \cdot (-12)$    38. $7 \cdot (-15)$    39. $-20 \cdot (-5)$

**30**  CHAPTER 1

**40.** You multiply $-3 \cdot (-1) \cdot (-2)$. Will the product be a positive integer or a negative integer?

**41.** You multiply $-2 \cdot (-3) \cdot (-8) \cdot (-10)$. Will the product be a positive integer or a negative integer?

**42.** In Exercises 40 and 41, do you have to multiply to find the actual product in order to answer the question? Explain.

## Connect and Extend

**Copy and complete this table.**

| | Situation | Math Symbols | Solution I | Opposite Situation | Math Symbols | Solution II |
|---|---|---|---|---|---|---|
| **43.** | Five times a debt of $100 | 5(−100) | −500 | Five times a profit of $100 | 5(100) | 500 |
| **44.** | Six times a depth of 15 meters below sea level | ? | ? | ? | ? | ? |
| **45.** | Twice a gain of 200 feet in altitude | ? | ? | ? | ? | ? |
| **46.** | Three times a loss of 4 yards in rushing | ? | ? | ? | ? | ? |

**47.** Compare Solution 1 and Solution 2 for each Exercise. Are they always opposites? Explain your answer.

**48.** If the product of two integers is less than zero, what can you say about the integers?

## Maintain Your Skills

**Write the answer only if it is between 100 and 300.**

**49.** $39 \cdot 6$  **50.** $56 + 48 + 96$  **51.** $9850 + 25$  **52.** $14 \cdot 14$
**53.** $86 + 75 + 36$  **54.** $1476 - 1258$  **55.** $15 \cdot 12$  **56.** $2025 \cdot 15$

**Write the opposite of each number. (Pages 2–6)**

**57.** $-9$  **58.** $-6$  **59.** $4$  **60.** $0$

**Subtract. (Pages 23–25)**

**61.** $36 - 45$  **62.** $-21 - (-16)$  **63.** $12 - (-18)$  **64.** $-42 - 42$

## ★ Math Team Problem

**65.** Eighteen students, numbered consecutively and evenly spaced, are standing in a circle. Which student is directly opposite student number 1?

1-8   Multiplying Integers

# LESSON 1-9

## Dividing Integers

**R**amon is hang gliding. He drops 30 yards in 5 minutes. What is the amount of change in altitude per minute?

You can represent Ramon's loss of altitude as $-30$. To find the amount of change per minute, you divide $-30$ by 5.

You can write this division problem in three ways.

$$5\overline{)-30} \qquad -30 \div 5 \qquad \frac{-30}{5}$$

In algebra, divisions are often written in fraction form. Recall that multiplication and division are inverse operations. You can check a division problem by multiplying.

$$\frac{30}{5} = 6 \qquad \text{because} \qquad 6 \cdot 5 = 30.$$

Because multiplication and division are inverse operations, the rules for multiplying and dividing integers are closely related.

$$\frac{-30}{5} = -6 \qquad \text{because} \qquad -6 \cdot 5 = -30.$$

$$\frac{30}{-5} = -6 \qquad \text{because} \qquad -6 \cdot (-5) = 30.$$

**32** CHAPTER 1

Since $\frac{-30}{5} = -6$, the amount of change in Ramon's altitude is $-6$ yards per minute.

> **Dividing Integers with Like Signs**
> 1. Find the quotient of the absolute values of the integers.
> 2. Write the quotient as a positive number.
>
> **Dividing Integers with Unlike Signs**
> 1. Find the quotient of the absolute values of the integers.
> 2. Write the quotient as a negative number.

**EXAMPLE** Find each quotient.

a. $\frac{-8}{-4}$  b. $\frac{-49}{7}$  c. $\frac{8}{-1}$

**Solution**

a. **Think:** Since the integers have like signs, the quotient will be positive.  $\frac{-8}{-4} = 2$

b. **Think:** Since the integers have unlike signs, the quotient will be negative.  $\frac{-49}{7} = -7$

c. **Think:** Since the integers have unlike signs, the quotient will be negative.  $\frac{8}{-1} = -8$

**Try This** Find each quotient.

1. $\frac{16}{-8}$  2. $\frac{-63}{-9}$  3. $\frac{-44}{11}$

What happens if you try to divide by zero?

You know $\frac{10}{2} = 5$ because $5 \cdot 2 = 10$. However, it is impossible to find this quotient if the divisor is zero.

$$\frac{5}{0} = \underline{\ ?\ } \text{ because } \underline{\ ?\ } \cdot 0 = 5.$$

To find $\frac{5}{0}$ you must find a number that you can multiply by zero to get 5. There is *no* such number!

Something even stranger happens when you try to divide *zero* by zero.

To find $\frac{0}{0}$, you look for a number that you can multiply by zero to get zero. You can see that there is no one answer to this. Every number works! So we say that division by zero is **not possible,** or that it is **not defined.**

Try dividing by zero on a calculator. What happens?

# EXERCISES

**Objective:** To divide integers

## Check Your Understanding

**Skill Check**  Write the letter of the correct answer.

**1.** Which of these does **not** have the same meaning?
  **a.** The quotient of 48 and $-16$
  **b.** $\frac{-16}{48}$  **c.** $\frac{48}{-16}$
  **d.** $48 \div (-16)$

**2.** Which statement is **not** correct?
  **a.** $5 \div 0 = 0$
  **b.** Division by 0 is not defined.
  **c.** Multiplication and division are inverse operations.
  **d.** $\frac{15}{-5} = -3$ because $-3 \cdot (-5) = 15$.

**3.** Which statement is **not** correct?
  **a.** $\frac{-27}{3} = \frac{81}{-9}$  **b.** $\frac{0}{6} > \frac{9}{-3}$
  **c.** $\frac{13}{-1} < \frac{2}{-2}$  **d.** $\frac{-54}{27} > \frac{-6}{-6}$

**4.** Find the quotient: $\frac{-144}{-4}$
  **a.** 31  **b.** 36
  **c.** $-31$  **d.** $-36$

Replace the  ?  with the word or words that make each statement true.
**5.** The product and quotient of two numbers having like signs is  ?  (positive or negative)
**6.** The product and quotient of two numbers having unlike signs is  ? . (positive or negative)
**7.** The product of any number and zero is  ? .
**8.** Dividing by zero  ?  (is or is not) defined.

## Practice and Apply

**Write two division statements for each multiplication statement.**
  **9.** $-3 \cdot (-11) = 33$    **10.** $11 \cdot (-12) = -132$    **11.** $-8 \cdot 12 = -96$

**Write P if the quotient will be positive. Write N if it will be negative.**
  **12.** $\frac{-48}{12}$    **13.** $\frac{50}{-10}$    **14.** $\frac{-48}{2}$    **15.** $\frac{-100}{-10}$

16. $\frac{96}{-8}$
17. $\frac{-84}{6}$
18. $\frac{153}{9}$
19. $\frac{-132}{4}$

**Find the quotient.**
20. $\frac{-9}{3}$
21. $\frac{-16}{-8}$
22. $\frac{12}{4}$
23. $\frac{-6}{-2}$
24. $\frac{-15}{-3}$
25. $\frac{-8}{-1}$
26. $\frac{0}{-5}$
27. $\frac{48}{6}$
28. $\frac{56}{7}$
29. $\frac{-63}{-9}$
30. $\frac{28}{-4}$
31. $\frac{-56}{-8}$
32. $\frac{-42}{-6}$
33. $\frac{27}{3}$
34. $\frac{-40}{5}$
35. $\frac{-54}{6}$
36. $\frac{-90}{6}$
37. $\frac{-100}{-25}$
38. $\frac{128}{-8}$
39. $\frac{-240}{-60}$

## Connect and Extend

**Replace the  ?  with the integer that makes each statement true.**
40. $-56 \div (-7) = $  ?
41.  ?  $\div 6 = -42$
42. $-35 \div $  ?  $= 7$
43.  ?  $\div (-6) = 0$
44. $120 \div $  ?  $= -60$
45. $225 \div (-15) = $  ?

46. The temperature dropped 5°F each hour for six hours. Which expression shows the total decrease in temperature?
    **a.** $-6 \cdot -6$   **b.** $6 \cdot -5$   **c.** $6 \cdot 5$   **d.** $-6 \cdot 5$

47. The temperature dropped 32°C over 8 hours. Which expression shows the average drop each hour?
    **a.** $-32 \div 8$   **b.** $-32 \div (-8)$   **c.** $32 \div 8$   **d.** $32 - 8$

**For Exercises 48–50, find the next number in each listing by dividing.**
48. $-40, -20, -10,$  ?
49. $40, -20, 10,$  ?
50. $96, -48, 24,$  ?

## Maintain Your Skills

**Write an inequality sign (< or >) or an equal sign (=) to compare the numbers.**
51. $12.6 \cdot 15$  ?  $126 \cdot 1.5$
52. $834 - 256$  ?  $600$
53. $10 \cdot 65.3$  ?  $100 \cdot 6.53$
54. $9 \cdot 16$  ?  $18 \cdot 8$

**Find the absolute value of each integer. (Pages 2–6)**
55. $-16$
56. $37$
57. $0$
58. $-85$

**Evaluate. (Pages 13–18, 23–25, 28–31)**
59. $19 - (-12)$
60. $8 \cdot -6$
61. $-14 + (-26)$
62. $-12 \cdot (-15)$
63. $-18 + 7$
64. $-21 \cdot 8$
65. $21 - (-43)$
66. $-349 \cdot 0$

# SUMMARY

## Key Terms

absolute value (p. 4)
ascending order (p. 7)
descending order (p. 7)
inequality sign (p. 8)
integer (p. 3)
opposite (p. 4)

## Key Ideas

**A.** The absolute value of an integer is the distance between the integer and zero on the number line.

**B.** Two different integers are opposites if they have the same absolute value.

**C.** When you compare two integers on the number line, the greater integer is to the right of the smaller.

**D.** To add two integers having like signs:
   **1.** Add the absolute values of the integers.
   **2.** The sign of the sum is the same as the sign of the integers.

**E.** To add two integers having unlike signs:
   **1.** Find the difference of the absolute values of the integers.
   **2.** The sign of the sum is the same as the sign of the integer having the greater absolute value.

**F.** To subtract an integer, add its opposite.

**G. Multiplying Integers**
   **1.** The product of integers having like signs is positive.
   **2.** The product of integers having unlike signs is negative.

**H. Dividing Integers**
   **1.** The quotient of integers having like signs is positive.
   **2.** The quotient of integers having unlike signs is negative.

**I.** Division by zero is not possible.

# REVIEW

**Write the absolute value of each integer. (Pages 2–6)**
1. $-30$
2. $-12$
3. $9$
4. $0$
5. $-14$

**Write the opposite of each integer. (Pages 2–6)**
6. $16$
7. $-4$
8. $0$
9. $-8$
10. $-1$

**Write an inequality sign (<, >) to compare the two integers. (Pages 7–10)**
11. $-8$ and $8$
12. $-9$ and $-6$
13. $-4$ and $3$

**Write in descending order. (Pages 7–10)**
14. $-9, -2, -6$
15. $-7, -4, 2$
16. $-8, -5, 1, -3$

**Add. (Pages 13–18)**
17. $-16 + 17$
18. $-5 + (-6)$
19. $9 + (-9)$
20. $12 + (-18)$
21. $5 + (-4)$
22. $7 + (-31)$
23. $-7 + (-32)$
24. $-10 + (-3)$

**Subtract. (Pages 23–25)**
25. $9 - 3$
26. $9 - (-3)$
27. $-3 - 9$
28. $-3 - (-9)$
29. $-8 - 7$
30. $-6 - (-6)$
31. $1 - (-1)$
32. $-1 - (-1)$

**Work with a partner to solve these problems. Compare your method of solution with those of your classmates. (Pages 26–27)**

The figure at the right is made up of 8 identical letter cubes. There is no cube in the center of the figure.

33. What letter is opposite the letter F?
34. What letter is opposite the letter P?
35. What letter is opposite the letter A?

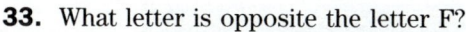

**Multiply. (Pages 28–31)**
36. $4 \cdot (-9)$
37. $-7 \cdot 3$
38. $-5 \cdot (-6)$
39. $-1 \cdot (-1)$
40. $-20 \cdot 3$
41. $-9 \cdot 0$
42. $8 \cdot (-10)$
43. $1 \cdot (-1)$

**Divide. (Pages 32–35)**
44. $\frac{-20}{5}$
45. $\frac{-50}{-25}$
46. $\frac{14}{-7}$
47. $\frac{-54}{9}$
48. $\frac{100}{-10}$
49. $\frac{-1500}{-1500}$
50. $\frac{1}{-1}$
51. $\frac{75}{-15}$

Chapter 1 Review

# TEST

**Find each answer.**
1. |−3|
2. |0|
3. |7|
4. |−7|
5. −(−10)
6. −(+9)
7. the opposite of 5

**Replace each ? with < or > to make the statement true.**
8. −7 ? 7
9. −7 ? −9
10. −7 ? −6
11. −7 ? 0

**Write in descending order.**
12. 1, −2, 3
13. −1, 2, −3
14. 0, −1, 3
15. 2, −2, 3

**Write P if the sum is positive, Z if it equals zero, and N if it is negative.**
16. −8 + 8
17. 23 + (−13)
18. −11 + (−7)
19. −3 + (−3)

**Find each answer.**
20. −3 + 2
21. −3 − 2
22. −3 + (−2)
23. −3 − (−2)
24. $\frac{-6}{2}$
25. $\frac{6}{-2}$
26. $\frac{-6}{-2}$
27. $\frac{0}{-2}$
28. −4 · 3
29. 3 · (−4)
30. −3 · 5
31. −3 · (−5)

32. Rewrite as a sum: −7 − (−3)

**For Exercises 33–35, write the operation (or operations) that you would use to solve the problem. Write +, −, ×, or ÷. Do not solve the problem.**

33. On three successive plays, a football team lost 5 yards, gained 15 yards, and gained 1 yard. If 10 yards are needed for a first down, did the team make a first down?

34. An airplane is at an altitude of 34,540 feet. A submarine directly below the plane is at an altitude of −150 feet. What is the distance between the plane and the submarine?

35. The product of two integers is 72. If one of the integers is −18, what is the other integer?

**38** CHAPTER 1 TEST

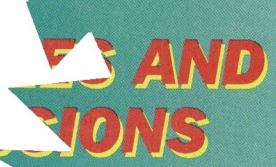

## LESSON 2-1

# Evaluating Expressions

**T**ops Record Store is having a sale. For one day only, you can buy cassette tapes for $8 each. This table shows how to write an algebraic expression to represent the total sales.

| Number of Cassettes | Dollars Received | |
|---|---|---|
| | Computation | Total |
| 1 | $8 \cdot 1$ | 8 |
| 2 | $8 \cdot 2$ | 16 |
| 3 | $8 \cdot 3$ | 24 |
| 4 | $8 \cdot 4$ | 32 |
| ⋮ | ⋮ | ⋮ |
| $n$ | $8 \cdot n$ | $8n$ |

To find the amount received after 5 cassettes are sold, you multiply 8 and 5. To find the amount received after $n$ cassettes are sold, you multiply 8 and $n$. The symbol $n$ is called a variable.

A **variable** is a letter or other symbol that can be replaced by any number or other object. When numbers and variables are combined, the result is an **algebraic expression**. If only numbers are used, the result is a **numerical expression**.

**Algebraic Expressions**  
$8n \quad x + 27 \quad y \div 19$

**Numerical Expressions**  
$8 \cdot 5 \quad -8 + 3 - 7$

This table will help you to translate word expressions to algebraic expressions. Notice that one algebraic expression can represent many word expressions.

| Word Expression | Algebraic Expression |
|---|---|
| 3 added to $a$<br>The sum of $a$ and 3<br>$a$ increased by 3<br>3 more than $a$ | $a + 3$ |
| 5 subtracted from $b$<br>$b$ decreased by 5<br>5 less than $b$<br>$b$ less 5 | $b - 5$ |
| $-2$ times $x$<br>The product of $-2$ and $x$<br>$-2$ multiplied by $x$ | $-2 \cdot x$ |
| 6 divided by $y$<br>The quotient of 6 and $y$ | $\dfrac{6}{y}$ |

**EXAMPLE 1** Translate each word expression to an algebraic expression.

|  | Word Expression | Algebraic Expression |
|---|---|---|
| a. | An amount, $a$, decreased by $35 | $a - 35$ |
| b. | $5 more than the cost, $c$ | $c + 5$ |
| c. | The number of days, $d$, divided by 7 | $\dfrac{d}{7}$ |

**Try This**   Translate each word expression to an algebraic expression.

1. The hours worked, $h$, times $5 per hour
2. The height, $t$, increased by $2\frac{1}{2}$ miles
3. The quotient of the number of days, $d$, and 7

Expressions such as $8 - 4$ and $\dfrac{-8}{-2}$ are **equivalent** because they name the same number.

$$8 - 4 = 4 \quad \text{and} \quad \dfrac{-8}{-2} = 4$$

Sometimes you have to simplify expressions to tell whether they are equivalent. An expression is **simplified** when no more operations, such as addition or multiplication, can be done.

When you simplify an expression containing parentheses, always do the operation within the parentheses first. This is called removing, or clearing, parentheses.

**EXAMPLE 2** Simplify each expression.

a. $5 \cdot (3 - 6)$
b. $\frac{(9 - 3)}{-6}$

**Solutions**
a. $5 \cdot (3 - 6) = 5 \cdot (-3)$
$= -15$

b. $\frac{(9 - 3)}{-6} = \frac{6}{-6}$
$= -1$

**Try This** Simplify each expression.

4. $(9 - 12) \cdot 6$
5. $\frac{42}{(4 - 10)}$
6. $7 - (12 - 5)$

Suppose that you want to find the cost of 13 tapes on sale at Tops Record Store. You can evaluate the algebraic expression $8n$ for $n = 13$.

Substitute 13 for $n$.   $8n = 8(13)$
Multiply.                      $= 104$

So the cost of 13 cassette tapes is $104.

To evaluate an algebraic expression for a given value of the variable, substitute that value for the variable. Then simplify.

**EXAMPLE 3** Evaluate each expression.

a. $7 + (m - 4)$, for $m = 2$
b. $\frac{(19 + c)}{-5}$, for $c = -4$

**Solutions**
a. Write the problem.
Substitute 2 for $m$.
Evaluate $(2 - 4)$.
Evaluate $7 + (-2)$.

$7 + (m - 4) = 7 + (2 - 4)$
$= 7 + (-2)$
$= 5$

b. Write the problem.
Substitute $-4$ for $c$.
Evaluate $19 + (-4)$.
Divide.

$\frac{(19 + c)}{-5} = \frac{[19 + (-4)]}{-5}$
$= \frac{15}{-5}$
$= -3$

**Try This** Evaluate each expression.

7. $(t + 8) \cdot (-3)$, for $t = 4$
8. $\frac{r - 7}{-4}$, for $r = 15$
9. $7 + (8 - b)$, for $b = 10$
10. $\frac{-30}{(m + 12)}$, for $m = -6$

# EXERCISES

**Objectives:** To translate word descriptions into algebraic descriptions
To evaluate algebraic expressions

## Check Your Understanding

**Skill Check** Write the letter of the correct answer.

1. Which expression is equivalent to $6 \cdot 8$?
   **a.** $6 + 8$  **b.** $8 \div 6$
   **c.** $\frac{96}{2}$  **d.** $\frac{24}{4}$

2. Which is the first step needed to simplify $3 + (4 - 6)$?
   **a.** $3 + (-2)$  **b.** $3 + 4 - 6$
   **c.** $1 - 6$  **d.** $3 + 2$

3. Which is the algebraic expression for "the quotient of $x$ and 7"?
   **a.** $\frac{x}{7}$  **b.** $\frac{7}{x}$
   **c.** $x - 7$  **d.** $7 - x$

4. Which is the algebraic expression of "8 less than $c$"?
   **a.** $8 - c$  **b.** $8 \div c$
   **c.** $c - 8$  **d.** $c \div 8$

5. Write an algebraic expression for "5 more than $m$."

6. Simplify: $9 - 2 - 5$

7. Evaluate $9 - (a + 2)$ for $a = 3$.

8. Write a word expression for this algebraic expression: $t - 5$

## Practice and Apply

Write each word expression as an algebraic expression.

9. The cost in cents of $s$ stamps at 25¢ each
10. The sum of a restaurant bill of $8.50 and the tip, $t$
11. The number of hours in $d$ days
12. The quotient of $x$ and 4
13. A school year of 180 days decreased by the number of days absent, $a$.
14. The cost for bowling $m$ games at $2 per game
15. The number of tickets you can buy for $30 if each ticket costs $d$ dollars
16. The amount, $a$, of tax withheld subtracted from a total wage of $100

Simplify.

17. $(4 + 7) - 9$
18. $(12 + 3) \cdot 2$
19. $\frac{-30}{(5 + 10)}$
20. $3 + (8 - 10)$
21. $6 \cdot (25 \div 5)$
22. $[56 \div (-7)] + 2$
23. $5 - (9 \cdot 2)$
24. $\frac{28}{(2 - 9)}$
25. $15 + [4 \cdot (-3)]$
26. $\frac{(-2 - 3)}{-1}$
27. $(81 \div 9) \cdot (-3)$
28. $(-3)(-7) + 7$

**Evaluate.**

29. $a + 8$, for $a = -2$
30. $9c$, for $c = -3$
31. $\frac{16}{g}$, for $g = 8$
32. $\frac{(x - 5)}{-8}$, for $x = 21$
33. $(s + 8) - 5$, for $s = 7$
34. $(t - 9) \cdot 5$, for $t = 6$
35. $-8 \cdot (17 - b)$, for $b = 10$
36. $(25 \div c) \cdot 6$, for $c = -5$
37. Beth weighs $w$ pounds. How much will she weigh after she gains 3 pounds?
38. Ted is $y$ years old. How old was he 4 years ago?
39. Sylvia earned $45 by washing cars. How much money will she have when she earns $d$ dollars more?
40. Karl has saved $x$ dollars. Carlos has saved 3 times as much. How much has Carlos saved?

## Connect and Extend

41. Write this algebraic expression in words in two different ways: $4 + a$
42. Write these words as algebraic expressions in two different ways: The sum of 3 and $b$

**Copy and complete the pattern.**

43. 2    3    4    5
    $n$    $n+1$    $n+2$    ?

44. 3    6    9    12
    $n$    $2n$    ?    ?

45. 20    15    10    5
    $n$    ?    ?    ?

46. 120    60    30    20
    $n$    ?    ?    ?

47. For what value of $y$ will $14 - (-y)$ equal 21?
48. For what value of $k$ will $k \cdot [8 \cdot (-2)]$ equal 16?
49. For what value of $r$ will $\frac{40}{(5 - r)}$ equal 10?

## Maintain Your Skills

**Copy and complete. (Pages 13–18, 23–25, 28–35)**

50. $-3 + (-5) = \underline{?}$
51. $\underline{?} \cdot (-7) = 35$
52. $15 - (-8) = \underline{?}$
53. $12 + (-18) = \underline{?}$
54. $\underline{?} \cdot (-6) = -54$
55. $11 - \underline{?} = -8$
56. $8 + \underline{?} = 5$
57. $\frac{?}{-6} = 8$
58. $-18 - (-2) = \underline{?}$

## ★ Math Team Problems

Write a paragraph about each of these topics. Give examples to illustrate the likenesses and differences.

59. How letters are like numbers
60. How letters are different from numbers

# LESSON 2-2
# Problem Solving Strategies Using a Step-by-Step Process

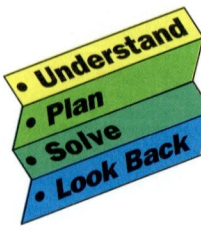

Many successful problem solvers do these things.

- Think about whether they **understand** a problem before trying to calculate an answer.

- **Plan** how they will use calculations to arrive at an answer.

- **Look back** at their work after they have solved a problem to be sure they have answered the question asked.

- Think about how they can **apply** what they have learned in other problem-solving situations.

Three students tried to solve this problem. Who solved it correctly? What mistakes were made?

**Problem:** One package of 2 videotapes is on sale for $6.35 and a special pack of 6 tapes is on sale for $12.98. Which is the better buy?

### Gerald's Solution

```
 6.35      12.98       77.88
 x  2      x   6      +12.70
 12.70     77.88       90.58
```
Answer: $90.58

### Katia's Solution

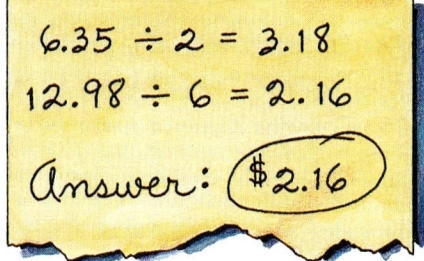

6.35 ÷ 2 = 3.18
12.98 ÷ 6 = 2.16
Answer: $2.16

### Lara's Solution

Plan: Find the price per tape.
Use estimation and division.

6.35 ÷ 2 is a little over 3.
12.98 ÷ 6 is a little over 2.
So the 6-pack is the better buy if I want to buy all 6 tapes and if I want to spend $13.
Answer: 6 tapes for $12.98 is cheaper if a person wants all 6 tapes and has $13 to spend.

You can see that Lara solved the problem correctly. She thought about what question was asked and how to answer it. She used estimation to find an approximate price per tape. Then she thought about whether her answer was reasonable.

Since Gerald did not use the correct operations, he probably did not understand what the problem was about.

Katia used the correct operations and her calculations were correct. However, she forgot to answer the question that was asked.

## EXERCISES

**Objective:** To follow a step-by-step process to solve problems

**Solve each problem.**

1. A board is 5 feet long. A piece $f$ feet long is sawed off.
   a. Draw and label a diagram showing the length of the board and the length of the piece that is sawed off.
   b. If a 2-foot piece is sawed off, how long is the remaining piece?
   c. If a 4-foot piece is sawed off, how long is the remaining piece?
   d. Write an algebraic expression for the length of the remaining piece. Use $f$ for the length of the piece sawed off.
   e. Let $f = 1$ foot and evaluate the algebraic expression you wrote for $f = 1$. Do you get the correct length for the remaining piece when 1 foot is sawed off? If not, go back to 1a and rethink your work.

2. At a Boy Scout Jamboree, Harry asks each person in a circle to name two numbers that, when multiplied together, will give a product of 24. No two answers can be exactly the same. How many people can answer before no more correct answers are possible? Explain.

3. Suppose that Harry (see Exercise 2) asks the same question and adds the condition that the numbers must be integers.
   a. How many different answers are possible?
   b. Does your answer assume that an answer of "2 and 12" is exactly the same as an answer of "12 and 2"?

4. Vera plans to fence in a pasture for her llamas. She starts at the outdoor water faucet and walks 20 yards due north. Then she walks due east for 5 yards, turns and walks 6 yards due north, turns due east and walks 6 yards. From there she turns due south and walks 26 yards, then turns due west and walks straight back to the water faucet where she began. If she builds a fence along the path she walked, what is the distance around the land she will enclose?

## LESSON 2-3

# Order of Operations

John and Tanu'e White called a plumber to repair a leaking pipe in the kitchen of their house. The plumber charges $30 per hour plus a service charge of $25. The two bills shown above show how the plumber and the Whites computed the cost for 4 hours of work. Who computed the cost correctly?

In mathematics, one expression, such as 4 · 30 + 55, cannot have two different values. Rules are needed to be sure that everyone simplifies an expression in a way that gives the same result. Mathematicians agree to use these rules for order of operations.

> **Order of Operations**
> 1. Perform all the operations enclosed in parentheses.
> 2. Perform all the operations in the numerator or denominator.
> 3. Perform all the multiplications and divisions, in order, from left to right.
> 4. Perform all additions and subtractions, in order, from left to right.

For expressions having multiplications and divisions only, do whichever comes first, in order, from left to right.

24 ÷ 2 · 3
12 · 3
36

For expressions having additions and subtractions only, do whichever comes first, in order, from left to right.

8 − 3 + 1
5 + 1
6

**46** CHAPTER 2

Now use order of operations to compute the plumber's bill at the top of page 46.

Multiply first. (Rule 3)    $4 \cdot 30 + 25 = 120 + 25$
Then add. (Rule 4)                              $= 145$

So the plumber computed the answer correctly.

When simplifying expressions remember to perform the operations inside parentheses first.

**EXAMPLE 1** Simplify each expression.
a. $7 + 9 - (6 + 4)$
b. $12 - 2 \cdot 5 + 3$

**Solutions** Follow the rules for order of operations.

a. $7 + 9 - (6 + 4)$
$7 + 9 - \underbrace{\quad 10 \quad}$
$\underbrace{16 \quad} - \quad 10$
$\quad\quad 6$

b. $12 - \underbrace{2 \cdot 5} + 3$
$12 - \underbrace{\quad 10 \quad} + 3$
$\quad\quad 2 \quad\quad + 3$
$\quad\quad 5$

**Try This** **Simplify each expression.**
1. $8 + 4 - (2 + 1)$
2. $15 - 5 \cdot 2 + 3$
3. $3 - 1 + 4 \div 2$

Some algebraic expressions contain more than one variable. To evaluate these expressions, substitute for each variable and follow the rules for order of operations.

**EXAMPLE 2** Evaluate $\frac{r + t}{3} + 3t$ for $r = 10$ and $t = -4$.

**Solutions**
Write the expression.                  $\frac{r + t}{3} + 3t$
Substitute 10 for $r$ and $-4$ for $t$.    $\frac{10 + (-4)}{3} + 3(-4)$
Simplify the numerator.               $\frac{6}{3} + 3(-4)$
Divide and multiply.                  $2 + (-12)$
Add.                                  $-10$

**Try This** **Evaluate each expression.**
4. $6 \div c + d$, for $c = 2$, $d = 1$
5. $\frac{p - q}{2}$, for $p = -3$, $q = 5$

If an expression uses the same variable more than once, substitute the same value for the variable each time. For example, to evaluate $(x + 3) \cdot (x - 1)$ for $x = -2$, write

$$(x + 3) \cdot (x - 1) = (-2 + 3) \cdot (-2 - 1)$$
$$= 1 \cdot (-3)$$
$$= -3$$

The dot indicating multiplication is usually omitted with parentheses. That is,

$(x + 3)(x - 1)$ means to multiply $(x + 3)$ and $(x - 1)$.
$5(x + 3)$ means to multiply 5 and $(x + 3)$.

You can also leave out the multiplication dot when you wish to show the product of two variables or the product of a number and a variable. The following expressions are **equivalent**.

$$c \cdot d \qquad (c)(d) \qquad c(d) \qquad (c)d \qquad cd$$

**EXAMPLE 3** Write the letter(s) of the equivalent expressions.
a. $3 \cdot x$  b. $3x$  c. $3 + x$  d. $3 \div x$  e. $3(x)$

**Solution** The expressions in **a, b,** and **e** are equivalent because each indicates the product of 3 and $x$.

**Try This** Copy each row of expressions. Circle equivalent expressions.
6. $3 + a \qquad 3 \cdot a \qquad 3 + b \qquad 3a \qquad 3(a)$
7. $xy \qquad x(y) \qquad x \cdot y \qquad y - x \qquad x - y$

Scientific calculators follow the rules for order of operations. When expressions have no parentheses, enter the numbers and operations in order from left to right.

| Expression | Calculator Keys | Answer |
|---|---|---|
| $8 + 15 \div 3$ | 8 [+] 15 [÷] 3 [=] | 13 |

Most scientific calculators have parentheses keys, [(] and [)]. Enter the parentheses or brackets where they appear in the problem. Remember, however, to press the [×] key to indicate "hidden" multiplication, such as $3 \times (18 - 6)$ in $96 + 3(18 - 6)$.

| Calculator Keys | Answer |
|---|---|
| 96 [+] 3 [×] [(] 18 [−] 6 [)] [=] | 132 |

# EXERCISES

**Objective:** To use the order of operations to simplify numerical expressions

## Check Your Understanding

*Skill Check* Write the letter of the correct answer.

1. Which is the first step in simplifying $16 \div 4 \cdot 2 + 6$?
   a. $16 \div 8 + 6$
   b. $4 \cdot 2 + 6$
   c. $16 \div 4 \cdot 8$
   d. $4 \cdot 8$

2. Which expression is equivalent to $5(a + b)$?
   a. $5 + (a + b)$
   b. $5 - (a + b)$
   c. $5 \cdot (a + b)$
   d. $5 \div (a + b)$

3. Which steps are used to simplify $120 \div (6 + 4) \cdot 3$?
   a. $120 \div 6 + 4 \cdot 3$
      $20 + 12$
   b. $120 \div 10 \cdot 3$
      $120 \div 30$
   c. $120 \div 6 + 4 \cdot 3$
      $20 + 4 \cdot 3$
      $24 \cdot 3$
   d. $120 \div 10 \cdot 3$
      $12 \cdot 3$

4. Which steps are used to simplify $26 - 12 \cdot 2 + 2$?
   a. $14 \cdot 2 + 2$
      $28 + 2$
   b. $26 - 24 + 2$
      $2 + 2$
   c. $26 - 24 + 2$
      $26 - 26$
   d. $24 \cdot 4$

5. Write four expressions that are equivalent to $7 \cdot c$.

**Evaluate each expression for $a = 8$ and $b = -2$.**

6. $b - a$
7. $2a + b$
8. $ab$
9. $\dfrac{a}{b}$
10. $\dfrac{a - b}{b - a}$

## Practice and Apply

**Use the order of operations to simplify each expression.**

11. $4 \cdot 6 + 8$
12. $3 \cdot (7 + 8)$
13. $9 + 8 \cdot 2$
14. $3 \cdot 5 - 20$
15. $20 - 6 \div 3$
16. $-20 + 9 - 5$
17. $25 + 14 - 20$
18. $5 \cdot 2 + 3 \cdot 8$
19. $6 \cdot 4 - 7 \cdot 5$
20. $16 - (4 - 2)$
21. $(16 - 4) - 2$
22. $16 - 4 - 2$
23. $9 \cdot 4 - (5 + 8)$
24. $(15 + 9) - (12 - 7)$
25. $-3 \cdot 8 - 4 \cdot 6$
26. $4 + 9(6 + 3)$
27. $(-5 - 4) \cdot (3 + 7)$
28. $9 \cdot (8 - 6) + 4$

29. Evaluate $\dfrac{m + a}{5}$ for $m = 2$ and $a = 8$.

30. Evaluate $ab + 9$ for $a = 7$ and $b = -2$.

31. Evaluate $\dfrac{3x}{2y}$ for $x = 8$ and $y = -6$.

32. Evaluate $\dfrac{2c - d}{5}$ for $c = 3$ and $d = -9$.

**33.** To evaluate 27 + 9(63 ÷ 7), Amy pressed these keys on her calculator.

27 [+] 9 [(] 63 [÷] 7 [)] [=]

Did she get the correct answer? If not, what error did she make?

## Connect and Extend

**34.** Use the numbers 2, 3, and 5 to write an expression that, when simplified, results in 17.

**35.** Use the numbers $-2$, 3, and 4 to write an expression that, when simplified, results in 14.

**36.** If $a - b = 20$, what is the value of $a - b + 5$?

**37.** If $c + d = -6$, what is the value of $c + d - 3$?

**38.** If $p + q = 9$, what is the value of $2(p + q)$?

**39.** If $x + y + 10 = 20$, what is the value of $x + y$?

## Maintain Your Skills

**Arrange in ascending order. (Pages 2–10)**

**40.** $-2, -4, 6, 0$  **41.** $10, -10, -20, -12$  **42.** $|-3|, -|-3|, -(-4), -4$

**Arrange in descending order. (Pages 2–10)**

**43.** $-6, -8, -10, -12$  **44.** $-5, |-4|, |3|, -|3|$  **45.** $|-10|, -|10|, |10|, |0|$

**Correct any wrong answers. (Pages 13–18, 23–25, 28–31)**

**46.** $-4 - (-3) = 1$  **47.** $(-9)(-3) = 27$  **48.** $5(-7) = 35$

## Mathematical Footnote

Over 3500 years ago in Egypt, a country in the northeast corner of Africa, a scribe named Ahmes copied some math problems that were already 300 years old. Ahmes wrote on papyrus, a paper made from tall reeds. In 1858, this papyrus roll, 18 feet long and 13 inches wide, was bought in an Egyptian shop by A. Henry Rhind, a Scottish scholar. The Rhind papyrus contains arithmetic problems, geometry problems, and problems about areas and volumes.

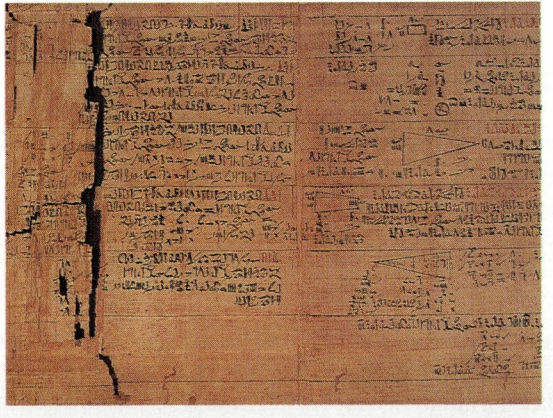

Write each word expression as an algebraic expression.
(Pages 39–43)
1. The cost of $n$ bottles of juice at 75¢ per bottle
2. The number of days in $h$ hours

Evaluate each expression. (Pages 39–43)
3. $b - 8$, for $b = -2$
4. $(p + 5) \cdot (-3)$, for $p = -10$

Evaluate each expression. (Pages 46–50)
5. $\frac{2d - t}{-5}$, for $d = 4$ and $t = -2$
6. $\frac{3x}{5y}$, for $x = 10$ and $y = -2$

Use the order of operations to simplify each expression.
(Pages 46–50)
7. $14 + 8 \div 2 - 1$
8. $3 \cdot (-6 + 4) + 5$
9. $3 \cdot 8 - (-4 \div 2)$
10. $(14 + 2) - (16 \div 4)$
11. $-9 \div 3 + 12 \div 6$
12. $(4 \cdot 5) + (6 \div 3)$

# Mathematical Footnote

Do research to find the math background required for each career. How is math background related to hourly pay? It pays to study math!

| Hourly Pay | Career |
|---|---|
| $4.18 | Bank teller |
| $7.21 | Secretary |
| $8.17 | Airline attendant |
| $9.33 | Social worker |
| $9.86 | Elementary teacher |
| $11.54 | Nurse |
| $13.94 | Photographer |
| $15.63 | House painter |
| $17.12 | Electrician |
| $21.63 | Optometrist |
| $31.68 | Dentist |

# LESSON 2-5
# Basic Properties and Mental Computation

Let $a = -9$ and $b = 10$. Substitute these values in the pairs of expressions. Then simplify.

$$
\begin{array}{cccc}
a + b & b + a & ab & ba \\
-9 + 10 & 10 + (-9) & (-9)(10) & (10)(-9) \\
1 & 1 & -90 & -90
\end{array}
$$

      **Same answer**              **Same answer**

You will get the same answer for each pair of expressions no matter what values you choose for $a$ and $b$. This illustrates the Commutative Properties of Addition and Multiplication.

> **Commutative Property of Addition**
> The order in which you add two numbers does not change the sum.
> $$a + b = b + a$$
>
> **Commutative Property of Multiplication**
> The order in which you multiply two numbers does not change the product.
> $$ab = ba$$

Now let $a = -3$, $b = 2$, and $c = 5$. Substitute these values in each pair of expressions. Then simplify.

$$
\begin{array}{cccc}
(a + b) + c & a + (b + c) & a \cdot (b \cdot c) & (a \cdot b) \cdot c \\
(-3 + 2) + 5 & -3 + (2 + 5) & -3 \cdot (2 \cdot 5) & (-3 \cdot 2) \cdot 5 \\
-1 \phantom{+} + 5 & -3 + \phantom{(}7\phantom{)} & -3 \cdot \phantom{(}10\phantom{)} & -6 \phantom{\cdot} \cdot 5 \\
4 & 4 & -30 & -30
\end{array}
$$

      **Same answer**              **Same answer**

You will get the same answer for each pair of expressions no matter what values you choose for $a$, $b$, and $c$. This illustrates the Associative Properties of Addition and Multiplication.

# LESSON 2-4
# Problem Solving Exploration
# Making Generalizations

A. **C**opy and complete the statements in each pattern.

**Pattern 1**
$3 + 6 = \underline{\ ?\ } + 3$
$7 + (-5) = -5 + \underline{\ ?\ }$
$\underline{\ ?\ } + 8 = 8 + (-2)$
$59 + \underline{\ ?\ } = -12 + 59$

**Pattern 2**
$4 \cdot \underline{\ ?\ } = 6 \cdot 4$
$3 \cdot (-2) = \underline{\ ?\ } \cdot 3$
$\underline{\ ?\ } \cdot 10 = 10 \cdot (-7)$
$-18 \cdot (-11) = -11 \cdot \underline{\ ?\ }$

1. Write one sentence that describes all the statements in Pattern 1.

2. Write one sentence that describes all the statements in Pattern 2.

3. How are the sentences you wrote alike? How are they different?

Each statement in Pattern 1 is a specific instance of a pattern. You can use variables to describe each pattern in general.

**Pattern 1**
$a + b = b + a$

**Pattern 2**
$a \cdot b = b \cdot a$

Using variables to describe a pattern is one way of making a **generalization** from the pattern.

Generalizations may be true or false. You can show that a generalization is false if you can find *only* one instance when it is false. This instance is called a **counterexample**.

For example, $a - b = b - a$ is false because it is not true for the instance when $a = 8$ and $b = 3$.

$8 - 3 \neq 3 - 8$ (The symbol $\neq$ is read as "does not equal.")

$8 - 3 \neq 3 - 8$ is called a counterexample for $a - b = b - a$.

4. Consider this generalization: $2 \cdot a = a \cdot a$
   If $a = 0$, this expression becomes $2 \cdot 0 = 0 \cdot 0$. Is the generalization true for $a = 0$?

5. Find another value for a that makes the expression true.

6. Find a value for a that makes the expression false.

7. Explain why $2 \cdot a = a \cdot a$ is not a true generalization.

8. The expression $a \cdot a = a + 0$ is true if $a$ has a value of one. Find a value for $a$ that makes the statement false.

9. Is $a \cdot a = a + 0$ a true or false generalization? Explain.

B. A generalization is true if *every instance* of the generalization is true.

**Copy and complete each statement.**

**Pattern 3**
$(2 + 3) + 1 = 2 + (\underline{?} + 1)$
$[5 + (-1)] + 4 = \underline{?} + (-1 + 4)$
$(-2 + 8) + 3 = -2 + (8 + 3)$

**Pattern 4**
$[6 \cdot (-3)] \cdot 4 = 6 \cdot (-3 \cdot \underline{?})$
$(8 \cdot 4) \cdot (-2) = \underline{?} [4 \cdot (-2)]$
$(-24 \cdot 20) \cdot 5 = -24 \cdot (\underline{?} \cdot 5)$

10. Write one sentence that describes all the statements in Pattern 3.

11. Write one sentence that describes all the statements in Pattern 4.

12. How are the sentences you wrote alike? How are they different?

13. Explain why the generalization $(a + b) + c = a + (b + c)$ can be used to describe Pattern 3. Is this a true generalization? Why?

14. Explain why the generalization $(ab)c = (bc)$ can be used to describe Pattern 4. Is this a true generalization? Why?

C. A generalization is false if one counterexample can be found.

A generalization is true if it is true for every instance.

**For each of the following, try several instances. If you think the description is a true generalization, answer yes. If not, give a counterexample.**

15. $a + 0 = a$

16. $a \div b = b \div a$

17. $a \cdot 1 = a$

18. $2a + a = 3a$

19. $a \cdot b + a = a(b + 1)$

20. $a - 0 = 0 - a$

21. $a \cdot (b + c) = a \cdot b + a \cdot c$

22. $a \cdot (b + c) = (b + c) \cdot a$

**Associative Property of Addition**

The way numbers are grouped for addition does not change the sum.

$$a + (b + c) = (a + b) + c$$

**Associative Property of Multiplication**

The way numbers are grouped for multiplication does not change the product.

$$a \cdot (b \cdot c) = (a \cdot b) \cdot c$$

Some special properties of numbers involve zero and one.

**Identity Property of Addition**

The sum of any number and 0 is that number.

$$a + 0 = a$$

**Identity Property of Multiplication**

The product of any number and 1 is that number.

$$a \cdot 1 = a$$

**EXAMPLE 1** Replace each ? with the number that makes the statement true. Write the name of the property shown.

a. $5 + \underline{?} = -3 + 5$ 
b. $3 + (4 \cdot 1) = 3 + \underline{?}$
c. $2 \cdot (6 + 4) = (6 + \underline{?}) \cdot 2$

**Solutions**
a. $5 + (-3) = -3 + 5$     Commutative Property of Addition
b. $3 + (4 \cdot 1) = 3 + 4$     Identity Property of Multiplication
c. $2 \cdot (6 + 4) = (6 + 4) \cdot 2$     Commutative Property of Multiplication

**Try This** Copy and replace each ? with the number that makes the statement true. Write the name of the property shown.

1. $(9 \cdot 5) \cdot 4 = 9 \cdot (\underline{?} \cdot 4)$
2. $9 \cdot (\underline{?} \cdot 8) = (9 \cdot 0) \cdot 8$
3. $(-7 \cdot \underline{?}) + 5 = [3 \cdot (-7)] + 5$

You can use the addition and multiplication properties to help you simplify expressions mentally.

2-5   Basic Properties

**EXAMPLE 2** Explain how you would use the addition and multiplication properties to help you simplify each expression mentally.
  a. $60 + (40 + 9)$
  b. $(18 + 12) + (-18)$

**Solutions**

a. Since it is easy to add 60 and 40 mentally, change the grouping to $(60 + 40)$. Then simplify.

Apply the Associative Property of Addition.
$$60 + (40 + 9) = (60 + 40) + 9$$
$$= 100 + 9$$
$$= 109$$

b. Since $-18 + 18 = 0$, change the position of $-18$.

Apply the Commutative Property of Addition.
$$(18 + 12) + (-18) = -18 + (18 + 12)$$

Apply the Associative Property of Addition.
$$= (-18 + 18) + 12$$

Apply addition of opposites.
$$= 0 + 12$$

Apply the Identity Property of Addition.
$$= 12$$

**Try This** Explain how you would use the various properties to simplify each expression mentally.

4. $(-9 - 6) + 9$
5. $(-11 + 12) \cdot 9$
6. $(25 \cdot 9) \cdot 4$

# EXERCISES

**Objective:** To use the Commutative, Associative, and Identity Properties to simplify numerical expressions mentally

## Check Your Understanding

**Skill Check** Write the letter of the correct answer.

1. Which property is shown by $(-2 \cdot 1) + 6 = -2 + 6$?
   a. Associative Property of Multiplication
   b. Commutative Property of Multiplication
   c. Identity Property of Addition
   d. Identity Property of Multiplication

2. Which expression shows the Commutative Property of Addition?
   a. $-3 \cdot (7 + 10) = (7 + 10) \cdot -3$
   b. $-3 \cdot (7 + 10) = (-3 \cdot 7) + 10$
   c. $-3 \cdot (7 + 10) = -3 \cdot (10 + 7)$
   d. $-3 \cdot (7 + 10) = -3 \cdot 7 + (-3) \cdot 10$

**3.** Which property will you use first to simplify $(16 + 25) + 25$ mentally?
  **a.** Associative Property of Addition
  **b.** Commutative Property of Multiplication
  **c.** Identity Property of Addition
  **d.** Identity Property of Multiplication

**4.** Which property will you use first to simplify $-8 \cdot 9 \cdot 5$ mentally?
  **a.** Associative Property of Addition
  **b.** Associative Property of Multiplication
  **c.** Commutative Property of Addition
  **d.** Commutative Property of Multiplication

**5.** Use the Associative Property of Multiplication to write an equivalent expression for $(a \cdot b) \cdot c$.

**6.** Give an example of the Identity Property of Addition.

**Explain how you would use the various properties to simplify each expression mentally.**

**7.** $(-6 \cdot 4) \cdot (-25)$

**8.** $(-31 + 12) + 31$

## Practice and Apply

**Replace each ? with the number that makes the statement true. Write the name of the property shown.**

**9.** $-3 \cdot 8 = \underline{?} \cdot -3$

**10.** $(-5 + 7) + \underline{?} = -5 + (7 + 2)$

**11.** $4 \cdot (-9 \cdot \underline{?}) = [4 \cdot (-9)] \cdot -6$

**12.** $8 + (-6) = \underline{?} + 8$

**13.** $[12 + (-3)] + 9 = 12 + (\underline{?} + 9)$

**14.** $-8 + (\underline{?} + 3) = -8 + (3 + 4)$

**15.** $[\underline{?} \cdot (-3)] \cdot (-5) = [-3 \cdot (-9)] \cdot (-5)$

**16.** $(-7 + 0) + \underline{?} = -7 - 6$

**17.** $(-16 \cdot 12) + 10 = 10 + (\underline{?} \cdot 12)$

**18.** $(57 + 43) \cdot \underline{?} = 57 + 43$

**19.** $(27 + 96) + 0 = \underline{?} + (27 + 96)$

**20.** $(-15 \cdot 1) \cdot 23 = \underline{?} \cdot 23$

**21.** $35 \cdot [-9 \cdot (-7)] = [\underline{?} \cdot (-7)] \cdot 35$

**22.** $[24 \cdot (-38)] + (-50) = -50 + (24 \cdot \underline{?})$

**Simplify each expression mentally.**

**23.** $(56 + 79) + 44$

**24.** $(45 \cdot 32) \cdot 0$

**25.** $(-96 + 156) + 96$

**26.** $(87 \cdot 5) \cdot 2$

**27.** $-78 + (116 + 78)$

**28.** $[25 \cdot (-72)] \cdot 4$

**29.** $(93 - 128) + 128$

**30.** $(138 - 64) - 138$

**31.** $(688 + 915) + 312$

**32.** $50 \cdot (118 \cdot 20)$

**33.** $(825 - 492) - 825$

**34.** $(914 + 816) - 914$

**Look up each of these words in a dictionary. Explain how the meaning of each word is related to a property of numbers that you have learned in this lesson.**

**35.** commuter

**36.** associate

**37.** identity

2-5 Basic Properties **57**

Write True or False for each statement. If a statement is false, give a counterexample.

**38.** If the product of two numbers is 0, then at least one of the numbers is 0.

**39.** If the sum of two numbers is 0, then the numbers are opposites of each other.

## Connect and Extend

**40.** Is putting on your shoes, then putting on your socks a commutative operation? Explain.

**41.** Is subtraction a commutative operation? Is it an associative operation? Give 3 examples to explain your answers.

**42.** Is division a commutative operation? Is it an associative operation? Give 3 examples to explain your answers.

Let * represent an operation that means double the first number and subtract the second number.

**43.** Evaluate $6 * y$, when $y = 3$.

**44.** Evaluate $y * 6$, when $y = 3$.

**45.** Is * a commutative operation? Explain your answer.

## Maintain Your Skills

Write $+$, $-$, $\times$ or $\div$ to make true statements.
(Pages 13–18, 23–25, 28–31, 32–35, 46–50)

**46.** $-5 \underline{\ ?\ } 1 = -4$

**47.** $35 \underline{\ ?\ } 7 \underline{\ ?\ } 3 = 15$

**48.** $-5 \underline{\ ?\ } 8 = -13$

**49.** $2 \cdot (6 + 6) \underline{\ ?\ } 8 = 3$

Translate each word expression into an algebraic expression.
(Pages 39–43)

**50.** The number $n$ divided by 8

**51.** The number of cents in $r$ nickels

Tell whether each pair of expressions has the same value when $a = -3$ and $b = 4$. (Pages 39–43)

**52.** $2a - b$ and $2(a - b)$

**53.** $ab + 18$ and $a(b - 6)$

Write the letter of the correct answer.

**54.** Sam gathered 12 buckets of scallops, Kirstin gathered 9 buckets, and Jose gathered 6 buckets. They sold the scallops for $4 a bucket and shared the profits equally. How much did each person earn?
 **a.** $27  **b.** $36
 **c.** $108  **d.** $30

**55.** Jennifer bought three sweaters at $27 each and six pairs of socks at $4 each. Which best describes how much she spent?
 **a.** $3 \times (\$27 + 6) \times \$4$
 **b.** $(3 + \$27) \times (6 + \$4)$
 **c.** $10 \times (\$27 + \$4)$
 **d.** $(3 \times \$27) + (6 \times \$4)$

## LESSON 2-6

# Using the Distributive Property

**M**s. Garcia earns $15 an hour tutoring math students. She tutored 4 hours on Friday and 6 hours on Saturday. Ms. Garcia asked her students to find two ways to compute her earnings. Here are the methods they used.

**Bill's Method**
Find the amount earned each day. Then add.

**Sam's Method**
Multiply the pay per hour and the total number of hours worked.

Fri. Pay + Sat. Pay = Total

$15 \cdot 4 \; + \; 15 \cdot 6 \; = 60 + 90$
$\qquad\qquad\qquad\qquad = 150$

Pay per Hour · Hours Worked = Total

$15 \;\; \cdot (4 + 6) = 15(10)$
$\qquad\qquad\qquad = 150$

You can see that Bill multiplied before adding, and that Sam added before multiplying. That is,

$15 \cdot 4 + 15 \cdot 6 = 150$ and $15 \cdot (4 + 6) = 150$.

So $15 \cdot 4 + 15 \cdot 6$ is equivalent to $15 \cdot (4 + 6)$,
and $15 \cdot 4 + 15 \cdot 6 = 15 \cdot (4 + 6)$.

2-6  Distributive Property  **59**

These patterns are instances of the Distributive Property of Multiplication over Addition.

> **Distributive Property of Multiplication over Addition**
> For all numbers $a$, $b$, and $c$,
> $a(b + c) = ab + ac$  $(b + c)a = ba + ca$
> $ab + ac = a(b + c)$  $ba + ca = (b + c)a$

To see how the Distributive Property of Multiplication over Addition got its name, look at the instance below. Notice how the factor, 25, is distributed (spread) to the two addends, 19 and 30.

$$25(19 + 30) = 25 \cdot 19 + 25 \cdot 30$$

**EXAMPLE 1**  Use the Distributive Property to complete each statement.
 a. $-5(9 + 6) = -5 \cdot (\underline{?}) + -5 \cdot (\underline{?})$
 b. $3 \cdot 2 + 6 \cdot 2 = (3 + \underline{?})2$
 c. $4 \cdot (-7 + 3) = \underline{?} \cdot (-7) + \underline{?}(3)$

**Solutions**  a. **9, 6**   b. **6**   c. **4, 4**

**Try This**  Use the Distributive Property to complete each statement.
 1. $8 \cdot 7 + 8 \cdot (-2) = 8(\underline{?} + \underline{?})$
 2. $(-4 + 9)(-3) = -4 \cdot (-3) + \underline{?} \cdot (-3)$
 3. $6 \cdot 2 + 14 \cdot 2 = (6 + 14) \cdot \underline{?}$

In the expression $4 \cdot x$, or more simply, $4x$, 4 is the **numerical coefficient** of $x$. In the expression $x$, the number 1 is the numerical coefficient of $x$, although the 1 is not usually written.

$$1 \cdot x = 1x = x$$

In the expression $-x$, $-1$ is the numerical coefficient of $x$.

$$-1 \cdot x = -1x = -x$$

The expression $3x + 4x$ is the sum of two terms, $3x$ and $4x$. The terms $3x$ and $4x$ are called **like terms** because they contain the same variable. You can use the Distributive Property of Multiplication over Addition to add like terms.

• Are $6x$ and $-5y$ like terms? Why or why not?

• Are $-12t$ and $-13$ like terms? Why or why not?

**EXAMPLE 2**  Add.
  a. $6x + 8x$     b. $7x + x$     c. $x + 4x$     d. $9x + 1.5x$

**Solutions**
  a. $6x + 8x = (6 + 8)x$     b. $7x + x = (7 + 1)x$
           $= \mathbf{14x}$                  $= \mathbf{8x}$
  c. $x + 4x = (1 + 4)x$     d. $9x + 1.5x = (9 + 1.5)x$
           $= \mathbf{5x}$                  $= \mathbf{10.5x}$

**Try This**  Add.
  **4.** $3x + 5x$   **5.** $15x + 4x$   **6.** $x + 9x$   **7.** $8x + 3.2x$

You can also use the Distributive Property to subtract like terms. First, rewrite the subtraction problem as an addition problem. Then use the Distributive Property.

**EXAMPLE 3**  Subtract.
  a. $12x - 8x$     b. $5x - x$

**Solutions**
  a. Write as an addition problem.   $12x - 8x = 12x + (-8x)$
     Use the Distributive Property.            $= [12 + (-8)]x$
     Simplify.                                 $= \mathbf{4x}$

  b. Write as an addition problem.   $5x - x = 5x + (-1x)$
     Use the Distributive Property.          $= [5 + (-1)]x$
     Simplify.                               $= \mathbf{4x}$

**Try This**  Subtract.
  **8.** $8x - 10x$   **9.** $7x - 4x$   **10.** $x - 4x$   **11.** $3x - x$

## EXERCISES

**Objectives:** To recognize instances of the Distributive Property
To simplify algebraic expressions using the Distributive Property

### Check Your Understanding

*Skill Check*  Write the letter of the correct answer.

1. Which is an equivalent expression for $6(x + 5)$?
   a. $6x + 5$      b. $6x + 30$
   c. $6 + 5x$      d. $30x$

2. Which is an equivalent expression for $4r + 12$?
   a. $16$          b. $(r + 12)4$
   c. $(r + 3)4$    d. $16r$

2-6  Distributive Property  **61**

# LESSON 2-7
# Formulas: Perimeter, Area, Average

**T**odd plans to back and frame a picture. He made a sketch to show how it will look when finished. Todd needs to buy framing to go around the picture and backing to go behind it. How much framing and how much backing does Todd need?

The **perimeter** of a rectangle is the distance around the rectangle.

The **area** of a rectangle is the number of square units needed to cover the rectangle.

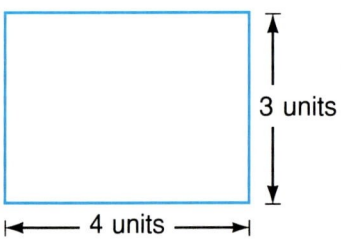

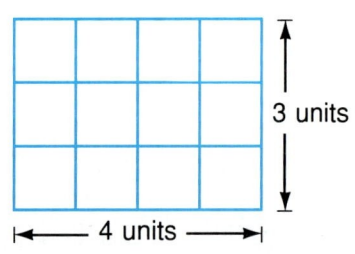

Perimeter = 2 · 4 + 2 · 3
= 8 + 6
= 14 units

Area = 3 · 4
= 12 square units

So Todd needs 14 feet of framing and 12 square feet of backing.

Copy and complete the table for the given length and width of each rectangle.

| Length | Width | Perimeter | Area |
|---|---|---|---|
| 5 | 4 | 2·5 + 2·4 | 5·4 |
| 7 | 2 | ? | ? |
| 3 | 6 | ? | ? |
| 8 | w | ? | ? |
| l | 9 | ? | ? |
| l | w | ? | ? |

You can use a **formula** to represent the pattern for the perimeter or for the area of any rectangle.

Consider a rectangle with length represented by $l$ and width represented by $w$. Let $P$ represent the perimeter and $A$ represent the area. Then the formulas for the perimeter and area of any rectangle will look like this.

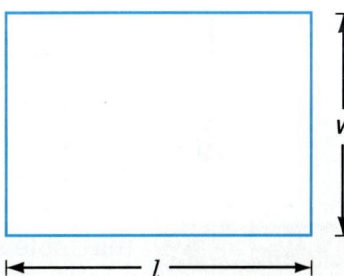

**Perimeter:** $P = 2l + 2w$, or $P = 2(l + w)$
**Area:** $A = lw$

A **square** is a rectangle with all four sides equal. Therefore,

$P = s + s + s + s$, or $P = 4s$
$A = s \cdot s$, or $A = s^2$

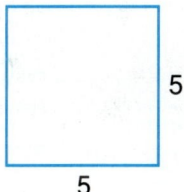

**EXAMPLE 1**   Find the perimeter and area of a rectangle with the given length and width.
   **a.** $l = 6$, $w = 5$     **b.** $l = 7$ meters, $w = 7$ meters

**Solutions**   Substitute for $l$ and $w$ in the perimeter and area formulas, then simplify.

**a.** Write the formula.         $P = 2(l + w)$      $A = lw$
   Substitute 6 for $l$ and 5 for $w$.   $P = 2(6 + 5)$      $A = 6 \cdot 5$
   Simplify.                   $P = 2 \cdot (11)$      $A = 30$
                               $P = 22$

The perimeter is **22 units** and the area is **30 square units.**

2-7   Formulas   **65**

**b.** Since the length and width are equal, the rectangle is a square.

| | | |
|---|---|---|
| Write the formula. | $P = 4s$ | $A = s^2$ |
| Substitute 7 for s. | $P = 4 \cdot 7$ | $A = 7 \cdot 7$ |
| Simplify. | $P = 28$ | $A = 49$ |

The perimeter is **28 meters** and the area is **49 square meters.**

**Try This**

**Find the perimeter and area of a rectangle having the given length and width.**

**1.** $l = 15$, $w = 15$     **2.** $l = 4$ inches, $w = 9$ inches

To find the **average** of a set of numbers, you find the sum, $s$, of the numbers and divide by $n$, where $n$ represents how many numbers are in the set. That is,

$$\text{Average} = \frac{s}{n}.$$

**EXAMPLE 2** This table shows the approximate length of a day on 5 planets in our solar system. What is the average length of a day for these 5 planets?

| Planet | Length of a Day in Earth Hours |
|---|---|
| Earth | 24 |
| Mars | 25 |
| Mercury | 1,417 |
| Neptune | 16 |
| Pluto | 153 |

**Solution**   Add to find $s$, the sum of the days.

$s = 24 + 25 + 1417 + 16 + 153$
$s = 1635$

Substitute 1635 for $s$ and 5 for $n$.    $\text{Average} = \frac{s}{n}$

Simplify.    $= \frac{1635}{5} = 327$

The average length of a day for the five planets is **327 hours.**

**Try This**

**3.** Rita's quiz scores on 5 different days are 92, 91, 86, 81, and 85. Is her average greater than 85?

How can you use a calculator to find the average of 83, 91, and 84?

|  | Calculator Keys | Answer |
|---|---|---|
| **Method 1:** | 83 [+] 91 [+] 84 [=] [÷] 3 [=] | 86 |
| **Method 2:** | [(] 83 [+] 91 [+] 84 [)] [÷] 3 [=] | 86 |

# EXERCISES

**Objectives:** To compute the perimeter and area of rectangles and squares
To find the average of a set of data

## Check Your Understanding

**Skill Check** Write the letter of the correct answer.

**1.** What is the perimeter of a rectangle with a length of 7 and a width of 2?
  **a.** 9 units  **b.** 14 units
  **c.** 16 units  **d.** 18 units

**2.** What is the area in square units of a rectangle with length 3 and width 4?
  **a.** 7  **b.** 12
  **c.** 14  **d.** 24

**3.** Which expression can be used to find the average of 42, 26, 39, and 42?
  **a.** $42 + 26 + 39 + 42$  **b.** $\frac{42 + 26 + 39}{3}$
  **c.** $\frac{42 + 26 + 39 + 42}{4}$  **d.** $\frac{42 + 26 + 39}{4}$

**4.** Which expression cannot be used to find the perimeter of a rectangle with length and width of 9 inches?
  **a.** $P = 2 \cdot 9 + 2 \cdot 9$  **b.** $P = 2(9 + 9)$
  **c.** $P = 9 + 9$  **d.** $P = 4 \cdot 9$

Tell whether these statements in a news item are talking about perimeter or talking about area.

**5.** In June, 1,500 acres in the Guadalupe Mountains National Park in Texas were blackened by a forest fire.

**6.** To control and stop the spread of the fire, a strip of land was plowed and cleared around the fire. This is called a firebreak.

## Practice and Apply

Find the perimeter and area of each rectangle or square only if the area is less than 100 square units.

**7.** $l = 8, w = 9$  **8.** $l = 11, w = 8$  **9.** $l = 9, w = 9$  **10.** $l = 9, w = 12$
**11.** $l = 14, w = 8$  **12.** $l = 15, w = 5$  **13.** $l = 30, w = 3$  **14.** $l = 21, w = 4$

Copy and complete.

**15.** Scores: 98, 64, 92, 82
Average: _?_

**16.** Scores: 91, _?_, 93
Average: 92

**17.** Scores: 96, _?_, 24
Average: 93

**18.** Explain how the two methods of finding an average by using a calculator differ. (See page 66.)

**19.** The low temperatures for 4 consecutive days in Houston are 75°, 76°, 76°, and 77°. Explain how to determine the average low temperature without adding the 4 numbers and dividing by 4.

The table lists the highest temperatures ever recorded in 6 different states.

20. What is the average high temperature for the six states?
21. For which two states will the average high temperature be greater than 123°?
22. For which two states will the average high temperature equal 113.5°?

**Find the length and width of the rectangle.**

23. Perimeter: 14; area: 10
24. Perimeter: 32; area: 64
25. Which of the figures in Exercises 23 and 24 is a square? Explain.

## Connect and Extend

Gayla and Jorge find the profit from their sales of school supplies with the formula $p = s - c$, where $p$ is the profit, $s$ is the price at which they sell supplies, and $c$ is the cost to them of the supplies. If $p$ is a negative number, they sell the supplies at less than they paid for them, that is, at a loss.

Complete the chart.

| | Profit $p$ | = | Price $s$ | − | Cost $c$ |
|---|---|---|---|---|---|
| 26. | ? | | 87 | | 23 |
| 27. | ? | | 56 | | 47 |
| 28. | ? | | 94 | | 104 |

29. Find the average profit, average price, and average cost for the 3 weeks.
30. For this three-week period, did the business make a profit or did it lose money?

## Maintain Your Skills

**Write an algebraic expression for each word expression. (Pages 39–43)**

31. $c$ subtracted from 4
32. 4 less than $c$
33. The quotient of 4 and $c$

**Correct any incorrect statements. (Pages 54–63)**

34. $3p + p = 3p$
35. $7c + 3d = 3d + 7c$
36. $3x - 7x = 4x$

37. Wallace found a shirt that costs $24.95. He received a $3.74 discount, and he paid $1.48 in sales tax. How much did he pay for the shirt?

# Area Formulas: Parallelograms, Triangles, and Trapezoids

**Y**ou can use what you know about the area of a rectangle to help you find the area of other geometric figures.

A **parallelogram** has 4 sides. Its opposite sides have the same length and are parallel. All of these figures are parallelograms.

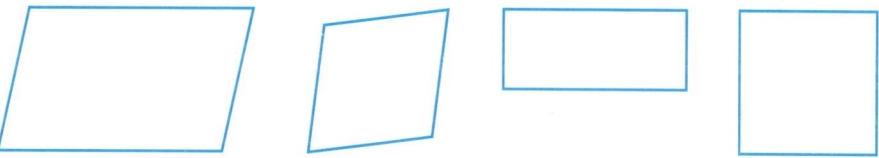

Draw a parallelogram on graph paper and cut it into two pieces as shown below. Rearrange the pieces to form a rectangle of the *same* area.

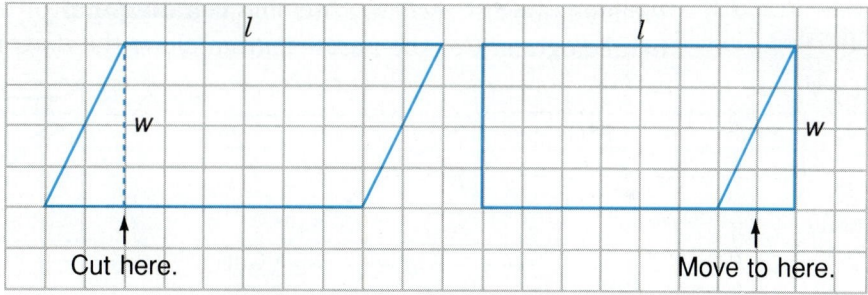

Count the number of square units in the rectangle.
- What is the area of the rectangle?
- What is the area of the parallelogram?

The length of a parallelogram is called its **base**, *b*. The width of a parallelogram is the distance between a pair of parallel sides. This is called the **height**, *h*, of the parallelogram.

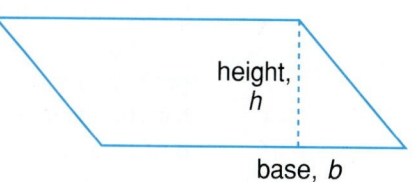

Since a rectangle and a parallelogram with equal lengths and equal widths have the same area, you can write these formulas for the area of a parallelogram.

$$A = lw \quad \text{or} \quad A = bh$$

2-8 Area Formulas **69**

**EXAMPLE 1**  The name plate on Maria's office has the shape of a parallelogram. It is 16 inches long and 6 inches high. Find the area.

**Solution**  Draw a diagram to help you picture the name plate.

| Write the formula. | $A = bh$ |
|---|---|
| Substitute for $b$ and $h$. | $A = 16 \cdot 6$ |
| Simplify. | $A = 96$ |

The area of the name plate is **96 square inches.**

**Try This**  **Find the area of each parallelogram.**
1. $b = 6$ inches, $h = 3$ inches
2. $b = 5$ feet, $h = 2$ feet

You can use what you learn about the area of a parallelogram to find the area of a triangle.

Draw a parallelogram on graph paper as shown below. Draw a line joining opposite corners. This line is a **diagonal** of the parallelogram. Cut the parallelogram along the diagonal.

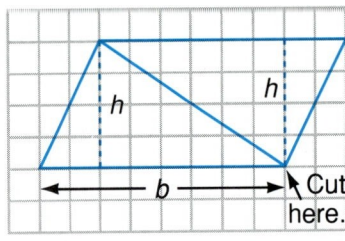  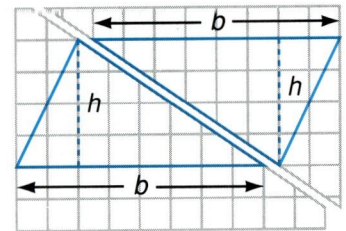

Compare the size and shape of each triangle.
• Is each triangle exactly half of the parallelogram?

Since each triangle covers exactly half the parallelogram, the formula for the area of a triangle is half that for the area of a parallelogram.

$$A = \frac{bh}{2}$$

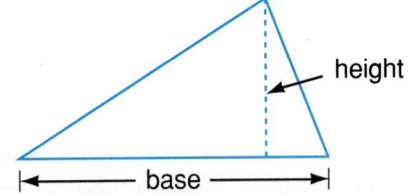

• The area of a parallelogram is 48 square inches. What is the area of a triangle having an equal base and height?

**EXAMPLE 2**  The gable of a roof has the shape of a triangle. It is 18 yards long and five yards high. What is its area?

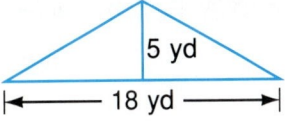

**Solution**

Write the formula.  $A = \dfrac{bh}{2}$

Substitute for b and h.  $A = \dfrac{5 \cdot \overset{9}{\cancel{18}}}{\underset{1}{\cancel{2}}}$

Simplify.  $A = 5 \cdot 9$
$A = 45$   The area is **45 square yards.**

**Try This**

**Find the area of each triangle.**

**3.** $b = 4$ feet, $h = 6$ feet   **4.** $b = 8$ yards, $h = 15$ yards

This shape is a **trapezoid.** It has one pair of opposite sides parallel. The parallel sides are not equal in length. You can use what you know about the area of a parallelogram to find the area of a trapezoid.

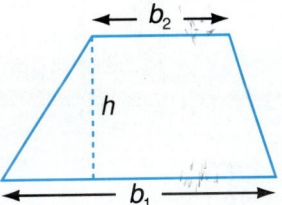

Draw a trapezoid as shown below. Represent the length of the parallel sides as $b_1$ and $b_2$ and the height as $h$. Cut out the trapezoid and trace around the cut-out figures. Trim the cut-out trapezoid so that it forms a parallelogram with the trapezoid you traced.

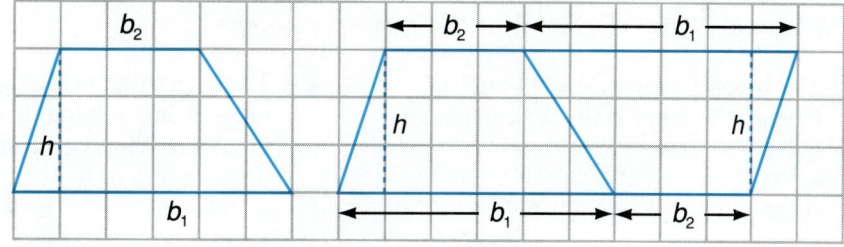

The area of the parallelogram is $(b_1 + b_2)h$.
The area of the trapezoid is half the area of the parallelogram. So the formula for the area of the trapezoid is

$$A = \dfrac{(b_1 + b_2)h}{2}.$$

**EXAMPLE 3** Find the area of this trapezoid.

**Solution**

$A = \dfrac{(b_1 + b)h}{2}$

$A = \dfrac{(3 + 5)9}{2}$

$A = \dfrac{(8 \cdot 9)}{2}$

$A = 36$     The area is **36 square yards.**

**Try This**  **Find the area of each trapezoid.**
5. $b_1 = 4$ feet, $b_2 = 5$ feet, $h = 2$ feet
6. $b_1 = 6$ feet, $b_2 = 2$ feet, $h = 5$ feet

This table shows the area formulas you have learned.

**Area Formulas**

Rectangle: $A = lw$
Square: $A = s^2$
Parallelogram: $A = bh$
Triangle: $A = \dfrac{bh}{2}$
Trapezoid: $A = \dfrac{(b_1 + b_2)h}{2}$

# EXERCISES

**Objective:** To find the areas of parallelograms, triangles, and trapezoids

## Check Your Understanding

**Skill Check**   Write the letter of the correct answer.

1. The base of a triangle is 14 and its height is 6. What is the area in square units?
   a. 84
   b. 42
   c. 35
   d. 21

2. The area of a parallelogram is 36 square units. A line joining its opposite corners is drawn. What is the area in square units of one of the triangles formed?
   a. 36
   b. 72
   c. 18
   d. 54

3. The area of a parallelogram is 56 square units and its height is 7. How long is the base?
   a. 7
   b. 392
   c. 63
   d. 8

4. The area of a trapezoid is 40 square units. The sum of the lengths of the parallel sides is 20. Which can be the height of the trapezoid?
   a. 4
   b. 8
   c. 2
   d. 40

**True or False? When a statement is false, give a reason.**

5. A trapezoid has one pair of parallel sides.
6. The parallel sides of a trapezoid have the same length.
7. If a parallelogram and a rectangle have equal bases and equal heights, then the parallelogram and the rectangle have equal areas.
8. If a parallelogram and a triangle have equal heights and equal bases, then the parallelogram and the triangle have equal areas.

## Practice and Apply

**Find the area of each triangle.**

9. $b = 7, h = 9$
10. $b = 23$ cm, $h = 16$ cm
11. $b = 14$ yd, $h = 11$ yd

**Find the area of each parallelogram.**

12. $b = 35, h = 23$
13. $b = 11$ km, $h = 16$ km
14. $b = 125$ in, $h = 83$ in

**Find the area of each trapezoid.**

15. $b_1 = 7, b_2 = 8, h = 6$
16. $b_1 = 27, b_2 = 15, h = 7$

**Refer to the figure at the right for Exercises 17–24.**

17. Figure ACEF is a ?.
18. Figure BCEF is a ?.
19. Figure CDE is a ?.
20. Figure BDEF is a ?.
21. The area of figure ACEF is ?.
22. The area of figure BCEF is ?.
23. The area of figure CDE is ?.
24. The area of figure BDEF is ?.

25. The length of a side of a square floor is 8 feet. How many square feet are there in the floor?
26. How many 1 foot-square tiles will it take to cover the floor in Exercise 25?
27. If tiles cost $2 each, what will it cost to tile the floor in Exercise 25?

**Refer to the figure at the right for Exercises 28–29.**

28. If the area of triangle WXY is 27 square inches, what is the area of figure WXYZ?
29. If the area of figure WXYZ is 72 square inches, what is the area of triangle WXZ?

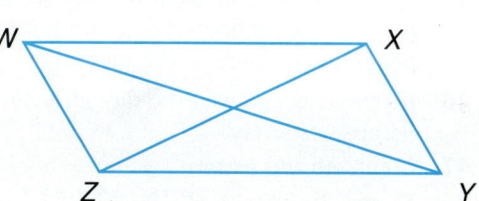

30. To find the area of a trapezoid, you can find the average of the lengths of the bases and multiply this average by the height of the trapezoid. Explain why this method works.

## Connect and Extend

31. What happens to the area of a triangle if you double the height?
32. Is one base of a trapezoid always shorter than the other base? Explain your answer.
33. Is a square a parallelogram? Explain your answer.
34. Is a trapezoid a parallelogram? Explain your answer.

## Maintain Your Skills

**In Exercises 35–37, write the answer only if it is less than 19. (Pages 23–31)**

35. $(-18)(17)$
36. $-54 - (-73)$
37. $-71 - 27$

38. The average monthly temperatures in Barrow Alaska were $-14°$, $-20°$, $-16°$, $-2°$, $19°$, $33°$, $39°$, $38°$, $31°$, $14°$, $-1°$, and $-13°$. Arrange the temperatures in ascending order. **(Pages 7–10)**

39. Which word expression has the greatest value if $n = 14$? **(Pages 39–43)**
    a. Sum of $n$ and 1
    b. Quotient of $n$ and 1
    c. Difference of $n$ and 1

**Simplify. (Pages 46–50, 54–58)**

40. $-19 + 56 \div 8$
41. $11f - 32f$
42. $(72 - 8) \div (4 \cdot 4)$

**For Exercises 43–44, the length of a side of a square is 6. The length of a rectangle is 9 and the width is 4. (Pages 64–68)**

43. Which figure has the greater perimeter, the square or the rectangle?
44. Which figure has the smaller area?

## ★ Math Team Problems

45. At a club meeting, each person present shakes hands with each other person exactly once. There are a total of 28 handshakes. How many people were there at the meeting?
46. If five days ago was the day after Saturday, what was the day before yesterday?
47. Complete the pattern.
    F ᴴ ꜔ T ? ꞁ N ? ?

# COMPUTER EXPLORATION

## Area Formulas

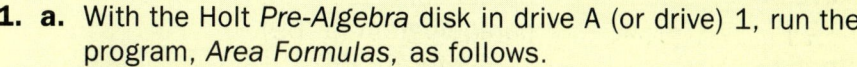

**Objective:** To explore area formulas using a computer

1. **a.** With the Holt *Pre-Algebra* disk in drive A (or drive) 1, run the program, *Area Formulas*, as follows.
   **IBM:** Type **a:b** and press **ENTER**.
   **Apple:** Turn on the computer. When the menu appears on the screen, press **B**.
   **b.** Read the opening messages of the program.

2. Press the **RETURN** key, and press **1** for parallelogram.
   **a.** What figure is shown? Press any key.
   **b.** What is the name of the shaded figure? Press any key.
   **c.** What is the name of the new figure formed?
   **d.** What is the area of the rectangle? of the parallelogram?

3. Press **2** for triangle. What figure is drawn? Press any key to continue.
   **a.** A diagonal divides the parallelogram into two triangles of equal area. What is the area of the parallelogram?
   **b.** What is the area of one triangle?

4. Press **3** for a trapezoid. Press any key.
   **a.** What new figure is formed on the screen when the two identical trapezoids are shown next to each other?
   **b.** What is the length of the base of the parallelogram? What is its height?
   **c.** How many trapezoids does it take to form one parallelogram?
   **d.** What is the area of one trapezoid?

5. Remove the disk and turn off the computer and the monitor. Return the disk to its storage place.

Computer Exploration **75**

# SUMMARY

## KEY TERMS

algebraic expression (p. 39)
area (p. 64)
average (p. 66)
counterexample (p. 51)
diagonal (p. 70)
equivalent expressions (p. 48)
formula (p. 65)
generalization (p. 51)
like terms (p. 60)

numerical coefficient (p. 60)
numerical expression (p. 39)
parallelogram (p. 69)
perimeter (p. 64)
rectangle (p. 64)
square (p. 65)
trapezoid (p. 71)
triangle (p. 70)
variable (p. 39)

## KEY IDEAS

**A.** To evaluate an algebraic expression for a given value of the variable, substitute the value for the variable. Then simplify.

**B. Order of Operations**
Simplify expressions in this order.
   1. Do the operations within parentheses.
   2. Do the operations in the numerator or denominator of a fraction.
   3. Do multiplications and divisions, in order, from left to right.
   4. Do additions and subtractions, in order, from left to right.

**C.**

| Properties of Numbers | | |
|---|---|---|
| Property | Addition | Multiplication |
| Commutative<br>Associative<br>Identity | $a + b = b + a$<br>$a + (b + c) = (a + b) + c$<br>$a + 0 = a$ | $ab = ba$<br>$a \cdot (b \cdot c) = (a \cdot b) \cdot c$<br>$a \cdot 1 = a$ |
| Distributive<br>Multiplication over<br>Addition | $a(b + c) = ab + ac$ and $(b + c)a = ba + ca$ | |

**D. Perimeter Formulas**   **Rectangle:** $P = 2l + 2w$   **Square:** $P = 4s$

**E. Area Formulas**   **Rectangle:** $A = lw$   **Square:** $A = s^2$
**Parallelogram:** $A = bh$   **Triangle:** $A = \dfrac{bh}{2}$
**Trapezoid:** $A = \dfrac{h(b_1 + b_2)}{2}$

# REVIEW

**Write each expression as an algebraic expression. (Pages 39–43)**
1. Brian's age 4 years ago if his present age is $r$ years
2. The cost of 7 quarts of oil if one quart costs $q$ cents

**Carlos plans to dig a vegetable garden with an area of 24 square feet. The lengths of the sides must be whole numbers. (Pages 44–45)**
3. What dimensions are possible for the garden?
4. What dimensions will give the garden the smallest perimeter?

**Simplify. (Pages 46–50)**
5. $12 - (-1 + 3)$
6. $144 \div 12 \cdot (54 - 48)$
7. $\dfrac{16 \div (9 - 5) \cdot 2}{19 - 21 + 4}$

**Evaluate each expression. (Pages 46–50)**
8. $\dfrac{a - 8}{-5}$ for $a = 3$
9. $-(y - w)$ for $y = 6m$ and $w = 2$

**Replace each ? with the numbers that make true statements. Write the name of the property shown. (Pages 54–58)**
10. $(-9 + 5)x = \underline{?}\,x + \underline{?}\,x$
11. $\underline{?} + 0 = 73$
12. $9 + (4 + \underline{?}) = (9 + \underline{?}) + 6$
13. $(4 \cdot 23) \cdot 5 = \underline{?} \cdot (\underline{?} \cdot 5)$

**Simplify. (Pages 59–63)**
14. $23p + (-22p)$
15. $19c + c$
16. $2w - w$
17. $x + 7x$

**The table on the right lists the four largest lakes in North America and their areas in square miles. (Pages 64–68)**
18. What is the average area per lake?
19. For which two lakes is the average area greater than 27,300 square miles?
20. For which two lakes is the average area less than 17,200 square miles?

**Find the area of each figure. (Pages 64–74)**
21. A triangle with a base of 10 and a height of 12
22. A square with a side of 6 inches
23. A parallelogram with a base of 7 and a height of 9
24. A trapezoid with bases of 5 feet and 9 feet and a height of 7 feet

Chapter 2 Review

# TEST

**Write each expression as an algebraic expression.**
1. Nine decreased by $t$
2. The number of months in $y$ years
3. The cost of one ear of corn if $r$ ears cost $2.34

**Evaluate each expression.**
4. $(4 - e)$, for $e = 9$
5. $3x - 2y$, for $x = -1$, $y = -3$
6. $a(b + 7)$, for $a = 4$, $b = -2$

**Write the name of the property shown.**
7. $a + (-1) = -1 + a$
8. $1x = x$
9. $(3 \cdot 7) \cdot s = 3(7s)$
10. $23ef + 0 = 23ef$
11. $6y - 6t = 6(y - t)$
12. $(4 + r) \cdot 5 = 5(4 + r)$

**Simplify, if possible.**
13. $25 \div (7 - 2) \cdot 3$
14. $7 - 2 \cdot 6 - 3 + 1$
15. $5z - 2z$
16. $(24 + 8) \div 4 + 7$
17. $8e + 21e$
18. $g - 3g$

**The table at the right lists the top hockey scorers and their points for one season.**

19. What is the average number of points per player?
20. Which two players have total points that average 133?
21. Which two players have total points whose average is less than 119?

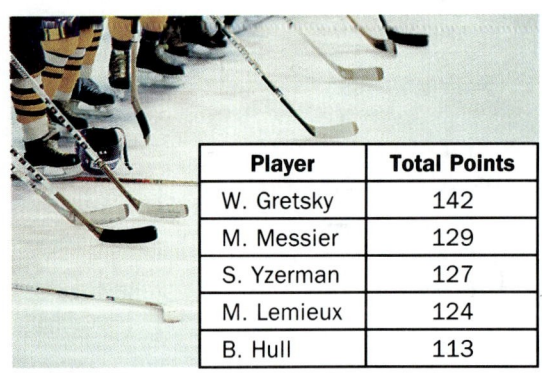

| Player | Total Points |
|---|---|
| W. Gretsky | 142 |
| M. Messier | 129 |
| S. Yzerman | 127 |
| M. Lemieux | 124 |
| B. Hull | 113 |

**Find the area for each figure.**
22. Triangle: $b = 8$; $h = 14$
23. Parallelogram: $b = 6$; $h = 11$
24. Triangle: $b = 24$; $h = 3$
25. Trapezoid: $b_1 = 18$; $b_2 = 14$; $h = 1$
26. Parallelogram: $b = 13$; $h = 4$
27. Trapezoid: $b_1 = 15$; $b_2 = 25$; $h = 12$

28. Find the perimeter and area of a sheet of paper having a width of 8 inches and a length of 11 inches.
29. Find the perimeter and area of a square window with a side of 7 meters.

**Find the opposite of each number. Then arrange the numbers in ascending order. (Pages 2–10)**

**1.** $-6, 181, 11, -17$     **2.** $-(-10), |-121|, -16$     **3.** $0, -(-5), |-4|, -3, -94$

**Find the altitude or depth change for each pair of readings. (Pages 23–25)**

**4.** 3000 meters to 800 meters

**5.** $-800$ meters to $-1500$ meters

**6.** $-5$ meters to 330 meters

**Find the difference between the pairs of successive monthly average temperatures for a recent year in Alaska. (Pages 23–25)**

**7.** $14°$ and $-1°$     **8.** $-2°$ and $19°$     **9.** $-20°$ and $-14°$

**Copy and complete. (Pages 28–35)**

**10.** $13 \cdot -4 = \underline{\ ?\ }$     **11.** $-54 \div \underline{\ ?\ } = 6$     **12.** $\underline{\ ?\ } \div 12 = -6$

**Write each word expression as an algebraic expression. (Pages 39–43)**

**13.** John's age $a$ decreased by 4 years     **14.** $16 increased by $t$ tips

**Copy and complete the following tables. (Pages 46–50)**

**15.**

| $b$ | $3b - 8$ | $-7b \cdot 2$ |
|---|---|---|
| 9 | ? | ? |

**16.**

| $w$ | $z$ | $-6(w - z)$ |
|---|---|---|
| $-7$ | $-5$ | ? |

**Insert grouping symbols in the expression $-5 + -5 \cdot 4 + 8$ so that it will have the given value when simplified. (Pages 46–50)**

**17.** $-120$     **18.** $-17$     **19.** $-32$     **20.** $-65$

**Simplify, if possible. (Pages 59–63)**

**21.** $15ad + 19ad$     **22.** $6h + 21 - h$     **23.** $7xy + 4x$

**Tell whether the problem is about perimeter or area. Then solve the problem. (Pages 64–68)**

**24.** How many 1-foot square tiles are needed to cover a floor 9 feet long and 8 feet wide?

**25.** How much fence is needed to enclose a square garden with each side 7 feet long?

**Solve. (Pages 69–74)**

**26.** A product label has the shape of a parallelogram and an area of 6 square inches. The height of the label is 2 inches. What is the width?

**27.** The area of a triangle is 24 square feet. The height of the triangle is 8 feet. What is the length of the base?

Cumulative Review: Chapters 1–2    **79**

## LESSON 3-1

# Equations and Inequalities

**R**on is saving for a class trip to Washington, D.C. So far he has saved $85. The cost of the trip per student is $125. To find how much more he needs to save, Ron lets $n$ represent the amount he needs. Then he wrote this equation:

$$125 = n + 85$$

An **equation** is a mathematical sentence with an equal sign. It says that the expressions on the left and right sides of the equation are equal to each other.

You cannot tell whether Ron's equation is true or false until you replace the variable, $n$, with a number. If the number replacement for $n$ makes the equation true, then that number is a **solution** of the equation.

Is $125 = n + 85$ true for $n = 35$?  Is $125 = n + 85$ true for $n = 40$?
$125 \stackrel{?}{=} 35 + 85$              $125 \stackrel{?}{=} 40 + 85$
$125 \neq 120$                             $125 = 125$

So 35 is not a solution.                   So 40 is a solution.

**EXAMPLE 1**   Which of the replacements, $-2$, 2, and 4, are solutions of $3x + 1 = -5$?

**Solution**   Replace the variable with each number.

a.
$$3x + 1 = -5$$
$$3(-2) + 1 \stackrel{?}{=} -5$$
$$-6 + 1 \stackrel{?}{=} -5$$
$$-5 = -5$$

Since $-2$ makes the equation true, **$-2$ is** a solution.

b.
$$3x + 1 = -5$$
$$3(2) + 1 \stackrel{?}{=} -5$$
$$6 + 1 \stackrel{?}{=} -5$$
$$7 \neq -5$$

Since 2 does not make the equation true, 2 is **not** a solution.

c.
$$3x + 1 = -5$$
$$3(4) + 1 \stackrel{?}{=} -5$$
$$12 + 1 \stackrel{?}{=} -5$$
$$13 \neq -5$$   So 4 is **not** a solution.

**Try This**   Which of the replacements, $-3$, $-4$, and $-5$, are solutions of these equations?

**1.** $2x + 15 = 5$   **2.** $-8x + 20 = 52$

The **replacement numbers** in Example 1 can be written as a **set** of numbers, $\{-2, 2, 4\}$. You use the symbols, $\{\ \}$, to indicate a set and you separate the **members,** or **elements,** of the set by commas. Usually a capital letter is used to name a set.

$R = \{0, 1, 2, 3\}$    Read: "The members of set $R$ are 0, 1, 2, and 3."

**EXAMPLE 2**   Find the solution set for $2x - 1 = -1$, if the replacement set, $R$, is $\{-1, 0, 1, 2\}$.

**Solution**   Replace $x$ in the equation with each member of set $R$.

a.
$$2x - 1 = -1$$
$$2(-1) - 1 \stackrel{?}{=} -1$$
$$-2 + (-1) \stackrel{?}{=} -1$$
$$-3 \neq -1 \quad \text{Not a solution}$$

b.
$$2x - 1 = -1$$
$$2(0) - 1 \stackrel{?}{=} -1$$
$$0 - 1 \stackrel{?}{=} -1$$
$$-1 = -1 \quad \textbf{Solution}$$

c.
$$2x - 1 = -1$$
$$2(1) - 1 \stackrel{?}{=} -1$$
$$2 - 1 \stackrel{?}{=} -1$$
$$1 \neq -1 \quad \text{Not a solution}$$

d.
$$2x - 1 = -1$$
$$2(2) - 1 \stackrel{?}{=} -1$$
$$4 - 1 \stackrel{?}{=} -1$$
$$3 \neq -1 \quad \text{Not a solution}$$

The solution set is $\{0\}$.

**Try This**    **Find the solution set for each equation if $R = \{-2, 0, 1, 2\}$.**

**3.** $\frac{r}{2} = 1$      **4.** $5 + 2q = 5$      **5.** $2 - t = 1$

An **inequality** is another kind of mathematical sentence. These sentences are examples of inequalities.

$$5 \neq 2 \qquad 3y < 8 \qquad 2x > 9.5$$

Any member of the replacement set that makes an inequality true is also a member of the solution set of the inequality.

**EXAMPLE 3**    Find the solution set of $3y < 8$ if $R = \{1, 2, 3\}$.

**Solution**    Replace the variable $y$ with each member of the solution set.

a.    $3y < 8$      b.    $3y < 8$      c.    $3y < 8$
     $3(1) \stackrel{?}{<} 8$          $3(2) \stackrel{?}{<} 8$          $3(3) \stackrel{?}{<} 8$
     $3 < 8$             $6 < 8$             $9 \not< 8$
     **1 is a solution.**      **2 is a solution.**      **Not a solution**

The solution set is $\{1, 2\}$.

**Try This**    **Find the solution set of each inequality if the replacement set is $\{-2, -1, 0, 1\}$.**

**6.** $-5y > 5$      **7.** $2b + 3 < 2$      **8.** $6 > 9y$

# EXERCISES

**Objective:** To find the solution of an equation or inequality

## Check Your Understanding

*Skill Check*    Write the letter of the correct answer.

**1.** For which equation is 4 a solution?
     **a.** $w - 32 = 28$    **b.** $6w - 6 = 18$
     **c.** $9w - 2 = 16$    **d.** $5w - 12 = 23$

**2.** Which mathematical sentence is an equation?
     **a.** $7 \neq 5 + 4$    **b.** $2r > 10$
     **c.** $6r + 1 = 7$    **d.** $5r - 8 < 1$

**3.** Which equation has a solution set of $\{-7\}$?
     **a.** $5x - 6 = -41$    **b.** $5x - 5 = 30$
     **c.** $5x + 6 = 41$    **d.** $4x + 5 = 33$

**4.** Which inequality has a solution set of $\{3, 4\}$?
     **a.** $3a - 6 < 4$    **b.** $2a + 8 > 16$
     **c.** $5a + 3 > 48$    **d.** $7a - 1 < 48$

Match each equation in Column A with its solution in Column B.

**Column A**
5. $49 - 3y = 40$
6. $3a - 1 = -100$
7. $4b + 8 = 88$
8. $-84 - 7z = -21$

**Column B**
a. $-33$
b. $-9$
c. $3$
d. $20$
e. $33$

Replace each __?__ with a whole number to make a true inequality.

9. $15 + \underline{\ ?\ } < 20$
10. $\underline{\ ?\ } \cdot 8 < 9$
11. $\underline{\ ?\ } - 4 > 6$
12. $3 \cdot 6 > \underline{\ ?\ }$

## Practice and Apply

Which of the replacements, $-4, -2, 2,$ and $4$, are solutions of the following equations?

13. $3x + 2 = 8$
14. $8r - 2 = -18$
15. $-17 = 7g - 3$
16. $\frac{-96}{e} = -24$

Find the solution set if the replacement set is $\{-5, 5, 6, 7\}$.

17. $9e + 7 = 52$
18. $-6d - 19 = -55$
19. $-4y - 2 = -30$
20. $6n + 8 = 50$
21. $-q = 2q + 15$
22. $2q = q + 7$
23. $4f - 1 = 23$
24. $6y - y - 1 = 24$

Find the solution set if $R = \{-5, -3, -1, 3, 5\}$.

25. $3s < 9$
26. $7r < -14$
27. $5x < 22$
28. $-4v > 7$
29. $4w < -10$
30. $-8a < 2$
31. $-6q > -15$
32. $4t + 1 > 20$
33. $-3a + 7 > 13$
34. $-2y - 7 < 3$
35. $9x + 7 > -5$
36. $-8c - 9 < 4$

Write an inequality for each word expression.

37. An amount of money, $a$, is more than $8.
38. The number of club members, $c$, is fewer than 20.
39. Cassette tapes are on sale for $6 each, including tax. Sara received $25 for her birthday from her grandparents and decided to buy some tapes. If Sara buys $t$ tapes, which mathematical sentence best describes the situation? Explain.
    a. $6t = 25$
    b. $6t < 25$
    c. $6t > 25$
    d. $6t < -25$

## Connect and Extend

For Exercises 40–41, refer to rectangle $ABCD$. Write an equation to represent each statement.

40. The perimeter of rectangle $ABCD$ is 60 units.
41. The area of rectangle $ABCD$ is 56 square units.

**42.** Find the solution set of $-7x > -28$ if the replacement set is all integers between $-7$ and $7$. Are all the members of the solution set less than or greater than a certain integer? Explain.

**43.** Tammy said that the sentences "a is not less than b" and "a is greater than or equal to b" are equivalent sentences. Jan disagreed. Who is right? Write an argument to prove your answer.

**44.** Write an equation involving multiplication and subtraction which has a solution set of $\{-2\}$.

**45.** Explain why the solution set for $-64x = 384$ does not have any elements if the replacement set contains only positive numbers.

## Maintain Your Skills

In Exercises 46–49, write the answer only if it is greater than $-98$.
(Pages 13–18, 23–25, 28–35)

**46.** $\frac{70}{-35} - 97$   **47.** $-73(12) + 27$   **48.** $3(25) + (-35)$   **49.** $(-4)(30) + 8(60)$

When a sports store orders jackets from a wholesale company, two options for ordering the jackets are available. The cost in dollars for each option is expressed in terms of $s$, the number of jackets.
(Pages 39–43)

Cost of Option 1:   Cost of Option 2:
$100 + 30s$           $35s$

**50.** Which option provides the better value when buying 10 jackets?

**51.** Which option provides the better value when buying 20 jackets?

**52.** Which option provides the better value when buying 30 jackets?

**Copy and complete.** (Pages 54–63)

**53.** $-8(46) + (-8)(4)$
$= -8(\underline{\ ?\ } + \underline{\ ?\ })$
$= \underline{\ ?\ }(50)$
$= \underline{\ ?\ }$

**54.** $(25 \cdot 38) \cdot 4 = 25 \cdot (38 \cdot \underline{\ ?\ })$
$= 25 \cdot (4 \cdot \underline{\ ?\ })$
$= (\underline{\ ?\ } \cdot \underline{\ ?\ }) \cdot 38$
$= 100 \cdot 38$
$= \underline{\ ?\ }$

**55.** $19t - 16t = [19 + (\underline{\ ?\ })]t$
$= \underline{\ ?\ }$

**56.** $6b + b = (6 + \underline{\ ?\ })b$
$= \underline{\ ?\ }$

# LESSON 3-2

## Solving Equations

In Friday's basketball game, Laura made 8 free throws. This brought her season total for free throws to 15. How many did Laura make before Friday's game?

How can you write an equation to represent this situation?

**Think:** I need to find how many free throws Laura made before Friday's game.

**Write:** Let $t$ represent the number made before Friday's game. Then $t + 8 = 15$.

You can use mental math to solve this equation.

**Think:** What number added to 8 will equal 15? Since $7 + 8 = 15$, the solution is 7.

So Laura made 7 free throws before Friday's game.

**EXAMPLE 1** Solve by using mental math.

  **a.** $r + 4 = -2$        **b.** $\frac{y}{5} = 20$

**Solutions**

**a. Think:** What number added to 4 equals $-2$?    Try $r = -6$.
Does $-6 + 4 = -2$?    Yes ✓
The solution is $-6$.

3-2 Solving Equations    **85**

**b. Think:** What number divided by 5 equals 20?   Try $y = 100$.

Does $\frac{100}{5} = 20$?   Yes ✓   The solution is **100.**

**Try This**   **Solve by using mental math.**
1. $m - 6 = 10$
2. $3c = -15$

Sometimes you can use guess-and-check to solve an equation. First, you make a reasonable guess about the value of the variable. Check your guess. If it is too high, lower the second guess. If it is too low, raise the second guess. Then check again. Repeat this procedure until you find the solution.

**EXAMPLE 2**   Solve by using guess-and-check.

**Solutions**   **a.** $\frac{p}{6} = 18$

**Guess 1:** 96

**Check 1:** $\frac{96}{6} = 16$   Too low

**Guess 2:** 108

**Check 2:** $\frac{108}{6} = 18$ ✓

The solution is **18.**

**b.** $35 - z = 115$

**Guess 1:** $-90$

**Check 1:** $35 - (-90) = 125$
Too low

**Guess 2:** $-75$

**Check 2:** $35 - (-75) = 110$
Too high

**Guess 3:** $-80$

**Check 3:** $35 - (-80) = 115$ ✓

The solution is **−80.**

**Try This**   **Solve by using guess-and-check.**
3. $m - 56 = 93$
4. $-9d = 135$

# EXERCISES

**Objectives:** To solve equations by using mental math
To solve equations by using guess-and-check

## Check Your Understanding

**Skill Check**   Write the letter of the correct answer.

**1.** What number added to $14 = -2$?
   **a.** 12   **b.** $-16$   **c.** $-12$   **d.** 16

**2.** For $7y = 98$, which guess is too high?
   **a.** 2   **b.** 5   **c.** 10   **d.** 17

**3.** For $t - 69 = 31$, which guess is too low?
  **a.** 30   **b.** 150
  **c.** 225   **d.** 1000

**4.** Solve: $\frac{-143}{c} = 11$
  **a.** $-12$   **b.** $-16$
  **c.** 13   **d.** $-13$

In Exercises 5–8, solve by using mental math. Write the solution only if it is negative.

**5.** $\underline{\ ?\ } - 7 = -3$   **6.** $\underline{\ ?\ } - 35 = 108$   **7.** $\underline{\ ?\ } + 12 = -4$   **8.** $\frac{\underline{\ ?\ }}{6} = 10$

## Practice and Apply

**Solve each equation by using mental math.**

**9.** $b + 3 = -7$   **10.** $c + 9 = -8$   **11.** $f - 4 = -2$   **12.** $g - 9 = -3$
**13.** $6j = 18$   **14.** $8x = -72$   **15.** $\frac{s}{-4} = 3$   **16.** $\frac{t}{7} = -8$

**Tell whether the guess is too high or too low. Then solve the equation by using guess-and-check.**

**17.** $8y = -64$   **18.** $1584 + y = 1262$   **19.** $\frac{x}{7} = 13$   **20.** $24 - w = -29$
   Guess: $-7$      Guess: $-302$      Guess: 98      Guess: 50

**Give a second guess for each solution. Explain your choice.**

**21.** $\frac{x}{9} = 29$   **22.** $128 + w = -75$   **23.** $-59y = -413$
   Guess: 252      Guess: $-200$      Guess: 8

**Solve by using guess-and-check.**

**24.** $87 - q = 12$   **25.** $49 - x = -30$   **26.** $y - 30 = -28$   **27.** $-5t = 115$
**28.** $6f = -72$   **29.** $\frac{s}{11} = 10$   **30.** $\frac{m}{4} = 47$   **31.** $\frac{-168}{u} = 14$

**Match each word expression in Column A with the correct equation in Column B.**

| Column A | Column B |
|---|---|
| **32.** Four added to a number $n$ is 16. | **a.** $4n = 16$ |
| **33.** Four times a number $n$ is 16. | **b.** $n - 16 = 4$ |
| **34.** Sixteen subtracted from a number $n$ is 4. | **c.** $\frac{n}{4} = 16$ |
| **35.** A number $n$ divided by 4 is 16. | **d.** $n + 4 = 16$ |
|  | **e.** $n - 4 = 16$ |

## Connect and Extend

In Exercises 36–37, write an equation to represent the situation.

**36.** [number line: total 29, segments $x$ and 12]

**37.** [number line: total 20, four equal segments each $c$]

**3-2 Solving Equations**

**Write a word expression for each equation.**

**38.** $a + 5 = 3$ **39.** $7f = 28$ **40.** $n - 4 = 2$ **41.** $\frac{t}{12} = 36$

**Solve each equation by using mental math or guess-and-check.**

**42.** $9r + 3 = 48$ **43.** $4p + 1 = -11$ **44.** $3q - 2 = 22$ **45.** $8s - 1 = -97$

## Maintain Your Skills

**Write the answer only if it is positive. (Pages 13–18, 23–25, 28–35)**

**46.** $(3)(64) + (-7)(61)$ **47.** $(5)(30) \div 3$ **48.** $6 + (-2) + 18 + (-41)$

**Copy each row of expressions. Circle the equivalent expressions. (Pages 59–63)**

**49.** $4p - pq$      $p(4 - pq)$      $q(4p - p)$      $p(4 - q)$
**50.** $5(-x + y)$      $-5x + 5y$      $-5x + y$      $(-x + y)5$
**51.** $4rs - 4cd$      $rs - cd$      $4rs + 4cd$      $(rs - cd)4$

**Find the area of the rectangle if the area is less than 100 square units. (Pages 64–68)**

**52.** $l = 49; w = 3$ **53.** $l = 15; w = 6$ **54.** $l = 19; w = 3$

## **M**athematical Footnote

Is there more than one zero point? There is if you are talking about temperature.

Water freezes at 32 degrees Fahrenheit, which is 0 degrees Celsius. On the other hand, 0 degrees Fahrenheit is about −18 degrees Celsius. Then there is "absolute zero." Absolute zero is −459.67 degrees Fahrenheit and −273.15 degrees Celsius. Scientists also use a Kelvin scale in which absolute zero is 0 degrees Kelvin.

In 1990, physicists at the University of Colorado produced the world's coldest temperature—within 1.1-millionth of a degree Celsius of absolute zero.

# LESSON 3-3

## Problem Solving Strategies
## Choosing Strategies

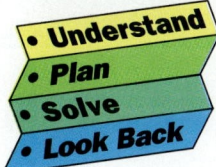
- Understand
- Plan
- Solve
- Look Back

Almost every problem can be solved in more than one way. This example shows several ways of solving the same problem.

### EXAMPLE

There are 10 hockey players on the ice at the end of a game. Each player shakes hands with each of the other players. What is the total number of handshakes?

**Strategy 1:** Try guess and check.

**Guess:** 10 handshakes

**Check:** Player 1 will shake hands with 9 people.
Player 2 will shake hands with 8 people. (The handshake with Player 1 has already happened.)

**Think:** There are already 17 handshakes, which is more than the first guess. Make a second guess and check again.

**Conclusion:** This strategy will work, but it will take time.

**Strategy 2:** Draw a diagram. Start with 10 dots.

Draw lines connecting each pair of dots. Each line represents one handshake.

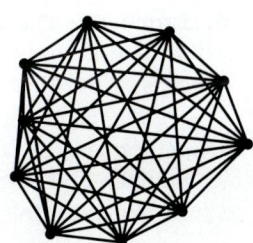

**Conclusion:** This strategy will work, but the lines may be difficult to count.

3-3 Problem Solving Strategies **89**

**Strategies 3 and 4:** Solve a simpler problem. Look for a pattern. Find the number of handshakes for 3 players, for 4 players, for 5 players, and so on. Find the pattern and use it to solve the problem.

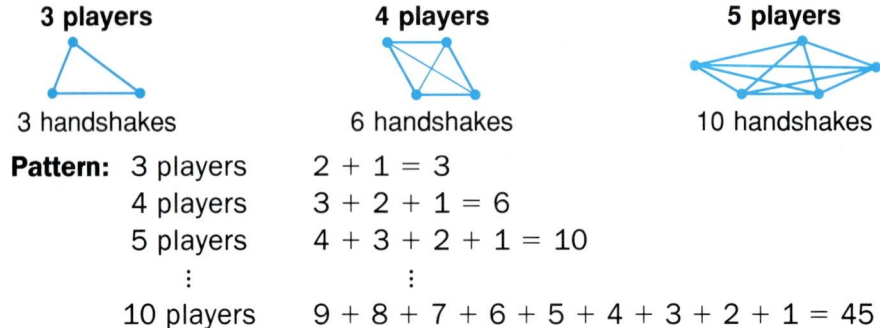

| 3 players | 4 players | 5 players |
|---|---|---|
| 3 handshakes | 6 handshakes | 10 handshakes |

**Pattern:**  3 players    $2 + 1 = 3$
4 players    $3 + 2 + 1 = 6$
5 players    $4 + 3 + 2 + 1 = 10$
⋮           ⋮
10 players   $9 + 8 + 7 + 6 + 5 + 4 + 3 + 2 + 1 = 45$

For 10 players, there will be **45** handshakes.

# EXERCISES

**Objective:** To use strategies in problem solving

Use any strategy or strategies to solve these problems.

1. Chris has $1.37 in his pocket.
   a. What is the smallest number of coins he can have?
   b. What is the greatest number of coins he can have?

2. Gina has 3 times as much money as Karl. If Gina had $18 more, she would have 4 times as much money as Karl. How much does Gina have?

3. Write a number that is more than 1.234 and less than 1.235.

4. a. Find an integer that makes this equation true: $(x - 3)(x + 2) = 0$
   b. Is there more than one answer?

5. The Gridiron Checkers Club has 7 members and the Hoopsters Checkers Club has 8 members. If each member of the Gridiron Club plays each member of the Hoopsters Club, how many games will they play?

6. Armin says that his phone number has the same first three digits as the other members of the Math Club. The last two digits are the same digit, and, when multiplied together, give the fourth digit of his phone number. The sum of the last four digits is 15. What are the last four digits?

# LESSON 3-4

# Solving Addition and Subtraction Equations

You have solved equations by substituting values from a replacement set, by using mental math, and by using guess-and-check. These methods may not help with equations involving larger numbers. A more general method for solving equations is based on a property of addition.

Begin with an equation.
If you add 6 to each side, the sums will be equal.

$$25 = 5 \cdot 5$$
$$25 + 6 = 5 \cdot 5 + 6$$
$$31 = 31$$

If you add $(-3)$ to each side, the sums will be equal.

$$25 + (-8) = 5 \cdot 5 + (-8)$$
$$17 = 17$$

These are two instances of the Addition Property of Equality. It says that the sides of an equation remain equal if you add the same number to each side.

> **Addition Property of Equality**
> For all numbers $a$, $b$, and $c$, if $a = b$,
> then $a + c = b + c$.

To use this property to solve $x + 3 = 10$, think: "$x$ is added to 3 on the left side of the equation. To get $x$ alone, add $(-3)$ to each side."

$$x + 3 = 10$$

Add $(-3)$ to each side. $\qquad x + 3 + (-3) = 10 + (-3)$
Apply the Property of Opposites. $\qquad x + 0 = 10 + (-3)$
Apply the Identity Property. $\qquad x = 7$

The key in solving equations such as these is to know what to add to each side.

**EXAMPLE 1** Solve and check each equation.
  **a.** $x + 7 = -9$   **b.** $t - 4 = 8$

**Solutions** **a.** **Think:** What number is added to x on the left side of the equation?
Add the opposite of that number to each side.

3-4 **Addition and Subtraction Equations** **91**

| | |
|---|---|
| Write the equation. | $x + 7 = -9$ |
| Add $-7$ to each side. | $x + 7 + (-7) = -9 + (-7)$ |
| Apply the Property of Opposites. | $x + 0 = -16$ |
| Apply the Identity Property. | $x = -16$ |

**Check**

| | |
|---|---|
| Substitute $-16$ for $x$ in the original equation. | $x + 7 = -9$ |
| | $-16 + 7 \stackrel{?}{=} -9$ |
| Are the two sides equal? | $-9 = -9$ ✓ |

Since $-16$ makes the equation true, the solution is **$-16$.**

**b. Think:** Since 4 is subtracted on the left side of the equation, add 4 to each side.

| | |
|---|---|
| Write the equation. | $t - 4 = 8$ |
| Apply the Addition Property of Equality. | $t - 4 + 4 = 8 + 4$ |
| Apply the Property of Opposites. | $t + 0 = 12$ |
| Apply the Identity Property. | $t = 12$ |

**Check**

| | |
|---|---|
| Substitute 12 for $t$ in the original equation. | $t - 4 = 8$ |
| | $12 - 4 \stackrel{?}{=} 8$ |
| | $8 = 8$ ✓ |

The solution is **12.**

**Try This**   **Solve and check.**
1. $x - 4 = -3$
2. $x + 9 = -5$

An equation can have the variable on either side. The procedure of solving the equation is the same whether the variable is on the right or the left of the equal sign.

$8 = x + 3$  has the same solution as  $x + 3 = 8$.
$-1 = 2 + r$  has the same solution as  $r + 2 = -1$.

Equations that have the same solution set are **equivalent.**

• Are $x - 5 = 10$ and $y + 3 = 7$ equivalent equations? Explain.

**EXAMPLE 2**  Solve and check.
a. $-5 = x + 2$
b. $9 = z - 3$

**Solutions**   **a. Think:** What number is added to $x$ on the right side of the equation? Since the number is 2, add $(-2)$ to each side.

$$-5 = x + 2$$
$$-5 + (-2) = x + 2 + (-2)$$
$$-7 = x + 0$$
$$-7 = x$$

Check: $-5 = x + 2$
$-5 \stackrel{?}{=} -7 + 2$
$-5 = -5$ ✓

The solution is **−7**.

**b. Think:** What number is subtracted from $z$ on the right side of the equation? Since the number is 3, add 3 to each side.

$$9 = z - 3$$
$$9 + 3 = z - 3 + 3$$
$$12 = z + 0$$
$$12 = z$$

Check: $9 = z - 3$
$9 \stackrel{?}{=} 12 - 3$
$9 = 9$ ✓

The solution is **12**.

**Try This**  Solve and check.

**3.** $-6 = 13 + s$   **4.** $9 = b - 13$   **5.** $-1 = t - 1$

# EXERCISES

**Objective:** To solve equations by using the Addition Property of Equality

## Check Your Understanding

**Skill Check**  Write the letter of the correct answer.

**1.** Which number must be added to each side of the equation to solve $e + 6 = 14$?
 **a.** 6   **b.** 14
 **c.** −6   **d.** −14

**2.** Which number should be added to each side of the equation to solve $y - 9 = 23$?
 **a.** 9   **b.** 23
 **c.** −9   **d.** −23

**3.** Which equation is equivalent to $x - 5 = 14$?
 **a.** $14 = x + 5$   **b.** $14 = x - 5$
 **c.** $x + 14 = 5$   **d.** $x + 14 = -5$

**4.** Which is the solution of $17 = k - 3$?
 **a.** −20   **b.** 14
 **c.** 17   **d.** 20

What number must be added to each side to solve the equation?

**5.** $f - 2 = 5$   **6.** $12 = g + 17$   **7.** $8 + x = -1$   **8.** $h - 9 = -16$

## Practice and Apply

Solve and check.

**9.** $d + 3 = -8$   **10.** $f + 2 = 7$   **11.** $13 = c + 8$
**12.** $x + 4 = -11$   **13.** $3 = r - 9$   **14.** $-4 = e + 2$

15. $x + 14 = -12$
16. $g - 7 = 15$
17. $y - 6 = -19$
18. $-22 = h - 10$
19. $s + 89 = 71$
20. $-31 = a - 98$
21. $38 = g - 4$
22. $-72 = -1 + z$
23. $-69 = b + 18$
24. $m - (-51) = 1$
25. $-64 = p + 57$
26. $q - 78 = 49$
27. $93 = c + 108$
28. $t - 129 = -56$
29. $x - 8 = 6 \cdot 3 + 4$

**Give a reason for each step.**

30. Equation: $x + 3 = 9$
    a. $x + 3 + (-3) = 9 + (-3)$   ?
    b. $x + 0 = 6$   ?
    c. $x = 6$   ?

31. Equation: $x - 5 = 7$
    a. $x - 5 + 5 = 7 + 5$   ?
    b. $x + 0 = 12$   ?
    c. $x = 12$   ?

## Connect and Extend

32. To solve $x + b = c$ for $x$, what should you add to each side of the equation?

33. To solve $y - d = f$ for $y$, what should you add to each side of the equation?

34. Write an equivalent equation of the form $x + a = b$ for the equation $x - 7 = 2$.

35. If $x + 6 = 13$, find the value of $8x$.

36. If $-16 = q - 8$, find the value of $9q$.

37. If $a - 15 = 0$, find the value of $a + 27$.

The table on the right can be used to evaluate each side of the equation, $2x - 4 = 8 - x$, for values of $x$ from 0 to 7.

38. Copy and complete the table. The first one is done for you.

39. *Complete:* When $x$ increases by 1, $2x - 4$ ? (increases/decreases) by ?.

40. *Complete:* When $x$ increases by 1, $8 - x$ ? (increases/decreases) by ?.

41. What is the solution for $2x - 4 = 8 - x$? How do you know?

| x | 2x − 4 | 8 − x |
|---|--------|-------|
| 0 | −4 | 8 |
| 1 | ? | ? |
| 2 | ? | ? |
| 3 | ? | ? |
| 4 | ? | ? |
| 5 | ? | ? |
| 6 | ? | ? |
| 7 | ? | ? |

## Maintain Your Skills

Write P or N to tell if the answer is positive (P) or negative (N). Write the answer if it is negative. (Pages 28–35)

42. $8 \cdot (-14)$
43. $\dfrac{48}{-12}$
44. $\dfrac{-93}{3}$
45. $(-36)(-10)$

**Find the solution set if the replacement set is {0, 5, 10, 15, 20}. Pages (80–84)**

46. $30 = 2t + 30$
47. $2y + 10 < 45$
48. $4p + 70 = 150$

## LESSON 3-5  Solving Multiplication and Division Equations

**F**ive members of the Math Team divide the profits from a car wash. Each person receives $18. What is the total profit?

You can write an equation to solve the problem.

| | |
|---|---|
| **What is the question?** | Find the total profit. |
| **Choose a variable** to represent what you need to find. | Let $p$ = the total profit. |

**In words:** Total profit divided by 5 is $18.

**In algebra:**   $p$   ÷   5 = 18,   or   $\frac{p}{5} = 18$

Since the variable, $p$, is divided by 5, the equation $\frac{p}{5} = 18$ is a division equation. You can use mental math to solve the equation.

**Think:** What number divided by 5 equals 18?
**Solution:** $p = 90$

**Check:** Does $\frac{90}{5} = 18$?   Yes ✓   So the solution is 90.

3-5   Multiplication and Division Equations   **95**

A more general method for solving division equations is based on a property of multiplication.

> **Multiplication Property of Equality**
> For all numbers $a$, $b$, and $c$, if $a = b$, then $ac = bc$.

To use this property to solve $\frac{z}{8} = 21$, think:

In the equation, $z$ is divided by 8. To get $z$ alone, multiply each side of the equation by 8.

Multiply each side by 8.     $8 \cdot \frac{z}{8} = 8 \cdot 21$

Simplify.     $z = 168$

The key in solving division equations such as these is to know by what number to multiply each side of the equation.

- If $x + 4 = 7$, what does $9(x + 4)$ equal?

**EXAMPLE 1**   Solve and check: $\frac{x}{4} = -7$

**Solution**   **Think:** What number divides $x$? Multiply each side by that number.

Write the equation.     $\frac{x}{4} = -7$

Multiply each side by 4.     $\frac{x}{4} \cdot 4 = -7 \cdot 4$

Simplify.     $x = -28$

Check in the original equation.     $\frac{x}{4} = -7$

$\frac{-28}{4} \stackrel{?}{=} -7$

$-7 = -7$ ✓

The solution is **−28.**

**Try This**   Solve and check.

1. $\frac{b}{-6} = -14$
2. $-6 = \frac{c}{20}$
3. $\frac{r}{12} = -11$
4. $-1 = \frac{t}{-7}$

You use the Division Property of Equality to solve a multiplication equation, such as $-5x = 125$. This property allows you to divide both sides of $-5x = 125$ by the same number, except for 0.

> **Division Property of Equality**
> For all numbers $a$, $b$, $and$ $c$, where $c \neq 0$,
> if $a = b$, then $\frac{a}{c} = \frac{b}{c}$.

- Why is this property not true for $c = 0$?

**EXAMPLE 2** Solve and check:    **a.** $-5n = 125$      **b.** $13 = \frac{-t}{8}$

**Solutions**

**a.** Think: What number multiplies $n$? Divide each side by that number.

| | |
|---|---|
| Write the equation. | $-5n = 125$ |
| Divide each side by $-5$. | $\frac{-5n}{-5} = \frac{125}{-5}$ |
| Simplify. | $n = -25$ |
| Check in the original equation. | $-5n = 125$ |
| | $-5(-25) \stackrel{?}{=} 125$ |
| | $125 = 125$ ✓ |

The solution is **$-25$.**

**b.** Think: What number divides $-t$? Multiply each side by that number.

| | |
|---|---|
| Write the equation. | $13 = \frac{-t}{8}$ |
| Multiply each side by 8. | $13 \cdot 8 = \frac{-t}{8} \cdot 8$ |
| | $104 = -t$ |
| Think: $-t = (-1)t$. Divide each side by $-1$. | $\frac{104}{-1} = \frac{(-1)t}{-1}$ |
| Simplify. | $-104 = t$ |

The check is left for you. The solution is **$-104$.**

**Try This**    Solve and check.    **5.** $6r = -78$      **6.** $-6 = \frac{-c}{20}$

3-5 Multiplication and Division Equations

# EXERCISES

**Objectives:** To solve multiplication equations
To solve division equations

## Check Your Understanding

**Skill Check** Write the letter of the correct answer.

**1.** Which is the first step in solving $28 = -7t$?
 **a.** $\frac{28}{28} = \frac{-7t}{28}$ **b.** $\frac{28}{-7} = \frac{-7t}{-7}$
 **c.** $\frac{28}{7} = \frac{-7t}{-7}$ **d.** $\frac{28}{-7} = \frac{-7t}{7}$

**2.** What is the first step in solving $\frac{m}{5} = 9$?
 **a.** $\frac{m}{5} \cdot 9 = 9 \cdot 9$ **b.** $\frac{m}{5} \cdot 5 = 9$
 **c.** $\frac{m}{5} \cdot 5 = 9 \cdot 5$ **d.** $\frac{m}{5} = 9 \cdot 5$

**3.** Which equation has the same solution as $r = 3$?
 **a.** $-r = 3$ **b.** $-2r = -8$
 **c.** $-r = -3$ **d.** $r = -3$

**4.** Which is the solution of $\frac{c}{-3} = 12$?
 **a.** $-36$ **b.** $12$
 **c.** $-12$ **d.** $36$

Write P if the solution to the equation is positive. Write N if the solution is negative.

**5.** $6g = 30$ **6.** $\frac{a}{-4} = 10$ **7.** $\frac{w}{7} = -3$ **8.** $-9t = -27$

## Practice and Apply

Write the number by which you would multiply or divide each side of the equation to solve for $x$. Write: "Multiply by (name a number)," or "Divide by (name a number)."

**9.** $10x = 120$ **10.** $\frac{x}{-5} = 4$ **11.** $-85 = 17x$ **12.** $\frac{x}{4} = -10$

Write E when the pair of equations is equivalent. Write N when they are not. When the equations are not equivalent, tell why.

**13.** $\frac{c}{-3} = 4$ and $\frac{c}{-3} \cdot 6 = 4 \cdot 6$ **14.** $2a = 8$ and $\frac{2a}{2} = \frac{8}{8}$

**15.** $-w = 6$ and $-6 = w$ **16.** $-7h = 2$ and $-7h \cdot 2 = 4$

Solve and check.

**17.** $6x = 60$ **18.** $9t = -63$ **19.** $\frac{c}{-4} = 4$ **20.** $\frac{x}{-8} = -7$

**21.** $-4 = \frac{h}{9}$ **22.** $-7z = -49$ **23.** $\frac{c}{7} = 11$ **24.** $-32 = 32y$

**25.** $-3r = -96$ **26.** $\frac{t}{-6} = -16$ **27.** $-75 = -5f$ **28.** $11 = \frac{x}{12}$

**29.** $-x = -13$ **30.** $\frac{-v}{7} = 18$ **31.** $\frac{-w}{4} = -31$ **32.** $127 = -x$

**Tell which algebraic equation represents each word description.**

**Equation A:** $4n = 12$  **Equation B:** $\frac{n}{4} = 12$

**33.** The product of 4 and $n$ is 12.
**34.** Four times a number $n$ is 12.
**35.** The quotient of a number $n$ and 4 is 12.
**36.** The number of peaches, $n$, divided by 4 is 12.

## Connect and Extend

**37.** To solve $ax = c$ for $x$, by what number would you divide each side?

**38.** To solve $\frac{y}{c} = h$ for $y$, by what number would you multiply each side?

**39.** Write a situation that can be described by the equation $9y = 54$.

**40.** Write a situation that can be described by the equation $\frac{x}{5} = 8$.

**For Exercises 41–43, write two equations that are equivalent to the given equation.**

**41.** $lp = w$   **42.** $-s = tu$   **43.** $\frac{x}{y} = -5$

## Maintain Your Skills

**Correct any wrong answers. (Pages 2–6, 13–18, 23–25, 28–35)**

**44.** $|-3| + (4)(2) = 5$   **45.** $(2)(-5) - |-7| = -3$   **46.** $(-3)(7) + |-21| = -42$

**Replace each  to make a true statement. (Pages 54–63)**

**47.** $[24 \cdot (-38)] + (-50) = -50 + (24 \cdot \underline{\ ?\ })$
**48.** $3x + (-3y) = 3(\underline{\ ?\ } - \underline{\ ?\ })$
**49.** The sail of a small sailboat is a triangle with a height of 6 meters and a width of 3 meters. What is the area of the sail? **(Pages 69–74)**
**50.** Write the letter of the equation that does not have the same solution as the other equations. **(Pages 91–94)**
  **a.** $13 + x = -7$  **b.** $-7 + x = 13$
  **c.** $x + 13 = -7$  **d.** $-7 = x + 13$

## ★ Math Team Problem

**51.** You have three piles of pennies. There are 10 pennies in the first pile, 18 pennies in the second pile, and a certain number in the third pile. If you double the number in each pile, you will have 70 pennies. How many pennies are there in the third pile?

Find the solution set for each equation or inequality if the replacement set is {−5, −3, −1, 0, 1, 3, 5}. (Pages 80–84)

1. $4x + 4 = -8$
2. $-4z > -10$
3. $12 = 7y - 23$
4. $9t < 20$
5. $2a + 5 < 0$
6. $10s - 6 = -6$

Solve each equation by using mental math or guess-and-check. (Pages 85–88)

7. $\frac{78}{m} = -26$
8. $-8y = 104$
9. $7t + 51 = 9$

Solve and check. (Pages 91–94, 95–99)

10. $5 + s = 91$
11. $y - 12 = -19$
12. $-28 = 33 + r$
13. $\frac{b}{-17} = 8$
14. $-405 = 9w$
15. $-7 = \frac{-y}{4}$

# Mathematical Footnote

Is zero something? Or is it nothing? Think about the following.

Below are some wooden blocks in the shape of numerals. Some of the numerals have been put inside open boxes.

1. What number is in Box A? How many numbers are there in Box A?
2. What number is in Box B? How many numbers are there in Box B?
3. Box C is empty. How many numbers are there in Box C?

If a box, or a set, has the number zero in it, is the set empty?

This set, { }, is empty. It has nothing in it.

This set, {0}, is not empty. It has one number, the number zero, in it.

So zero really is something!

# LESSON 3-6

## Inverse Operations

The Teen-T Shoppe sells personalized T-shirts. They sold 63 T-shirts this week. This is 3 more than 5 times the number sold last week. How many were sold last week?

**What is the question?** How many shirts were sold last week?
**Choose a variable.** Let $t$ represent the number sold last week.

**Plan:** Write an equation.

**In words:** 5 times the number sold plus 3 equals 63.

**In algebra:**      $5t$      $+$   $3$   $=$   $63$

To solve $5t + 3 = 63$, you can perform operations to "undo" what was done to write the equation. Study these charts.

**Writing an Equation**

| Start with $t$. | $t$ → | Multiply by 5. | $5t$ → | Add 3. | $5t + 3$ → | Equation: $5t + 3 = 63$ |

| Write the solution. $t = 12$ | ← $\frac{5t}{5} = \frac{60}{5}$ | Divide each side by 5. | ← $5t = 60$ | Subtract 3 from each side. | ← | Start: $5t + 3 = 63$ |

**Solving an Equation**

The top four boxes in the figure above (read from left to right) show what was done to $t$ to write the equation. The next four boxes (read from right to left) show how to "undo" what was done in order to solve for $t$.

Since $t = 12$, Teen-T Shoppe sold 12 T-shirts last week.

Addition and subtraction are called inverse operations. Inverse operations "undo" each other. Multiplication and division are also inverse operations. Division and multiplication "undo" each other.

**EXAMPLE 1** Solve and check: $\frac{x}{-3} - 2 = 4$

**Solution** First, apply the inverse of subtracting 2. Then simplify.
Then apply the inverse of dividing by $-3$. Then simplify.

3-5 Inverse Operations **101**

| | |
|---|---|
| Write the equation. | $\frac{x}{-3} - 2 = 4$ |
| Add 2 to each side. | $\frac{x}{-3} - 2 + 2 = 4 + 2$ |
| Simplify. | $\frac{x}{-3} = 6$ |
| Multiply each side by $(-3)$. | $\frac{x}{-3} \cdot (-3) = 6(-3)$ |
| Simplify. | $x = -18$ |
| Check by substituting $-18$ for $x$ in the original equation. | $\frac{x}{-3} - 2 = 4$ |
| Simplify. | $\frac{-18}{-3} - 2 \stackrel{?}{=} 4$ |
| | $6 - 2 \stackrel{?}{=} 4$ |
| | $4 = 4$ ✓ |

The solution is $-18$.

**Try This**   **Solve and check.**

1. $\frac{t}{4} - 3 = -5$        2. $\frac{x}{-2} + 3 = 1$

3. $-11 = \frac{q}{8} + 9$        4. $-10 = \frac{z}{7} + 6$

When you solve an equation using more than one operation, you also apply the Properties of Equality.

**EXAMPLE 2**   Solve and check: $17 = 2w + 5$

**Solution**

| | |
|---|---|
| Write the equation. | $17 = 2w + 5$ |
| Add $-5$ to each side. | $17 + (-5) = 2w + 5 + (-5)$ |
| Apply the Division Property of Equality. | $\frac{12}{2} = \frac{2w}{2}$ |
| | $6 = w$ |
| Check in the original equation. | $17 = 2w + 5$ |
| | $17 \stackrel{?}{=} 2(6) + 5$ |
| | $17 \stackrel{?}{=} 12 + 5$ |
| | $17 = 17$ ✓ |

The solution is **6.**

**Try This**   **Solve and check.**

5. $7w - 5 = -47$        6. $-29 = 12d + 7$

7. $3y + 9 = 6$        8. $44 = -13k - 21$

# EXERCISES

**Objective:** To solve equations by using inverse operations

## Check Your Understanding

**Skill Check**  Write the letter of the correct answer.

1. Which pattern does **not** show inverse operations?
   a. $-5 + 5 + 6 = 6$
   b. $12 \div 3 + 3 = 7$
   c. $(17 \cdot 4) \div 4 = 17$
   d. $1 + 3 - 3 = 1$

2. You start with $-16$ and perform an operation. Then you perform the inverse operation. What is the final result?
   a. $x$   b. $-16x$
   c. $\frac{x}{-16}$   d. $-16$

3. What is the first thing to do to *both sides* of
$$19 = -5x + 4$$
in order to solve it?
   a. Subtract 19.   b. Divide by $-5$.
   c. Add 4.   d. Subtract 4.

4. What is the first thing to do to *both sides* of
$$\frac{x}{5} - 8 = 7$$
in order to solve it?
   a. Add 8.   b. Multiply by 5.
   c. Subtract 7.   d. Subtract 8.

**Solve and check.**

5. $3y - 9 = 12$

6. $5 + 2x = 19$

7. $\frac{x}{3} + 2 = 10$

## Practice and Apply

Replace each  ?  with the name of the property used or the name of the operation done in each step.

8. Solve: $-7x + 3 = -25$
   a. $-7x + 3 + (-3) = -25 + (-3)$ _?_
   b. $\phantom{aaaa}-7x + 0 = -25 + (-3)$ _?_
   c. $\phantom{aaaaaaaa}-7x = -28$ _?_
   d. $\phantom{aaaaaaaaa}\frac{-7x}{-7} = \frac{-28}{-7}$ _?_
   e. $\phantom{aaaaaaaaaa}1x = 4$ _?_
   f. $\phantom{aaaaaaaaaaa}x = 4$ _?_

9. Solve: $5 = 4x + 21$
   a. $5 + (-21) = 4x + 21 + (-21)$ _?_
   b. $5 + (-21) = 4x + 0$ _?_
   c. $\phantom{aaaaa}-16 = 4x$ _?_
   d. $\phantom{aaaaa}\frac{-16}{4} = \frac{4x}{4}$ _?_
   e. $\phantom{aaaaaa}-4 = 1x$ _?_
   f. $\phantom{aaaaaa}-4 = x$ _?_

**Solve and check.**

10. $8 + 5y = -32$

11. $7 + 3t = -20$

12. $-4 + 11n = -92$

13. $\frac{p}{4} - 3 = 15$

14. $1 + \frac{r}{3} = -7$

15. $\frac{s}{7} + 8 = -3$

16. $8 + 9q = -64$

17. $-4 + 11x = -92$

18. $14 = 2 + 3a$

19. $6 + \frac{b}{5} = 7$

20. $13 = \frac{b}{5} + 2$

21. $27 = 3 + \frac{z}{6}$

3-6  Inverse Operations

## Connect and Extend

**22.** Write an equation of the form $ax + b = c$, where $a$, $b$, and $c$ are integers, that has a solution, $x$, that is also an integer.

**23.** Write an equation of the form $ax - b = c$, where $a$, $b$, and $c$ are integers, that has a solution, $x$, that is also an integer.

An equation of the form $ax + b = c$ can be solved with a simple computer program.

```
10 PRINT "For the equation ax + b = c,"
20 PRINT "What are a, b, and c?"
30 PRINT "(Type the numbers separated
   by commas.)"
40 INPUT A,B,C
50 LET X=(C-B)/A
60 PRINT "x= ";X
70 PRINT
80 PRINT "Any more equations?
   (Type y or n.)"
90 INPUT Z$
100 IF Z$="y" OR Z$="Y" THEN 10
110 IF Z$="n" OR Z$="N" THEN 130
120 GOTO 80
130 END
```

**Use the program to solve each equation.**

**24.** $-7x + 34 = -71$     **25.** $22x - 37 = 139$    **26.** $-115x - 147 = 1348$

**27.** $11 + 25x = -1214$    **28.** $39 - 16x = -201$    **29.** $145 = 12x + 1$

## Maintain Your Skills

**Simplify. Write the answer only if it is negative. (Pages 46–50)**

**30.** $(32)(40) - (51)(-11)$    **31.** $(412)(-10) \div 1$    **32.** $75 \div (-3) + 30$

**33.** $-81 \div (-9) - 8$    **34.** $9 - 6 \cdot 2 - 7$    **35.** $2 - 3 \cdot 4 + 9$

**Insert parentheses in the equation to make it true. (Pages 46–50)**

**36.** $-40 + 2 - 10 \cdot (-4) = -8$    **37.** $12 \div 3 \cdot 2 - 10 \cdot 3 + 1 = -38$

**Find the solution set if the replacement set is $\{-25, -3, 0, 3, 25\}$. (Pages 80–84)**

**38.** $-4x + 25 = 37$    **39.** $-3x + 5 = 5$    **40.** $-5x - 8 > -20$

## ★ Math Team Problem

**41.** Copy the figure at the right. Connect all points in the figure by drawing exactly four line segments without lifting your pencil off the paper. Do not draw through any one point more than once.

# LESSON 3-7
# Solving Multi-Step Equations

To reduce inventory, Galactic Games offers single game cartridges at a $4-off special. Sol buys 3 of the games on sale for $48. What is the original price of a game when it is not on sale?

You can use an equation to find the original price. Let $x$ represent the original price. Then $x - 4$ represents the price Sol paid for one game. Since Sol bought 3 games, this equation represents the situation.

$$3(x - 4) = 48$$

To solve an equation having a variable inside the parentheses, you begin by using the Distributive Property to remove the parentheses.

| | |
|---|---|
| Write the original equation. | $3(x - 4) = 48$ |
| Use the Distributive Property. | $3x - 12 = 48$ |
| Add 12 to each side. | $3x - 12 + 12 = 48 + 12$ |
| Simplify. | $3x = 60$ |
| Divide each side by 3. | $\dfrac{3x}{3} = \dfrac{60}{3}$ |
| Simplify. | $x = 20$ |

To check your answer, replace $x$ with 20 in the *original* equation.

$$3(20 - 4) \stackrel{?}{=} 48$$
$$3(16) \stackrel{?}{=} 48$$
$$48 = 48 \checkmark$$

The original price of a game is $20.

**EXAMPLE 1** Solve and check: $3(x + 3) = 18$

**Solution** Use the Distributive Property.

| | |
|---|---|
| Write the equation. | $3(x + 3) = 18$ |
| Use the Distributive Property. | $3x + 9 = 18$ |
| Subtract 9 from each side. | $3x = 9$ |
| Divide each side by 3. | $x = 3$ |

**Check** Substitute 3 for $x$ in the original equation.

$$3(x + 3) = 18$$
$$3(3 + 3) \stackrel{?}{=} 18$$
$$9 + 9 \stackrel{?}{=} 18$$
$$18 = 18 \checkmark$$

The solution is **3**.

**Try This** **Solve and check.**

**1.** $5(x + 2) = 12$   **2.** $-3(x - 2) = 6$   **3.** $-2(x - 5) = 8$

When an equation has the variable in more than one term, combine these terms before you use the Properties of Equality.

**EXAMPLE 2** Solve and check: $4(x + 2) + 3x = -6$

**Solution** Use the Distributive Property first.

| | |
|---|---|
| Write the equation. | $4(x + 2) + 3x = -6$ |
| Use the Distributive Property. | $4x + 8 + 3x = -6$ |
| Combine like terms. | $7x + 8 = -6$ |
| Add $(-8)$ to each side. | $7x = -14$ |
| Divide each side by 7. | $x = -2$ |

**Check** Substitute $-2$ for $x$ in the original equation.

$$4(x + 2) + 3x = -6$$
$$4[(-2) + 2] + 3(-2) \stackrel{?}{=} -6$$
$$[4 \cdot 0] + 3(-2) \stackrel{?}{=} -6$$
$$0 + -6 \stackrel{?}{=} -6$$
$$-6 = -6 \checkmark$$

The solution is **−2**.

**Try This** **Solve and check.**

**4.** $3(x - 2) + 2x = 1$   **5.** $4x + 2(x + 1) = -10$

Taking the opposite of an expression is the same as multiplying the expression by $-1$. Study these examples carefully.

$$-x = -1 \cdot x = -x$$
$$-(3x) = -1 \cdot 3x = -3x$$
$$-(ab) = -1 \cdot ab = -ab$$

If an expression contains a sum, taking the opposite also includes using the Distributive Property.

$$\begin{aligned}-(a + b) &= (-1)(a + b) \\ &= (-1 \cdot a) + (-1 \cdot b) \\ &= -a + (-b), \text{ or } -a - b\end{aligned}$$

If an expression contains a difference, first rewrite the subtraction as the addition of the opposite. Then use the Distributive Property. For example,

$$\begin{aligned}-(2x - 5) &= -[2x + (-5)] \\ &= (-1)[2x + (-5)] \\ &= (-1)(2x) + (-1)(-5) \\ &= -2x + 5\end{aligned}$$

**EXAMPLE 3**  Solve and check: $3x - (5x + 1) = 9$

**Solution**  Use the definition of subtraction to rewrite $3x - (5x + 1)$.

| | |
|---|---|
| Write the equation. | $3x - (5x + 1) = 9$ |
| Use the definition of subtraction. | $3x + [-(5x + 1)] = 9$ |
| Use the Distributive Property. | $3x + (-1)(5x) + (-1)(1) = 9$ |
| Multiply. | $3x - 5x - 1 = 9$ |
| Combine like terms. | $-2x - 1 = 9$ |
| Add 1 to each side. | $-2x = 10$ |
| Divide each side by $-2$. | $x = -5$ |

**Check**  Substitute $-5$ for $x$ in $3x - (5x + 1) = 9$.

$$3(-5) - [5(-5) + 1] \stackrel{?}{=} 9$$
$$-15 - (-24) \stackrel{?}{=} 9$$
$$-15 + 24 \stackrel{?}{=} 9$$
$$9 = 9 \checkmark$$

The solution is **−5**.

**Try This**  **Solve and check.**

**6.** $2x - (3x + 2) = 1$    **7.** $4x - (2x + 1) = -5$

3-7  Solving Multi-Step Equations

# EXERCISES

**Objective:** To solve equations using a multi-step process

## Check Your Understanding

**Skill Check** Write the letter of the correct answer.

**1.** To solve $3x + 7 = 10$, first
   a. add 7 to each side.
   b. add $-7$ to each side.
   c. multiply each side by $-7$.
   d. divide each side by $-7$.

**2.** To solve $4(x + 3) = 9$, first
   a. add $-3$ to each side.
   b. multiply each side by 4.
   c. use the Distributive Property.
   d. use the Subtraction Property.

**3.** Which equation is equivalent to $2x - 3 = 5$?
   a. $2x = 8$     b. $2x = 2$
   c. $2x = 7$     d. $x = 3$

**4.** Which expression is equivalent to $-(x + y)$?
   a. $-x + y$     b. $x - y$
   c. $-x - y$     d. $x + y$

Rewrite each expression as a sum or difference.

**5.** $3(x - 2)$     **6.** $-4(x + 5)$     **7.** $-2(2x - 5)$     **8.** $5(x + 3) - 2x$

## Practice and Apply

**Solve and check.**

**9.** $4(y + 2) = 28$
**10.** $2(t + 4) = -4$
**11.** $6(x - 1) = -12$
**12.** $9(a - 1) = 27$
**13.** $-3(b + 5) = 6$
**14.** $-5(r - 6) = 10$
**15.** $-4(n - 3) = 8$
**16.** $-13(g + 1) = 39$
**17.** $-20 = -5(a + 7)$
**18.** $18 = 6(4 - z)$
**19.** $2(n - 6) + 4n = -12$
**20.** $5(p - 6) + 3p = -6$
**21.** $-2(a + 3) - 4a = 12$
**22.** $-4(q + 1) - 6q = 6$
**23.** $9(h + 3) - 3h = 39$
**24.** $-d - 2(d + 3) = 0$
**25.** $4z - 3(7 + 3z) = -1$
**26.** $15 = -5c + 7(c + 5)$
**27.** $0 = -2(4y - 9) - y$
**28.** $-2(6 + a) - a = -18$
**29.** $c - 3(c + 4) = -6$
**30.** $2t - (3t + 7) = 9$
**31.** $12r - (r + 1) = 21$
**32.** $5m - (m - 7) = -1$
**33.** $8a - (3 - a) = -30$
**34.** $-6(2x - 1) - (3x + 5) = 31$
**35.** $7(x + 6) - (1 - x) = 33$
**36.** $16 = 5(p - 6) - (p + 2)$
**37.** $1 = 3(s - 4) - (s + 11)$
**38.** $0 = 7t - (t + 12)$

## Connect and Extend

Let the replacement set be $R = \{-3, -2, 0, 1, 2, 3\}$. Find the solution set for each sentence.

**39.** $-(x - 3) = -x + 3$     **40.** $-x - 2 = -x + (-2)$

**41.** Give 5 examples to show that the opposite of the sum of ten numbers equals the sum of their opposites.

**42.** Give 5 examples to show that the opposite of the difference of two numbers equals the difference of their opposites.

Here is a computer program to solve equations of the form $a(x + b) = c$.

```
100 PRINT "For the equation a * (x + b) = c,"
110 PRINT "What are a, b, and c? Type the numbers sepa-
           rated by commas."
120 INPUT A,B,C
130 LET X = (C - A * B)/A
140 PRINT "x = ";X
150 PRINT
160 PRINT "Any more equations? (Y = yes, N = no)
170 INPUT Z$
180 IF Z$ = "y" OR Z$ = "Y" THEN 100
190 END
```

Use the program to solve these equations.

**43.** $-2(x + 9) = 12$    **44.** $12(x + 243) = -96$    **45.** $-5(x - 89) = 120$
**46.** $7(x + 87) = 56$    **47.** $-15(x - 678) = -225$    **48.** $42(x + 876) = 882$

## Maintain Your Skills

**Solve only the problems having answers greater than zero.**
**(Pages 13–18, 23–25, 28–35)**

**49.** $-68 + (-12)$    **50.** $26 + (-12)$    **51.** $4 - (-9)$    **52.** $-(-6) + 6$
**53.** $-9(-1)$    **54.** $80(-20)$    **55.** $-21 \div 7$    **56.** $-(-50)$

**Find the area. (Pages 64–74)**

**57.** A triangle with $b = 8$, $h = 17$
**58.** A square with side $s = 13$ inches

## ★ Math Team Problems

Numbers written in a particular order form a **sequence**. The sequence below has a pattern. Each term, starting with the second term, is 3 more than the previous term.

$$1, 4, 7, 10, 13, \ldots$$

**For each sequence, find the pattern and write the next three terms.**

**59.** 8, 16, 24, 32, …    **60.** $-50, -45, -40, -35, \ldots$    **61.** 1, 2, 4, 8, …
**62.** 1, 8, 27, 64, …    **63.** 1, 1, 2, 3, 5, …    **64.** $3, -6, 9, -12, \ldots$

**65.** Explain how the pattern works in Exercises 62 and 63.

# Translating Word Expressions to Algebraic Expressions

Translating word expressions to algebraic expressions is an essential skill in solving word problems. This table shows that the *same* operation symbol can be used to translate *more than one* word expression. For example, the symbol for addition, +, translates 7 different word expressions in this table.

| | Word Expression | Algebraic Expression |
|---|---|---|
| **Addition Expressions** | The **sum** of a number, $t$, and $-3$ | $t + (-3)$ |
| | $z$ **plus** 1.9 | $z + 1.9$ |
| | **Add** $7\frac{1}{2}$ to $k$. | $k + 7\frac{1}{2}$ |
| | $50 **more than** $d$ dollars | $d + 50$ |
| | $-\frac{7}{8}$ **increased by** $n$ | $-\frac{7}{8} + n$ |
| | **Increase** the quantity $(2t + 4)$ by 6. | $(2t + 4) + 6$ |
| | The **total** of 16 and $-r$ | $16 + (-r)$ |
| **Subtraction Expressions** | $x$ **minus** 8 | $x - 8$ |
| | The **difference of** $60 and $b | $60 - b$ |
| | **Subtract** 7.8 from $n$. | $n - 7.8$ |
| | The quantity $(3t - 9)$ **decreased by** $-8$ | $(3t - 9) - (-8)$ |
| | $y°$ **less** $5°$ | $y - 5$ |
| | $y°$ **less than** $5°$ | $5 - y$ |
| **Multiplication Expressions** | 1.7 **times** $z$ | $1.7z$ |
| | $b$ **multiplied by** 23 | $b \cdot 23$ or $23b$ |
| | The **product of** 5 and the quantity $(2a - 9)$ | $5(2a - 9)$ |
| | **Twice** $m$ centimeters | $2m$ |
| | **Triple** the width, $w$ | $3w$ |
| **Division Expressions** | The cost, $c$ **divided by** 12 | $c \div 12$, or $\frac{c}{12}$ |
| | The **quotient of** the quantity $(2a - 9)$ and 15 | $\frac{2a - 9}{15}$ |
| | The **quotient of** 15 and $y$ | $15 \div y$, or $\frac{15}{y}$ |

Notice how these two expressions differ.

$y$ less 5    **is not the same as**    $y$ less than 5.

$y - 5$    **is not the same as**    $5 - y$.

Terms enclosed in parentheses are treated as one quantity. Thus, 5(a + b) can be read "5 times the quantity a + b."

To represent a word expression by algebraic symbols, perform these steps.

1. Choose a variable. Tell what it represents.
2. Identify the key word or words that tell which *operation* to use.
3. Write the word expression in *symbols*.

**EXAMPLE 1** Use algebraic symbols to translate each word expression.
   a. 15 seconds more than Amy's record time
   b. The amount of a car loan divided by 36 payments

**Solutions**
   a. Let t = Amy's record time.
      15 seconds <u>more than</u> t        t + 15

   b. Let a = the amount of the car loan.
      a <u>divided</u> by 36        a ÷ 36

**Try This** **Use algebraic symbols to translate the word expression.**
   1. 2 seconds less than Karl's record time, t
   2. The distance traveled, d, divided by 3 hours

In the table on page 110, "b multiplied by 23" translates exactly into b · 23. You can then use the Commutative Property to rewrite this as 23b.

**EXAMPLE 2** Use algebraic symbols to translate the words.
   a. The sum of x and 12         b. The product of 3y and 7

**Solutions**
   a. x + 12 is the exact translation.
      (x + 12) equals (12 + x) by the Commutative Property.

   b. 3y · 7 is the exact translation.
      You can simplify this to 7 · 3y or 21y.

**Try This** **Use algebraic symbols to translate the words.**
   3. 6 less y                    4. Subtract z from 6.
   5. y more than 9               6. Triple the length, l

3-8  Translating Word Expressions to Algebraic Expressions

# EXERCISES

**Objective:** To translate word expressions to algebraic expressions

## Check Your Understanding

*Skill Check*  Write the letter of the correct answer.

1. Which is the algebraic translation for "the total of $-23$ and $s$"?
   a. $-23 - s$   b. $-23s$
   c. $-23 + s$   d. $-23 \div s$

2. Which is the algebraic translation for "the quotient of 10 and $r$"?
   a. $\frac{r}{10}$   b. $\frac{10}{r}$
   c. $r - 10$   d. $10 - r$

3. Which is the algebraic translation for "subtract 3 from $g$"?
   a. $3 - g$   b. $3 \div g$
   c. $g \div 3$   d. $g - 3$

4. Which is the algebraic translation for "21 less $t$"?
   a. $t - 21$   b. $21 - t$
   c. $t + 21$   d. $21t$

For each word expression, choose a variable and tell what it represents. Use algebraic symbols to translate the expression.

5. The cost of the coat plus $4

6. Four times the width of a rectangle

## Practice and Apply

In Exercises 7–11, use the following information.

John is $n$ years old. Catalina is $n - 4$ years old. Tasha is $n + 4$ years old. Walt is $4n$ years old. Sam is $\frac{n}{4}$ years old.

7. What variable is used to represent John's age?
8. Who is 4 years older than John?
9. Who is 4 times as old as John?
10. Suppose John is 16 years old. How old is Catalina?
11. Suppose John is 20 years old. How old is Sam?

Replace each <u>?</u> with the word or words that make the word expression match the algebraic expression.

12. $k - 3$ can be read as "3 <u>?</u> $k$".
13. $k \div 3$ can be read as "3 <u>?</u> $k$".
14. $k + 3$ can be read as "3 <u>?</u> $k$".
15. $k \cdot 3$ can be read as "$k$ <u>?</u> 3".

**Use this information for Exercises 16–18.**
Christine is $8k$ years old. Tom is $8k + 2$ years old. Kathy is $8k - 1$ years old. Marc is $\frac{8k}{2}$ years old.

16. Who is the oldest?
17. Who is the youngest?
18. Which two people are closest in age?

**Write two word expressions for each algebraic expression.**

19. $s + 7.5$
20. $m - 13$
21. $\frac{21}{n + 4}$
22. $2(r - 15)$

**Use algebraic symbols to translate each word expression.**

23. The product of $-3$ and $x$
24. Seven increased by $c$
25. The total of $j$, $3k$, and 5
26. Eight decreased by the quantity $(z + 2)$
27. Triple the sum of $g$ and $h$
28. Seven divided by the quantity $(9 + q)$
29. Six less than the sum of $s$ and $t$
30. The quotient of the quantity $(8 + d)$ and $f$

**You know that Stefani is $s$ years old. Write an algebraic expression to complete each statement.**

31. If Mikki is 14 years younger than Stefani, then ? is Mikki's age.
32. If Stefani's age is 4 years more than twice Mikki's, then ? is Stefani's age.

## Connect and Extend

**For each expression, choose a variable and tell what it represents. Then write an algebraic expression for each word expression.**

33. Five centimeters more than the width of a rectangle
34. Triple the perimeter of the base of the Great Pyramid of Egypt
35. The product of the distance traveled and 3 hours
36. Four times the perimeter of a parking lot less 16
37. Five hundred less than the quotient of the number of pages in a telephone directory and 9

## Maintain Your Skills

**Which equations have the same solution? (Pages 91–99)**

38. $y + 5 = 181$
39. $-7r = 168$
40. $\frac{t}{-8} = -22$
41. $x - 185 = -9$

**Solve and check. (Pages 101–109)**

42. $7g + 8 = 1$
43. $-10x + 81 = 51$
44. $7(3q - 1) = 35$

# LESSON 3-9: Problem Solving: Writing an Equation

Sanjay deposited some money in his savings account. Maggie deposited one dollar more than Sanjay, and Daren deposited one dollar more than Maggie. Altogether they deposited $258. How much did each person deposit?

## UNDERSTAND THE PROBLEM

What is the question?   How much did each person deposit?

What is given?   Maggie's deposit: $1 more than Sanjay's.
Daren's deposit: $1 more than Maggie's.
Total Deposit: $258.

**Think: Consecutive numbers** follow one another. Each is one more than the number before it.

**Consecutive Numbers**
56, 57, 58, ...

## DEVELOP A PLAN

**Strategy:** Write an equation to represent the situation.

## CARRY OUT THE PLAN

Represent the unknowns.

Let $a$ = the smallest amount.
Then $a + 1$ = the middle amount.
Then $a + 2$ = the largest amount.

Write an equation. The sum of the three amounts is 258.

$a + (a + 1) + (a + 2) = 258$
$a + a + 1 + a + 2 = 258$
$3a + 3 = 258$
$3a = 255$
$a = 85$

Remember to find $a + 1$ and $a + 2$.

$a + 1 = 86$
$a + 2 = 87$

## LOOK BACK

**Check:** Do the deposits differ by $1?   Yes ✓
Does $85 + $86 + $87 = $258?   Yes ✓

Sanjay deposited $85, Maggie deposited $86, and Daren deposited $87.

Identifying conditions in a problem will help you in translating from word expressions to algebraic expressions and in writing an equation to represent the situation.

**EXAMPLE 1** A marathon had 448 more local participants than visiting runners (Condition 1). The total number in the race was 3640 (Condition 2). Find the number of local runners and visiting runners.

**Solution**

**UNDERSTAND THE PROBLEM**
**Find:** The number of local runners and visiting runners
**Given:** There are 448 more local runners than visiting runners.
(Condition 1)
There are 3640 runners in all.
(Condition 2)

**DEVELOP A PLAN**
Use Condition 1 to represent the unknowns.
Use Condition 2 to write an equation.

**CARRY OUT THE PLAN**
**Represent the unknowns.** Let $r$ = number of visiting runners.
Then $r + 448$ = number of local runners.

Write an equation to "connect" Condition 1 and Condition 2.

**Think:** Number of Visiting Runners + Number of Local Runners = 3640

**Translate:** $r$ + $r + 448$ = 3640

| | |
|---|---:|
| Write the equation. | $r + r + 448 = 3640$ |
| Add ($-448$) to each side. | $2r + 448 = 3640$ |
| Divide each side by 2. | $2r = 3192$ |
| | $r = 1596$ |
| Don't forget to find $r + 448$. | $r + 448 = 2044$ |

**LOOK BACK**

Check Condition 1: Does $2044 = 1596 + 448$?  Yes ✓
Check Condition 2: does $1596 + 2044 = 3640$?  Yes ✓

There were **1,596 visiting runners** and **2,044 local runners.**

3-9 Problem Solving: Writing an Equation

**Try This**

1. Susan weighs 43 pounds less than Jim (Condition 1).
   Their total weight is 261 pounds (Condition 2).
   Find Susan's weight and Jim's weight.

Sometimes Condition 2 is given in the first statement of the problem.

**EXAMPLE 2** Two movie studios have provided a total of 104 science fiction movies (Condition 2). Studio B has produced 20 fewer science fiction movies than Studio A (Condition 1). How many has Studio B produced?

**Solution:** **Find:** The number of science fiction movies Studio B produced
**Given:** Together both studios produced 104 science fiction movies.

Represent the unknowns.

Let $x$ = the number Studio A produced.
Then $x - 20$ = the number Studio B produced.

**Think:** Number for Studio A + Number for Studio B = Total
$$x \quad + \quad x - 20 \quad = 104$$

Solve the equation.
$$x + x - 20 = 104$$
$$2x - 20 = 104$$
$$2x - 20 + 20 = 104 + 20$$
$$2x = 124$$
$$\frac{2x}{2} = \frac{124}{2}$$

Don't forget to find $x - 20$.
$$x = 62 \leftarrow \text{Studio A}$$
$$x - 20 = 62 - 20$$
$$x - 20 = 42 \leftarrow \text{Studio B}$$

Check Condition 1: Does $42 = 62 - 20$?  Yes ✓
Check Condition 2: Does $62 + 42 = 104$?  Yes ✓

Studio B produced **42** science fiction movies.

**Try This**

2. Two companies sold a total of 13,574 shares of their stock (Condition 2). Company A sold 3,260 fewer shares than Company B (Condition 1). How many shares of stock did each company sell?

# EXERCISES

**Objective:** To solve word problems by writing equations

## Check Your Understanding

**Skill Check** Write the letter of the correct answer.

1. Which group of algebraic expressions represents three consecutive integers?
   a. $j, j + 2, j + 4$
   b. $j + 1, j + 3, j + 5$
   c. $j, j + 5, j + 10$
   d. $j, j + 1, j + 2$

2. Dick has 3 times as many quarters as dimes. He has 220 dimes and quarters altogether. Which equation represents this situation?
   a. $d + 3d = 220$   b. $3d - d = 220$
   c. $3d = 220$   d. $3d = d$

3. A coat costs $65 more than a dress. Together they cost $113. How much does the dress cost?
   a. $24   b. $89
   c. $49   d. $65

4. The sum of Janet's and Mischa's ages is 31. Mischa is a year younger than Janet. How old is Mischa?
   a. 30   b. 18
   c. 16   d. 15

For Exercises 5–6, refer to the check in Example 1 on page 115.

5. Explain why $2044 = 1596 + 448$ checks Condition 1.

6. Explain why $1596 + 2044 = 3640$ checks Condition 2.

## Practice and Apply

Write an equation to represent each situation. Then solve the equation and check the results.

7. Jane scored 14 points less than the points Natalie scored (Condition 1). Together they scored 46 points (Condition 2). How many points did each girl score?

8. A Doberman weighs twice as much as a cocker spaniel (Condition 1). If together the dogs weigh 135 pounds (Condition 2), how much does each dog weigh?

9. A hospital has 1550 patients (Condition 2). If there are 128 more female patients than male patients (Condition 1), find the number of female and male patients.

10. Tina threw a discus triple the distance that Kim threw it (Condition 1). Together they threw the discus a total distance of 280 feet (Condition 2). How far did each throw the discus?

11. Together Bob and Dan earned $80. Bob earned 4 times as much as Dan. How much did each boy earn?

12. TKY High School sold 842 tickets to their carnival. They sold 512 more children's tickets than adult tickets. How many of each type did they sell?

3-9   Problem Solving: Writing an Equation

**13.** Find three consecutive integers whose sum is 138.

**14.** Find four consecutive integers whose sum is −66.

**15.** In one long-distance phone call, Amy talked to her parents for twice as long as her brother talked. Her sister talked for 12 minutes longer than Amy. If the phone call was 62 minutes long, how long did each person talk on the phone?

**16.** Cal is 4 inches taller than Kay. Jeremy is 2 inches taller than the sum of Cal's and Kay's heights. The sum of their three heights is 158 inches. How tall is each person?

## Connect and Extend

**Write a question that fits the situation and the given answer.**

At a sale, a store reduced the price of T-shirts from $12 to $8 and reduced the price of shorts from $15 to $12.

**17.** Answer: $3

**18.** Answer: $7

The distance from Buda to Crull is 54 miles. Buda is the same distance from Adler as from Crull. The distance from Crull to Ennis is 36 miles less than the distance from Adler to Delco. The total distance from Adler to Ennis is 216 miles.

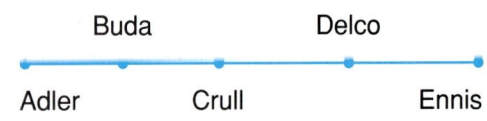

**19.** Copy the diagram at the right above. Use the information to label the diagram.

**20.** How far is Crull from Delco?

**21.** How far is it from Ennis to Buda?

## Maintain Your Skills

**Write an integer to represent each amount. (Pages 2–6)**

**22.** 9 floors down

**23.** A profit of $76

**24.** A tax of $1.50

**Write +, −, ×, or ÷ to identify the operation to be done first. (Pages 46–50).**

**25.** $9 \div 9 \cdot 4$

**26.** $[(73 - 8) \div 5] + (-3)$

**27.** $(49 - 4 \cdot 4) \div 9 - 3$

**Use the Distributive Property to write an equivalent expression. (Pages 59–63)**

**28.** $9b + (-7b)$

**29.** $-8c + 4c$

**30.** $63f - 7f$

# SUMMARY

## KEY TERMS
elements of a set (p. 81)
equation (p. 80)
equivalent equations (p. 92)
inequality (p. 82)
inverse operations (p. 101)

members of a set (p. 81)
replacement set (p. 81)
set (p. 81)
solution (p. 80)
solution set (p. 81)

## KEY IDEAS

**A. Addition Property of Equality:** Adding the same number to each side of an equation results in an equivalent equation.

**B. Multiplication Property of Equality:** Multiplying each side of an equation by the same nonzero number results in an equivalent equation.

**C. Division Property of Equality:** Dividing each side of an equation by the same nonzero number results in an equivalent equation.

**D.** To solve equations involving more than one operation, use the Addition Property of Equality first. Then use the Multiplication or Division Properties of Equality.

**E.** To represent a word expression by an algebraic expression, follow these steps.
  1. Choose a variable. Tell what it represents.
  2. Identify the key word or words that tell which operation to use.
  3. Write the word expression in symbols.

**F. Steps in Problem Solving**
  1. Understand the problem.
  2. Develop a plan.
  3. Carry out the plan.
  4. Look back.

# REVIEW

**Find the solution set if the replacement set is {−3, −2, −1, 0, 1, 2, 3}. (Pages 80–84)**

1. $7b - 4 = -25$
2. $8 < -3c + 1$
3. $-4t - 5 < -10$
4. $3 = -7m + 17$

**Find the solution set if the replacement set is {−25, −2, 0, 2, 25}. (Pages 80–84)**

5. $-2s - 5 = -55$
6. $7n - 16 = -2$
7. $5p + 15 < 25$
8. $9y + 25 > -2$

**Solve each equation by using mental math. (Pages 85–88)**

9. $g + 7 = -3$
10. $7y = -56$
11. $\frac{s}{-9} = -6$
12. $-10 = -4 - k$

**Solve by using guess-and-check. (Pages 85–88)**

13. $75 = 121 + g$
14. $-157 = 93 - x$
15. $-18d = 324$
16. $\frac{r}{-6} = -39$

**Solve and check. (Pages 91–99, 101–104)**

17. $v - 17 = 63$
18. $8m = -120$
19. $\frac{h}{-4} = 45$
20. $19 = d + 53$
21. $6x + 19 = 7$
22. $-80 = w - 90$
23. $-56 = d + 21$
24. $128 = -4y$
25. $-20n = -320$
26. $104 = 161 + u$
27. $\frac{-w}{36} = -8$
28. $\frac{y}{5} + 4 = -18$
29. $\frac{z}{-9} - 7 = -20$
30. $x - (-8) = 30$
31. $-g = 424$
32. $-23 = 26z - 205$

**Solve and check. (Pages 105–109)**

33. $5(c + 1) = -20$
34. $-6(y + 2) = 12$
35. $4(r - 3) = 36$
36. $-7(s - 6) = 0$
37. $3(a - 3) - 4a = -1$
38. $4q - (q - 5) = 50$

**Use algebraic symbols to translate each word expression. (Pages 110–113)**

39. The number of apples picked, $n$, plus 10
40. The quotient of the quantity, $(a - 6)$, and 5
41. Evelyn's age, less 6
42. Three times the weight, $w$

**Write an equation to represent the situation. Then solve the equation and check the results. (Pages 114–118)**

43. Marco traveled 1258 fewer miles than Chris traveled. If they traveled a total of 3854 miles, how far did each person travel?
44. Together Cynthia and Joe lifted 75 pounds. If Joe lifted twice the weight Cynthia lifted, how much did each person lift?
45. Carol Soong's scores for 3 games of bowling were consecutive whole numbers. The sum of her scores was 315. Find her scores.
46. The ages of Jason, Amy, and Eric are consecutive whole numbers. Jason is the oldest and Eric is the youngest. If the sum of their ages is 63, how old is Amy?

# TEST

**Find the solution set if the replacement set is {−40, −2, −1, 0, 2, 40}.**

1. $12b - 53 = -29$
2. $7y + 9 < 2$
3. $6m + 8 > -2$
4. $3a + 153 = 33$
5. $30 < 15 - 7n$
6. $12 = 2 - \frac{x}{4}$

**Solve and check.**

7. $j + 9 = -5$
8. $90 = -6k$
9. $p - 34 = -72$
10. $\frac{t}{-4} = 24$
11. $87v = 261$
12. $\frac{-u}{-9} = -144$
13. $-7y - 54 = -40$
14. $x + 47 = -29$
15. $12x = -132$
16. $11(f + 3) = 88$
17. $44 = 13k - 21$
18. $-x = -23$

19. Write the letter of the equation that could represent this statement. Jerry has two quarters less than Susan.
    a. $j + s = 4$
    b. $j = 2 - s$
    c. $j = s - 2$
    d. $j - 2 = s$

20. Which of these choices best describes what $j$ represents in the equation for Exercise 19?
    a. Jerry
    b. The number of quarters Jerry has
    c. The quarters
    d. Jerry's age

21. What does s represent in the equation in Exercise 19?

**Use algebraic symbols to translate each word expression.**

22. Triple the quantity $(p - q)$
23. The quantity $(5e - f)$ divided by 6
24. The sum of $w$ and $x$ plus $z$
25. 18 less $a$
26. The product of Steve's age, $a$, and 4
27. Eleven pounds less than Lori's weight, $w$

**Write an equation to represent the situation. Then solve the equation and check the results.**

28. Tina is two years older than Rose. The sum of their ages is 32. Find each girl's age.

29. Together, Rudolfo and his cat weigh 135 pounds. Rudolfo weighs 8 times as much as his cat. How much does the cat weigh?

30. Julia's test score was one point lower than Roberta's score. The sum of their scores was 187. What was Roberta's score?

31. Kurt and Melissa were born in consecutive years. Kurt is one year younger than Melissa. Find their present ages if the sum of their ages is 35.

# REVIEW: CHAPTERS 1–3

**Replace each ? with <, >, or = to make a true statement. (Pages 2–10)**

1. $-(-7)$ ? $(-7)$
2. $-(-5)$ ? $|-6|$
3. $|-4|$ ? $-(-4)$
4. $-(2)$ ? $-|-1|$

**Write an algebraic expression for each word expression. Then find the answer. (Pages 2–6, 13–18, 23–25, 32–35)**

5. A $64 loss plus a $27 profit
6. Nine degrees below zero minus 20 degrees above zero
7. A temperature drop of 15 degrees from $-1°$.
8. Eight years ago divided by two years
9. On three successive plays, a football team gained 2 yards, lost 8 yards, and gained 15 yards. If 10 yards are needed for a first down, did the team make a first down? Explain your answer. **(Pages 7–10)**
10. The temperature was recorded at 18° Celsius. It then rose 7°, fell 5°, fell 21°, and rose 14°. What was the temperature after these changes? **(Pages 13–18)**

**Write an algebraic expression for each word expression. Use the following information. (Pages 39–43)**

Beth earned $d$ dollars by babysitting. How much will she have in each situation?

11. After she earns $10 more
12. After she triples the amount
13. After she decreases the amount by $7
14. After she divides the amount by 4

The sum of the first 100 natural numbers can be found by adding the numbers one by one or by using the formula, $\frac{n(n+1)}{2}$, where $n$ = the number of numbers. **(Pages 39–43)**

15. Find the sum of the first 6 natural numbers.
16. Find the sum of the first 10 natural numbers.
17. Find the sum of the first 108 natural numbers.

**Which expression in each pair has the greater value? (Pages 46–50)**

18. $24 - 6 - 4$ or $24 - (6 - 4)$
19. $(-3)(-8) + 8$ or $(-3)(-8 + 8)$

**Replace each ? with +, −, ×, or ÷ to make a true statement. (Pages 46–50)**

20. $8 + 4$ ? $12 = 0$
21. $5 \times 5$ ? $5 \times 5 = 50$
22. $35$ ? $7$ ? $3 = 15$

**23.** Write the letter of the expression that has the smallest value. (Pages 46–50)
  **a.** (20 ÷ 2) ÷ 5 + 5   **b.** (20 ÷ 5) ÷ 4 + 5   **c.** (20 ÷ 2) ÷ (5 + 5)

Give two examples of each property. (Pages 54–63)

**24.** Commutative Property of Addition

**25.** Distributive Property

**26.** Identity Property of Multiplication

**27.** Associative Property of Multiplication

**28.** The number of rooms in the four largest U.S. hotels is 4032, 3529, 3174, and 3049. Find the average number of rooms per hotel. (Pages 64–68)

Find each area. (Pages 69–74)

**29.** 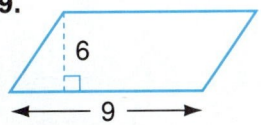    **30.** 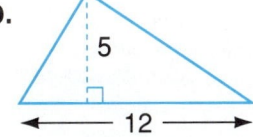    **31.**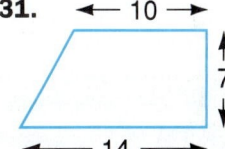

Find the solution set if the replacement set is {−6, −3, 0, 3, 6}. (Pages 80–84)

**32.** $3t - 6 = -15$   **33.** $2y + 10 < 15$   **34.** $8n + 22 = 70$

Solve and check. (Pages 91–99, 101–104)

**35.** $t - 15 = -55$   **36.** $12 + e = -90$   **37.** $-15 = \frac{c}{7}$

**38.** $-4a = 144$   **39.** $7(v - 4) = -119$   **40.** $44 = -13k - 21$

Write an equation to represent each situation. Then solve and check. (Pages 110–118)

**41.** Julie's test score was three points lower than Jack's score. The sum of their scores was 189. Find their scores.

**42.** The sum of Willie's three diving scores was 24. The scores were consecutive numbers. Find his scores.

**43.** Dena weighs 4 times as much as her younger sister, Flor. If together the girls weigh 160 pounds, how much does each girl weigh?

**44.** Ana is 5 inches shorter than Chico. Felicia is 3 inches taller than Chico. The sum of their total heights is 190 inches. How tall is each person?

# UNIT 2

### PROJECT 1
The Chamber of Commerce is sponsoring a Make a Circus Day at a local park. Your club volunteered to run the sandwich booth as a money-making project. You are on the planning committee that will decide on those kinds of sandwiches, what special names you will give them, how much can you spend on ingredients, and how much you will charge for each kind of sandwich. Work as a team to determine the costs and the profits you intend to make.

### PROJECT 2
A clothing shop owner asks if you will create some designs on oversized denim jackets, which she will sell in her boutique. You agree to decorate six jackets for her to sell. You will need to plan each of the jacket themes, buy the jackets, determine what materials to use to carry out the theme, and estimate the cost of making each of these one-of-a-kind jackets. The shop owner will loan you enough money for supplies to get started.

# Rational Numbers

## CONTENTS

### 4 Number Theory

Factors and Multiples
Tests for Divisibility
Problem Solving Exploration:
   Factors and Patterns
Prime Numbers
Prime Factorization
Problem Solving Strategies:
   Solving a Simpler Problem
Exponents
LCM and GCF

### 5 Fractions

Equivalent Fractions
Problem Solving Exploration:
   Modeling Multiplication
Multiplying Fractions
Multiplying Mixed Numbers
Problem Solving Strategies:
   Organizing Information
Using Reciprocals to Solve
   Equations
Dividing Fractions and
   Mixed Numbers
Fractions with Like Denominators
Adding and Subtracting Fractions:
   Unlike Denominators
Subtracting Mixed Numbers

## PROJECT 3

Your family has a five-year-old car that needs new tires and new brakes. Work as a team to determine if it is worthwhile to repair this car in order to drive it three more years or whether it would be better to buy a new car.

# LESSON 4-1

## Factors and Multiples

If there are 24 students in your class, in how many different ways can the class line up at a fire drill so that each row has the same number of students?

**Plan:** Make a drawing to show all the different ways.

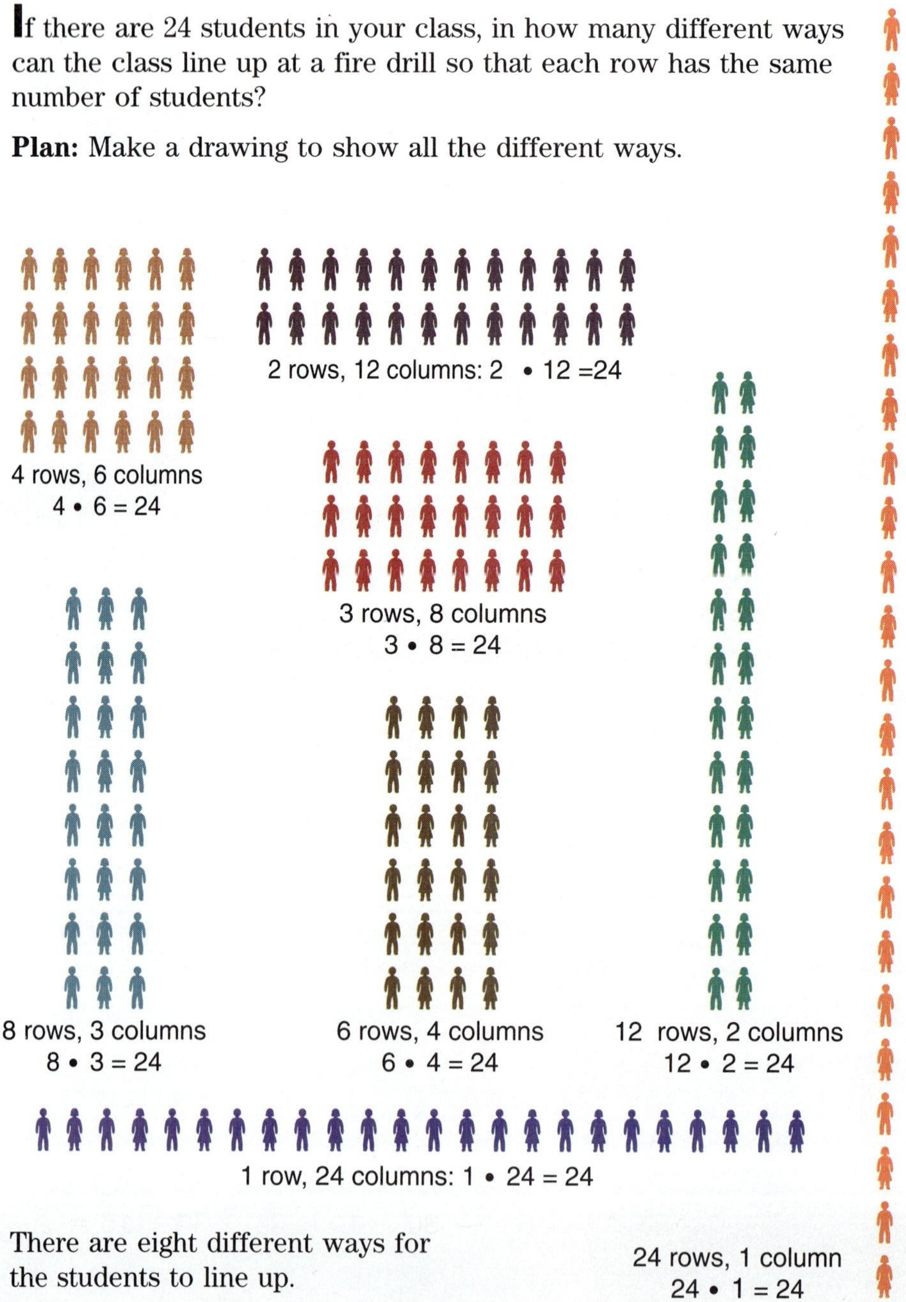

4 rows, 6 columns
4 • 6 = 24

2 rows, 12 columns: 2 • 12 = 24

3 rows, 8 columns
3 • 8 = 24

8 rows, 3 columns
8 • 3 = 24

6 rows, 4 columns
6 • 4 = 24

12 rows, 2 columns
12 • 2 = 24

1 row, 24 columns: 1 • 24 = 24

There are eight different ways for the students to line up.

24 rows, 1 column
24 • 1 = 24

When two or more natural numbers (such as 1, 2, 3, and so on) are multiplied to form a product, each number is a **factor** of the product. The product is a **multiple** of each factor.

$$\left. \begin{array}{rcl} \text{Factor} \cdot \text{Factor} &=& \text{Product} \\ 3 \cdot 8 &=& 24 \\ 4 \cdot 6 &=& 24 \\ 12 \cdot 2 &=& 24 \end{array} \right\} \text{24 is a multiple of 2, 3, 4, 6, 8, and 12.}$$

The numbers 1, 2, 3, 4, 6, 8, 12, and 24 are factors of 24. That is, the set of all natural number factors of 24 is

$$F = \{1, 2, 3, 4, 6, 8, 12, 24\}.$$

The number, 24, is a multiple of each of its factors.

A multiple of a number is divisible by each of the factors of that number. A number **is divisible by** a natural number if the quotient is a natural number and the remainder is 0. For example, 24 is divisible by 3 because $24 \div 3$ is the natural number, 8, and the remainder is 0.

**EXAMPLE 1** Write **True** or **False**. Give a reason for your answer.
a. 18 is a multiple of 6.
b. 4 is a factor of 16.
c. 23 is divisible by 5.
d. 8 and 7 are factors of 56.

**Solutions**
a. **True,** because $18 = 6 \cdot 3$.
b. **True,** because $16 = 4 \cdot 4$.
c. **False,** because $23 \div 5 = 4$ R3.
d. **True,** because $56 = 8 \cdot 7$.

**Try This** Write True or False. Give a reason for your answer.
1. 43 is divisible by 9.
2. 3 is a factor of 36.

It is important to remember that *factor of, multiple of,* and *is divisible by* are related terms.

24 is divisible by 6.          24 is divisible by 24.
24 is a multiple of 6.         24 is a multiple of 24.
6 is a factor of 24.           24 is a factor of 24.

**EXAMPLE 2** List all the factors of 30.

**Solution** Mentally divide 30 by each natural number from 1 to 30. If there is no remainder, the quotient is a factor of 30.

$30 \div 1 = 30 \rightarrow 30 \div 2 = 15 \rightarrow 30 \div 3 = 10 \rightarrow 30 \div 5 = 6$
$30 \div 6 = 5 \rightarrow 30 \div 10 = 3 \rightarrow 30 \div 15 = 2 \rightarrow 30 \div 30 = 1$

The factors of 30 are **1, 2, 3, 5, 6, 10, 15,** and **30.**

4-1 Factors and Multiples

**Try This**  3. List all the factors of 36.

**EXAMPLE 3**  List the first five multiples of 8.

**Solution**  Starting with 1, multiply each natural number up to 5 by 8. Each product is a multiple of 8.

$1 \cdot 8 = 8 \quad 2 \cdot 8 = 16 \quad 3 \cdot 8 = 24 \quad 4 \cdot 8 = 32 \quad 5 \cdot 8 = 40$

The first five multiples of 8 are **8, 16, 24, 32,** and **40.**

**Try This**  4. Write the first five multiples of 7.

How can you show all the multiples of 8? Since the multiples of 8 go on forever, they form an **infinite set.** The set of multiples of 8 is {8, 16, 24, 32, 40, ⋯}. The three dots mean that the multiples continue without end.

# EXERCISES

**Objective:** To list the factors of a number
To list the multiples of a number

## Check Your Understanding

**Skill Check**  Write the letter of the correct answer.

1. Which lists all the factors of 48?
   a. 1, 2, 4, 6, 8, 12, 24, 48
   b. 48, 96, 144, 192
   c. 1, 2, 3, 4, 6, 8, 12, 16, 24, 48
   d. 1, 2, 3, 4

2. Which list shows multiples of 12 only?
   a. 1, 2, 6, 8, 12, 24
   b. 2, 3, 4, 6
   c. 1, 2, 3, 4, 6, 12
   d. 12, 24, 36, 48, 60, 72

3. Which statement is **not** true?
   a. 16 is divisible by 4.
   b. 21 is divisible by 4.
   c. 5 is divisible by 5.
   d. 20 is divisible by 2 and by 10.

4. If $x$ is a factor of $y$, which statement is true?
   a. $y$ is divisible by $x$.
   b. $y$ is a factor of $x$.
   c. $x$ is a multiple of $y$.
   d. $x$ is divisible by $y$.

Replace each ? with one number that makes the sentence true. There may be more than one right answer.

5. 42 is divisible by ?.
6. ? is a factor of 12.
7. 63 is a multiple of ?.
8. ? is a multiple of 11.

## Practice and Apply

**Write True or False for each statement. Give a reason for your answer.**

9. 54 is a multiple of 9.
10. 30 is divisible by 6.
11. 23 is divisible by 3.
12. 8 is a factor of 48.
13. 16 is a factor of 4.
14. 13 is a multiple of 13.
15. 5 is a multiple of 15.
16. 9 is a factor of 181.

**List the numbers from the given set that are factors of the given number.**

17. 15; {1, 2, 3, 15, 30}
18. 9; {3, 6, 9, 18, 27}
19. 12; {2, 4, 6, 8, 10, 12}
20. 16; {1, 2, 4, 6, 8, 10, 12}

**List all the factors of each number.**

21. 28
22. 17
23. 23
24. 54
25. 51

**List the first five multiples of each number.**

26. 2
27. 3
28. 11
29. 10
30. 25

**Write a true statement using each given word with the pair of numbers. Example: 3 is a factor of 18.**

31. factor; 3 and 18
32. multiple; 6 and 24
33. divisible; 8 and 16
34. factor; 12 and 2
35. multiple; 2 and 10
36. divisible; 1 and 18

37. How many different factors does 49 have?
38. How many different factors does 72 have?

**This pictograph shows the number of people who attended orchestra concerts for four months.**

39. How many people attended the concerts in September?
40. How many more people attended the concerts in December than in June?
41. Between which consecutive months did attendance at the concerts increase?

| ORCHESTRA CONCERT ATTENDANCE ||
|---|---|
| March | ♪♪♪♪ |
| June | ♪♪♪♪♪ |
| September | ♪♪♪ |
| December | ♪♪♪♪♪♪ |

Key: Each ♪ represents 125 people.

4-1 Factors and Multiples

## Connect and Extend

**42.** Write six different numbers that have exactly two factors.

**43.** Explain why every even natural number greater than 2 has at least 3 factors.

**44.** The number 102 is divisible by 17 and by 6. Without doing any calculation, name two other numbers that divide into 102 evenly.

**45.** Does a larger number always have more factors than a smaller number? Explain your answer.

**46.** If a number is divisible by a second number, is the first number also divisible by the factors of the second number? Give an example to illustrate your answer.

**47.** A number has 2, 3, and 5 as factors. Which of the following numbers are also factors?

1, 4, 6, 8, 9, 10, 12, 14, 15, 18, 20, 25, 30

## Maintain Your Skills

Match each equation in Column A with the value or values of the variables in Column B that make the equation true. (Pages 39–43, 46–50)

| Column A | Column B |
|---|---|
| **48.** $6e + (e \div 3) = -57$ | **a.** $e = 6$ and $f = -1$ |
| **49.** $3e + f = 17$ | **b.** $e = -4$ |
| **50.** $30 - (e - f) = 23$ | **c.** $e = -9$ |
| **51.** $-e + (12 \div e) - 2 = -1$ | **d.** $e = 2$ and $f = -5$ |

Solve. (Pages 46–50, 59–63)

**52.** Harold gathered 3 baskets of peaches, Roseanne gathered 4 baskets of peaches, and Kacy gathered 5 baskets of peaches. Their grandmother paid them $3 a basket and they split the profit equally. How much did each person get?

Solve and check. (Pages 105–109)

**53.** $6x + 15 = 57$     **54.** $4x - 8 = 20$

**55.** $-11(x - 7) = 22$    **56.** $5 - 2x = 15$

**57.** Bonnie and four friends go to lunch. The total cost of lunch is $30. Write an equation to describe the situation if each person pays the same amount, $a$. (Pages 110–113)

## LESSON 4-2

# Tests for Divisibility

Sheila has saved $123 to buy gifts for five friends. Can the $123 be divided into five equal whole-dollar amounts?

One way for Sheila to solve this problem is to divide 123 by 5. However, she can also use mental math and the tests for divisibility to determine whether 123 is divisible by 5. Divisibility tests are shortcuts for determining whether one number is divisible by another.

Make a list of the multiples of 5 and 10. Look for patterns.

**Multiples of 5:** 5, 10, 15, 20, 25, 30, 35, ...
**Multiples of 10:** 10, 20, 30, 40, 50, 60, 70, ...

The list for 5 shows that if a number ends in 0, or in 5, it is divisible by 5. The list for 10 shows that if a number ends in 0, it is divisible by 10.

**EXAMPLE 1** Which numbers are divisible by 5? by 10?
    **a.** 7235      **b.** 60      **c.** 874

**Solutions**    **a.** Since 7235 ends in 5, it is divisible by 5.
    **b.** Since 60 ends in 0, it is divisible by 5 and by 10.
    **c.** Since 874 does not end in 0 or 5, it is divisible neither by 5 nor by 10.

**Try This**    Which numbers are divisible by 5? by 10?
    **1.** 980      **2.** 645      **3.** 6791      **4.** 23,190

- Can the $123 that Sheila saved be divided into five equal whole-dollar amounts? Why or why not?

### Tests for Divisibility by 3 and 9

Each of these numbers is divisible by 3.

42   87   120   225   609   1683

Add the digits of each number. What do you notice? If the sum of the digits of a number is divisible by 3, the number is divisible by 3.

The test for divisibility by 9 is similar to the test for 3. If the sum of the digits of a number is divisible by 9, the number is divisible by 9.

**EXAMPLE 2** Which numbers are divisible by 3? by 9?
  a. 37,218          b. 4,347

**Solutions**
a. Add the digits: 3 + 7 + 2 + 1 + 8 = 21
   21 is divisible by 3. So, 37,218 is divisible **by 3.**

b. Add the digits: 4 + 3 + 4 + 7 = 18
   18 is divisible by 3 and by 9. So, 4,347 is divisible **by 3** and **by 9.**

**Try This** Test each number for divisibility by 3 and by 9.
  **5.** 7,215     **6.** 8,271     **7.** 3,474     **8.** 58,243

### Tests for Divisibility by 2 and 6

**Even numbers** are multiples of 2. That is, the set of even numbers can be written as shown below.

| 2 · 0 | 2 · 1 | 2 · 2 | 2 · 3 | 2 · 4 | 2 · 5 | 2 · 6 ... |
|---|---|---|---|---|---|---|
| 0 | 2 | 4 | 6 | 8 | 10 | 12 ... |

Since even numbers are multiples of 2, they are all divisible by 2. Notice that the last digit of an even number is even. This suggests a quick test for divisibility by 2.

A number is divisible by 2 if its last digit is divisible by 2.

Numbers that are not divisible by 2 are **odd numbers.**

To test for divisibility by 6, first test for divisibility by 2. Then test for divisibility by 3.

A number that is divisible by 2 *and* by 3 is divisible by 6.

**EXAMPLE 3** Which numbers are divisible by 2? by 6?
  **a.** 236    **b.** 1,140    **c.** 3,041

**Solutions**

| Last digit even? | Sum of digits divisible by 3? | Divisible by 2? | Divisible by 6? |
|---|---|---|---|
| **a.** Yes | No | Yes | No |
| **b.** Yes | Yes | Yes | Yes |
| **c.** No | No | No | No |

**Try This** Which numbers are divisible by 2? by 6?
  **9.** 108    **10.** 393    **11.** 766    **12.** 7,860

A number is divisible by 4 if the number formed by its last two digits is divisible by 4. To see why this happens, notice that any number with three or more digits can be written in this way.

$$228 = 2 \cdot 100 + 28 \quad \text{Divisible by 4}$$
$$972 = 9 \cdot 100 + 72 \quad \text{Divisible by 4}$$
$$5{,}495 = 54 \cdot 100 + 95 \quad \text{Not divisible by 4}$$
$$37{,}568 = 375 \cdot 100 + 68 \quad \text{Divisible by 4}$$

Since the product of any number and 100 is divisible by 4, you need only to check the last two digits for divisibility by 4.

Can you also test the last two digits for divisibility by 8? No, because some multiples of 100 are not divisible by 8. However, 1000 is divisible by 8. This means that you can apply the same kind of reasoning, using 1000 in place of 100, to discover this test.

If the last three digits of a number form a number that is divisible by 8, then the original number is divisible by 8.

**EXAMPLE 4** Which numbers are divisible by 4? by 8?
  **a.** 5064    **b.** 7376    **c.** 7316

**Solutions**

| Number formed by last 2 digits | Divisible by 4? | Number formed by last 3 digits | Divisible by 8? |
|---|---|---|---|
| **a.** 64 | Yes | 064 | Yes |
| **b.** 76 | Yes | 376 | Yes |
| **c.** 16 | Yes | 316 | No |

**Try This**   **Which numbers are divisible by 4? by 8?**
**13.** 196   **14.** 2976   **15.** 5700   **16.** 12,096

> **Summary of Divisibility Tests**
> A number is divisible by
>   2, if the last digit is even.
>   3, if the sum of its digits is divisible by 3.
>   4, if the number formed by the last two digits is divisible by 4.
>   5, if the last digit is 0 or 5.
>   6, if it is divisible by both 2 and 3.
>   8, if the number formed by the last three digits is divisible by 8.
>   9, if the sum of its digits is divisible by 9.
>   10, if the last digit is 0.

# EXERCISES

**Objective:** To apply the tests for divisibility by 2, 3, 4, 5, 6, 8, 9, and 10

## Check Your Understanding

**Skill Check** Write the letter of the correct answer.

**1.** For which number do you have to look at the last three digits to test for divisibility by that number?
  **a.** 8    **b.** 6
  **c.** 4    **d.** 2

**2.** If a number is divisible by 2 and by 3, then the number is divisible by what other number?
  **a.** 8    **b.** 6
  **c.** 5    **d.** 4

**3.** A number that ends in 0 is always divisible by which numbers?
  **a.** 2, 4    **b.** 4, 8
  **c.** 2, 5, 10  **d.** 4, 5

**4.** Which number is divisible by 3 and by 9?
  **a.** 275    **b.** 1,014
  **c.** 4,302   **d.** 102

Replace each ? with a digit so that the number is divisible by 4.
  **5.** 74_?_    **6.** 1,31_?_    **7.** 2,00_?_    **8.** 11,4_?_2

## Practice and Apply

**Which numbers are divisible by 2? by 5? by 10?**
  **9.** 235   **10.** 523   **11.** 730   **12.** 2,895   **13.** 5,912

**Which numbers are divisible by 4? by 8?**

**14.** 1,728    **15.** 2,500    **16.** 7,318    **17.** 27,912    **18.** 60,992

**Which numbers are divisible by 3? by 6? by 9?**

**19.** 108    **20.** 3,507    **21.** 51,726    **22.** 3,346    **23.** 4,266

**24.** Explain how you can determine if a number is divisible by 15.

**Which of these numbers are divisible by 15?**

**25.** 205    **26.** 195    **27.** 375    **28.** 140    **29.** 270

**30.** Can 21 whole apples be divided evenly among 7 people? Explain your answer.

**31.** Can $172 be divided evenly among 4 employees? Explain your answer.

**32.** There are 17 music books. If 2 people share each book, are there enough books for 10 people? Explain your answer.

**33.** A club has 26 members. Can everybody sit at 6 square tables that seat 4 people each? Explain your answer.

**34.** Give two ways that 35 people can be divided into equal teams.

**35.** Draw two different rectangles, each with an area of 12 square units.

**36.** Draw two different rectangles, each with an area of 20 square units.

List the members of set $S = \{2, 3, 4, 5, 6, 8, 9\}$ that are factors of the given number.

**37.** 540    **38.** 588    **39.** 1176    **40.** 2401    **41.** 44,328

## Connect and Extend

**Write a number that is not divisible by any of the given numbers.**

**42.** 7, 8    **43.** 2, 5, 6    **44.** 2, 3, 4    **45.** 5, 6, 7

**Show that the statement is true by writing a logical argument.**

**46.** Any number divisible by 12 is also divisible by 2, by 3, by 4, and by 6.

**47.** Zero is divisible by any nonzero number.

**48.** Any even number divisible by 9 is also divisible by 6.

**49.** What is the smallest number that is divisible by 2, 3, 4, and 8?

**50.** What is the smallest number that is divisible by 3, 5, 7, and 9?

## Maintain Your Skills

**Write the perimeter of a rectangle with the given width and length only if the perimeter is greater than 25. (Pages 64–68)**

**51.** $w = 4, l = 8$  **52.** $w = 3, l = 10$  **53.** $l = 7, w = 7$

**Write the area of a rectangle with the given width and length only if the area is less than 200. (Pages 64–68)**

**54.** $w = 11, l = 20$  **55.** $w = 9, l = 9$  **56.** $l = 10, w = 3$

**Write an equation to represent each situation. Then solve the equation and check the results. (Pages 95–99)**

**57.** Lonnie scored 3 times the number of goals that Jamie did. If Lonnie scored 12 goals, how many did Jamie score?

**58.** Milan worked 52 hours last week and earned $416. What was his rate of pay per hour?

**Solve. (Pages 126–130)**

**59.** Which number has the most factors, 12, 16, or 17?

**60.** List the multiples of 13 that are greater than 25 and less than 75.

##  Math Team Problems

**61.** Steve forgot his locker number at school. The school secretary reminded him that each digit in his locker number is an odd number, and that the sum of the digits is 11. There are 120 lockers. What was Steve's locker number?

**62.** Katia wanted to find the difference of the sum of all even numbers from 2 through 1,000 and the sum of all the odd numbers from 1 through 999. She arranged her work as follows.

$$\begin{array}{r} 2 + 4 + 6 + \cdots + 996 + 998 + 1{,}000 \\ -(1 + 3 + 5 + \cdots + 995 + 997 + \phantom{0}999) \\ \hline 1 + 1 + 1 + \cdots + \phantom{00}1 + \phantom{00}1 + \phantom{000}1 \end{array}$$

What is the difference?

**63.** In the addition problem at the right, the letters P, K, and N represent three different digits. Identify the digits.

$$\begin{array}{r} P\,K \\ +\,K\,N \\ \hline K\,N\,K \end{array}$$

# LESSON 4-3
## Problem Solving Exploration
## Factors and Patterns

This activity builds models with unit squares. A **unit square** is a square that has an area of one square unit.

Start with 6 unit squares like this.

Then build all possible rectangles made up of 6 squares.

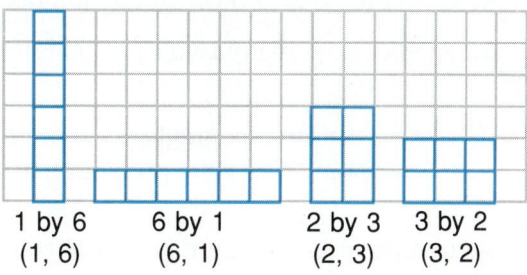

1 by 6    6 by 1    2 by 3    3 by 2
(1, 6)    (6, 1)    (2, 3)    (3, 2)

You can build exactly four different rectangles made up of 6 unit squares. Notice that a rectangle that is 1 by 6 is not the same as a rectangle that is 6 by 1. You can represent the length and width of each rectangle by ordered pairs of numbers.

These pairs of numbers are called **ordered pairs** because the order of the numbers makes a difference. In this example, the horizontal (right and left) dimension is always given first and then the vertical (up and down) dimension.

1. Use the given numbers of unit squares to build all possible rectangles. Copy and complete this table through 25 unit squares.

| Number of Unit Squares | Dimensions of Possible Rectangles | Number of Rectangles Formed |
|---|---|---|
| 1 | (1,1) | 1 |
| 2 | (1,2), (2,1) | 2 |
| 3 | (1,3), (3,1) | 2 |
| 4 | (1,4), (2,2), (4,1) | 3 |
| 5 | (1,5), (5,1) | 2 |
| 6 | (1,6), (6,1), (2,3), (3,2) | 4 |
| 7 | ? | ? |

Study the entries in the "Number of Rectangles" column. Notice that exactly one rectangle can be built from 1 unit square. Call the number 1 a **one-rectangle number.**

2. Are there any other one-rectangle numbers in this activity?

Since exactly two rectangles can be built from 2 unit squares, call the number 2 a **two-rectangle number.** The numbers 3 and 5 are also examples of two-rectangle numbers.

3. List all the two-rectangle numbers from 2 through 25.

4. Find the next two-rectangle number after 25.

5. Predict how many two-rectangle numbers there are between 1 and 100.

6. List all the two-rectangle numbers from 1 through 100 and count them. Was your prediction correct?

The number 4 is an example of a **three-rectangle number.**

7. List all the three-rectangle numbers from 1 through 25.

8. Find the next three-rectangle number after 25.

9. Predict how many three-rectangle numbers there are between 1 and 100.

10. List all the three-rectangle numbers from 1 through 100 and count them. Was your prediction correct?

11. Describe the set of three-rectangle numbers in your own words.

# LESSON 4-4  Prime Numbers

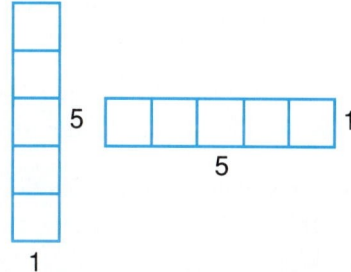

In the previous Exploration, you found that exactly two different rectangles can be built by using 5 unit squares. One rectangle is 5 by 1 and the other rectangle is 1 by 5. However, both rectangles have the same area, 5 square units.

How many ways can you multiply two natural numbers to get 5? There are two ways, 1 · 5 or 5 · 1. Another way to say this is that the number 5 has exactly 2 different factors, 5 and 1. Numbers that have exactly two different factors are called *prime numbers*.

A **prime number** is a natural number with exactly two different factors, itself and one.

**EXAMPLE**  Write **True** or **False**. Give a reason for your answer.
- **a.** 8 is a prime number.
- **b.** 1 is a prime number.
- **c.** 17 is a prime number.
- **d.** 0 is a prime number.

**Solutions**
- **a.** False
  The number 8 has four factors, 1, 2, 4, and 8.
- **b.** False
  The number 1 has only one factor, 1.
- **c.** True
  The number 17 has exactly two factors, 1 and 17.
- **d.** False
  The number 0 has many factors.

**Try This**  Write True or False. Give a reason for your answer.
1. 91 is a prime number.
2. 105 is a prime number.
3. 35 is a prime number.
4. 51 is a prime number.

Natural numbers that have more than two factors are called **composite numbers**. For example, 10 is a composite number because it has 4 factors 1, 2, 5 and 10.

Every natural number except 1 is either prime or composite.

• Why is the number 1 not a composite number?

4-4  Prime Numbers  **139**

# EXERCISES

**Objective:** To determine whether a number is prime or composite

## Check Your Understanding

*Skill Check* Write the letter of the correct answer.

**1.** How many different factors does 9 have?
   **a.** 1   **b.** 2
   **c.** 3   **d.** 4

**2.** Which number is a composite number?
   **a.** 17   **b.** 15
   **c.** 13   **d.** 11

**3.** Which number is a prime number?
   **a.** 2   **b.** 4
   **c.** 6   **d.** 8

**4.** How many different factors does 23 have?
   **a.** 1   **b.** 2
   **c.** 3   **d.** 4

Write True or False for each statement. Give a reason for your answer.

**5.** 77 is a prime number.

**6.** 23 is a prime number.

**7.** 3 is a composite number.

**8.** 39 is a composite number.

**9.** Can a number be both prime and composite? Explain your answer.

## Practice and Apply

Tell whether each of the numbers is prime, composite, or neither. Give a reason for your answer.

**10.** 61   **11.** 57   **12.** 185   **13.** 1   **14.** 372   **15.** 819

List the prime numbers between each pair of numbers.

**16.** 1 and 9   **17.** 9 and 15   **18.** 16 and 28   **19.** 28 and 40

**20.** List all the composite numbers less than 50 that have exactly three different factors.

**21.** List all the prime numbers that are even.

**22.** List all the composite numbers less than 25 that have exactly four different factors.

**23.** How many rectangles with an area of 7 square units can you draw?

**24.** How many different ways can 16 compact discs be arranged in rows so there are the same number in each row?

## Connect and Extend

25. Find two prime numbers whose sum is a prime number.
26. Find two composite numbers whose difference is a prime number.
27. **Twin primes** are consecutive primes, such as 3 and 5, that have a difference of 2. List three more pairs of twin primes.
28. One definition of the word composite is "formed of distinct parts, compound." Explain how this definition applies to composite numbers.
29. What do you notice about the ones digit of the 2-digit primes between 10 and 75?
30. **Reversal primes** are prime numbers, such as 37 and 73, in which the digits are reversed. Find two other pairs of reversal primes.

Suppose that one of your classmates claims that every even integer greater than 2 can be written as the sum of two prime numbers. Check out this claim by finding two prime numbers that have these sums.

**31.** 12  **32.** 32  **33.** 60  **34.** 84

35. Do you think that your classmate's claim is correct? Why or why not?
36. Do research to explain how the Sieve of Eratosthenes can be used to find the prime numbers from 1 through 100.

## Maintain Your Skills

Find the solution set if $\{-3, -2, -1, 0, 1, 2, 3\}$ is the replacement set. (Pages 80–84)

**37.** $-3v > 3$  **38.** $-5t < 10$  **39.** $3s + 7 = 1$  **40.** $-24 = -3 - 7w$

Ted is $3n$ years old. Ned is 5 years younger than Ted. Fred is 1 year older than Ted. (Pages 110–113)

**41.** Who is the oldest?  **42.** Who is the youngest?

43. Write an algebraic expression for Ned's age.
44. Write an algebraic expression for Fred's age.

List the members of $S = \{2, 3, 4, 5, 6, 8, 9, 10\}$ that are factors of each number. (Pages 126–136)

**45.** 44  **46.** 81  **47.** 455  **48.** 78  **49.** 10

**List the factors of each number. (Pages 126–130)**
1. 15
2. 19
3. 27
4. 42
5. 50

**List the first five multiples of each number. (Pages 126–130)**
6. 30
7. 12
8. 28
9. 41
10. 100

**True or False? Give a reason for each answer. (Pages 131–136)**
11. 210 is divisible by 3.
12. 6 and 3 are divisible by 18.

**List the members of set $S = \{2, 3, 4, 5, 6, 8, 9, 10\}$ that are factors of each number. (Pages 131–136)**
13. 72
14. 18
15. 84
16. 270
17. 745

18. Sue has 40 baseball cards. Can she give the same number of cards to each of her five friends? Explain. (Pages 131–136)

19. Louis scored the same number of runs in each of four games. Can his total runs for the four games be 15? Explain. (Pages 131–136)

**Tell whether each number is prime, composite, or neither. Give a reason for your answer. (Pages 139–141)**
20. 79
21. 1
22. 244
23. 83
24. 381

# Mathematical Footnote

Two mathematicians, working with hundreds of others, recently used a thousand computers and 275 years of computer time to break an "unbreakable" code. To do this, they factored a number having 155 digits into its prime factors. This is about 50 digits longer than the previous record.

Banks, corporations, and governments use codes based on these huge numbers. Code makers multiply several large numbers to create an even larger one whose factors are kept secret. Decoding requires knowing the factors. Now numbers with more than 200 digits will be required for security.

# Prime Factorization

**To factor** a natural number is to express the number as a product of natural numbers. For example, you can factor 36 by expressing it as a product of 4 and 9.

$$36 = 4 \cdot 9$$

However, each of the factors, 4 and 9, can also be factored.

$$36 = 4 \cdot 9$$
$$36 = 2 \cdot 2 \cdot 3 \cdot 3$$

When you express 36 as the product of $2 \cdot 2 \cdot 3 \cdot 3$, you cannot factor any further (except by using 1). Now that each factor is a prime number, 36 is factored completely. To **factor a number completely** means to express the number as the product of prime numbers only.

$2 \cdot 2 \cdot 3 \cdot 3$ is called the **prime factorization** of 36.

The prime factors of 36 can be arranged in several different orders, but there is only one set of prime factors for any number. A diagram, called a **factor tree,** is sometimes used to show factors.

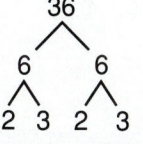 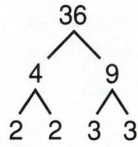 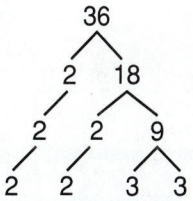

The factor trees show 3 different ways to find the prime factors of 36. However, the end result is always the same. The prime factorization of a number is usually written with the factors in ascending order.

When you cannot see immediately how to begin factoring a number, try this method.

1. If the number is even, start with 2 as a factor.

2. If the number is odd, test prime numbers beginning with 3, 5, 7, and so on. Use divisibility tests to eliminate some possibilities.

**EXAMPLE 1** Write the prime factorization of 105.

**Solution**
Step 1 Start with the smallest prime number, 2.
Think: Is 105 divisible by 2? No, 105 is an odd number.
Step 2 Think: Is 105 divisible by 3? Yes. Divide 105 by 3.
$105 = 3 \cdot 35$
Step 3 Think: 3 is a prime number. Find the prime factors of 35.
$35 = 5 \cdot 7$
Step 4 Since 3, 5, and 7 are prime numbers, write the prime factorization.

Prime factorization: **3 · 5 · 7**

**Try This**
1. Write the prime factorization of 180.

**EXAMPLE 2** Write the prime factorization of 100. Write the prime factors in ascending order.

**Solution**
$100 = 2 \cdot 50$
$= 2 \cdot 2 \cdot 25$
$= 2 \cdot 2 \cdot 5 \cdot 5$       Prime factorization: **2 · 2 · 5 · 5**

**Try This** Write the prime factorization of each number. Write the factors in ascending order.
2. 16    3. 87    4. 126    5. 363

# EXERCISES

**Objective:** To write the prime factorization of a number

## Check Your Understanding

**Skill Check** Write the letter of the correct answer.

1. How many prime factors does 64 have?
   a. 1    b. 2
   c. 4    d. 6

2. What is the remaining prime factor of 90 if three of the factors are 2, 3, and 3?
   a. 2    b. 3
   c. 5    d. 7

3. What is the prime factorization of 24?
   a. 4 · 6    b. 2 · 2 · 2 · 3
   c. 3 · 8    d. 2 · 3 · 4

4. If the prime factors of a number are 2, 2, 3, and 5, what is the number?
   a. 60    b. 19    c. 16    d. 12

Copy and complete each factor tree. Then write the prime factorization of each number.

5.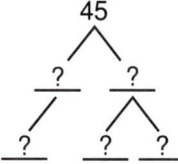

6.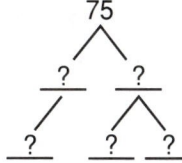

7.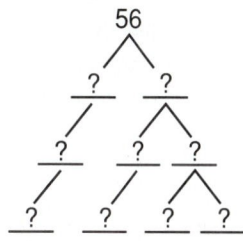

## Practice and Apply

Draw a factor tree to show the prime factors of each number. Circle the prime factors.

**8.** 28  **9.** 42  **10.** 30  **11.** 52  **12.** 84  **13.** 150

Write the prime factorization of each number.

**14.** 48  **15.** 99  **16.** 40  **17.** 86  **18.** 125  **19.** 340

Write True or False for each statement. When a statement is false, explain why.

**20.** The prime factorization of 18 is $3 \cdot 6$.

**21.** The prime factorization of 32 is $2 \cdot 2 \cdot 2 \cdot 4$.

**22.** The prime factorization of 112 is $2 \cdot 2 \cdot 2 \cdot 7$.

**23.** The prime factorization of 182 is $2 \cdot 91$.

Replace each ? with the number that makes a true sentence.

**24.** The prime factors of 63 are 3, 3, and ?.

**25.** The prime factors of 330 are 2, 3, ?, and ?.

**26.** The prime factors of 236 are 2, ?, and ?.

**27.** The prime factorization of 2093 is $? \cdot 13 \cdot ?$.

Find the number whose prime factorization is given.

**28.** $3 \cdot 3 \cdot 5 \cdot 5$   **29.** $5 \cdot 7$   **30.** $2 \cdot 2 \cdot 3 \cdot 3 \cdot 3 \cdot 7$

Lana made an interesting discovery when she was studying prime numbers and patterns. Work through Exercises 31–36 to find out what Lana observed.

Use the expression $a \cdot 3 \cdot 37 \cdot 91$, where $a$ = the age of a person in years.

**31.** Evaluate the given expression with Lana's age of 15.

**32.** Evaluate the given expression for $a = 27$.

**33.** Evaluate the given expression for $a = 43$.
**34.** Evaluate the given expression for $a = 75$.
**35.** What do you notice about the patterns in Exercises 31–34?
**36.** If Lana used her grandmother's age and got an answer of 818,181, how old is her grandmother?

## Connect and Extend

**37.** Write all possible factors of 136.
**38.** Write all possible factors of 216.
**39.** Find the smallest number whose prime factorization is seven different primes.

**Solve each equation to find $p$.**
**40.** $2 \cdot 2 \cdot 7 \cdot p = 644$
**41.** $1705 = 5 \cdot 11 \cdot p$
**42.** $p \cdot 2 \cdot 19 = 874$
**43.** $2697 = p \cdot 29 \cdot 3$

## Maintain Your Skills

**Solve and check. (Pages 101–109)**
**44.** $8 + 3c = -19$
**45.** $-70 = 2 - 8s$
**46.** $(2 + x) - (3x + 5) = -3$

**Tell whether each number is divisible by 2, 3, 4, 5, 6, 8, 9, and 10. (Pages 131–136)**
**47.** 819
**48.** 670
**49.** 26,104
**50.** 928
**51.** 1,584
**52.** 935

**Solve. (Pages 131–136)**
**53.** Ed earns $3 per hour mowing lawns. He made $407 during the summer. Did he work an even number of hours? How do you know?

**54.** Juan Garcia wishes to give $70,622 in equal amounts to four different charities.
  **a.** Can $70,622 be divided in four equal whole-dollar amounts?
  **b.** If not, what is the smallest number of dollars that must be added so that this can be done?

# COMPUTER EXPLORATION

## Prime Factorization

**Objective:** To write the prime factorization of a number by using a computer

1. With the Holt *Pre-Algebra* disk in drive 1 (or drive A), run the program, *Prime Factorization*, as follows.

   **IBM:** Type **a : c** and press **ENTER.**

   **Apple:** Turn on the computer. When the menu appears on the screen, type **C.**

2. Read the opening messages and press **RETURN.**
   a. The question, **WHAT NUMBER DO YOU WANT TO FACTOR?,** will appear on the screen. Type 31. Before pressing **RETURN,** write the prime factors of 31 on a piece of paper. If 31 is a prime number, write Prime. Then press **RETURN.**
   b. Compare your answer with the answer displayed on the screen. Was your answer correct?
   c. To the question, **WANT TO FACTOR ANOTHER (Y/N)?,** type **Y.**

3. a. To find the prime factorization of 36, type 36 and press **RETURN.**
   b. Type a prime factor of 36, such as 2, and press **RETURN.** Since $36 = 2 \cdot 18$, the computer will then ask for a prime factor of 18.
   c. Enter a prime factor of 18, such as 3. Since $18 = 3 \cdot 6$, the computer will ask for a prime factor of 6.
   d. Enter a prime factor of 6 (either 2 or 3) to complete the factorization: $36 = 2 \cdot 3 \cdot 2 \cdot 3$, or $2 \cdot 2 \cdot 3 \cdot 3$.

   NOTE: If you enter a number that is not a prime factor of the number you are factoring, the computer will print a message such as **3 IS NOT A PRIME FACTOR OF 350.** Press any key to continue and correct the error by entering a prime factor.

Use the program to find the prime factorization of each number. If a number is prime, write **PRIME.**

**4.** 87   **5.** 350   **6.** 385   **7.** 240   **8.** 233   **9.** 374

# Problem Solving Strategies
## Solving a Simpler Problem

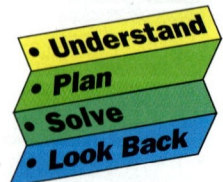

Problems that seem difficult can often be made simpler by writing and solving a simpler, related problem or by using smaller numbers. Then you can apply what you have learned to solving the original problem.

### EXAMPLE 1

Is 2345 divisible by 26?

**Solution**

Think of the problem as $\frac{2345}{2 \cdot 13}$.

**Think:** If 2345 is divisible by 26, then it is divisible by both 2 and 13. But 2345 is an odd number, and odd numbers are not divisible by 2.

Therefore, 2345 is **not divisible** by 26.

Notice that you do not have to actually divide to solve this problem.

### EXAMPLE 2

How many rectangles are there in the figure at the right?

**Solution**

Simplify the problem. Find the number of rectangles of each size in the figure.

| Size | Number of Rectangles |
|---|---|
| | 4 |
| | 3 |
| | 2 |
| | 1 |

So the total number of rectangles in the given figure is **10.**

# EXERCISES

**Objective:** To apply strategies in problem solving

**Solve each problem.**

1. Think of a number. Add 6 to it, double the result, subtract 12, and then divide by 2. The result is the original number. Explain why this works.

2. How many squares are there in this figure?

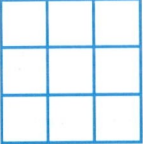

3. Find a set of numbers that are all multiples of 7 and which add to a sum of 150. Explain how you solved this problem.

4. The number 12 has three prime factors, 2, 2, and 3. Find a number less than 101 that has 6 prime factors.

5. Is there only one number that satisfies the conditions of Exercise 4? Explain your answer.

6. The Todd family had dinner at the Eat Well Diner. They spent $20.75 for the meal. How many of the Todd family are under 12 years of age?

7. The average of seven numbers is 43. What is the new average when 1 is added to the first number, 2 is added to the second number, 3 is added to the third number, and so on through the seventh number?

8. Jim is climbing a 20-yard rope. Every 30 seconds he climbs 3 yards. In the next 30 seconds, he slips back 2 yards. How long will it take Jim to climb the rope?

9. Write a problem that can be solved by solving a simpler problem first. Have one of your classmates solve the problem.

## LESSON 4-7

# Exponents

**E**xponents are sometimes used in writing prime factorizations.

$$\underbrace{2 \cdot 2 \cdot 2 \cdot 2}_{\text{Prime factorization}} = 2\underset{\text{Base}}{\overset{\leftarrow \text{Exponent}}{^4}}$$

An **exponent** shows how many times a base is used as a factor.

$2^4$ is a **power.** You read $2^4$ as "two to the fourth power." Some powers are given special names.

$8^2$ is read as "eight squared."     $a^2$ is read as "$a$ squared."

$6^3$ is read as "six cubed."     $c^3$ is read as "$c$ cubed."

When simplifying an expression involving exponents, clear parentheses first. Then simplify the exponents. Finally, follow the rules for the order of operations.

**EXAMPLE 1** Simplify each expression.
 a. $3^5$      b. $3^2 - 6$      c. $(8 + 2)^3$      d. $4 + 5^2$

**Solutions**

a. $3^5 = 3 \cdot 3 \cdot 3 \cdot 3 \cdot 3$
    $= \mathbf{243}$

b. $3^2 - 6 = 3 \cdot 3 - 6$
    $= 9 - 6$
    $= \mathbf{3}$

c. $(8 + 2)^3 = (10)^3$
    $= 10 \cdot 10 \cdot 10$
    $= \mathbf{1{,}000}$

d. $4 + 5^2 = 4 + 5 \cdot 5$
    $= 4 + 25$
    $= \mathbf{29}$

**Try This**  Simplify each expression.
 **1.** $12^3$     **2.** $3 + 4^2$     **3.** $(2 + 3)^2$     **4.** $5^2 - 4^2$

The expression $3a^2$ means $3 \cdot a \cdot a$. Note that only the $a$ is squared. In the expression $(3a)^2$, note that the base, $3a$, is in parentheses.

$$(3a)^2 = (3a)(3a)$$
$$= 3 \cdot a \cdot 3 \cdot a$$
$$= 3 \cdot 3 \cdot a \cdot a$$
$$= 9a^2$$

Therefore, $3a^2 \neq (3a)^2$.

**EXAMPLE 2** Evaluate $-4a^2b$ for $a = -2$ and $b = 3$.

**Solution**  Substitute $-2$ for $a$ and 3 for $b$.   $-4a^2b = -4(-2)^2(3)$
Evaluate $(-2)^2$ first.                              $= -4(4)(3)$
                                                       $= -48$

**Try This**  Evaluate each expression.
5. $(-4c)^2d$, for $c = 3$, $d = 2$   6. $4m^3 + 3m$, for $m = -3$

**EXAMPLE 3** You can use the formula $d = 16t^2$ to find how many feet, $d$, an object falls in $t$ seconds. How far does an object fall in 5 seconds?

**Solution**  
Write the formula.        $d = 16t^2$
Substitute 5 for $t$.     $d = 16(5)^2$
Evaluate $(5)^2$ first.   $d = 16(25)$
                          $d = 400$

It falls **400 feet** in 5 seconds.

**Try This**  
7. The formula, $V = e^3$ gives the volume of a cube where $e$ is the length of an edge (side). Find the volume of a cube that has an edge of 6 units.

Scientific calculators have a $y^x$ or an $x^y$ key that can be used to evaluate expressions with exponents. For example, to evaluate $9^5$, press these keys.

$9 \; \boxed{y^x} \; 5 \; \boxed{=}$    The result is 59,049.

The key sequence may be different on some calculators. Check your calculator manual.

# EXERCISES

**Objectives:** To simplify expressions containing exponents
To evaluate expressions containing exponents

## Check Your Understanding

**Skill Check**  Write the letter of the correct answer.

1. What is the base of the expression $2^3$?
   a. 2    b. 3    c. 6    d. 8

2. In $3^4$ what is the exponent?
   a. 3    b. 4    c. 18    d. 81

**3.** Which expression is equivalent to $4 \cdot 4 \cdot 4$?
  **a.** $3^4$      **b.** $4 + 4 + 4$
  **c.** $4^3$      **d.** $3 \cdot 4$

**4.** Which is the first step in simplifying $(7 - 9)^3 \cdot 5$
  **a.** $7^3 - 9^3 \cdot 5$     **b.** $2^3 \cdot 5$
  **c.** $7 - 9^3 \cdot 5$       **d.** $(-2)^3 \cdot 5$

**Simplify each expression.**
  **5.** $(-7)^4$     **6.** $-9 + 12^2$     **7.** $(14 - 3)^3$     **8.** $5 - 8^2$

**Evaluate each expression.**
  **9.** $(5q)^2 r$, for $q = 3$ and $r = -5$
  **10.** $7t^2 - s$, for $t = 6$ and $s = 9$

## Practice and Apply

**Identify the base and the exponent in each expression.**
  **11.** $2^8$     **12.** $4^3$     **13.** $-(-6)^2$     **14.** $(-3)^6$     **15.** $-(7^9)$

**Simplify.**
  **16.** $(-4)^4$     **17.** $-5^3$     **18.** $(4 - 8)^3$     **19.** $15 - 2^2$
  **20.** $(2 - 3)^3$  **21.** $(4 + 9)^2$  **22.** $7^2 - (3 + 1)^3$  **23.** $(3 + 4)^2 - 4^3$

**Evaluate for $w = 2$, $x = -2$, $y = 3$, and $z = -3$.**
  **24.** $4x^4$        **25.** $6x^5$        **26.** $9x^3 y$      **27.** $-7x^5 y$      **28.** $-6yz^3$
  **29.** $(-5z)^2 w$   **30.** $25 - 4z^2$   **31.** $17 - 6x^5$   **32.** $12 + 9x^3 z$  **33.** $54 + 3xz^3$

**Use the formula $16t^2 = d$ to find the distance, in feet $d$, that an object will fall in the given time in seconds, $t$.**
  **34.** $t = 20$ seconds     **35.** $t = 30$ seconds     **36.** $t = 98$ seconds

**Use the formula $V = e^3$ to find the volume, $v$, in cubic inches for the given length of an edge, $e$.**
  **37.** $e = 5$ inches     **38.** $e = 7$ inches     **39.** $e = 87$ inches

**40.** How do the expressions, $6a^3$ and $(6a)^3$, differ?

## Connect and Extend

**Write True or False for each statement. If a statement is false, explain why.**
  **41.** $-5^2 = (-5)^2$     **42.** $2 - x^4 = 2 + (-x)^4$     **43.** $7 + (-4)x^2 = 7 + (-4x^2)$

**44.** What is the general rule for evaluating $(-1)^n$ when $n$ is an odd number? (Hint: Try replacing $n$ with odd numbers.)

**45.** What is the general rule for evaluating $(-1)^n$ when $n$ is an even number? (Hint: Try replacing $n$ with even numbers.)

**152**   CHAPTER 4

**46.** Write the word expression "the fifth power of a number, $t$," as an algebraic expression.

## Maintain Your Skills

**47.** Eric earned $16,848 last year. What was his average monthly salary? **(Pages 64–68)**

**Find the largest number that is a factor of each number in each pair. (Pages 126–130)**

**48.** 36 and 54   **49.** 24 and 48   **50.** 21 and 59

**51.** Find the first seven multiples of 2 and of 3. What multiples do they have in common? **(Pages 126–130)**

**52.** Find the first seven multiples of 4 and of 6. What multiples do they have in common? **(Pages 126–130)**

**Write the largest prime factor of each number. (Pages 143–146)**

**53.** 147   **54.** 2730   **55.** 5040   **56.** 5130

# Mathematical Footnote

The kite, which dates back to China more than 3,000 years ago, is the world's oldest aircraft. Kite flying is also a sport which attracts people of all ages and from all walks of life.

Stunt kiting is a definite "in" with American kiters. In Olympic-like competitions, kiters put their stunters through clockwise and counterclockwise loops, figure-eights, multiple spirals, deep dives, and free-form maneuvers.

Most kites are built on a frame that is then covered with some type of fabric. Some kites are triangular in shape, others look like boxes with many sides. Flexible kites, such as the parafoil, take their shape from the wind.

4-7 Exponents

## LESSON 4-8

# LCM and GCF

**T**wo cars start off together around a racetrack. The first car completes one lap every 30 seconds. The second car completes one lap every 40 seconds. In how many seconds will both cars be at the starting point again?

One way to solve this problem is to list some multiples of 30 and 40.

**Multiples of 30:** 30, 60, 90, **120**, 150, 180, 210, **240**, 270, ...
**Multiples of 40:** 40, 80, **120**, 160, 200, **240**, 280, 320, 360, ...

Notice that 120 is a multiple of 30 *and* a multiple of 40. That is, 120 is a common multiple of 30 and 40. Another common multiple of 30 and 40 is 240. Since 120 is the smallest, or *least common multiple* of 30 and 40, the two cars will both be at the starting point in 120 seconds, or 2 minutes.

The **least common multiple (LCM)** of two or more natural numbers is the smallest natural number that is divisible by each of the given numbers.

What is the least common multiple of 5 and 7?
Since 5 and 7 are prime numbers, the smallest number divisible by both 5 and 7 is their product, 5 · 7, or 35.

What is the least common multiple of 4 and 9? Since 4 and 9 are not prime numbers, first find their prime factors.

$$4 = 2 \cdot 2 \qquad 9 = 3 \cdot 3$$

Since 4 and 9 have no prime factors in common, their least common multiple is $4 \cdot 9$, or 36.

What is the least common multiple of 6 and 8? First find the prime factors of each number.

$$6 = 2 \cdot 3 \qquad 8 = 2 \cdot 2 \cdot 2, \text{ or } 2^3$$

Notice that 6 has one factor of 2 and 8 has 3 factors of 2.

You can find the least common multiple by multiplying the factors with the greatest exponent in either prime factorization.

| | |
|---|---|
| Write the prime factorization of each number, using exponents. | $6 = 2 \cdot 3 \qquad 8 = 2^3$ |
| For each prime factor, write the base with the greatest exponent. | $3 \cdot 2^3$ |
| Multiply. | $3 \cdot 8 = 24$ |

The least common multiple of 6 and 8 is **24**.

**EXAMPLE 1** Find the LCM of 12, 15, and 18.

**Solution**

Write the prime factorization of each number using exponents.

$$12 = 2^2 \cdot 3$$
$$15 = 3 \cdot 5$$
$$18 = 2 \cdot 3^2$$

For each prime factor, write the base with the greatest exponent. Then multiply.

$$2^2 \cdot 3^2 \cdot 5 = 4 \cdot 9 \cdot 5$$
$$= 36 \cdot 5$$
$$= 180$$

The LCM of 12, 15, and 18 is **180**.

**Try This**   Find the LCM.

1. 4 and 8
2. 8 and 12
3. 6, 9, and 15

**Greatest Common Factor**

The numbers 12 and 18 have several common factors.

Factors of 12:  1, 2, 3, 4, 6, 12
Factors of 18:  1, 2, 3, 6, 9, 18

**Common factors:**  1, 2, 3, 6

4-8   LCM and GCF   **155**

The *greatest common factor* of 12 and 18 is 6. This means that 6 is the greatest number that is a factor of each number.

The **greatest common factor (GCF)** of two or more natural numbers is the greatest number that is a factor of each number.

**EXAMPLE 2** Find the GCF of 8, 12 and 20.

**Solution**  Write the prime factorization of each number using exponents.

$8 = 2^3 \qquad 12 = 2^2 \cdot 3$
$20 = 2^2 \cdot 5$

Write the product of the prime factors common to the factorizations. Choose the smallest exponent for each factor.

GCF: $2^2 = 4$

The GCF of 8, 12, and 20 is **4**.

**Try This**  Find the GCF.

**4.** 12 and 30  **5.** 54 and 108  **6.** 24, 36, and 60

# EXERCISES

**Objective:** To find the LCM of two or more numbers
To find the GCF of two or more numbers

## Check Your Understanding

**Skill Check**  Write the letter of the correct answer.

**1.** If the prime factors of two numbers are $2^5 \cdot 5 \cdot 7$ and $2^3 \cdot 5^2$, what is the least common multiple of the numbers?
  **a.** $2^3 \cdot 5$   **b.** $2^3 \cdot 5 \cdot 7$
  **c.** $2^5 \cdot 52$  **d.** $2^5 \cdot 5^2 \cdot 7$

**2.** If the prime factors of two numbers are $3^2 \cdot 7 \cdot 11$ and $3^3 \cdot 5 \cdot 7$, what is the greatest common factor of the numbers?
  **a.** $3^2 \cdot 5 \cdot 7 \cdot 11$   **b.** $3^2 \cdot 7$
  **c.** $3^3 \cdot 5 \cdot 7 \cdot 11$   **d.** $3^3 \cdot 7$

**3.** If the greatest common factor of two numbers is 12, which of these could be one of the numbers?
  **a.** 76   **b.** 63
  **c.** 12   **d.** 6

**4.** What is the least common multiple of two prime numbers?
  **a.** The sum of the numbers
  **b.** The product of the numbers
  **c.** The difference of the numbers
  **d.** The quotient of the numbers

Write prime factorizations for each pair of numbers. Find the LCM.

**5.** 6 and 10   **6.** 9 and 15   **7.** 20 and 25   **8.** 6, 12, and 18

**9.** Find the GCF of 24 and 40.   **10.** Find the GCF of 48 and 64.

## Practice and Apply

**Write the prime factorization of each number, using exponents.**
**11.** 14   **12.** 60   **13.** 90   **14.** 84   **15.** 100   **16.** 75

**Find the LCM.**
**17.** 4 and 10   **18.** 30 and 70   **19.** 18 and 60   **20.** 8, 10, and 12

**Find the GCF.**
**21.** 42 and 44   **22.** 30 and 105   **23.** 48 and 72   **24.** 28, 42, and 90

Use the following information for Exercises 25–28.
$$a = 2^3 \cdot 3^2 \cdot 5 \qquad b = 2^2 \cdot 3^4 \cdot 5^2$$

**25.** List two common factors of $a$ and $b$.
**26.** What is the GCF of $a$ and $b$?
**27.** List two common multiples of $a$ and $b$.
**28.** What is the LCM of $a$ and $b$?

**Find the GCF and LCM.**
**29.** 48 and 96   **30.** 180 and 270
**31.** 26, 78, and 104   **32.** 90, 225, and 315

**33.** Brian, Mary, and Eugene deliver newspapers. The newspapers are packaged in bundles with the same number of papers in each bundle. Brian delivers 252 papers. Mary delivers 105 papers, and Eugene delivers 168 papers. What is the greatest number of papers there could be in each bundle?

**34.** Donny wears out a tennis racquet every 6 months. A pair of tennis shoes lasts for 8 months. He buys a new shirt every 4 months. How often does he buy a racquet, tennis shoes, and a shirt at the same time?

## Connect and Extend

**Find the LCM and GCF of each pair of expressions.**
**35.** $16a^2b^3c$ and $20ab^2c^3$
**36.** $12xy^2z$ and $18x^3y$

If the GCF of two numbers is 1, then the numbers are said to be *relatively prime*. Determine if the following pairs of numbers are relatively prime. If they aren't, find the GCF.

**37.** 90 and 189   **38.** 26 and 51   **39.** 165 and 182

**40.** Find two numbers whose LCM is 180.   **41.** Find two numbers whose GCF is 15.

## Maintain Your Skills

Replace each  ?  with the number that makes a true sentence. Write the name of the property shown. (Pages 54–58)

**42.** $(\underline{\ ?\ } - 3) \cdot (-2) = -2(-4 - 3)$   **43.** $(-15 + 4) + \underline{\ ?\ } = -15 + (4 + 6)$

Solve each equation. Check your answer. (Pages 101–109)

**44.** $3(x - 1) - 2(x + 4) = -15$   **45.** $7y(8) - 4y = -104$

**46.** $2(m + 9) + (-3)(7 - m) = 2$   **47.** $\frac{n}{4} + 9 = 7$

This pictograph shows a baseball player's hits over one month. (Pages 126–130)

**48.** How many of the player's hits were home runs?

**49.** How many hits were doubles?

**50.** How many more singles than doubles did the player hit?

**51.** How many more doubles than triples did the player hit?

**52.** What was the total number of hits for this player over one month?

Evaluate each pair of expressions for $a = 2$ and $b = 5$. (Pages 150–153)

**53.** $3a^2$ and $(3a)^2$   **54.** $a^2 + b^2$ and $(a + b)^2$   **55.** $b^3$ and $(-b)^3$

## ★ Math Team Problem

**56.** A pet store owner wanted to separate his fish so that the same number of fish were in each tank. When he separated the fish by twos, by threes, by fours, by fives, or by sixes, there was always one fish left over. What is the least number of fish the storeowner could have had?

# SUMMARY

## KEY TERMS

base (p. 150)
composite number (p. 139)
divisible (p. 127)
even number (p. 132)
exponent (p. 150)
factor (p. 127)
factor tree (p. 143)
greatest common factor (p. 156)

infinite set (p. 128)
least common multiple
 (p. 154)
multiple (p. 127)
odd numbers (p. 132)
power (p. 150)
prime factorization (p. 143)
prime number (p. 139)

## KEY IDEAS

**A.** To factor a number means to write it as a product.
**B.** A **multiple** of a whole number, *n*, is the product of *n* and any whole number.
**C.** The set of natural numbers is N = {1, 2, 3, 4, ⋯}.
**D.** A summary of the divisibility tests is given on page 134.
**E.** In the prime factorization of a number, the number is expressed as a product of prime factors.
**F.** To find the least common multiple (LCM) of two or more numbers:
  1. Write the prime factorization of each number.
  2. Write a product using each prime factor only once.
  3. For each factor, write the highest exponent used in any of the prime factorizations.
  4. Multiply these factors.
**G.** To find the greatest common factor (GCF) of two or more numbers:
  1. Write the prime factorization of each number. Identify the common factors.
  2. Write the common factors as a product.

# CHAPTER REVIEW

Write *True* or *False* for each statement. Give a reason for your answer. (Pages 126–130)

1. 63 is a multiple of 7.
2. 9 is a factor of 9.
3. 40 is divisible by 3.
4. 24 is a factor of 8.
5. 54 is divisible by 6.
6. 8 is a multiple of 64.

**List all the factors of each number. (Pages 126–130)**
**7.** 39    **8.** 76    **9.** 35    **10.** 26    **11.** 77    **12.** 53

**List the first five multiples of each number. (Pages 126–130)**
**13.** 5    **14.** 6    **15.** 9    **16.** 2    **17.** 19    **18.** 26

**List the members of set S = {2, 3, 4, 5, 6, 8, 9, 10} that are factors of each number. (Pages 126–130)**
**19.** 108    **20.** 462    **21.** 730    **22.** 867    **23.** 375    **24.** 17,328

**Solve.**
**25.** Sandy has 148 books. Will the books evenly fill shelves that hold 12 books each? Explain your answer. (Pages 131–136)

**26.** Mr. Smith wants to give his employees a bonus. Can $1664 be divided evenly among his 8 employees? Explain your answer. (Pages 131–136)

**Tell whether each of the numbers is prime, composite, or neither. Give a reason for your answer. (Pages 139–141)**
**27.** 696    **28.** 2    **29.** 47    **30.** 273    **31.** 0    **32.** 101

**Draw a factor tree for each number. Then write the prime factorization of the number. (Pages 143–146)**
**33.** 36    **34.** 60    **35.** 40    **36.** 54    **37.** 32    **38.** 140

**Solve each problem. (Pages 148–149)**
**39.** Find numbers such that each is a multiple of 8 and whose sum is 180. Explain how you solved the problem.

**40.** Think of a number. Add 5 to it, triple the result, subtract 15, and then divide by 3. What is the result? Explain why this works.

**Simplify each expression. (Pages 150–153)**
**41.** $-4^4$    **42.** $(-6)^3$    **43.** $(2-19)^2$    **44.** $17+(-3)^4$
**45.** $18-(3+2)^2$    **46.** $30-(-4)^2$    **47.** $(1-3)^5+70$    **48.** $-9^2+10^2$

**Evaluate each expression for $a = 1$, $b = -2$, $c = 3$, and $d = -3$. (Pages 148–153)**
**49.** $-5c^2d$    **50.** $9b^3$    **51.** $12+2ac^2$    **52.** $25-ab^2$
**53.** $a^2d-44$    **54.** $10+2cd^3$    **55.** $54-bd^3$    **56.** $8c-a^3b^2$

**Find the LCM of each group of numbers. (Pages 154–158)**
**57.** 18 and 72    **58.** 36 and 54    **59.** 16 and 120    **60.** 63 and 180
**61.** 135 and 54    **62.** 216 and 45    **63.** 15, 25, and 30    **64.** 2, 27, and 18

**Find the GCF of each group of numbers. (Pages 154–158)**
**65.** 16 and 60    **66.** 45 and 75    **67.** 80 and 100    **68.** 45 and 105
**69.** 42 and 84    **70.** 48 and 144    **71.** 16, 20, and 24    **72.** 18, 27 and 36

# TEST

Write *True* or *False* for each statement. Give a reason for your answer.

**1.** 9 is a factor of 72.
**2.** 91 is divisible by 7.
**3.** 18 is a multiple of 1.
**4.** 39 is a factor of 13.

List all the factors of each number.

**5.** 16    **6.** 28    **7.** 34    **8.** 69    **9.** 63

List the first four nonzero multiples of each number.

**10.** 8    **11.** 4    **12.** 20    **13.** 18    **14.** 22

List the members of set S = {2, 3, 4, 5, 6, 8, 9, 10} that are factors of each number.

**15.** 342    **16.** 240    **17.** 252    **18.** 912    **19.** 532

Tell whether each of the numbers is prime, composite, or neither. Give a reason for your answer.

**20.** 57    **21.** 17    **22.** 0    **23.** 98    **24.** 101

**25.** Write the prime factorization of 126.

**26.** Write the prime factorization of 150.

Simplify each expression.

**27.** $-3^3$    **28.** $(-2)^6$    **29.** $(10 - 15)^5$    **30.** $(9 - 13)^2 + 2$

Evaluate each expression for $c = 2$, $d = -3$, and $f = -6$.

**31.** $4c^3d^2$    **32.** $c^4f - 21$    **33.** $41 + d^3f$    **34.** $400 + cf^3$

Find the LCM of each group of numbers.

**35.** 8 and 18    **36.** 27 and 36    **37.** 9, 10, and 15

Find the GCF of each group of numbers.

**38.** 55 and 121    **39.** 63 and 84    **40.** 12, 14, and 20

**41.** Both Walter and Susan start a new job on the same day. Walter works every fourth day and Susan works every sixth day. How many days will it be before they will again be working on the same day?

**1.** The actual temperature is 5°F. The wind chill temperature is −31°F. What is the difference of the two temperatures? **(Pages 23–25)**

**2.** The lowest temperature ever recorded in Nebraska was −47°F. The lowest temperature ever recorded in Florida was −2°F. Find the difference of the two temperatures. **(Pages 23–25)**

**Solve and check each equation. (Pages 91–99, 101–109)**

**3.** $p + 27 = 62$  **4.** $-47 = c - 29$  **5.** $-z = 0$  **6.** $6(y - 12) - 4y = -100$

**7.** $\frac{f}{-5} - 7 = 9$  **8.** $102 = -12x + 18$  **9.** $-15 = \frac{c}{9}$  **10.** $-2(-6b - 4) = 44$

**Write an equation to represent the situation. Then solve and check. (Pages 114–118)**

**11.** The ages of Lynne, Jack and Kelli are consecutive numbers. Lynne is the oldest and Jack is the youngest. The sum of their ages is 69. Find their ages.

**12.** Tim has twice as much money as Nell. Dora has $7 less than Nell. Together they have $41. How much money does each person have?

**13.** How many different factors does 88 have? **(Pages 126–131)**

**14.** What is the sixth nonzero multiple of 23? **(Pages 126–131)**

**15.** Replace _?_ with a digit that will make 4_?_,631 divisible by 9. **(Pages 131–136)**

**16.** Replace _?_ with a digit that will make 11,_?_12 divisible by 8. **(Pages 131–136)**

**17.** Find five numbers between 67 and 115 that are divisible by 6. **(Pages 131–136)**

**18.** Write the prime numbers between 36 and 49. **(Pages 139–141)**

**19.** How many numbers between 1 and 25 have exactly five factors? **(Pages 139–141)**

**20.** If the numbers 2, 2, 3, and 11 are at the bottom of a factor tree, what number is at the top? **(Pages 143–146)**

**Simplify. (Pages 46–50, 54–64, 150–153)**

**21.** $18(9 - 12) - 2$  **22.** $(-5)^4$  **23.** $84 \div 3 \div (113 - 117)$  **24.** $30 - 7^2$

**25.** $6f - 7f$  **26.** $(3 - 9)^3$  **27.** $9^2 - (2 - 10)^2$  **28.** $-4g - 3g$

**For each pair of expressions, write the answer of the expression with the larger value when $c = 7$ and $d = 5$. (Pages 39–44, 46–50, 150–153)**

**29.** $(-c + d) \cdot 2$ and $[c + (-d)] \cdot 2$  **30.** $\frac{c + 3}{2}$ and $\frac{-(d - 3)}{2}$  **31.** $(-4d)^2$ and $-4d^2$

**32.** Find the GCF and LCM of 16, 24, and 40. **(Pages 154–158)**

# CHAPTER 5 / FRACTIONS

## LESSON 5-1  Equivalent Fractions

**T**im and Tami each earned $84 mowing lawns last week. Tim spent $\frac{1}{2}$ of his earnings and Tami spent $\frac{3}{4}$ of her earnings. Who spent less?

**Think:** Is $\frac{1}{2} = \frac{3}{4}$?   Study the drawings below.

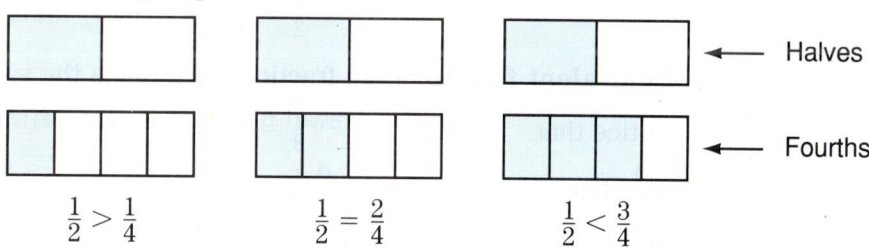

Since $\frac{1}{2} < \frac{3}{4}$, Tim has spent less of the $84 than Tami.

Numbers such as $\frac{1}{2}$ and $\frac{3}{4}$ are fractions. You can think of a **fraction** as a number written in the form $\frac{a}{b}$, where $a$ is a whole number and $b$ is a natural number. Every fraction has a numerator and a denominator.

$$\frac{1}{2} \; \substack{\longleftarrow \text{Numerator} \longrightarrow \\ \longleftarrow \text{Denominator} \longrightarrow} \; \frac{a}{b}$$

5-1  Equivalent Fractions   **163**

You can also express a fraction in lowest terms by dividing the numerator and denominator of the fraction by a common factor more than once. For example,

$$\frac{12}{18} = \frac{12 \div 2}{18 \div 2} = \frac{6}{9}$$ ⟵ Not in lowest terms.   Common factor: 3

$$\frac{6}{9} = \frac{6 \div 3}{9 \div 3} = \frac{2}{3}$$ ⟵ Lowest terms.

Using the GCF, however, gives the fraction in lowest terms after the first division is completed.

- How many equivalent fractions are there for $\frac{2}{3}$?

# EXERCISES

**Objective:** To show that two fractions are equivalent
To write fractions in lowest terms

## Check Your Understanding

**Skill Check**  Write the letter of the correct answer.

1. Which fraction is equivalent to $\frac{2}{3}$?
   a. $\frac{4}{9}$   b. $\frac{2}{9}$
   c. $\frac{4}{6}$   d. $\frac{4}{5}$

2. Which fraction is equivalent to $\frac{8}{10}$?
   a. $\frac{4}{10}$   b. $\frac{2}{5}$
   c. $\frac{6}{8}$   d. $\frac{4}{5}$

3. Which fraction is expressed in lowest terms?
   a. $\frac{9}{12}$   b. $\frac{4}{6}$
   c. $\frac{2}{5}$   d. $\frac{18}{45}$

4. Which fraction is **not** in lowest terms?
   a. $\frac{5}{16}$   b. $\frac{15}{21}$
   c. $\frac{8}{9}$   d. $\frac{7}{12}$

Replace each  ?  with the word that makes the sentences true.

5. When the numerator of a fraction is less than the denominator, the fraction is a  ?  fraction.

6. The denominator, $b$, of any fraction, $\frac{a}{b}$, represents any  ?  number.

## Practice and Apply

Make a drawing to show which is the larger fraction in each of these pairs.

7. $\frac{1}{2}$ and $\frac{3}{8}$
8. $\frac{1}{8}$ and $\frac{1}{4}$
9. $\frac{1}{2}$ and $\frac{5}{8}$
10. $\frac{2}{3}$ and $\frac{5}{6}$

**Show that each pair of fractions is equivalent.**

11. $\frac{3}{5}$ and $\frac{9}{15}$
12. $\frac{18}{24}$ and $\frac{3}{4}$
13. $\frac{4}{9}$ and $\frac{16}{36}$
14. $\frac{66}{36}$ and $\frac{22}{12}$
15. $\frac{7}{12}$ and $\frac{56}{96}$
16. $\frac{32}{64}$ and $\frac{1}{2}$
17. $\frac{12}{18}$ and $\frac{6}{9}$
18. $\frac{9}{24}$ and $\frac{81}{216}$

**Write the GCF of the numerator and denominator of each fraction.**

19. $\frac{30}{54}$
20. $\frac{42}{63}$
21. $\frac{18}{90}$
22. $\frac{16}{80}$
23. $\frac{18}{72}$

**Express each fraction in lowest terms.**

24. $\frac{8}{24}$
25. $\frac{6}{18}$
26. $\frac{5}{45}$
27. $\frac{7}{42}$
28. $\frac{108}{126}$
29. $\frac{42}{50}$
30. $\frac{48}{54}$
31. $\frac{21}{27}$
32. $\frac{12}{15}$
33. $\frac{36}{12}$

34. Explain how you know when a fraction is in lowest terms.
35. The numerator of a fraction is a multiple of the denominator. Is the fraction a proper or an improper fraction? Explain.
36. The numerator and denominator of a fraction are even numbers. Is the fraction in lowest terms? Explain.

## Connect and Extend

**Rewrite each pair of fractions so that they have the same denominator.**

37. $\frac{4}{5}$ and $\frac{24}{40}$
38. $\frac{5}{7}$ and $\frac{28}{49}$
39. $\frac{12}{14}$ and $\frac{25}{35}$
40. $\frac{9}{15}$ and $\frac{28}{35}$

41. Write five fractions, each having a numerator and denominator that is relatively prime.
42. Suppose that you need to measure $\frac{3}{5}$ of an inch, but your ruler is marked every $\frac{1}{3}$ of an inch. How could you come close to, or approximate, $\frac{3}{5}$ on your ruler?
43. Which is greater, a proper or an improper fraction? Explain your answer.

## Maintain Your Skills

**Name the integer in each pair that is farthest from 0 on the number line. (Pages 7–10)**

44. $-25$ and $-12$
45. $3$ and $-4$
46. $-3$ and $0$
47. $-1$ and $1$

48. Taylor has 24 yards of fencing. What should be the length and width, in yards, to enclose the most area for a rectangular garden? **(Pages 69–74)**
49. How many different factors does 52 have? **(Pages 126–130)**
50. What is the largest number that evenly divides 80, 220, and 460? **(Pages 154–158)**

# COMPUTER EXPLORATION

## Equivalent Fractions

1. With the Holt *Pre-Algebra* disk in drive A (or drive 1), run the program, *Equivalent Fractions*, as follows.
   a. **IBM:** Type **a:d** and press **RETURN**.
      **Apple:** Turn on the computer. When the menu appears on the screen, type **D**.
   b. Read the opening messages of the program and press **RETURN**. Type **1** to see the demonstrations in the program.

2. a. The computer asks whether $\frac{2}{3}$ is equivalent to $\frac{4}{6}$, is greater than $\frac{4}{6}$, or is less than $\frac{4}{6}$. Type **1, 2,** or **3** to indicate your choice. Press any key to continue.
   b. What fraction does the computer illustrate? Press any key to continue.
   c. What is the next fraction the computer illustrates?
   d. Explain how the computer drawings show that the fractions are equivalent.
   e. The computer tells you whether your answer is correct.

3. The computer repeats steps a, b, c, and e for these pairs of fractions.

       **i.** $\frac{3}{5}$ and $\frac{6}{7}$     **ii.** $\frac{10}{13}$ and $\frac{14}{17}$     **iii.** $\frac{4}{9}$ and $\frac{4}{25}$

   a. Which pairs of fractions are equivalent?
   b. If not, which fraction in each pair is greater?

4. The computer asks whether you wish to enter some fractions. Type **Y** for yes. Then run the program for these pairs of fractions.
   a. Enter $\frac{7}{14}$ and $\frac{8}{15}$. Which fraction is greater?
   b. Enter $\frac{6}{9}$ and $\frac{12}{18}$. Which fraction is smaller?
   c. Enter $\frac{11}{18}$ and $\frac{10}{19}$. Which fraction is greater?
   d. Enter other pairs of fractions of your choice as long as you have time at the computer. Then press **N** (for no) when the computer asks, **WANT TO ENTER ANOTHER FRACTION (Y/N)**.

# LESSON 5-2
## Problem Solving Exploration
## Modeling Multiplication

**A.** In this activity you use a paper strip, an ordinary pencil, a red pencil and a blue pencil.

Fold the paper strip into fourths and open it. Mark over the fold lines with the ordinary pencil and draw red horizontal lines in the first section.

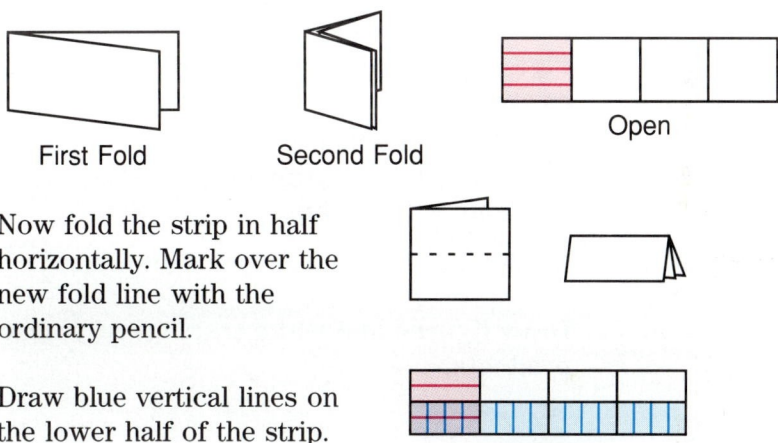

Now fold the strip in half horizontally. Mark over the new fold line with the ordinary pencil.

Draw blue vertical lines on the lower half of the strip.

**Figure 1**

Write a fraction to answer Exercises 1–4. Refer to Figure 1.

1. What part of the figure has red lines?
2. What part of the figure has blue lines?
3. What part of the figure has both red and blue lines?
4. What part of the figure is one-half of one-fourth?
5. How do you write "one-half of one-fourth" in mathematical symbols?
6. Explain how this activity shows that $\frac{1}{2} \cdot \frac{1}{4} = \frac{1}{8}$.

**Fold other paper strips in different ways to find each of these products.**

7. $\frac{1}{4} \cdot \frac{1}{2}$
8. $\frac{1}{2} \cdot \frac{1}{5}$
9. $\frac{1}{3} \cdot \frac{1}{4}$

10. Write a rule that tells how to find mathematically the product of any two fractions represented by $\frac{1}{a} \cdot \frac{1}{b}$.

**B.** This activity uses tracing paper, a red pencil, and a blue pencil.

Trace Figure 2 and draw red lines as shown.

$\frac{3}{4}$ of this figure is red.

**Figure 2**

Trace Figure 3 and draw blue lines as shown.

$\frac{1}{2}$ of this figure is blue.

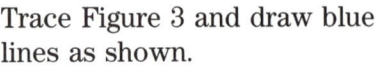

**Figure 3**

Use the tracings to place Figure 3 over Figure 2 as shown.

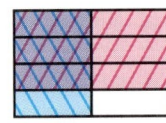

**Figure 4**

11. Write a fraction to tell what part of Figure 4 is both red and blue.

12. Explain how the activity shows that $\frac{1}{2} \cdot \frac{3}{4} = \frac{3}{8}$.

**Draw figures and make tracings of your own to find these products.**

13. $\frac{1}{3} \cdot \frac{1}{5}$   14. $\frac{1}{4} \cdot \frac{2}{3}$   15. $\frac{1}{2} \cdot \frac{5}{6}$

16. Write a rule that tells how to find mathematically the product of two fractions represented by $\frac{1}{a} \cdot \frac{b}{c}$.

# Mathematical Footnote

Where is the oldest and deepest lake on our planet? It is Lake Baikal in Russia, just north of Mongolia. It contains about one-fifth of the fresh water on the earth's surface! Its area is more than that of Maryland and Delaware combined. It has a depth of 4,259 feet below sea level.

Lake Baikal is surrounded by mountains, some of which tower 6,560 feet above the lake. There are 336 rivers and streams that flow into the lake. Lake Baikal is home to more than 1,000 species of plants and animals that are not found in other parts of the globe.

## LESSON 5-3    Multiplying Fractions

**T**wo-thirds of Grant City Park is covered with grass. The Director of City Parks plans to replace $\frac{1}{5}$ of this grass with shrubs. What part of the park will contain shrubs?

If you use a folded paper strip to model this problem, you first fold the strip into thirds and then fold it the other way into fifths. This suggests that you can write the answer using a multiple of 3 and 5.

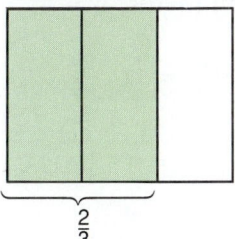
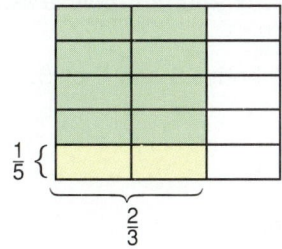

15 equal parts: 2 of the 15 will have shrubs.

Grass covers $\frac{2}{3}$ of the park. One-fifth of the two-thirds covered by grass will be replaced by shrubs.

$\frac{1}{5}$ of $\frac{2}{3}$ will be shrubs.     $\frac{1}{5} \cdot \frac{2}{3} = \frac{1 \cdot 2}{5 \cdot 3} = \frac{2}{15}$.

So $\frac{2}{15}$ of the park will contain shrubs.

One way to multiply fractions is to multiply the numerators and multiply the denominators.

> **Multiplication Rule for Fractions**
> If $\frac{a}{b}$ and $\frac{c}{d}$ are fractions, then $\frac{a}{b} \cdot \frac{c}{d} = \frac{a \cdot c}{b \cdot d}$.

**EXAMPLE 1** Find each product. Write answers in lowest terms.

a. $\frac{2}{3} \cdot \frac{7}{10}$
b. $\frac{3}{16} \cdot 5$

**Solutions**

a. $\frac{2}{3} \cdot \frac{7}{10} = \frac{2 \cdot 7}{3 \cdot 10}$

$= \frac{14}{30}$

$= \frac{14 \div 2}{30 \div 2}$

$= \frac{7}{15}$

b. $\frac{3}{16} \cdot 5 = \frac{3}{16} \cdot \frac{5}{1}$

$= \frac{3 \cdot 5}{16 \cdot 1}$

$= \frac{15}{16}$

**Try This** Find each product. Write answers in lowest terms.

1. $\frac{1}{3} \cdot \frac{9}{10}$
2. $3 \cdot \frac{4}{6}$
3. $\frac{3}{4} \cdot \frac{16}{12}$

It is sometimes easier to divide a numerator and denominator by a common factor before multiplying.

**EXAMPLE 2** Multiply. Write answers in lowest terms.

a. $\frac{5}{7} \cdot \frac{4}{15}$
b. $\frac{2}{9} \cdot \frac{3}{4}$

**Solutions**

a. **Think:** 5 and 15 have a common factor, 5.

Divide both 15 and 5 by 5. $\quad \frac{5}{7} \cdot \frac{4}{15} = \frac{\overset{1}{\cancel{5}}}{7} \cdot \frac{4}{\underset{3}{\cancel{15}}}$

$= \frac{1}{7} \cdot \frac{4}{3}$

Multiply. $= \frac{4}{21}$

b. **Think:** 2 and 4 have a common factor, 2.
3 and 9 have a common factor, 3.

Divide both 2 and 4 by 2.
Divide both 3 and 9 by 3. $\quad \frac{2}{9} \cdot \frac{3}{4} = \frac{\overset{1}{\cancel{2}}}{\underset{3}{\cancel{9}}} \cdot \frac{\overset{1}{\cancel{3}}}{\underset{2}{\cancel{4}}}$

$= \frac{1}{3} \cdot \frac{1}{2}$

Multiply. $= \frac{1}{6}$

**Try This** Multiply. Write answers in lowest terms.

4. $\frac{3}{4} \cdot \frac{8}{15}$
5. $\frac{15}{16} \cdot \frac{4}{5}$
6. $\frac{5}{9} \cdot \frac{12}{15}$

# EXERCISES

**Objective:** To find the product of fractions

## Check Your Understanding

**Skill Check** Write the letter of the correct answer.

1. Which is the product of $\frac{2}{3} \cdot \frac{4}{5}$?
   a. $\frac{6}{8}$
   b. $\frac{3}{4}$
   c. $\frac{8}{15}$
   d. $\frac{8}{8}$

2. Which is the product, in lowest terms, of $2 \cdot \frac{3}{8}$?
   a. $\frac{3}{4}$
   b. $\frac{6}{8}$
   c. $\frac{3}{16}$
   d. $\frac{5}{8}$

3. Which is the product, in lowest terms, of $\frac{1}{4} \cdot \frac{1}{4}$?
   a. $\frac{2}{8}$
   b. $\frac{2}{16}$
   c. $\frac{1}{8}$
   d. $\frac{1}{16}$

4. Which is the product, in lowest terms, of $\frac{3}{4} \cdot \frac{8}{15}$?
   a. $\frac{2}{5}$
   b. $\frac{11}{19}$
   c. $\frac{8}{20}$
   d. $\frac{1}{3}$

5. Write the Multiplication Rule for Fractions in your own words.

## Practice and Apply

**Multiply. Write answers in lowest terms.**

6. $\frac{1}{2} \cdot \frac{3}{4}$
7. $\frac{1}{2} \cdot \frac{1}{2}$
8. $\frac{5}{4} \cdot \frac{7}{12}$
9. $\frac{9}{12} \cdot \frac{10}{15}$
10. $6 \cdot \frac{5}{8}$
11. $\frac{7}{10} \cdot 12$
12. $\frac{10}{12} \cdot \frac{8}{12}$
13. $\frac{7}{8} \cdot \frac{8}{21}$
14. $\frac{32}{25} \cdot \frac{15}{36}$
15. $\frac{2}{3} \cdot \frac{3}{5}$
16. $\frac{5}{8} \cdot 8$
17. $7 \cdot \frac{2}{3}$

**Use this information for Exercises 18–21.**

Elias' favorite recipe calls for $\frac{1}{2}$ cup of flour and $\frac{3}{4}$ cup of milk.

18. How much flour will Elias need to make $\frac{2}{3}$ of the recipe?

19. How much flour will he need to double the recipe?

20. How much milk will he need to triple the recipe?

21. How much milk will he need to make $\frac{1}{2}$ of the recipe?

5-3 Multiplying Fractions

22. If a 32-ounce pitcher is $\frac{5}{8}$ full, how much liquid does it contain?
23. If a 12-gallon gas tank is $\frac{3}{4}$ empty, how much gas is in the tank?
24. Four-fifths of the students taking a computer programming course are passing. Of these students, $\frac{1}{3}$ have an A-average. What fraction of the students taking the course have an A-average?
25. When you multiply two fractions that are greater than 0 and less than 1, will the product be greater than or less than each of the factors? Explain.

## Connect and Extend

**A 9-piece pizza is shared by 4 people.**

26. If each person gets the same amount, how many whole pieces does each person get?

27. After the whole pieces are eaten, the rest of the pizza is divided equally. How much of the remaining pizza does each person get?

28. If one person eats 4 pieces and the others share the rest, how much pizza will each of the other 3 persons get? Include fractions of a piece.

29. If one person eats only one piece and the others share what remains equally, how much does each of the other 3 persons get? Include fractions of a piece.

**Write a multiplication problem with fractions to fit each description.**

30. One factor is an improper fraction and the product is $\frac{1}{2}$.
31. The factors are proper fractions and the product is $\frac{1}{3}$.
32. The factors are improper fractions and the product is 7.

## Maintain Your Skills

**Write an integer to represent each situation. (Pages 2–6)**

33. A 26 billion dollar deficit
34. A $514 profit
35. A $20 deposit
36. 42 feet below sea level

**Find the solution set if the replacement set is $\{-5, -3, -1, 0, 1, 3, 5\}$. (Pages 80–84)**

37. $-25s > 50$
38. $8t > 24$
39. $3r < -12$

**Write an equation to represent the situation. Then solve the equation. (Pages 114–118)**

40. If Harold had $13 less than the amount of money he has now, he would have $52. How much money does he have now?
41. The sum of three consecutive numbers is 87. What are the numbers?
42. The number of persons entering Sealy's Department store at 10 A.M. one day was 235 more than at 10 A.M. the next day. The total for both days was 885. How many persons entered the store at 10 A.M. each day?

## LESSON 5-4
# Multiplying Mixed Numbers

**E**very week, the members of Easton High Drill Team have 3 practice sessions of $2\frac{1}{2}$ hours each. How many hours is this per week?

One way to solve this problem is to use a number line to show 3 sessions of $2\frac{1}{2}$ hours each.

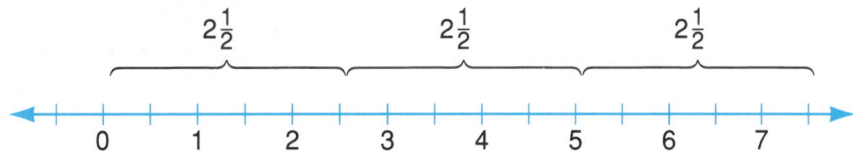

The number line shows that

$$2\frac{1}{2} + 2\frac{1}{2} + 2\frac{1}{2}, \text{ or } 3 \cdot 2\frac{1}{2}, = 7\frac{1}{2}.$$

So the drill team practices $7\frac{1}{2}$ hours per week.

The number $2\frac{1}{2}$ is a mixed number. A mixed number is another way to represent an improper fraction. You can think of a **mixed number** as the sum of a whole number and a fraction.

$$\frac{5}{2} = \frac{4}{2} + \frac{1}{2} = 2 + \frac{1}{2}, \text{ or } 2\frac{1}{2} \qquad \text{or} \qquad 2\overline{)5}^{\,2\frac{1}{2}}$$

5-4  Multiplying Mixed Numbers  **175**

**Method 1:** You can use this idea and the Distributive Property to find the product of a whole number and a mixed number mentally.

**EXAMPLE 1** Find each product mentally.

a. $3 \cdot 2\frac{1}{2}$  b. $3\frac{1}{4} \cdot 8$

**Solutions**

|  | a. $3 \cdot 2\frac{1}{2}$ | b. $3\frac{1}{4} \cdot 8$ |
|---|---|---|
| Write the mixed number as a sum. | $3 \cdot \left(2 + \frac{1}{2}\right)$ | $\left(3 + \frac{1}{4}\right)8$ |
| Use the Distributive property and multiply. | $3 \cdot 2 + \frac{3}{1} \cdot \frac{1}{2}$ | $3 \cdot 8 + \frac{1}{4} \cdot \frac{\cancel{8}^2}{1}$ |
| Add. | $6 + \frac{3}{2}$ | $24 + 2$ |
| Think: $\frac{3}{2} = \frac{2}{2} + \frac{1}{2}$ | $6 + 1\frac{1}{2}$ | **26** |
|  | $7\frac{1}{2}$ |  |

**Try This** Find each product mentally.

1. $6 \cdot 4\frac{1}{3}$  2. $3\frac{1}{3} \cdot 9$  3. $10 \cdot 2\frac{3}{5}$  4. $5\frac{3}{4} \cdot 12$

Another method for finding the product of a whole number and a mixed number involves renaming the mixed number as a fraction. For example, to write a fraction for $3\frac{1}{2}$:

Write $3\frac{1}{2}$ as a sum.   $3\frac{1}{2} = 3 + \frac{1}{2}$

Think: How many halves in 3?   $= \frac{6}{2} + \frac{1}{2}$

Add.   $= \frac{7}{2}$

**EXAMPLE 2** Write a fraction for each mixed number.

a. $5\frac{3}{4}$  b. $2\frac{1}{8}$

**Solutions**

a. $5\frac{3}{4} = 5 + \frac{3}{4}$   b. $2\frac{1}{8} = 2 + \frac{1}{8}$

$= \frac{20}{4} + \frac{3}{4}$   $= \frac{16}{8} + \frac{1}{8}$

$= \frac{23}{4}$   $= \frac{17}{8}$

**Method 2:** In Example 3, you rename both the mixed number and the whole number as fractions before multiplying to find the product.

**EXAMPLE 3** Find each product.

a. $4 \cdot 2\frac{1}{3}$      b. $1\frac{1}{8} \cdot 6$

**Solutions**

|  | a. $4 \cdot 2\frac{1}{3}$ | b. $1\frac{1}{8} \cdot 6$ |
|---|---|---|
| Rewrite as fractions. | $\frac{4}{1} \cdot \frac{7}{3}$ | $\frac{9}{8} \cdot \frac{6}{1}$ |
| Divide by common factors, if possible. | $\frac{4 \cdot 7}{1 \cdot 3}$ | $\frac{9}{\underset{4}{8}} \cdot \frac{\overset{3}{6}}{1}$ |
| Multiply. | $\frac{28}{3} \leftarrow \frac{27}{3} + \frac{1}{3}$ | $\frac{27}{4} \leftarrow 4\overline{)27}^{\,6\frac{3}{4}}$ |
| Write as a mixed number. | $9\frac{1}{3}$ | $6\frac{3}{4}$ |

**Try This** Find each product.

5. $4 \cdot 5\frac{1}{3}$      6. $3\frac{1}{6} \cdot 9$      7. $1\frac{7}{8} \cdot 56$

You can also use Method 2 to find the product of two mixed numbers.

**EXAMPLE 4** Find each product. Check whether each answer is reasonable.

a. $2\frac{1}{4} \cdot 3\frac{1}{3}$      b. $4\frac{1}{5} \cdot 1\frac{3}{7}$

**Solutions**

|  | a. $2\frac{1}{4} \cdot 3\frac{1}{3}$ | b. $4\frac{1}{5} \cdot 1\frac{3}{7}$ |
|---|---|---|
| Rewrite as fractions. | $\frac{9}{4} \cdot \frac{10}{3}$ | $\frac{21}{5} \cdot \frac{10}{7}$ |
| Divide by common factors, if possible. | $\frac{\overset{3}{9}}{\underset{2}{4}} \cdot \frac{\overset{5}{10}}{\underset{1}{3}}$ | $\frac{\overset{3}{21}}{\underset{1}{5}} \cdot \frac{\overset{2}{10}}{\underset{1}{7}}$ |
| Multiply. | $\frac{15}{2}$ | $\frac{6}{1}$ |
| Write as a mixed number. | $7\frac{1}{2}$ | $6$ |

5-4   Multiplying Mixed Numbers

**Check**

a. Is $7\frac{1}{2}$ a reasonable answer for $2\frac{1}{4} \cdot 3\frac{1}{3}$?

**Estimate:** $2\frac{1}{4}$ is close to 2. $3\frac{1}{3}$ is more than 3 and less than 4. The product of a number close to 2 and a number between 3 and 4 will be greater than 6 and less than 8. So $7\frac{1}{2}$ is a reasonable answer.

b. Is 6 a reasonable answer for $4\frac{1}{5} \cdot 1\frac{3}{7}$?

**Estimate:** $4\frac{1}{5}$ is close to 4. $1\frac{3}{7}$ is about $1\frac{1}{2}$.

Since $4 \cdot 1\frac{1}{2}$ is 6, 6 is a reasonable answer.

**Try This** Find each product. Check whether the answer is reasonable.

8. $1\frac{1}{4} \cdot 4\frac{4}{5}$
9. $2\frac{1}{3} \cdot 4\frac{1}{8}$
10. $6\frac{3}{4} \cdot 11\frac{1}{9}$

# EXERCISES

**Objective:** To find the product of mixed numbers

## Check Your Understanding

**Skill Check** Write the letter of the correct answer.

1. Which fraction is equivalent to $5\frac{3}{4}$?
   a. $\frac{23}{5}$
   b. $\frac{23}{4}$
   c. $\frac{19}{4}$
   d. $\frac{23}{3}$

2. Which is the product of $4\frac{1}{3} \cdot 5$?
   a. $4\frac{1}{3}$
   b. 6
   c. $21\frac{2}{3}$
   d. $20\frac{1}{3}$

3. Which is the product of $1\frac{1}{3} \cdot 3\frac{3}{4}$?
   a. $3\frac{1}{4}$
   b. $4\frac{1}{4}$
   c. $\frac{60}{12}$
   d. 5

4. Which is the best estimate for $1\frac{5}{6} \cdot 6\frac{1}{3}$?
   a. About 6
   b. About 7
   c. About 8
   d. About 12

5. Write a mixed number for $\frac{25}{4}$.

6. Write a fraction for $7\frac{2}{3}$.

## Practice and Apply

**Write a fraction for the mixed number.**

7. $3\frac{4}{5}$
8. $2\frac{2}{3}$
9. $9\frac{7}{10}$
10. $8\frac{5}{6}$
11. $9\frac{3}{8}$

**Write a mixed number for the improper fraction.**

12. $\frac{17}{12}$
13. $\frac{13}{10}$
14. $\frac{78}{8}$
15. $\frac{27}{10}$
16. $\frac{61}{5}$

**Use the Distributive Property (see Example 1) to find each product.**

**17.** $7 \cdot 2\frac{3}{5}$   **18.** $5 \cdot 3\frac{1}{8}$   **19.** $8 \cdot 9\frac{3}{4}$   **20.** $6 \cdot 10\frac{5}{12}$

**21.** $8\frac{7}{12} \cdot 9$   **22.** $6\frac{3}{4} \cdot 7$   **23.** $112 \cdot 4\frac{3}{8}$   **24.** $32 \cdot 2\frac{3}{4}$

**Find each product. Use estimates to show that your answer is reasonable.**

**25.** $8\frac{1}{5} \cdot 7\frac{1}{2}$   **26.** $2\frac{2}{5} \cdot 4\frac{1}{6}$   **27.** $2\frac{5}{8} \cdot 5\frac{1}{4}$   **28.** $1\frac{5}{6} \cdot 3\frac{1}{9}$

**29.** $8\frac{1}{4} \cdot 15\frac{3}{5}$   **30.** $7\frac{2}{3} \cdot 12\frac{1}{8}$   **31.** $2\frac{1}{6} \cdot 11\frac{3}{9}$   **32.** $3\frac{1}{5} \cdot 8\frac{2}{3}$

**33.** Explain why the method of Example 1 (using the Distributive Property) is not the best method to use when finding the product of two mixed numbers.

**34.** Explain why this statement is true.

The product of $2\frac{4}{5}$ and $2\frac{3}{8}$ will be greater than 4 and less than 9.

## Connect and Extend

**35.** The method shown at the right is often used to write a fraction for a mixed number. Explain why it works.

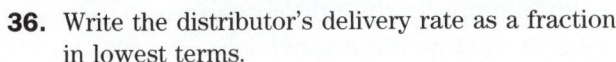

**Rates,** such as miles per hour and cost per item, can be written as fractions.

**Examples:** 60 miles per hour = $\frac{60}{1}$; 120 miles per 2 hours = $\frac{120}{2}$

A newspaper distributor can deliver 72 papers in 15 minutes.

**36.** Write the distributor's delivery rate as a fraction in lowest terms.

**37.** At this rate, how many papers can the distributor deliver in 10 minutes?

**38.** At this rate, how many papers can the distributor deliver in 45 minutes?

## Maintain Your Skills

**Find the divisor that makes each statement true. (Pages 32–35)**

**39.** $64 \div \underline{?} = -8$   **40.** $-75 \div \underline{?} = -15$   **41.** $-126 \div \underline{?} = -9$

**Solve each equation. (Pages 95–99)**

**42.** $8b = 1872$   **43.** $\frac{x}{6} = 36$   **44.** $\frac{c}{27} = 21$   **45.** $65k = 1495$

**46.** Darryl spent $\frac{5}{6}$ of an hour practicing basketball. He spent $\frac{1}{3}$ of that time practicing lay-up shots. What fraction of the hour did he spend practicing lay-up shots? **(Pages 171–174)**

# Problem Solving Strategies
# Organizing Information

Organizing information in a table can help you to see how different pieces of given information are related and how to use logical reasoning to solve the problem.

### EXAMPLE

Jeff, Cathy, and Amy work on the school newspaper. One is the editor, one is a reporter, and one is a word processor. Use the clues below to find each person's job.

**CLUE A**  Amy's only exercise is jogging.

**CLUE B**  Jeff and the editor play tennis together.

**CLUE C**  Amy and the reporter are cousins.

### Solution

Make a table to show all the possibilities. Use an X to show that a possibility cannot be true. Use a ✓ when you are certain that a possibility is true.

**1** Read Clue A. Can you reach any conclusion about Amy's job?

**2** Read Clue B. Is Jeff the editor?
Place an X next to Jeff's name in the editor column.
Read Clue A and Clue B. Is Amy the editor?
Place an X next to Amy's name in the editor column.

Who must be the editor?
Place a ✓ next to Cathy's name in the editor column.
Place X's next to Cathy's name in the reporter and word processor columns.

|  | Editor | Reporter | Word Processor |
|---|---|---|---|
| Jeff | X |  |  |
| Cathy | ✓ | X | X |
| Amy | X |  |  |

**3** Read Clue C. Is Amy the reporter?
Place an X in the reporter column and a ✓ in the processor column next to Amy's name. Who must be the reporter?

|  | Editor | Reporter | Word Processor |
|---|---|---|---|
| Jeff | X | ✓ | X |
| Cathy | ✓ | X | X |
| Amy | X | X | ✓ |

Jeff: reporter      Cathy: editor      Amy: word processor

Check the answer with the clues.

# EXERCISES

**Objective:** To apply strategies in problem solving

**Copy the table. Use the clues to complete the table and to solve the problem.**

1. Alice, Nathan, and Marie play in the school band. One plays the drum, one plays the saxophone, and one plays the flute.

   **CLUE A** Alice is a senior.

   **CLUE B** Alice and the saxophone player practice together after school.

   **CLUE C** Nathan and the flute player are sophomores.

   |        | Drum | Saxophone | Flute |
   |--------|------|-----------|-------|
   | Alice  |      |           |       |
   | Nathan |      |           |       |
   | Marie  |      |           |       |

   Who plays each instrument?

2. Brenda has three cats named Tiki, Moby, and Copper. One is a Persian, one is a Siamese, and one is a Himalayan.

   **CLUE A** Copper's favorite food is Fish Treats.

   **CLUE B** The Siamese will only eat Liver Bites.

   **CLUE C** Tiki will not eat with the Himalayan.

   **CLUE D** The Himalayan does not have a favorite food.

   What is the name of Brenda's Siamese cat?

3. Alex, Dee, Rod, and Sue are athletes. One is on the baseball team, one is on the soccer team, one is on the track team, and one is on the golf team.

   **CLUE A** Alex is taller than the soccer player.

   **CLUE B** Dee and Rod do not know how to play golf.

   **CLUE C** Neither Alex nor Dee has time to play baseball.

   **CLUE D** Sue's sport does not use a ball.

   Who plays on the soccer team?

5-5  Problem Solving Strategies   **181**

**Show that the two fractions in each pair are equivalent. (Pages 163–167)**

1. $\frac{4}{7}$ and $\frac{8}{14}$
2. $\frac{12}{54}$ and $\frac{2}{9}$
3. $\frac{4}{14}$ and $\frac{8}{28}$
4. $\frac{4}{10}$ and $\frac{16}{40}$

**Express each fraction in lowest terms. (Pages 163–167)**

5. $\frac{40}{52}$
6. $\frac{68}{76}$
7. $\frac{66}{102}$
8. $\frac{78}{112}$

**Multiply. Write answers in lowest terms. (Pages 171–174)**

9. $\frac{1}{5} \cdot \frac{1}{5}$
10. $\frac{1}{9} \cdot 6$
11. $\frac{10}{21} \cdot \frac{14}{15}$
12. $\frac{3}{8} \cdot \frac{8}{9}$

13. If a 15-gallon gas tank is $\frac{4}{5}$ empty, how much gas is in the tank? (Pages 171–174)

**Find each product. (Pages 175–179)**

14. $2\frac{1}{2} \cdot 7$
15. $2\frac{2}{3} \cdot 1\frac{1}{2}$
16. $2\frac{2}{5} \cdot 3\frac{1}{3}$
17. $9\frac{3}{8} \cdot 5\frac{1}{5}$

18. Judi does $4\frac{1}{2}$ hours of volunteer work every Saturday. How many hours does she work in a month having four Saturdays? (Pages 175–179)

## Mathematical Footnote

The Vietnam War Memorial in Washington, D.C., and the Civil Rights Memorial in Montgomery, Alabama, were both designed by Maya Lin.

Studying mathematics where she grew up in Ohio after leaving China, Maya Lin learned how to approach a problem and solve it. She learned how to express her ideas exactly so that others could carry them out with precision. Maya Lin used mathematics as she studied architecture and she uses it now as she creates stone memorials that impress all who visit them.

# LESSON 5-6
## Using Reciprocals to Solve Equations

Study these multiplication statements. Notice that each product is 1.

$$\frac{2}{3} \cdot \frac{3}{2} = 1 \qquad \frac{8}{5} \cdot \frac{5}{8} = 1 \qquad \frac{5}{6} \cdot \frac{6}{5} = 1 \qquad \frac{1}{2} \cdot \frac{2}{1} = 1$$

If the product of two numbers is 1, the numbers are called **reciprocals** of each other. The reciprocal of a number is also called its **multiplicative inverse**.

$\frac{4}{5}$ and $\frac{5}{4}$ are reciprocals because $\frac{4}{5} \cdot \frac{5}{4} = 1$.

3 and $\frac{1}{3}$ are reciprocals because $\frac{3}{1} \cdot \frac{1}{3} = 1$.

$\frac{1}{8}$ and 8 are reciprocals because $\frac{1}{8} \cdot \frac{8}{1} = 1$.

These suggest a generalization called the *Multiplication Property of Reciprocals*.

> **Multiplication Property of Reciprocals**
> For all nonzero numbers $a$ and $b$,
> $$\frac{a}{b} \cdot \frac{b}{a} = 1.$$

Notice that zero does not have a reciprocal. There is no number that can be multiplied by zero so that the product equals 1.

• Why is the number 1 its own reciprocal?

**EXAMPLE 1** Name the reciprocal of each number.

a. $\frac{1}{20}$      b. $6\frac{1}{4}$      c. $\frac{a}{15}$, $a \neq 0$

**Solution**

a. $\frac{20}{1}$, or **20**

b. Since $6\frac{1}{4} = \frac{25}{4}$, its reciprocal is $\frac{4}{25}$.

c. $\frac{15}{a}$

• If a number is a proper fraction, will its reciprocal be greater than 1 or less than 1? Why?

You can use reciprocals to solve multiplication equations.

**EXAMPLE 2** Solve and check.

a. $\frac{1}{3}x = \frac{5}{8}$
b. $\frac{3}{4}x = 4\frac{1}{2}$

**Solutions**

a. Write the equation.
Think: $\frac{1}{3} \cdot \underline{\phantom{?}} = 1$

Since $\frac{1}{3} \cdot 3 = 1$, multiply each side by 3.

$\frac{1}{3}x = \frac{5}{8}$

$3\left(\frac{1}{3}x\right) = 3\left(\frac{5}{8}\right)$

$\left(\frac{3}{1} \cdot \frac{1}{3}\right)x = \frac{3}{1} \cdot \frac{5}{8}$

Simplify.

$1x = \frac{15}{8}$

$x = \frac{15}{8}$, or $1\frac{7}{8}$

**Check** Replace $x$ with $\frac{15}{8}$ in the original equation.

$\frac{1}{3}x = \frac{5}{8}$

$\frac{1}{\cancel{3}}\left(\frac{\cancel{15}^{5}}{8}\right) \stackrel{?}{=} \frac{5}{8}$

$\frac{5}{8} = \frac{5}{8}$ ✓

The solution is $\frac{15}{8}$, or $1\frac{7}{8}$.

b. Write the equation.
Think: $\frac{3}{4} \cdot \underline{\phantom{?}} = 1$
Since $\frac{3}{4} \cdot \frac{4}{3} = 1$, multiply each side by $\frac{4}{3}$.

$\frac{3}{4}x = 4\frac{1}{2}$

$\frac{4}{3}\left(\frac{3}{4}x\right) = \frac{4}{3}\left(\frac{9}{2}\right)$

$\frac{\cancel{4}^{1}}{\cancel{3}_{1}}\left(\frac{\cancel{3}^{1}}{\cancel{4}_{1}}x\right) = \frac{\cancel{4}^{2}}{\cancel{3}_{1}}\left(\frac{\cancel{9}^{3}}{\cancel{2}_{1}}\right)$

Simplify.

$x = 6$

**Check** Replace $x$ with 6.

$\frac{3}{4}x = 4\frac{1}{2}$

$\frac{3}{\cancel{4}_{2}} \cdot \cancel{6}^{3} \stackrel{?}{=} 4\frac{1}{2}$

$\frac{9}{2} = \frac{9}{2}$ ✓

The solution is **6**.

## LESSON 5-8

# Fractions with Like Denominators

The table at the right above shows the Walk-a-Thon results in miles for Sherri, Mike, and Jenna. Organizing data in a table makes it easier to make comparisons, formulate questions, draw conclusions, and make decisions about the data.

This lesson considers three questions about the data in the table.

**Question 1:** How far did each person walk during the two days?

To find how far Sherri walked during the two days, add to find the total distance.

Write the problem.

$$3\frac{1}{10} + 4\frac{3}{10} = \left(3 + \frac{1}{10}\right) + \left(4 + \frac{3}{10}\right)$$

Apply the Commutative and Associative Properties.

$$= (3 + 4) + \left(\frac{1}{10} + \frac{3}{10}\right)$$

Add the whole numbers.

$$= 7 + \frac{1}{10} + \frac{3}{10}$$

Add the fractions.

$$= 7\frac{4}{10}, \text{ or } 7\frac{2}{5}$$

Sherri's total distance was $7\frac{2}{5}$ miles.

**194** CHAPTER 5

## Connect and Extend

Refer to this diagram for Exercises 32–36.

$$\text{Divisor} \rightarrow a\overline{)b} \begin{matrix} \leftarrow \text{Quotient} \\ \leftarrow \text{Dividend} \end{matrix}$$

Write True or False for each statement. When a statement is false, give a counterexample.

32. The quotient equals the dividend multiplied by the reciprocal of the divisor.
33. If the divisor is less than the dividend, then the quotient is always greater than 1.
34. If the divisor equals the dividend, then the quotient is zero.
35. The quotient and the divisor are factors of the dividend, if there is no remainder.
36. The quotient of an improper fraction divided by an improper fraction is always a whole number.

## Maintain Your Skills

In each group of numbers, name the number that is closest to 0 on the number line. (Pages 7–10)

37. 5, −2, or −3
38. 1, −4, or −3
39. $\frac{4}{2}, -\frac{1}{2}, \frac{3}{2}$

Write the name of the property shown. (Pages 54–58)

40. $-\frac{1}{4}(8 + \frac{1}{2}) = -\frac{1}{4}(\frac{1}{2} + 8)$
41. $\frac{1}{3}xy^2 = \frac{1}{3}y^2x$
42. $(\frac{2}{5} + \frac{3}{5}) + \frac{1}{5} = \frac{2}{5} + (\frac{3}{5} + \frac{1}{5})$

Which of the following are composite numbers? (Pages 139–141)

43. 91
44. 234
45. 51
46. 89
47. 117

Write two fractions that are equivalent to the given fraction. (Pages 163–167)

48. $\frac{1}{2}$
49. $\frac{2}{5}$
50. $\frac{4}{3}$
51. $\frac{5}{6}$
52. $\frac{9}{4}$

## ★ Math Team Problem

53. An ant climbs up a pole five feet during the day and slides back four feet during the night. The pole is 30 feet high. How many days will it take the ant to climb from ground level to the top of the pole?

**3.** Which is the quotient for $\frac{5}{8} \div \frac{3}{4}$?

a. $\frac{15}{32}$    b. $\frac{5}{6}$

c. $1\frac{1}{5}$    d. $\frac{3}{8}$

**4.** Which is the quotient for $9\frac{3}{4} \div 1\frac{1}{2}$?

a. $14\frac{5}{8}$    b. $10\frac{1}{8}$

c. $6\frac{1}{2}$    d. $3\frac{1}{6}$

**Refer to the diagrams to find each quotient.**

**5.**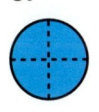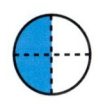

$1\frac{1}{2} \div \frac{1}{4}$

**6.**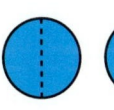

$3\frac{1}{2} \div \frac{1}{2}$

**7.**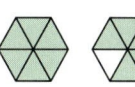

$2\frac{5}{6} \div \frac{1}{6}$

## Practice and Apply

**Write each division as a multiplication expression.**

**8.** $\frac{3}{4} \div \frac{3}{8}$    **9.** $\frac{2}{5} \div \frac{1}{10}$    **10.** $\frac{2}{3} \div \frac{8}{9}$    **11.** $2\frac{1}{7} \div \frac{5}{2}$

**Write each division as a multiplication expression. Then simplify.**

**12.** $3 \div \frac{1}{3}$    **13.** $\frac{2}{3} \div 6$    **14.** $\frac{2}{5} \div 8$    **15.** $8 \div \frac{4}{5}$

**16.** $\frac{2}{3} \div \frac{7}{8}$    **17.** $\frac{4}{5} \div \frac{9}{10}$    **18.** $\frac{2}{5} \div \frac{3}{4}$    **19.** $\frac{7}{10} \div \frac{1}{2}$

**20.** $3\frac{1}{5} \div \frac{4}{15}$    **21.** $2\frac{1}{3} \div 1\frac{5}{6}$    **22.** $4\frac{1}{2} \div 2\frac{7}{10}$    **23.** $\frac{9}{10} \div 1\frac{1}{5}$

**24.** $2\frac{1}{6} \div 3\frac{1}{2}$    **25.** $\frac{5}{9} \div 1\frac{2}{3}$    **26.** $7\frac{1}{2} \div 1\frac{1}{3}$    **27.** $3\frac{1}{2} \div 4\frac{1}{4}$

**Write a division expression for each situation. Use the expression to answer the question.**

**28.** Mr. Barker has $\frac{1}{3}$ of a pound of raisins to distribute equally to his 4 children. How much will each child get?

**29.** Christine is carving a model airplane from a piece of wood that is 5 yards long. How many models can she carve if each model is $\frac{5}{6}$ of a yard long?

**30.** A baker has $17\frac{1}{2}$ cups of flour on hand. If it takes $2\frac{1}{3}$ cups to make a loaf of bread, how many loaves can the baker make?

**31.** When dividing whole numbers by a fraction less than 1, will the quotient be less than or greater than the whole number? Explain.

**EXAMPLE 2** Find each quotient.

a. $\frac{3}{4} \div \frac{9}{10}$

b. $2\frac{1}{4} \div 1\frac{1}{2}$

**Solutions**

a. Rewrite as multiplication. $\quad \frac{3}{4} \div \frac{9}{10} = \frac{3}{4} \cdot \frac{10}{9}$

Divide by common factors. $\quad = \frac{\overset{1}{\cancel{3}}}{\underset{2}{\cancel{4}}} \cdot \frac{\overset{5}{\cancel{10}}}{\underset{3}{\cancel{9}}}$

Multiply. $\quad = \frac{5}{6}$

b. Rewrite as improper fractions. $\quad 2\frac{1}{4} \div 1\frac{1}{2} = \frac{9}{4} \div \frac{3}{2}$

Rewrite as multiplication. $\quad = \frac{9}{4} \cdot \frac{2}{3}$

Divide by common factors. $\quad = \frac{\overset{3}{\cancel{9}}}{\underset{2}{\cancel{4}}} \cdot \frac{\overset{1}{\cancel{2}}}{\underset{1}{\cancel{3}}}$

Multiply. $\quad = \frac{3}{2}$, or $1\frac{1}{2}$

**Try This** Find each quotient.

4. $\frac{2}{3} \div \frac{5}{6}$

5. $8\frac{1}{3} \div \frac{5}{9}$

6. $5\frac{1}{4} \div 5\frac{5}{6}$

- How many $1\frac{1}{4}$'s are there in 60?
- How many $1\frac{1}{5}$'s are there in 120?

# EXERCISES

**Objective:** To divide fractions and mixed numbers

## Check Your Understanding

**Skill Check** Write the letter of the correct answer.

**1.** Which is the first step in finding this quotient?

$$\frac{4}{9} \div 4$$

a. $\frac{4}{9} \cdot \frac{4}{1}$
b. $\frac{9}{4} \cdot \frac{4}{1}$
c. $\frac{9}{4} \cdot \frac{1}{4}$
d. $\frac{4}{9} \cdot \frac{1}{4}$

**2.** Which is the quotient for $8 \div \frac{1}{4}$?

a. 2
b. $\frac{1}{32}$
c. $\frac{1}{2}$
d. 32

5-7 Dividing Fractions and Mixed Numbers

**Method 2:** Use a mathematical model.
Rewrite the question: How many halves are there in the 6 square sections?

Draw a number line.
Count the number of halves in 6.

$6 \div \frac{1}{2} = 12$

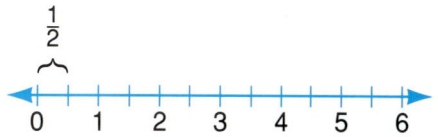

**Method 3:** Use inverse operations.
You learned in the previous lesson that dividing by a number gives the same result as multiplying by the reciprocal of that number. That is,

$2 \div \frac{1}{2} = 2 \cdot 2 = 4$   $\qquad$   $4 \div 2 = 4 \cdot \frac{1}{2} = 2$   $\qquad$   $6 \div \frac{1}{2} = 6 \cdot 2 = 12$

This suggests the Division Rule for Fractions.

> **Division Rule for Fractions**
> To divide by a number, multiply by the reciprocal (multiplicative inverse) of the number.

**EXAMPLE 1** Use the Division Rule for Fractions to find each quotient.

**a.** $6 \div \frac{2}{3}$   $\qquad$   **b.** $\frac{8}{15} \div 12$

**Solutions** Write each division problem as a multiplication problem. Then simplify.

Rewrite as multiplication.   **a.** $6 \div \frac{2}{3} = 6 \cdot \frac{3}{2}$   **b.** $\frac{8}{15} \div 12 = \frac{8}{15} \cdot \frac{1}{12}$

Divide by common factors.   $= \frac{\overset{3}{\cancel{6}}}{1} \cdot \frac{3}{\cancel{2}}$   $= \frac{\overset{2}{\cancel{8}}}{15} \cdot \frac{1}{\cancel{12}}$
$\qquad\qquad\qquad\qquad\qquad\qquad\qquad 1 \qquad\qquad\qquad\qquad 3$

Multiply.   $= \frac{9}{1}$, or **9**   $= \frac{2}{45}$

**Try This** Use the Division Rule for Fractions to find each quotient.

**1.** $8 \div \frac{2}{3}$   **2.** $9 \div \frac{6}{15}$   **3.** $15 \div \frac{5}{8}$

You can also use the Division Rule for Fractions to divide mixed numbers.

**190**   CHAPTER 5

## LESSON 5-7

# Dividing Fractions and Mixed Numbers

**S**am and Paul are planning a design for the tile floor of a hallway. There are six square sections along the length of the hall. Each pattern covers one-half of each square section. How many times will the pattern occur?

Sam says to take $\frac{1}{2}$ of 6 to find the answer. Paul argues that $\frac{1}{2}$ of 6 is 3, and he knows that more than 3 patterns will be needed. So Paul concludes that they will have to divide 6 by $\frac{1}{2}$ to get the answer. Sam wants to know why dividing 6 by $\frac{1}{2}$ results in 12, a number greater than 6.

Here are three ways Paul might use to convince Sam that the answer is 12.

**Method 1:** Act it out.
Trace and cut out the pattern enough times to cover the six square sections. Count how many are needed.

## Connect and Extend

A **complex fraction** has a fraction either in its numerator or in its denominator or in both the numerator and denominator.

$$4 \div \tfrac{1}{2} \text{ means } \underbrace{\dfrac{\overbrace{4}^{\text{Dividend}}}{\underbrace{\tfrac{1}{2}}_{\text{Divisor}}}}_{} \longleftarrow \text{This is a complex fraction.}$$

To simplify a complex fraction, multiply by the reciprocal of the divisor as shown below. This will result in a divisor (denominator) of 1.

$$\dfrac{4}{\tfrac{1}{2}} = \dfrac{4 \cdot 2}{\tfrac{1}{2} \cdot 2} = \dfrac{8}{1} = 8$$

**Complete.**

**34.** $\dfrac{25}{\tfrac{1}{5}} = \dfrac{25 \cdot ?}{\tfrac{1}{5} \cdot 5} = \underline{\ ?\ }$

**35.** $\dfrac{9}{\tfrac{3}{5}} = \dfrac{9 \cdot ?}{\tfrac{3}{5} \cdot \tfrac{5}{3}} = \underline{\ ?\ }$

**36.** $\dfrac{12}{\tfrac{4}{3}} = \dfrac{12 \cdot ?}{\tfrac{4}{3} \cdot ?} = \underline{\ ?\ }$

**Simplify each complex fraction.**

**37.** $\dfrac{15}{\tfrac{1}{4}}$

**38.** $\dfrac{2}{\tfrac{4}{5}}$

**39.** $\dfrac{18}{\tfrac{2}{9}}$

**40.** $\dfrac{1}{\tfrac{5}{8}}$

## Maintain Your Skills

**Find the area of the rectangle *only* if the area is *less than* 100 square miles. (Pages 64–68)**

**41.** $w = 6, l = 19$

**42.** $w = 8, l = 13$

**43.** $w = 7, l = 14$

**44.** Find the average of the numbers 91, 96, 94, 95, and 99. **(Pages 64–68)**

**Write an algebraic expression for each word expression. (Pages 110–113)**

**45.** The quotient of $y$ and the sum of $-7$ and 5.

**46.** Six times the difference of $a$ and $b$.

**47.** Sixteen centimeters more than 5 times the width, $w$, of a board

**48.** Twice the number of hours, $h$, decreased by 15

**49.** Four times the number of artists, $a$, divided by 15

**50.** The total of bank deposits, $t$, less $1200

**Write the LCM and the GCF. (Pages 154–158)**

**51.** 97 and 89

**52.** 20 and 35

**53.** 30, 45, and 60

**54.** Write a fraction for $2\tfrac{4}{9}$. **(Pages 175–179)**

## Practice and Apply

**Write the reciprocal of each number.**

**6.** $\frac{3}{5}$  **7.** $\frac{9}{4}$  **8.** $2\frac{3}{5}$  **9.** 25  **10.** $\frac{8}{3}$

**Solve and check each equation.**

**11.** $\frac{1}{5}x = \frac{2}{3}$  **12.** $\frac{3}{8}t = \frac{3}{4}$  **13.** $\frac{5}{8}a = \frac{10}{4}$  **14.** $\frac{5}{12}p = \frac{3}{5}$

**15.** $\frac{3}{4}s = 5\frac{1}{2}$  **16.** $2\frac{1}{3} = 5y$  **17.** $2\frac{5}{6} = 3r$  **18.** $2\frac{7}{8}p = 2\frac{3}{10}$

**19.** $4 = 4\frac{2}{3}z$  **20.** $1\frac{2}{3}q = 5\frac{7}{9}$  **21.** $\frac{3}{10}n = 12$  **22.** $\frac{8}{9}h = 24$

**23.** $\frac{1}{2} = \frac{1}{24}y$  **24.** $\frac{4}{5}s = 3\frac{1}{5}$  **25.** $1\frac{5}{9}n = \frac{11}{15}$  **26.** $3\frac{1}{8}x = 21\frac{1}{2}$

**Write an equation that represents the situation. Then solve the equation.**

**27.** Nick works 16 hours and completes $\frac{3}{4}$ of the job. How long will it take him to do the entire job?

**28.** Mandy ran $5\frac{1}{2}$ miles at a steady rate of $\frac{1}{8}$ mile per minute. How many minutes did Mandy run? (Hint: Use the formula, rate · time = distance.)

**29.** A bag of Kool Kat holds 160 ounces of cat food. If each serving is $1\frac{1}{2}$ ounces, how many servings does the bag hold?

**30.** A full package of Green-Gro has 60 teaspoons of fertilizer. If Marty sprinkles $\frac{2}{3}$ of a teaspoon of fertilizer on a plant each week, how many weeks will it take Marty to use a full package of Green-Gro?

**31.** The distance a nut moves on a bolt in one turn is $\frac{3}{16}$ of an inch. How many turns are needed to make the nut move $2\frac{3}{8}$ inches?

**32.** A road worker painting a line down the side of a road paints $53\frac{1}{3}$ feet per minute. How many minutes will it take her to paint one mile? (1 mile = 5280 feet)

**33.** If a snail can crawl $1\frac{7}{8}$ inches per minute, how many minutes will it take the snail to crawl $5\frac{5}{8}$ inches?

**Solution**   Choose a variable to represent the unknown. Use the given information to write an equation.

Let $p$ represent the total length.
The piece cut off is $\frac{3}{4}$ of the total length.

$$2\frac{1}{2} = \frac{3}{4} \cdot p$$

Solve for $p$. Think: $\frac{3}{4} \cdot \underline{\ ?\ } = 1$

Multiply each side by $\frac{4}{3}$.

$$\frac{4}{3}\left(\frac{5}{2}\right) = \frac{4}{3}\left(\frac{3}{4}p\right)$$

$$\frac{\cancel{4}}{3} \cdot \frac{5}{\cancel{2}} = \left(\frac{4}{3} \cdot \frac{3}{4}\right)p$$

Simplify.

$$\frac{10}{3} = p$$

$$p = 3\frac{1}{3}$$

The length of the piece of lumber is $3\frac{1}{3}$ feet.

**Try This**   4. Teresita spent $\frac{2}{3}$ of her money on tapes. If she spent $24, how much money did she have originally?

# EXERCISES

**Objectives:** To find the reciprocal of a number
To solve multiplication equations using the Multiplication Property of Reciprocals

## Check Your Understanding

**Skill Check**   Write the letter of the correct answer.

1. Which is the reciprocal of $1\frac{1}{2}$?
   a. $\frac{1}{2}$   b. $\frac{3}{2}$   c. $\frac{2}{3}$   d. 3

2. Which number has no reciprocal?
   a. $\frac{0}{8}$   b. $\frac{1}{8}$   c. $\frac{8}{5}$   d. $\frac{6}{6}$

3. Which is the first step you would use to solve the equation $\frac{2}{3}b = 18$?
   a. Multiply each side by 3.
   b. Divide each side by 2.
   c. Multiply each side by $\frac{3}{2}$.
   d. Divide each side by $\frac{3}{2}$.

4. Which is the first step you would use to solve the equation $\frac{1}{9}a = 21$?
   a. Divide each side by 9.
   b. Multiply each side by 81.
   c. Multiply each side by 3.
   d. Multiply each side by 9.

5. *Complete:* When the product of two numbers is 1, the numbers are called __?__ of each other.

**Try This**    **Solve each equation.**

1. $\frac{1}{5}x = \frac{7}{10}$    2. $\frac{2}{3}x = 5\frac{1}{3}$    3. $5x = \frac{7}{8}$

You can solve the equation $4x = 8$ in two ways.

| Method 1 | Method 2 |
|---|---|
| Divide each side by 4. | Multiply each side by $\frac{1}{4}$. |
| $4x = 8$ | $4x = 8$ |
| $\frac{4x}{4} = \frac{8}{4}$ | $\frac{1}{4}(4x) = \frac{1}{4}(8)$ |
| $x = 2$ | $x = 2$ |

You can see that dividing by 4 gives the same result as multiplying by $\frac{1}{4}$. In other words, dividing by a number is the same as multiplying by its reciprocal. So the Division Property of Equality (see page 96) can be combined with the Multiplication Property of Equality in this way.

> **Multiplication Property of Equality**
> For all numbers $a$, $b$, and $c$, where $c \neq 0$, if $a = b$, then $ac = bc$, and if $ac = bc$, then $a = b$.

The Multiplication Property of Equality is useful in solving word problems.

**EXAMPLE 3**    Sam cuts $2\frac{1}{2}$ feet from a piece of lumber. The piece he cuts off is $\frac{3}{4}$ of the length of the total piece.

How long was the original piece of lumber?

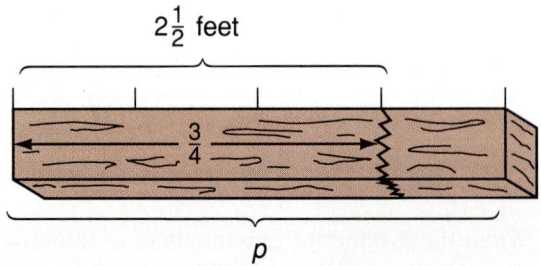

5-6 Using Reciprocals to Solve Equations

# SUMMARY

## KEY TERMS

Conjecture (p. 202)
Complex fraction (p. 188)
Denominator (p. 163)
Equivalent fractions (p. 164)
Fraction (p. 163)
Improper fraction (p. 164)
Least common denominator (p. 201)
Lowest terms (p. 165)
Mixed number (p. 175)
Multiplicative inverse (p. 183)
Numerator (p. 163)
Proper fraction (p. 164)
Reciprocal (p. 183)

## KEY IDEAS

**A.** To find an equivalent fraction, multiply or divide both the numerator and denominator of the given fraction by the same nonzero number.

**B.** A fraction is in lowest terms if the only common factor of its numerator and denominator is 1. To express a fraction in lowest terms, divide both the numerator and the denominator of the fraction by their greatest common factor (GCF).

**C. Multiplication Rule for Fractions:** To multiply fractions, multiply the numerators and multiply the denominators.

**D. Multiplication Property of Reciprocals:** For all nonzero numbers $a$ and $b$, $\frac{a}{b} \cdot \frac{b}{a} = 1$.

**E.** When solving equations, dividing by a number is the same as multiplying by the reciprocal of the number.

**F. Division Rule for Fractions:** To divide by a number, multiply by the reciprocal (multiplicative inverse) of the number.

**G.** To add, subtract, or compare fractions with unlike denominators, first rewrite the fractions so that they have like denominators.

**Find the value of $n$.**

**23.** $17\frac{1}{8} - 12\frac{1}{2} = n$
**24.** $12 - n = 11\frac{1}{5}$
**25.** $n + 6\frac{1}{2} = 4\frac{2}{3}$

## Connect and Extend

How many fractions are there between $\frac{1}{3}$ and $\frac{1}{2}$?

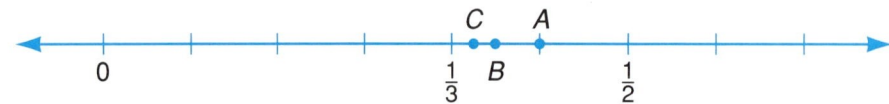

**26.** To solve this problem, find a fraction halfway between $\frac{1}{3}$ and $\frac{1}{2}$. This is the same as finding the average of $\frac{1}{3}$ and $\frac{1}{2}$. (HINT: $\left(\frac{1}{3} + \frac{1}{2}\right) \div 2 = \underline{\ ?\ }$) This is point $A$ on the number line above.

**27.** Find the fraction halfway between $\frac{1}{3}$ and the coordinate of point $A$. This is point $B$ on the number line above.

**28.** Find the fraction halfway between $\frac{1}{3}$ and the coordinate of point $B$. This is point $C$ on the number line above.

**29.** How many fractions do you think there are between $\frac{1}{3}$ and $\frac{1}{2}$?

**30.** How many fractions do you think there are between $0$ and $\frac{1}{3}$?

**31.** Complete this generalization. It is called the **density property** of fractions.

Between any two fractions, no matter how close they may be, there is an $\underline{\ ?\ }$ number of other fractions.

## Maintain Your Skills

**Find the solution set if the replacement set is $\{0, 3, 6, 9, 12\}$. (Pages 80–84, 171–174)**

**32.** $\frac{1}{6}c + 2 = 4$
**33.** $\frac{2}{6}k + 3 = 6$
**34.** $\frac{1}{3}g - 1 < 0$

**Solve and check. (Pages 105–109)**

**35.** $-14 = \frac{c}{3} - 6$
**36.** $7n + 101 = 17$
**37.** $2(x - 5) - 4 = 60$

**38.** A can holds $10\frac{3}{4}$ ounces of soup. If each serving is 4 ounces, how many servings does the can hold? **(Pages 189–193)**

**39.** Mrs. Hudson has $\frac{2}{5}$ of a pound of chocolate to distribute equally among 6 people. How much will each person get? **(Pages 189–193)**

# REVIEW

**Show that the two fractions in each pair are equivalent. (Pages 163–167)**

1. $\frac{7}{8}$ and $\frac{35}{40}$
2. $\frac{4}{12}$ and $\frac{1}{3}$
3. $\frac{15}{27}$ and $\frac{5}{9}$
4. $\frac{12}{38}$ and $\frac{24}{76}$

**Express each fraction in lowest terms. (Pages 163–167)**

5. $\frac{21}{24}$
6. $\frac{17}{68}$
7. $\frac{56}{70}$
8. $\frac{91}{13}$
9. $\frac{96}{104}$

**Multiply. Write the answers in lowest terms. (Pages 171–174)**

10. $\frac{2}{3} \cdot \frac{9}{10}$
11. $\frac{3}{2} \cdot 7$
12. $8 \cdot \frac{5}{12}$
13. $\frac{32}{25} \cdot \frac{15}{36}$

**Find each product. Write the answers in lowest terms. (Pages 175–179)**

14. $5 \cdot 3\frac{2}{5}$
15. $4\frac{3}{8} \cdot 3\frac{1}{7}$
16. $6\frac{2}{9} \cdot 7\frac{1}{8}$
17. $1\frac{3}{8} \cdot 26$

18. Lewis, Shannon, Ana and Keith live in the towns Saratoga, Kilgore, Lawton, and Ardmore.
    - **CLUE A** No one lives in a town that has the same first letter as his or her name.
    - **CLUE B** Neither Lewis nor Keith has ever been to Saratoga.
    - **CLUE C** Shannon has spent all of her life in Ardmore.
    
    Where does each person live? **(Pages 180–181)**

**Solve and check. (Pages 183–188)**

19. $\frac{1}{6}d = \frac{3}{4}$
20. $1\frac{2}{3}e = 1\frac{1}{5}$
21. $1\frac{2}{9} = 18f$
22. $\frac{7}{10} = 5\frac{3}{12}g$

23. A cook needs $\frac{1}{3}$ hour cooking time for each pound of turkey. How big is the turkey if it is cooked for $4\frac{1}{2}$ hours? **(Pages 183–188)**

**Find each quotient. (Pages 189–193)**

24. $\frac{6}{7} \div 12$
25. $\frac{9}{10} \div \frac{3}{5}$
26. $\frac{4}{5} \div 2\frac{2}{3}$
27. $4\frac{3}{5} \div 2\frac{3}{10}$

**Add or subtract. Write the answers in lowest terms. (Pages 194–199)**

28. $8\frac{3}{10} + 4\frac{1}{10}$
29. $7\frac{5}{12} - 4\frac{1}{12}$
30. $10\frac{3}{16} - 4\frac{1}{16}$
31. $6\frac{9}{11} + 3\frac{3}{11}$

**Add or subtract. Write the answers in lowest terms. (Pages 200–209)**

32. $9\frac{2}{3} - 7\frac{1}{5}$
33. $6\frac{1}{7} + 2\frac{2}{3}$
34. $5\frac{3}{8} - 3\frac{2}{7}$
35. $12\frac{5}{6} + 7\frac{7}{8}$

# CHAPTER TEST

**Show that the two fractions in each pair are equivalent.**

1. $\frac{36}{54}$ and $\frac{2}{3}$
2. $\frac{9}{21}$ and $\frac{27}{63}$
3. $\frac{14}{63}$ and $\frac{2}{9}$
4. $\frac{18}{72}$ and $\frac{1}{4}$

**Express each fraction in lowest terms.**

5. $\frac{56}{80}$
6. $\frac{49}{140}$
7. $\frac{65}{75}$
8. $\frac{72}{150}$
9. $\frac{96}{108}$

**Find each product. Write the answers in lowest terms.**

10. $\frac{2}{9} \cdot \frac{4}{5}$
11. $\frac{4}{5} \cdot \frac{7}{12}$
12. $\frac{3}{10} \cdot 3$
13. $2\frac{2}{5} \cdot 3\frac{1}{3}$
14. $12 \cdot 2\frac{3}{8}$
15. $4\frac{2}{7} \cdot \frac{5}{6}$
16. $2\frac{4}{3} \cdot 2\frac{3}{4}$
17. $6 \cdot \frac{3}{20}$

**Solve and check.**

18. $\frac{1}{4}t = \frac{5}{7}$
19. $5 = \frac{4}{5}m$
20. $6x = \frac{4}{3}$
21. $\frac{12}{9} = \frac{1}{6}k$
22. $\frac{3}{4} = 1\frac{3}{8}y$
23. $4\frac{3}{5}g = 8$
24. $4\frac{3}{10} = 2\frac{3}{5}h$
25. $18j = 1\frac{2}{9}$

**Find each quotient.**

26. $\frac{5}{8} \div \frac{3}{2}$
27. $\frac{14}{81} \div 2\frac{12}{18}$
28. $6\frac{2}{3} \div 5$
29. $1\frac{1}{2} \div \frac{3}{4}$
30. $6 \div 2\frac{1}{4}$
31. $5\frac{2}{3} \div 1\frac{3}{4}$
32. $3\frac{1}{10} \div 100$
33. $\frac{1}{6} \div 3\frac{1}{2}$

**Add or subtract. Write the answers in lowest terms.**

34. $7\frac{2}{3} + 4\frac{2}{3}$
35. $6\frac{2}{5} - 4\frac{1}{5}$
36. $4\frac{3}{5} + 2\frac{3}{5}$
37. $10\frac{1}{2} - 4\frac{3}{8}$
38. $3\frac{5}{6} - 2\frac{1}{8}$
39. $4\frac{3}{10} + 7\frac{3}{4}$
40. $9\frac{1}{6} - 4\frac{5}{8}$
41. $8\frac{4}{7} + 6\frac{4}{5}$

**Solve.**

42. A tank holds $12\frac{3}{4}$ gallons of water. If the tank is $\frac{4}{5}$ full, how many gallons of water are in the tank?

43. A chef has $9\frac{3}{8}$ cups of chopped bell peppers. If each serving of the daily special requires $\frac{1}{4}$ cup of peppers, how many orders can he make?

**In a doubles round of disc golf, Lanna threw her disc $238\frac{1}{12}$ feet. Brandi threw her disc $142\frac{7}{12}$ feet.**

44. What was the total distance of Lanna's and Brandi's throws?
45. How much further was Lanna's throw than Brandi's throw?

# EXERCISES

**Objectives:** To subtract mixed numbers

## Check Your Understanding

**Skill Check**  Write the letter of the correct answer.

1. Which expression is equivalent to $24\frac{1}{3}$?
   a. $\frac{63}{3}$
   b. $23\frac{3}{4}$
   c. $23\frac{2}{3}$
   d. $23\frac{4}{3}$

2. Which expression is equivalent to $4\frac{12}{8}$?
   a. $4\frac{2}{3}$
   b. $5\frac{1}{2}$
   c. $5\frac{1}{8}$
   d. $5\frac{3}{8}$

3. Subtract: $3\frac{1}{8} - 1\frac{2}{3}$
   a. $\frac{11}{24}$
   b. $1\frac{8}{24}$
   c. $1\frac{11}{24}$
   d. $1\frac{17}{24}$

4. Subtract: $10 - 3\frac{1}{6}$
   a. $6\frac{5}{6}$
   b. $7\frac{1}{6}$
   c. 7
   d. $13\frac{1}{6}$

## Practice and Apply

**Subtract. Express your answer in lowest terms.**

5. $7\frac{1}{12} - 3\frac{2}{3}$
6. $8\frac{1}{4} - 2\frac{5}{8}$
7. $10\frac{1}{5} - 5\frac{2}{3}$
8. $8\frac{3}{10} - 2\frac{5}{6}$
9. $2 - 1\frac{1}{16}$
10. $15\frac{1}{3} - 10\frac{5}{8}$
11. $14 - 3\frac{7}{10}$
12. $5\frac{2}{3} - 3\frac{1}{2}$
13. $21\frac{1}{2} - 13\frac{9}{10}$
14. $8 - 6\frac{3}{5}$
15. $3\frac{1}{4} - 2\frac{1}{2}$
16. $5\frac{3}{8} - 4\frac{1}{2}$

17. Is $20\frac{1}{3} - 4\frac{2}{3}$ greater than or less than 15? How much more or less?

18. Is $50\frac{1}{4} - 39\frac{3}{4}$ greater than or less than 20? How much more or less?

**Solve.**

19. Tom worked $3\frac{3}{4}$ hours on Thursday and $4\frac{1}{2}$ hours on Friday. How much longer did he work on Friday?

20. A 21-pound turkey weighed $17\frac{5}{16}$ pounds after it was cooked. What was the weight loss?

21. A machinist cuts $7\frac{1}{16}$ inches off a pipe that is 18 inches long. How many inches are left?

22. There are $16\frac{1}{3}$ yards of material on a bolt. If $5\frac{1}{2}$ yards are needed, how much is left on the bolt?

**Try This**  Rename as indicated.

1. $9\frac{5}{6} = \frac{?}{6}$   2. $4\frac{3}{5} = 3\frac{?}{5}$   3. $7\frac{6}{8} = 6\frac{?}{8}$

Now use Example 1 to find a more exact weight for Apollo by subtracting $101\frac{1}{3}$ from $120\frac{1}{4}$.

Find the LCM of 3 and 4.   LCM: 12

$$120\frac{1}{4} = 120\frac{3}{12}$$
$$-101\frac{1}{3} = -101\frac{4}{12}$$

Since $\frac{4}{12} > \frac{3}{12}$, you cannot subtract. Rename 120 as in Example 1.

$$120\frac{3}{12} = 119\frac{15}{12}$$
$$-101\frac{4}{12} = -101\frac{4}{12}$$
$$\phantom{-101\frac{4}{12} = }\ 18\frac{11}{12}$$

Apollo weighs $18\frac{11}{12}$ pounds.

Since $18\frac{11}{12}$ is close to the estimate, the answer is reasonable.

**EXAMPLE 2**   Subtract:  **a.** $24\frac{1}{3} - 15\frac{3}{5}$   **b.** $26 - 6\frac{5}{9}$

**Solutions**   **a.**  $24\frac{1}{3} = 24\frac{5}{15} = 23\frac{20}{15}$
$\phantom{a.\ }-15\frac{3}{5} = -15\frac{9}{15} = -15\frac{9}{15}$
$\phantom{a.\ \ \ \ \ \ \ \ \ \ \ \ \ \ \ \ \ \ \ \ \ \ \ \ \ \ \ \ \ \ \ \ \ }8\frac{11}{15}$

**b.** Rename 26 as $25\frac{9}{9}$.

$26 = 25\frac{9}{9}$
$-6\frac{5}{9} = -6\frac{5}{9}$
$\phantom{-6\frac{5}{9} = }\ \mathbf{19\frac{4}{9}}$

**Try This**   Subtract.

4. $16\frac{1}{5} - 3\frac{1}{4}$   5. $8 - 2\frac{1}{10}$   6. $9\frac{1}{2} - 4\frac{2}{3}$

## LESSON 5-10

# Subtracting Mixed Numbers

Together, Gina and her dog, Apollo, weigh $120\frac{1}{4}$ pounds. Gina alone weighs $101\frac{1}{3}$ pounds. How much does Apollo weigh?

Estimate first.   $120\frac{1}{4}$ is closer to 120 than to 121.

$101\frac{1}{3}$ is closer to 101 than to 102.

$120 - 101 = 19$

**Estimate:** Apollo weighs about 19 pounds.

When subtracting mixed numbers, you may have to rename a whole number.

**EXAMPLE 1**   Rename: $120\frac{3}{12} = 119\frac{?}{12}$

**Solution**   Think: $120\frac{3}{12} = 119 + 1 + \frac{3}{12}$

$= 119 + \frac{12}{12} + \frac{3}{12}$

$= \mathbf{119\frac{15}{12}}$

**206**   CHAPTER 5

**34.** What kind of problem can you solve by running this program?

```
10 PRINT "TO FIND E/F + G/H, TYPE E,F,G,H"
20 INPUT E,F,G,H
30 LET N=(E*H)+(G*F): D=F*H
40 PRINT "UNREDUCED ANSWER IS ";N;"/";D
50 END
```

**35.** Run the program for several different sets of values of E, F, G, and H.

**36.** In Line 30, what does N represent?   **37.** In Line 40, what does D represent?

## Connect and Extend

The diagram below shows that $\frac{1}{3} + \frac{3}{4} = \frac{13}{12}$, or $1\frac{1}{12}$.

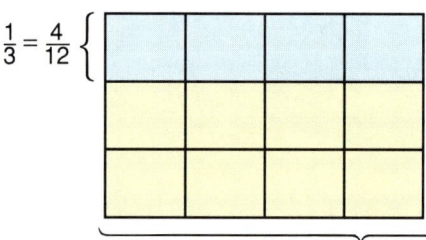

**Draw a figure to represent each sum.**

**38.** $\frac{1}{2} + \frac{5}{8}$     **39.** $\frac{7}{8} + \frac{3}{4}$     **40.** $\frac{2}{3} + \frac{3}{5}$     **41.** $1\frac{2}{3} + 1\frac{3}{4}$

**Write an expression for adding mixed numbers or fractions that has the given sum.**

**42.** $\frac{11}{24}$     **43.** $\frac{13}{24}$     **44.** $\frac{13}{15}$     **45.** $\frac{17}{20}$     **46.** $\frac{43}{72}$

## Maintain Your Skills

**47.** Which has a smaller value, $\frac{5}{8}xy$ or $\frac{2}{3}yz$ for $x = \frac{8}{9}$, $y = \frac{3}{2}$, and $z = \frac{1}{6}$? **(Pages 39–43)**

**Simplify. (Pages 150–153)**

**48.** $(6 - 13)^3$     **49.** $12^2 - 11^2$     **50.** $(-51 + 31)^4$     **51.** $(5 + 8)^2 - 12^2$

**52.** What is the improper fraction for $6\frac{5}{6}$? **(Pages 175–179)**

**Solve and check. (Pages 105–109)**

**53.** $2b + 9 = -17$     **54.** $-4x - 3 = 17$     **55.** $-31 = 2e + 7$
**56.** $-32 - 30t = -287$     **57.** $125 = 10c - 25$     **58.** $42 = 4x + 6$

5. Write three fractions that might be rounded to 1.
6. Write three fractions that might be rounded to 0.

## Practice and Apply

**Rewrite the fractions in each pair so that they have common denominators.**

7. $\frac{5}{12}$ and $\frac{1}{6}$
8. $\frac{1}{4}$ and $\frac{3}{5}$
9. $\frac{5}{8}$ and $\frac{2}{3}$
10. $\frac{7}{12}$ and $\frac{2}{5}$
11. $\frac{1}{2}, \frac{3}{10},$ and $\frac{4}{5}$
12. $\frac{2}{3}, \frac{5}{12},$ and $\frac{3}{8}$

**Estimate each sum or difference.**

13. $3\frac{4}{5} + 2\frac{5}{8} + 7\frac{1}{6}$
14. $5\frac{3}{12} - 2\frac{8}{9}$
15. $7\frac{13}{16} - 3\frac{4}{5}$

**Add or subtract. Write your answer in lowest terms.**

16. $3\frac{1}{3} + 4\frac{1}{6}$
17. $7\frac{3}{5} - 2\frac{3}{10}$
18. $8\frac{2}{3} - 5\frac{1}{6}$
19. $3\frac{2}{5} + 4\frac{1}{3}$
20. $4\frac{7}{8} - 1\frac{1}{2}$
21. $9\frac{3}{5} + 7\frac{1}{3}$
22. $12\frac{1}{2} + 11\frac{1}{8}$
23. $7\frac{7}{10} - \frac{2}{5}$
24. $8\frac{7}{8} - 6\frac{3}{4}$
25. $8\frac{2}{5} + 3\frac{1}{6} - 1\frac{3}{10}$
26. $8\frac{2}{3} - 2\frac{1}{4} - 3\frac{1}{6}$
27. $2\frac{1}{5} + 5\frac{3}{4} - 1\frac{5}{6}$
28. $9\frac{3}{4} - 6\frac{1}{5} - 1\frac{1}{2}$

Five classes held a contest to see which class could collect the most paper to recycle. The table at the right below shows the collections in pounds of paper for the three days.

29. Which class (or classes) collected the greatest amount of paper in three days?
30. Which class (or classes) collected the least amount of paper in 3 days?
31. List the total number of pounds of paper collected by each class in order from least to greatest.
32. How much paper did the 5 classes together collect on Wednesday?
33. On Wednesday, how much more paper was collected by Class 2 than by Class 5?

|  | Mon | Tues | Wed |
|---|---|---|---|
| Class 1 | $18\frac{1}{3}$ | $17\frac{5}{12}$ | $16\frac{1}{6}$ |
| Class 2 | $17\frac{1}{4}$ | $16\frac{1}{6}$ | $18\frac{1}{3}$ |
| Class 3 | $18\frac{1}{2}$ | $16\frac{1}{2}$ | $17\frac{1}{6}$ |
| Class 4 | $16\frac{1}{6}$ | $17\frac{2}{3}$ | $18\frac{1}{12}$ |
| Class 5 | $16\frac{1}{2}$ | $17\frac{1}{12}$ | $18\frac{1}{4}$ |

$$5\frac{3}{4} \qquad 5\frac{9}{12}$$
$$-5\frac{2}{3} = -5\frac{8}{12}$$
$$\phantom{-5\frac{2}{3}} = \phantom{-5}\frac{1}{12} \qquad \text{Compare } \frac{1}{12} \text{ and } \frac{1}{2}. \text{ Since } \frac{1}{2} = \frac{6}{12}, \frac{1}{12} < \frac{1}{2}.$$

The second conjecture is also **true**.

**Try This**

6. Which is greater, $7\frac{3}{4}$ or $7\frac{5}{6}$? Explain.

Once you rewrite fractions so that they have a common denominator, you can add, subtract, or compare them.

> **Summary**
> To add, subtract, or compare fractions:
> 1. Rewrite the fractions so that they have like denominators.
> 2. Apply the rule for fractions that have like denominators.

# EXERCISES

**Objective:** To add, subtract and compare unlike fractions

## Check Your Understanding

**Skill Check** Write the letter of the correct answer.

1. Which expression is equivalent to $\frac{5}{8} + \frac{2}{3}$?
   a. $\frac{2}{11} + \frac{2}{11}$
   b. $\frac{5}{24} + \frac{2}{24}$
   c. $\frac{15}{24} + \frac{16}{24}$
   d. $\frac{15}{24} + \frac{2}{24}$

2. $9\frac{7}{12} - 3\frac{3}{8} = \underline{\phantom{?}}$
   a. $6\frac{4}{16}$
   b. $6\frac{5}{24}$
   c. $6\frac{1}{4}$
   d. $12\frac{23}{24}$

3. $5\frac{3}{6} + 3\frac{1}{10} = \underline{\phantom{?}}$  Add. Express the answer in lowest terms.
   a. $8\frac{5}{60}$
   b. $8\frac{1}{12}$
   c. $8\frac{3}{5}$
   d. $2\frac{23}{30}$

4. Which expression lists the fractions in order from least to greatest?
   a. $\frac{3}{5}, \frac{11}{15}, \frac{2}{3}$
   b. $\frac{2}{3}, \frac{3}{5}, \frac{11}{15}$
   c. $\frac{2}{3}, \frac{11}{15}, \frac{3}{5}$
   d. $\frac{3}{5}, \frac{2}{3}, \frac{11}{15}$

5-9 Adding and Subtracting Fractions: Unlike Denominators

**b.** Find the LCM of 8 and 12.  $8 = 2^3$   $12 = 2^2 \cdot 3$
LCM: $2^3 \cdot 3 = 8 \cdot 3$, or 24

$$8\frac{3}{8} = 8\frac{9}{24}$$
$$+ 6\frac{5}{12} = + 6\frac{10}{24}$$
$$\overline{\phantom{+ 6\frac{5}{12} = }14\frac{19}{24}}$$

**Try This**   Add. Express answers in lowest terms.

**4.** $6\frac{1}{4} + 5\frac{7}{8}$   **5.** $3\frac{1}{6} + 2\frac{4}{9}$

Kareem and Pierre look at the map to plan their trip. They make two conjectures. A **conjecture** is a statement that is believed to be true but has not been proved.

**Conjecture 1:** The distance from Caram to Stokes is greater than the distance from Stokes to Lindburgh.

**Conjecture 2:** The difference of these distances is less than $\frac{1}{2}$ mile.

Here's how Kareem and Pierre checked whether their conjectures were true.

**EXAMPLE 3**   **a. Conjecture 1:** Is $5\frac{3}{4}$ greater than $5\frac{2}{3}$?

**b. Conjecture 2:** Is $5\frac{3}{4} - 5\frac{2}{3}$ less than $\frac{1}{2}$?

**Solutions**   **a.** Since the whole-number parts of $5\frac{3}{4}$ and $5\frac{2}{3}$ are the same, write equivalent fractions having a common denominator of 12 for $\frac{3}{4}$ and $\frac{2}{3}$.

$$\frac{3}{4} = \frac{9}{12} \qquad \frac{2}{3} = \frac{8}{12}$$

Since $9 > 8$, $\frac{9}{12} > \frac{8}{12}$, and $5\frac{9}{12} > 5\frac{8}{12}$.

Thus, $5\frac{3}{4} > 5\frac{2}{3}$, and the first conjecture is **true.**

**b.** Write equivalent fractions having a common denominator of 12. Subtract the fractions and subtract the whole numbers.

**Try This**  Find a whole-number estimate for each sum or difference.

1. $2\frac{7}{8} + 1\frac{4}{5} + 9\frac{1}{4}$
2. $7\frac{1}{8} - 3\frac{5}{6}$
3. $15\frac{7}{8} - 1\frac{11}{12}$

Kareem plans to meet Pierre at Coram and bicycle with him from Coram to Stokes to Lindburgh. How far will they travel together? That is,

$$5\frac{3}{4} + 5\frac{2}{3} = \underline{\ ?\ }$$

Notice that $\frac{3}{4}$ and $\frac{2}{3}$ have unlike denominators. To add or subtract fractions having unlike denominators, you write equivalent fractions having a least common denominator. The **least common denominator** (LCD) is the least common multiple of the denominators.

Find the LCM of 4 and 3.   $4 = 2^2 \quad 3 = 3 \cdot 1 \quad$ LCM: $2^2 \cdot 3 = 12$

Change $\frac{3}{4}$ to twelfths.   $\frac{3}{4} \cdot \frac{3}{3} = \frac{9}{12}$

Change $\frac{2}{3}$ to twelfths.   $\frac{2}{3} \cdot \frac{4}{4} = \frac{8}{12}$

So $5\frac{3}{4} + 5\frac{2}{3} = 5\frac{9}{12} + 5\frac{8}{12} = (5 + 5) + \left(\frac{9}{12} + \frac{8}{12}\right)$.

$= 10 + \frac{17}{12}$

$= 10 + \frac{12}{12} + \frac{5}{12} = 11\frac{5}{12}$

The distance from Stokes to Lindburgh is $11\frac{5}{12}$ miles.

**EXAMPLE 2**  Add. Express answers in lowest terms.

a. $4\frac{3}{8} + 2\frac{3}{4}$
b. $8\frac{3}{8} + 6\frac{5}{12}$

**Solutions**   a. Find the LCM of 8 and 4.   $8 = 2^3 \quad 4 = 2^2 \quad$ LCM: $2^3$, or 8

$4\frac{3}{8} = \phantom{+}4\frac{3}{8}$
$+ 2\frac{3}{4} = + 2\frac{6}{8}$
$\overline{\phantom{+ 2\frac{3}{4}}}$
$\phantom{+ 2\frac{3}{4}}6\frac{9}{8} = 6 + \frac{8}{8} + \frac{1}{8}$

$= 7\frac{1}{8}$

5-9 Adding and Subtracting Fractions: Unlike Denominators

# LESSON 5-9
## Adding and Subtracting Fractions: Unlike Denominators

Pierre plans to bicycle from Newton to Lincoln to Coram to Stokes. He uses mental math to estimate the total distance to the nearest mile.

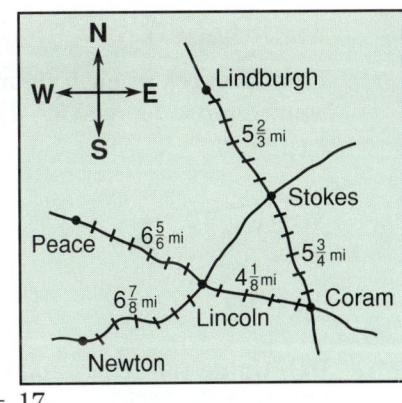

To estimate to the nearest mile, think:

$6\frac{7}{8}$ is closer to 7 than to 6.

$4\frac{1}{8}$ is closer to 4 than to 5.

$5\frac{3}{4}$ is closer to 6 than to 5.

**Estimate** (nearest mile): $7 + 4 + 6 = 17$

**EXAMPLE 1** Find a whole number estimate for each sum or difference.

  **a.** $2\frac{7}{8} + 1\frac{3}{4} + 1\frac{1}{8}$    **b.** $16\frac{7}{9} - 12\frac{3}{5}$    **c.** $15\frac{1}{8} - 12\frac{2}{3}$

**Solutions**

**a.** Think: $2\frac{7}{8}$ is about 3. $1\frac{3}{4}$ is about 2. $1\frac{1}{8}$ is about 1.
Estimate: $3 + 2 + 1 = $ **6**

**b.** Think: $16\frac{7}{9}$ is about 17. $12\frac{3}{5}$ is about 13.
Estimate: $17 - 13 = $ **4**

**c.** Think: $15\frac{1}{8}$ is about 15. $12\frac{2}{3}$ is about 13.
Estimate: $15 - 13 = $ **2**

## Connect and Extend

You use whole numbers to represent dollars and you can represent parts of a dollar by fractions. For example, 1 cent can be represented by $\frac{1}{100}$, a dime can be represented by $\frac{10}{100}$, or $\frac{1}{10}$ and so on.

**Write fractions or mixed numbers in lowest terms to represent the amounts.**

**31.** 25¢    **32.** 5¢    **33.** 75¢    **34.** $1.25    **35.** $1.75

**Rewrite each statement as the sum or difference of mixed numbers. After finding the sum or difference, use the $ symbol to write it as dollars and cents.**

**36.** Twelve dollars and 80 cents minus 4 dollars and 35 cents.
**37.** Eight dollars and 22 cents less than 10 dollars and 58 cents.
**38.** Nine dollars and 41 cents subtracted from 20 dollars and 95 cents.
**39.** Write a mixed number that shows how many yards equal 7 feet. (1 yard = 3 feet).
**40.** Write a mixed number that shows how many pints equal 5 cups. (1 pint = 2 cups).

## Maintain Your Skills

**Find the area of each figure *only* if the area is greater than 75 square units. (Pages 69–74)**

**41.** 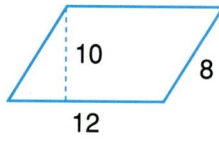    **42.** 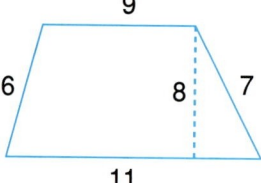    **43.**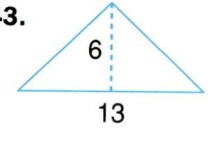

**Determine the remaining prime factors of each number. (Pages 143–146)**

**44.** Number: 70
Prime factor: 2

**45.** Number: 500
Prime factors: 2, 5

**46.** Number: 693
Prime factors: 3

**Find the LCM. (Pages 154–158)**

**47.** 24 and 5    **48.** 20 and 75    **49.** 12, 18, and 42

**50.** Maria traveled by plane from Oregon to Kentucky. Only $\frac{5}{6}$ of the 234 seats were occupied. How many people, excluding airline personnel, were on the plane? **(Pages 171–174)**

5-8  Fractions with Like Denominators  **199**

3. Which statement is **not** true?
   a. $\frac{5}{7} > \frac{3}{7}$    b. $2\frac{7}{10} < 1\frac{9}{10}$
   c. $3\frac{1}{6} < 3\frac{5}{6}$    d. $4\frac{7}{8} > 4\frac{5}{8}$

4. For which statement does $n = 3\frac{3}{5}$?
   a. $1\frac{1}{3} + 2 = n$    b. $7\frac{5}{6} - 5\frac{1}{6} = n$
   c. $n = 1\frac{4}{15} + 2\frac{5}{15}$    d. $n = 9\frac{9}{12} - 6\frac{6}{12}$

5. Which is greater?
   $9\frac{7}{8} - 2\frac{1}{8}$ or $2\frac{3}{8} + 4\frac{5}{8}$?

6. Which is less?
   $7\frac{5}{12} + 5\frac{6}{12}$ or $15\frac{11}{12} - 3\frac{1}{12}$?

## Practice and Apply

**Add or subtract. Write answers in lowest terms.**

7. $3\frac{1}{3} + 4\frac{1}{3}$    8. $6\frac{1}{7} + 4\frac{3}{7}$    9. $5\frac{3}{8} - 2\frac{1}{8}$    10. $4\frac{5}{9} - 3\frac{1}{9}$

11. $8\frac{5}{12} - 2\frac{1}{12}$    12. $6\frac{4}{15} + 8\frac{1}{15}$    13. $9\frac{3}{5} - 4\frac{2}{5}$    14. $3\frac{1}{6} + 7\frac{3}{6}$

15. $8\frac{3}{10} - 4\frac{1}{10}$    16. $9\frac{3}{8} + 3\frac{4}{8}$    17. $2\frac{7}{10} + 11\frac{3}{10}$    18. $9\frac{7}{8} - 5\frac{3}{8}$

19. $17\frac{3}{8} + 18\frac{2}{8} + 15\frac{1}{8}$    20. $8\frac{3}{10} + 13\frac{1}{10} - 6\frac{2}{10}$

The table below shows the results of three players' practice throws of a baseball. Distances are given in feet.

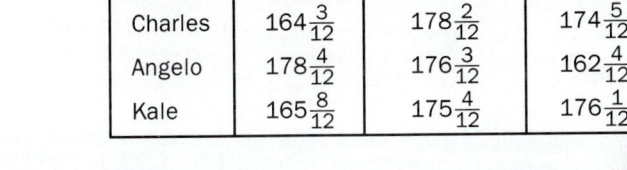

| Players | Throws in Feet | | |
|---|---|---|---|
| | First Try | Second Try | Third Try |
| Charles | $164\frac{3}{12}$ | $178\frac{2}{12}$ | $174\frac{5}{12}$ |
| Angelo | $178\frac{4}{12}$ | $176\frac{3}{12}$ | $162\frac{4}{12}$ |
| Kale | $165\frac{8}{12}$ | $175\frac{4}{12}$ | $176\frac{1}{12}$ |

21. Which player had the greatest combined distance for the three throws?

22. Which player had the smallest combined distance for the three throws?

23. On the first try, how much farther did Kale throw than Charles?

24. On the third try, how much farther did Charles throw than Angelo?

25. How much farther was Angelo's second throw than Kale's second throw?

**Find two mixed numbers that have the given sum.**

26. $3\frac{2}{3}$    27. $4\frac{11}{12}$    28. $7\frac{5}{8}$    29. $5\frac{4}{9}$    30. $6\frac{5}{8}$

198   CHAPTER 5

To subtract fractions having like denominators, subtract the numerators. Write the difference over the like denominator. Express the answer in lowest terms.

> **Subtracting Fractions with Like Denominators**
> If $\frac{a}{b}$ and $\frac{c}{b}$ are fractions, then $\frac{a}{b} - \frac{c}{b} = \frac{a-c}{b}$.

**EXAMPLE 3** Subtract $6\frac{7}{10}$ from $7\frac{8}{10}$ to find the answer to Question 3.

**Solution**

| Write the problem. | $7\frac{8}{10}$ |
| Subtract the fractions. | $-6\frac{7}{10}$ |
| Subtract the whole numbers. | $1\frac{1}{10}$ |

The longest two-day walk (Jenna's) is $1\frac{1}{10}$ **miles** farther than the shortest two-day walk (Mike's).

**Try This** Subtract. Express answers in lowest terms.

5. $15\frac{9}{10} - 12\frac{3}{10}$

6. $21\frac{4}{5} - 14\frac{3}{5}$

# EXERCISES

**Objective:** To add, subtract, and compare fractions having like denominators

## Check Your Understanding

**Skill Check** Write the letter of the correct answer.

1. Which statement is **not** true?
   a. $2\frac{1}{5} + 6\frac{3}{5} = 8\frac{4}{5}$
   b. $1\frac{1}{8} + 9\frac{5}{8} = 10\frac{3}{4}$
   c. $6\frac{1}{4} + 7\frac{3}{4} = 15$
   d. $3 + 5\frac{11}{12} = 8\frac{11}{12}$

2. Which statement is **not** true?
   a. $12\frac{5}{10} - 3\frac{5}{10} = 9$
   b. $9\frac{5}{8} - 3\frac{3}{8} = 6\frac{1}{4}$
   c. $1\frac{4}{5} - 1\frac{3}{5} = 1\frac{1}{5}$
   d. $7\frac{3}{4} - 5 = 2\frac{3}{4}$

Since Sherri and Jenna each walked a fraction more than 7 miles, compare the fractional part of a mile each person walked. Use a number line.

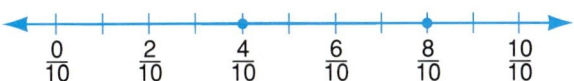

Since $\frac{8}{10}$ is to the right of $\frac{4}{10}$, $\frac{8}{10}$ is greater than $\frac{4}{10}$. So $7\frac{8}{10}$ is greater than $7\frac{4}{10}$ and Jenna walked farther than Sherri.

Therefore, Jenna walked the farthest of the three persons.

Comparing fractions on the number line suggests that to compare fractions with like denominators, you compare the numerators.

**EXAMPLE 2** Replace each ? with < (is less than) or > (is greater than) to make a true sentence.

a. $\frac{5}{16}$ ? $\frac{3}{16}$      b. $\frac{7}{12}$ ? $\frac{9}{12}$

**Solutions** Think: Each pair of fractions has like denominators. Compare the numerators.

a. $\frac{5}{16}$ ? $\frac{3}{16}$      b. $\frac{7}{12}$ ? $\frac{9}{12}$

Since 5 > 3,      Since 7 < 9,

$\frac{5}{16} > \frac{3}{16}.$      $\frac{7}{12} < \frac{9}{12}.$

**Try This** Replace each ? with < or > to make a true sentence.

3. $\frac{8}{12}$ ? $\frac{11}{12}$      4. $\frac{21}{25}$ ? $\frac{19}{25}$

**Question 3:** How much farther was the longest two-day walk than the shortest two-day walk?

**Think:** To find "how much farther," subtract the smallest total distance from the greatest total distance walked (see the answer to Question 2).

$$7\frac{8}{10} - 6\frac{7}{10} = \underline{\ ?\ }$$

To subtract mixed numbers, first subtract the fractions. Then subtract the whole numbers.

You can write a generalization for adding fractions with like denominators.

> **Adding Fractions with Like Denominators**
> If $\frac{a}{b}$ and $\frac{c}{b}$ are fractions, then $\frac{a}{b} + \frac{c}{b} = \frac{a+c}{b}$.

You can use mental math to find Mike's and Jenna's total distances.

**EXAMPLE 1**  **a.** How far did Mike walk during the two days?
**b.** How far did Jenna walk during the two days? Express this distance in lowest terms.

**Solutions**

**a.** Mike's total distance:
$$3\tfrac{5}{10} + 3\tfrac{2}{10} = (3 + 3) + \left(\tfrac{5}{10} + \tfrac{2}{10}\right)$$
$$= 6\tfrac{7}{10}$$

Mike's total distance was **$6\tfrac{7}{10}$ miles.**

**b.** Jenna's total distance:
$$4\tfrac{6}{10} + 3\tfrac{2}{10} = (4 + 3) + \left(\tfrac{6}{10} + \tfrac{2}{10}\right)$$
$$= 7 + \tfrac{8}{10}$$
$$= 7\tfrac{8}{10}, \text{ or } 7\tfrac{4}{5}$$

Jenna's total distance was **$7\tfrac{4}{5}$ miles.**

**Try This**  Add. Express answers in lowest terms.

1. $3\tfrac{5}{8} + 6\tfrac{1}{8}$
2. $4\tfrac{3}{15} + 6\tfrac{7}{15}$

**Question 2:** Which person walked the farthest for the two days combined?

To compare the total distances, express them in tenths so the denominators will be the same.

Since Sherri and Jenna each walked more than 7 miles, both walked farther than Mike.

5-8   Fractions with Like Denominators

# CUMULATIVE REVIEW CHAPTERS 1–5

**Determine whether the numbers in each group are written in ascending order, descending order, or neither. Explain your answer. (Pages 2–10)**

1. $-(-1),\ -|-2|,\ -3,\ |-4|$
2. $-1\frac{1}{2},\ -\frac{1}{2},\ 0,\ \frac{1}{2}$
3. $|-6|,\ 2\frac{1}{2},\ 1,\ -\frac{1}{2}$

**Replace each ? with the correct number. (Pages 13–18, 23–25)**

4. $22 - \underline{\ ?\ } = -36$
5. $\underline{\ ?\ } + 12 = -62$
6. $-98 \div \underline{\ ?\ } = 7$
7. $\underline{\ ?\ } \div -4 = 18$

**Use positive and negative numbers to represent each situation by a multiplication problem. Then find the product. (Pages 28–31)**

8. Three penalties of 5 yards each
9. Two debts of $78 each

**Simplify each expression. (Pages 46–50, 59–63)**

10. $-3 \cdot 4 \cdot 2 \div 6$
11. $9xy + 4 + xy$
12. $56 + 4 \div (6 - 10)$

13. What is the solution set of $5y - 1 > -1$ if the replacement set is $\{-4, -1, 0, 1, 4\}$. **(Pages 80–84)**

14. How many different factors does 84 have? **(Pages 126–130)**

15. Four consecutive multiples of a number are 28, 42, 56, and 70. What is the number? **(Pages 126–130)**

16. How many *prime* factors does 84 have? **(Pages 143–146)**

17. Copy and complete the pattern of squares. The first one is done for you. **(Pages 148–149)**
    a. $22^3 = 10{,}648$
    b. $3^4$
    c. $4^5$
    d. $12^2$
    e. $17^3$

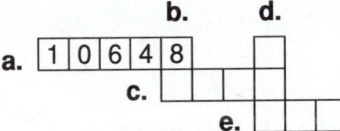

18. Write four fractions that are equivalent to $\frac{24}{36}$. **(Pages 163–167)**

**Find each answer. (Pages 171–179, 189–209)**

19. $\frac{9}{10} \cdot \frac{8}{15}$
20. $5\frac{7}{8} + 3\frac{3}{8}$
21. $9\frac{11}{12} - 4\frac{3}{12}$
22. $2\frac{1}{4} \div 3\frac{3}{8}$
23. $12\frac{3}{5} + 8\frac{2}{3}$
24. $3\frac{2}{3} \cdot 3\frac{2}{3}$
25. $22\frac{1}{7} - 14\frac{3}{8}$
26. $2\frac{2}{30} \div \frac{18}{21}$
27. $1\frac{3}{10} \div \frac{1}{5}$

28. Juan is putting molding around the ceiling of a room that is $15\frac{3}{4}$ feet long and $9\frac{2}{5}$ feet wide. How many feet of molding should he buy?

Cumulative Review: Chapters 1–5

# UNIT 3

## PROJECT 1

You have heard the saying "Time is running out." So is water. When was the last time you saw a dripping faucet? Did you turn it off?

There are many things that you can do every day to save the earth's resources. For this project, you will explore where you have seen water wasted, observe and calculate current water consumption in and around your home, and investigate ways to save both water and dollars.

## PROJECT 2

What is the best investment you can make now with the $1,500 you have saved from doing chores and jobs, from your allowance, and from birthday and holiday gifts? In this project, you will explore what kinds of investments are available, and which are the best bets for your money. After all, you and your family are probably making plans for your education or job training after high school.

# Using Ratios

## PROJECT 3

The lyrics in some old songs tell you about things that you can always count on: "... sun in the morning and the moon at night," or "three strikes and you're out at the old ball game." There's also a familiar headline: "Prices are on the increase!" Working as a team, you will research the costs of some of your favorite things,— what they cost ten years ago and what they cost now.

## CONTENTS

### 6 Probability
Ratio and Proportion
Problem Solving: Using Proportions
Ratio and Measurement
Problem Solving Strategies:
   Collecting Data
Problem Solving Exploration:
   Recording Chances
Probability
Independent and Dependent
   Events
Making Choices
Permutations

### 7 Decimals
Rational Expressions
Rational Numbers
Decimals and Fractions
Problem Solving Strategies:
   Drawing a Diagram
Repeating Decimals
Estimating Sums and Differences
Problem Solving Exploration:
   Powers of Ten
Scientific Notation
Problem Solving Exploration:
   The Metric System
Estimating Products and Quotients
   of Decimals

### 8 Percent
The Meaning of Percent
Decimals and Percents
Estimating the Percent of a
   Number
Problem Solving Strategies:
   Deciding on Estimates
Finding the Percent of a Number
Interest
Discount
Percent Equations and Proportions
Percent Increase and Decrease

## LESSON 6-1
# Ratio and Proportion

**S**hown above are the results of a survey of three high schools to determine which of two TV shows students preferred, the Jack Silver Comedy Show or Quiz for $. At Lochmore High, 3 out of 5 students voted for the Jack Silver Comedy Show.

You can use **ratios** to compare the data from the schools. For example, you can write a ratio for "3 out of 5 students" in these ways.

$$3 \text{ to } 5 \qquad 3:5 \qquad \frac{3}{5}$$

The ratio $\frac{3}{5}$ is in lowest terms. The **first term** of the ratio is 3; the **second term** is 5. Just as with fractions, there is an unlimited (infinite) number of ratios that are equivalent to $\frac{3}{5}$.

$$\frac{3}{5} = \frac{6}{10} = \frac{9}{15} = \frac{12}{20} = \frac{18}{30} = \ldots$$

When you use a ratio in lowest terms to compare two numbers, think carefully about the meaning.

Look at this table. It shows the results of the survey for Lochmore High and Jipsom High.

| School | Prefer Comedy Show | Ratio in Lowest Terms | Prefer Quiz Show | Ratio in Lowest Terms |
|---|---|---|---|---|
| Lochmore | 3 out of 5 | $\frac{3}{5}$ | 2 out of 5 | $\frac{2}{5}$ |
| Jipsom | 12 out of 20 | $\frac{12}{20}$, or $\frac{3}{5}$ | 8 out of 20 | $\frac{8}{20}$, or $\frac{2}{5}$ |

The ratios are the same for both high schools. This does not mean, however, that both high schools have the same number of students, or that the same number of students in both schools prefer the comedy show, or that the same number prefer the quiz show.

For some ratios, the terms do represent actual data. For example, if you make 3 out of 5 shots in a basketball game, you have taken 5 shots. If you make 6 out of 10 shots, you have taken 10 shots. In both cases, the ratio, shots made to shots taken, is the same.

$$\frac{\text{Shots made}}{\text{Shots taken}} = \frac{3}{5} = \frac{6}{10}$$

A statement that two ratios are equal is called a **proportion**. Since a proportion is made up of two ratios, it has four terms. Two of these are called **means** (the middle terms), and two are called **extremes** (the first and last terms).

$$\text{Extreme} \rightarrow \frac{3}{5} = \frac{6}{10} \leftarrow \text{Mean}$$
$$\text{Mean} \rightarrow \phantom{\frac{3}{5}} \phantom{=} \phantom{\frac{6}{10}} \leftarrow \text{Extreme}$$

**In a proportion, the product of the means equals the product of the extremes.** These products are often called **cross-products**.

$$\frac{3}{5} \times \frac{6}{10}$$
$$3 \cdot 10 = 5 \cdot 6$$
$$30 = 30$$

> **Property of Proportions**
> If $\frac{a}{b} = \frac{c}{d}$, $b \neq 0$, $d \neq 0$, then $ad = bc$.

When you know any three terms of a proportion, you can find the remaining term by using the Property of Proportions.

6-1 Ratio and Proportion

**EXAMPLE 1** Solve and check each proportion.

a. $\frac{3}{8} = \frac{n}{24}$

b. $\frac{48}{p} = \frac{12}{2}$

**Solutions** Write the problem.

a. $\frac{3}{8} \times \frac{n}{24}$

b. $\frac{48}{p} \times \frac{12}{2}$

Cross multiply.
Simplify.

$3 \cdot 24 = 8 \cdot n$
$72 = 8n$
$\frac{72}{8} = \frac{8n}{8}$
$9 = n$

$12 \cdot p = 48 \cdot 2$
$12p = 96$
$\frac{12p}{12} = \frac{96}{12}$
$p = 8$

**Checks**

a. Does $\frac{3}{8} = \frac{9}{24}$?
$\frac{3}{8} = \frac{3}{8}$ ✓

b. Does $\frac{48}{8} = \frac{12}{2}$?
$6 = 6$ ✓

**Try This** Solve and check each proportion.

1. $\frac{p}{6} = \frac{12}{9}$

2. $\frac{8}{r} = \frac{2}{10}$

3. $\frac{12}{36} = \frac{4}{n}$

Some problems involving proportions can be solved by using mental math.

**EXAMPLE 2** Suppose that there are 360 students at Cabot Central. If 18 out of 27 students preferred Quiz for $, how many students was this?

**Solution** Write a ratio in lowest terms for 18 out of 27.

$\frac{18}{27} = \frac{2}{3}$

**Think:** 2 out of 3 students preferred Quiz for $, and s out of 360 students preferred Quiz for $.

Write a proportion.

$\frac{2}{3} = \frac{s}{360}$

**Think:** Since $3 \cdot 120 = 360$, multiply 2 by 120.

$\frac{2 \cdot 120}{3 \cdot 120} = \frac{240}{360} \leftarrow s$

There were **240** students who preferred Quiz for $.

**Try This**

4. Suppose that there were 1800 students at Jipsom High. How many students preferred the Jack Silver Comedy Show?

• Explain how to use mental math to solve Try This Exercise 1 above.

# EXERCISES

**Objectives:** To make comparisons by using ratios
To solve proportions

## Check Your Understanding

**Skill Check** Write the letter of the correct answer.

1. If the ratio of nutmeg to cinnamon in a recipe is 2 : 3, how many teaspoons of cinnamon are used for every 2 teaspoons of nutmeg?
   **a.** 6   **b.** 3   **c.** 4   **d.** 5

2. The Tiger baseball team finished the season with 18 wins and 2 losses. What is the ratio of losses to total games?
   **a.** 9 to 1   **b.** 18 to 20
   **c.** 2 to 20   **d.** 2 to 18

3. Which pair of ratios can be written as a proportion?
   **a.** $\frac{3}{8}; \frac{12}{16}$   **b.** $\frac{3}{2}; \frac{18}{12}$
   **c.** $\frac{27}{30}; \frac{3}{10}$   **d.** $\frac{14}{10}; \frac{7}{4}$

4. Which is the first step in solving $\frac{6}{5} = \frac{30}{p}$ for $p$?
   **a.** $6 \cdot 30 = 5p$   **b.** $6 \cdot 5 = 30 \cdot p$
   **c.** $6 \cdot p = 5 \cdot 30$   **d.** $\frac{5}{6} = \frac{p}{30}$

**True or False?** When a statement is false, give a reason.

5. The ratio of 6 months to 18 months can be written as 1 : 3 or as 3 : 1.

6. If two football teams each have a record of winning 2 out of every 3 games played, then each team played the same number of games.

## Practice and Apply

**Write each ratio as a fraction in lowest terms.**

7. 3 out of 8   8. 21 : 12   9. 24 to 15   10. 18 : 21

**Determine whether each pair of ratios can be written as a proportion. Write Yes or No.**

11. $\frac{2}{4}$ and $\frac{10}{30}$   12. $\frac{5}{6}$ and $\frac{15}{2}$   13. $\frac{2}{5}$ and $\frac{14}{35}$

**Solve and check each proportion.**

14. $\frac{2}{3} = \frac{n}{36}$   15. $\frac{t}{72} = \frac{5}{6}$   16. $\frac{q}{20} = \frac{27}{45}$

17. $\frac{84}{60} = \frac{14}{r}$   18. $\frac{18}{n} = \frac{108}{48}$   19. $\frac{24}{a} = \frac{54}{18}$

20. $\frac{12}{16} = \frac{p}{24}$   21. $\frac{3}{a} = \frac{81}{135}$   22. $\frac{b}{13} = \frac{10}{65}$

6-1 Ratio and Proportion **219**

**The ratio of red carnations to yellow carnations in a florist's bouquet is 2 : 5.**

23. If the bouquet has 20 yellow carnations, how many red carnations does it have?

24. How many yellow carnations are there in the bouquet if it has 6 red carnations?

**Write a proportion for each situation. Solve the proportion and answer the question.**

25. The ratio of juniors to seniors on a football team is 3 : 2. If there are 24 juniors on the team, how many seniors are there?

26. At Grand High School, 3 out of every 10 students has blue eyes. One student says that, of the 480 students at the school, 124 have blue eyes. Is the student correct?

27. The ratio of canoes to sailboats on Lake Winnepesaukee is 5 to 2. If there are 22 sailboats, how many canoes are there?

28. Explain the difference between a ratio and a proportion.

## Connect and Extend

29. If $\frac{a}{b} = \frac{c}{d}$, does $\frac{b}{a} = \frac{d}{c}$? Explain.

30. If $\frac{a}{b} = \frac{c}{d}$, does $\frac{d}{a} = \frac{c}{b}$? Explain.

31. If $\frac{2}{3} = \frac{8}{12}$, does $\frac{2+3}{3} = \frac{8+12}{12}$?

32. Write a generalization for the property of proportions shown in Exercise 31.

**In a nut mixture, 2 out of 12 nuts are cashews, the ratio of almonds to cashews is 1 : 1, the ratio of filberts to almonds is 1 : 2, and the ratio of almonds to peanuts is 1 : 2.**

33. The mixture also includes Brazil nuts. How many nuts out of each 12 are Brazil nuts?

34. About how many peanuts are there in a mixture of 100 nuts?

## Maintain Your Skills

**Solve and check. (Pages 95–99)**

35. $10r = -90$
36. $-4a = 144$
37. $-12r = -132$
38. $-x = -5$

**Determine whether each fraction is in lowest terms. Write Yes or No. When the answer is No, write the fraction in lowest terms. (Pages 163–167)**

39. $\frac{6}{9}$
40. $\frac{12}{48}$
41. $\frac{9}{25}$
42. $\frac{24}{31}$
43. $\frac{16}{28}$
44. $\frac{21}{52}$

# LESSON 6-2 Problem Solving Using Proportions

**P**roportions can be used to solve many different types of problems. Since a proportion involves equivalent ratios, it is helpful to write the known ratio in words and in numbers before setting up the proportion.

**EXAMPLE 1** A manufacturer produces 24 wooden baseball bats for every 30 aluminum bats. If the factory produces 60 wooden bats each day, how many aluminum bats does it produce per day?

**Solution** Let $a$ = the number of aluminum bats produced per day.

Write the given ratio. $\qquad \dfrac{\text{Wooden bats}}{\text{Aluminum bats}} = \dfrac{24}{30} = \dfrac{4}{5}$

Use $\dfrac{4}{5}$ to write a proportion. Then find the cross-products. $\qquad \dfrac{4}{5} = \dfrac{60}{a} \;\begin{matrix}\leftarrow \text{Wooden bats}\\ \leftarrow \text{Aluminum bats}\end{matrix}$

Simplify. $\qquad 4 \cdot a = 5 \cdot 60$

$$\dfrac{\overset{1}{\cancel{4}a}}{\underset{1}{\cancel{4}}} = \dfrac{\overset{75}{\cancel{300}}}{\underset{1}{\cancel{4}}}$$

$$a = 75$$

The factory produces **75** aluminum bats per day.

**Try This**

1. Six out of every ten juniors voted for Louise for class president. There are 500 juniors. How many voted for Louise?

To use a calculator to solve a proportion, first rewrite the proportion with the variable alone on the left side.

$$\dfrac{9}{n} = \dfrac{57}{152}$$

$$57 \cdot n = 9 \cdot 152$$

$$n = \dfrac{9 \cdot 152}{57}$$

$9 \;\boxed{\times}\; 152 \;\boxed{\div}\; 57 \;\boxed{=}\; 24 \quad \leftarrow n = 24$

6-2 Problem Solving: Using Proportions **221**

Proportions are also useful in solving some problems involving quality control.

**EXAMPLE 2** A technician found that 2 out of every 75 light bulbs tested were defective. About how many defective bulbs can you expect in a shipment of 120,000?

**Solution**  Let $b$ = the number of defective bulbs expected.

Write the given ratio. $\quad \dfrac{2}{75} \leftarrow$ Defective bulbs
$\phantom{\text{Write the given ratio.}\quad \dfrac{2}{75}} \leftarrow$ Total number

Write a proportion. $\quad \dfrac{b}{120000} = \dfrac{2}{75}$

Solve the proportion. $\quad 75 \cdot b = 120{,}000 \cdot 2$

$$b = \dfrac{120{,}000 \cdot 2}{75}$$

$120{,}000 \boxed{\times} 2 \boxed{\div} 75 \boxed{=} 3200$

The expected number of defective bulbs is **3200**.

**Try This**
2. A sample of 800 toasters contains 39 that are defective. About how many defective toasters can you expect in a shipment of 36,000?

A **rate** is a comparison of numbers that have different units. Here are some examples of rates.

89¢ for 6 oz
60 miles per hour
28 points in 5 games

Since rates compare numbers, you can express them as ratios.

$\dfrac{89¢}{6 \text{ ounces}} \qquad \dfrac{60 \text{ miles}}{1 \text{ hour}}$

$\dfrac{28 \text{ points}}{5 \text{ games}}$

The rate, $\dfrac{60 \text{ miles}}{1 \text{ hour}}$, is a **unit rate** because it has 1 unit as its second term.

- What is the difference between a rate and a ratio?

- What is the meaning of the word "per" in the expression, 60 miles per hour?

**EXAMPLE 3** A jet airplane travels 1256 miles in 4 hours. At this rate, how far will it travel in 6 hours?

**Solution** Let $d$ = the distance traveled in 6 hours.

Write a proportion. $\quad \dfrac{1256 \text{ miles}}{4 \text{ hours}} = \dfrac{d \text{ miles}}{6 \text{ hours}}$

Solve the proportion. $\quad \dfrac{1256}{4} \times \dfrac{d}{6}$

$$4d = 1256 \cdot 6$$
$$d = \dfrac{1256 \cdot 6}{4}$$
$$d = \dfrac{7536}{4}$$
$$d = 1884$$

The plane will travel **1884 miles** in 6 hours.

**Check** Is the answer reasonable?
**Think:** 1256 ÷ 4 is about 300 miles per hour, and 6 · 300 = 1800 miles. So the answer is reasonable.

**Try This** 3. An automobile travels 300 miles in 6 hours. At this rate, how far will it travel in 15 hours?

# EXERCISES

**Objective:** To use proportions to solve problems

## Check Your Understanding

*Skill Check* Write the letter of the correct answer.

**1.** Which proportion can be used to solve this problem?

A painter is mixing blue and red paint in the ratio 3 : 4. How many gallons, $p$, of red paint are needed to mix with 9 gallons of blue paint?

a. $\dfrac{9}{p} = \dfrac{3}{4}$     b. $\dfrac{9}{p} = \dfrac{4}{3}$

c. $\dfrac{9}{4} = \dfrac{3}{p}$     d. $\dfrac{3}{4} = 9p$

**2.** Which proportion can be used to solve this problem?

Carlos typed 364 words in 7 minutes. At this rate, how many minutes, $m$, will it take him to type a book report of 1092 words?

a. $\dfrac{7}{364} = \dfrac{1}{m}$     b. $\dfrac{7}{364} = \dfrac{m}{1092}$

c. $\dfrac{364}{7} = \dfrac{m}{1092}$     d. $\dfrac{364}{m} = \dfrac{7}{1092}$

6-2 Problem Solving: Using Proportions

**3.** David ran 6 miles in 42 minutes. At this rate, how many hours will it take him to run 10 miles?
  **a.** $1\frac{3}{7}$   **b.** $1\frac{1}{6}$
  **c.** 70   **d.** $1\frac{1}{10}$

**4.** Which ratio shows a unit rate?
  **a.** $\frac{154 \text{ mi}}{22 \text{ gal}}$   **b.** $\frac{\$45.50}{7 \text{ hours}}$
  **c.** $\frac{28 \text{ in}}{4 \text{ in}}$   **d.** $\frac{2\frac{1}{2}¢}{1 \text{ ounce}}$

## Practice and Apply

**Suppose that Hiero drives 135 miles in 3 hours.**

**5.** Write a proportion you can use to find how far Hiero can travel in 5 hours if he drives at the same rate.

**6.** Solve the proportion you wrote for Exercise 5. How far can Hiero drive in 5 hours?

**7.** What is Hiero's rate expressed in miles per hour?

**8.** How many hours will it take Hiero to travel 90 miles if he drives at the same rate?

**Express each of the following as a unit rate.**

**9.** Kara is paid $35 for 5 hours of work.

**10.** Vernon saved $360 in 8 months.

**Write a proportion to represent the situation. Solve the proportion and answer the question.**

**11.** Devon can type 40 words in one minute. At this rate, how long will it take him to type 900 words?

**12.** If your pulse rate is 18 beats in 15 seconds, how many beats is this per minute?

**13.** A photograph that is 4 inches wide and 7 inches long is enlarged to have a width of 8 inches. What is the length of the enlarged photograph?

**14.** A magazine ad claims that 3 out of 5 dentists in a survey use Extra Gleem toothpaste. If there are 15,000 dentists in the survey, how many do **not** use Extra Gleem?

**15.** On a recent vacation, Caitlin found that her car traveled an average of 35 miles per gallon of gasoline. How many gallons can she expect to use on a trip of 2,625 miles?

**16.** A farmer takes a random sample of 75 ears of corn from a cornfield and finds that five of them are diseased. If the field contains 48,000 ears of corn, how many can the farmer expect to be diseased?

## Connect and Extend

**Solve.**

**17.** John receives $450 for working 18 hours and Clayton receives $1690 for working 30 hours. Whose pay rate is higher?

**18.** Melinda buys a 24-ounce box of cereal for $2.64 at Quick Stop. Janet buys a 12-ounce box of the same cereal for $1.56 at True Price. Who pays less per ounce, Melinda or Janet?

**19.** A caution light flashes 3 times every 5 seconds. At this rate, how many times will it flash per hour?

Ms. Liola stated in her will that her son is to receive $3 for each $5 in the estate. Her niece will receive the rest.

**20.** How much of an estate worth $83,200 will her niece receive?

**21.** If the niece receives $51,000, how much was the estate worth?

**Find the ratio of $x$ to $y$. (Hint: Use inverse operations.)**

**22.** $2x = 3y$

**23.** $5x = 6y$

**24.** $7y = 10x$

## Maintain Your Skills

**Evaluate each expression. Write the answer in lowest terms. (Pages 39–43, 175–179)**

**25.** $\frac{3}{4}c$, for $c = 4\frac{1}{2}$

**26.** $\frac{9}{4}g$, for $g = 6$

**27.** $4\frac{5}{6}k$, for $k = 2\frac{2}{3}$

**Solve and check. (Pages 183–188)**

**28.** $\frac{1}{4}g = \frac{5}{7}$

**29.** $\frac{2}{3} = 3b$

**30.** $8 = 4\frac{3}{5}a$

**31.** $1\frac{2}{3}x = 1\frac{1}{5}$

**Solve. (Pages 216–220)**

**32.** The ratio of hamsters to parakeets in a store is 4:3. If there are 28 hamsters, how many parakeets are there?

**33.** The ratio of women to men in an office is 6:5. If there are 30 women in the office, how many men are in the office?

## ★ Math Team Problems

**34.** Trace the figure at the right. Then draw lines that divide it into four regions that have the same size and shape.

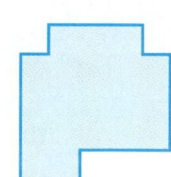

**35.** Four people are running in a marathon. Bob is 30 blocks behind Mary. Mary is 75 blocks ahead of Kathy. Kathy is 20 blocks behind Tim. How far behind Bob is Tim?

**36.** Suppose that a person was born on the 25th day of the year 35 B.C. and died on the 25th day of the year 15 A.D. How many years did the person live?

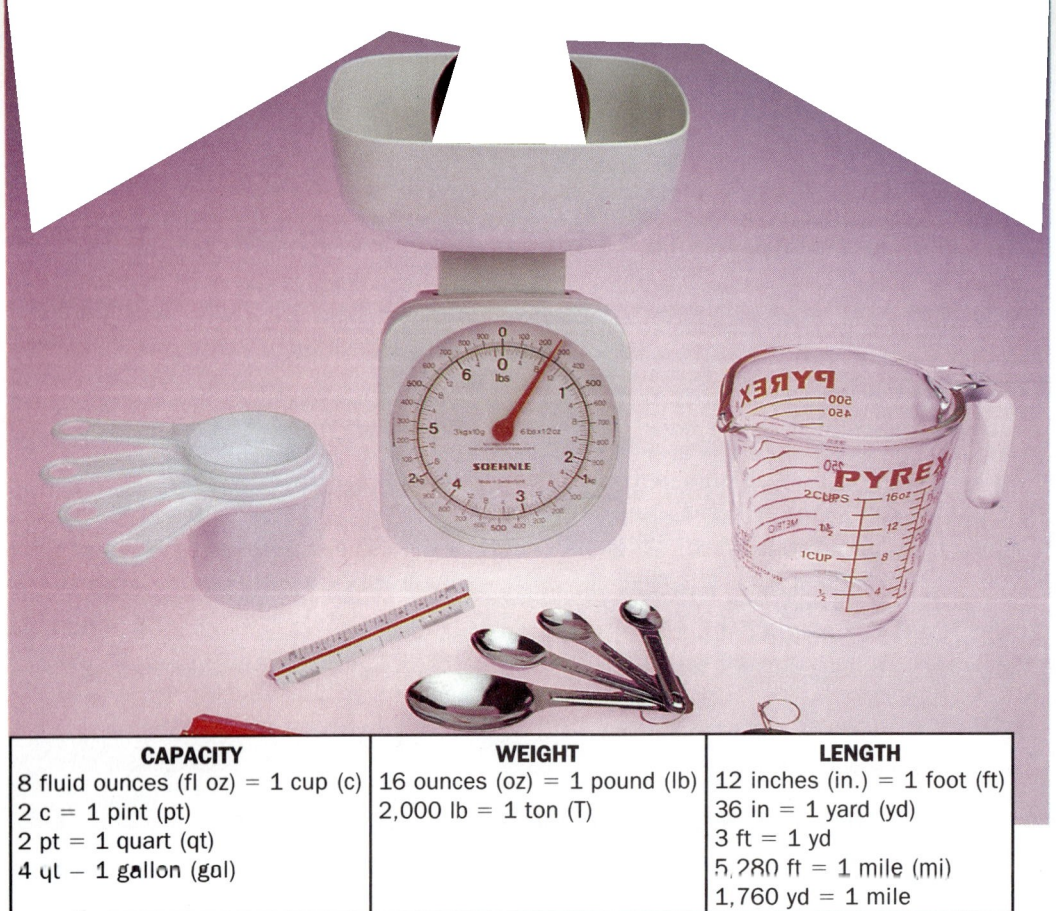

| CAPACITY | WEIGHT | LENGTH |
|---|---|---|
| 8 fluid ounces (fl oz) = 1 cup (c) | 16 ounces (oz) = 1 pound (lb) | 12 inches (in.) = 1 foot (ft) |
| 2 c = 1 pint (pt) | 2,000 lb = 1 ton (T) | 36 in = 1 yard (yd) |
| 2 pt = 1 quart (qt) | | 3 ft = 1 yd |
| 4 qt = 1 gallon (gal) | | 5,280 ft = 1 mile (mi) |
| | | 1,760 yd = 1 mile |

## LESSON 6-3 Ratio and Measurement

The tables above show the basic units of length, capacity, and weight most commonly used in the United States.

You can use ratios and the relationships in the tables to change a measurement from one unit to another. For example, to change 7 feet to yards, first look for the relationship in the table for lengths.

$$3 \text{ feet} = 1 \text{ yard}$$

Express the relationship as a ratio.   $\dfrac{3 \text{ ft}}{1 \text{ yd}}$

Write a proportion. Let $n$ represent the number of yards in 7 feet.

Feet → $\dfrac{3}{1} = \dfrac{7}{n}$ ← Feet
Yards →         ← Yards

Solve the proportion for $n$.

$$\dfrac{3}{1} = \dfrac{7}{n}$$
$$3n = 7$$
$$n = \dfrac{7}{3}, \text{ or } 2\dfrac{2}{3}$$

So 7 feet = $2\dfrac{2}{3}$ yards.

**EXAMPLE 1** To mix a liquid plant food, Tanya has to find the number of ounces in $4\frac{1}{2}$ pounds. How many ounces are there?

**Solution** Use this ratio: $\frac{16 \text{ ounces}}{1 \text{ pound}}$

**Write a proportion. Let s represent the number of ounces.**

$$\text{oz} \rightarrow \frac{16}{1} = \frac{s}{4\frac{1}{2}} \leftarrow \text{oz}$$

**Solve for s.**

$$s = 4\frac{1}{2} \cdot 16$$

$$s = \frac{9}{2} \cdot \frac{\overset{8}{\cancel{16}}}{1}$$

$$s = 72$$

So $4\frac{1}{2}$ pounds = 72 ounces.

**Try This**  **Complete.**

**1.** $3\frac{1}{2}$ cups = __?__ ounces    **2.** 45 inches = __?__ feet

You can use a ratio to compare map distances to actual distances. A map of the Florida Panhandle uses this scale to show the ratio.

5  15  25 Miles

$\frac{1}{2}$ inch represents 25 miles.

**EXAMPLE 2**  **a.** If the map distance from Pensacola to Panama City is $2\frac{1}{16}$ inches, what is the actual distance?

**b.** If the actual distance from Prosperity to Mexico Beach is 73 miles, what is the map distance?

**Solutions**  **a.** Write a proportion. Use ratios of $\frac{\text{inches}}{\text{miles}}$.

Let d represent the actual distance.

**Write a proportion.**  $\frac{\frac{1}{2}}{25} = \frac{2\frac{1}{16}}{d}$

**Cross-multiply.**  $\frac{1}{2}d = 25 \cdot 2\frac{1}{16}$

**Multiply each side by 2.**  $\overset{1}{\underset{1}{\cancel{2}}}(\frac{1}{\cancel{2}}d) = \frac{1}{2} \cdot 25 \cdot \frac{33}{\underset{8}{\cancel{16}}}$

**Solve for d.**  $d = \frac{825}{8}$, or $103\frac{1}{8}$ **miles**

6-3 Ratio and Measurement

**b.** Write a proportion. Let *w* represent the map distance.

$$\frac{\frac{1}{2}}{25} = \frac{w}{73}$$

Cross-multiply.  $25w = \frac{1}{2} \cdot 73$

Solve for *w*.  $w = \frac{1}{25} \cdot \frac{1}{2} \cdot 73$

$w = \frac{73}{50} = 1\frac{23}{50}$

The map distance is about $1\frac{1}{2}$ inches.

**Try This**

**3.** If the actual distance from Prosperity to Mexico Beach is 75 miles, what is the map distance?

**4.** If the map distance from Pensacola to Bascom is $2\frac{3}{4}$ inches, what is the actual distance?

• How can you use mental math to determine whether the answer to Example 2a is reasonable?

# EXERCISES

**Objective:** To convert customary units of measure
To find lengths using scale drawings

## Check Your Understanding

**Skill Check**  Write the letter of the correct answer.

**1.** Which proportion can be used to find the number of pints in 5 quarts?

a. $\frac{2}{1} = \frac{p}{5}$   b. $\frac{1}{2} = \frac{p}{5}$
c. $\frac{2}{1} = \frac{5}{p}$   d. $\frac{2}{5} = \frac{p}{1}$

**2.** Which proportion can be used to find the number of cups in 68 fluid ounces?

a. $\frac{1}{8} = \frac{c}{68}$   b. $\frac{8}{1} = \frac{c}{68}$
c. $\frac{1}{8} = \frac{68}{c}$   d. $\frac{c}{1} = \frac{8}{68}$

**3.** On a drawing, the length of a living room is 3 inches. The actual length is 24 feet. Which ratio represents the scale (inches to feet) used?

a. $\frac{1}{6}$   b. $\frac{1}{12}$
c. $\frac{8}{1}$   d. $\frac{1}{8}$

**4.** On a map, 1 inch represents 600 miles. Find the actual distance between two cities if they are $2\frac{1}{2}$ inches apart on the map.

a. 1200 mi   b. 1500 mi
c. 240 mi    d. 300 mi

5. Is 16 fluid ounces equal to, greater than, or less than 1 pint?
6. Is 7 quarts equal to, greater than, or less than 2 gallons?

## Practice and Apply

**Write a proportion to find each amount.**

7. 5 ft = _?_ in
8. 3 yd = _?_ in
9. _?_ ft = 4 miles
10. 7 c = _?_ pt
11. _?_ c = 78 fl oz
12. _?_ gal = 51 qt
13. $3\frac{1}{2}$ T = _?_ lb
14. _?_ oz = $5\frac{3}{4}$ lb
15. 153 in = _?_ yd

16. A recipe calls for $1\frac{1}{2}$ pints of water. How many cups is this?
17. You need to measure a length of 42 inches. How many yards is this?
18. The ratio of Omar's weight to Ricardo's weight is 120 lb to 150 lb. What is this ratio in ounces?
19. A recipe calls for 6 cups of milk for every 2 cups of water. What is the ratio of milk to water in fluid ounces?
20. The ratio of the gasoline used by a truck and gasoline used by a car along the same route is 64 gallons to 32 gallons. What is this ratio in quarts?
21. The ratio of one cargo's weight to another cargo's weight is 3 to 2. If the smaller cargo weighs $2\frac{1}{2}$ tons, what is the weight of the larger cargo?
22. One inch on a model airplane represents 12 feet on the actual plane. If the wingspan of the model is $12\frac{1}{2}$ inches, what is the actual wingspan of the plane?

**Make a scale drawing of each figure. Use the scale at the right.**

Scale: $\frac{1}{2}$ in = 1 ft

23. A square with sides of 3 feet
24. A rectangle with a length of $3\frac{1}{2}$ feet and a width of $1\frac{1}{2}$ feet.

**The length of this drawing of a shark is about 3 inches.**

25. Use the scale shown to find the actual length of the shark.
26. A whale has an actual length of about 12 yards. If the ratio of the length of the whale to the length of a drawing of the whale is 3 yd = 1 in, how long is the drawing?

Scale: 1 in = 13 ft

## Connect and Extend

The ratios of the lengths of the two rectangles at the right is the same as the ratio of their widths. That is,

$$\frac{\text{length of larger rectangle}}{\text{length of smaller rectangle}} = \frac{\text{width of larger rectangle}}{\text{width of smaller rectangle}}.$$

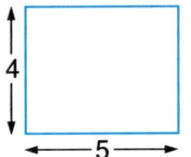

27. Show that the ratios listed above are equal.

28. Find this ratio: $\frac{\text{area of larger rectangle}}{\text{area of smaller rectangle}}$. Write the ratio in lowest terms.

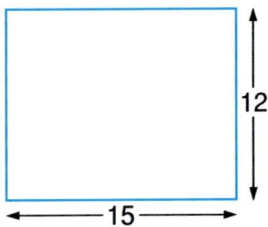

29. Compare the ratios in Exercises 27 and 28. What do you notice?

30. The length of a third rectangle is 45 and its width is 36. Test whether the ratios, length : length and width : width are the same for this rectangle and the smaller rectangle at the right above.

31. What is the ratio of the area of the third rectangle in Exercise 30 to the area of the smaller rectangle at the right above?

32. If the lengths and widths of two rectangles have the ratio 3 : 2, what is the ratio of their areas?

33. You are traveling at a rate of 44 feet per second on a road with a speed limit of 35 miles per hour. Could you get a speeding ticket?

## Maintain Your Skills

**Write True or False for each statement.
(Pages 126–136)**

34. 936 is divisible by 4.
35. 35,226 is divisible by 9.
36. 8 is a factor of 657.
37. 3 is a factor of 372,646.

**Give three examples of each of the following numbers.
(Pages 131–136, 139–141)**

38. An even number
39. An odd number
40. A prime number
41. A composite number

42. What is the smallest prime factor of 119? **(Pages 143–146)**

43. The prime factorization of a number is $2^3 \cdot 3 \cdot 7^2$. What is the number?

## LESSON 6-4

# Problem Solving Strategies
# Collecting Data

Suppose that the owner of a new music store near your school hires you as a consumer consultant. The owner wants to be sure that the store carries the kind of music that students will buy. Your first job is to prepare a one-page report showing what you think will sell best.

### Discussion 1
• What information will you need?

Think about the questions you will need to answer before you can prepare the report.

1. Only students who buy music (or may do so) will be interested in the store. Which students in your school buy music now? How might other students become interested in buying?
2. What form of music do the student buyers choose, sheet music, records, tapes, or compact discs? Is the trend changing? If so, in what direction and how fast is it changing?
3. What kind of music, rock, new age, country, classical, and so on, do student buyers purchase?
4. What established artists and what up-and-coming artists are student buyers interested in?

### Discussion 2
• What questions will you ask the people you interview?

Before interviewing people to gather information, think about the best way to ask each question and about how you will use the answer.

Here are two forms of the same question. Which will give you the most useful answer? Why?

**A.** What kind of music do you buy? _____

**B.** Write 1, 2, or 3 to show your first, second and third choices of the music you buy the most often.
    **a.** Country _____     **b.** Rock _____     **c.** New Age _____
    **d.** Rhythm and Blues _____     **e.** Classical _____

• How will you choose the people to interview?

Will you interview students in your school and people who live close to the store, or people who live on the other side of town? How many people will you interview?

• How will you present the report?

What will you do with the data you gather in your interview? Will you try to make a chart to summarize the data? Will you suggest decisions that the store owner might make based on your findings?

## EXERCISES

**Objective:** To apply strategies in problem solving

**TRYOUT INTERVIEW**

1. Write four interview questions you can use in the music survey.
2. Interview three friends. Use the four interview questions you wrote.
3. Make a chart that summarizes your findings.

**REVISED QUESTIONS AND INTERVIEW**

4. Revise your questions based on what you learned as you interviewed your friends and prepared the report.
5. Try out your three new questions with three new people. Make a second chart to summarize your findings.
6. Write a short paragraph describing what you have learned about collecting and analyzing data.

The total surface area of the earth is about 200 million square miles. Oceans cover about 140 million square miles. Use this information to express each ratio in lowest terms. (Pages 216–220)

1. Ocean area to total area
2. Non-ocean area to ocean area

Solve and check each proportion. (Pages 216–220)

3. $\frac{3}{7} = \frac{a}{56}$
4. $\frac{54}{63} = \frac{6}{n}$
5. $\frac{25}{c} = \frac{5}{15}$

Write a proportion to represent the situation. Solve the proportion. (Pages 221–225)

6. Mr. Fowler drove 144 miles in 3 hours. At this rate, how long will it take him to travel 240 miles?

7. Peggy drove 198 miles on 18 gallons of gasoline. How far can she travel on 32 gallons of gasoline?

A map of the Southern United States is drawn to a scale of 1 inch = 150 miles. (Pages 226–230)

8. If the distance between a town in Indiana and one in Illinois is 450 miles, how far apart are the towns on the map?

9. If two towns in Alabama are $1\frac{1}{5}$ inches apart on the map, how far apart are the towns in miles?

# Mathematical Footnote

Consider these questions.

- How many moves are there in the shortest possible checker game?
- Why is 2 the "oddest" of all the primes?

For more than twenty years, Martin Gardner asked questions such as these in his column of recreational mathematics, *Mathematical Games*, published in the magazine *Scientific American*. In his column, Gardner writes about games, puzzles, and other playful approaches to mathematics.

Martin Gardner's last formal mathematics class was in high school. As a professional writer, he published more than 30 books. His publications include a book on magic and several books for children.

# LESSON 6-5 Problem Solving Exploration
## Recording Chances

In this activity, you throw a pair of number cubes and keep a record of the numbers on top of each cube. Use cubes of different colors that have sides numbered from 1 through 6. Call one of the cubes the *first* cube and the other cube the *second* cube. Use an *ordered pair* of numbers to record each throw.

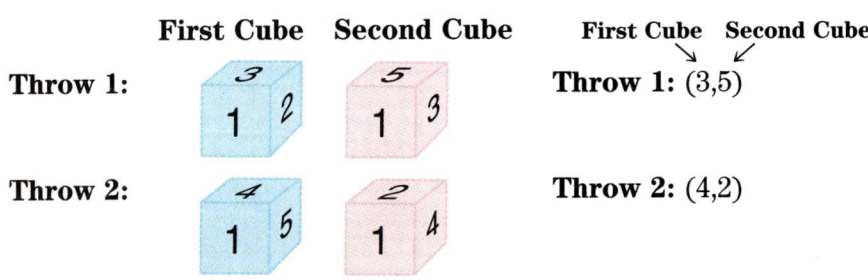

1. Throw a pair of cubes 50 times. For each throw, write the result as an ordered pair.

| Throw | (First Cube, Second Cube) |
|---|---|
| 1 | ( , ) |
| 2 | ( , ) |
| 3 | ( , ) |
| ⋮ | |
| 49 | ( , ) |
| 50 | ( , ) |

2. Now make a summary table such as the one shown below. In this table, each tally mark shows the result of one throw.

   For example, the tally mark in the "3" column and the "5" row represents a throw of (3,5). The tally mark in the "4" column and the "2" row represents a throw of (4,2).

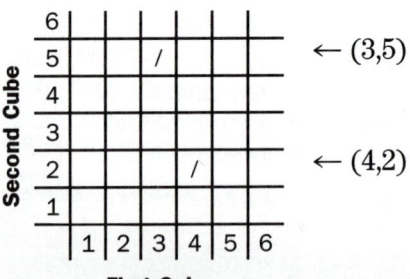

**234** CHAPTER 6

3. Do you think the chances of throwing one particular ordered pair are better than the chances of throwing any of the others?

4. Compare your summary table with that of your classmates. If you recorded all the throws you and your classmates made in one summary table, how do you think the results would compare with the results in your summary table? Explain.

5. How many different ordered pairs are possible? What do you think are your chances of throwing a (3,5)? Do you think that the chances of throwing one particular ordered pair might be 1 out of 36? Why or why not?

6. If you throw the number cubes just once, what do you think are the chances of throwing (1,1)? of throwing (1,2)?

7. If you throw the number cubes just once, and throwing a (1,1) or a (1,2) is counted as a successful throw, what are the chances of making a successful throw?

8. How many columns are there in your summary table? What do you think are the chances that, in one throw of the number cube, you will throw a pair that you record in the first column?

9. Write a paragraph that compares the chances, on one throw of a number cube, of throwing an ordered pair beginning with 1 to the chances of throwing an ordered pair that begins with 2. Will the chances be different for a pair that begins with a 3, or a 4, or a 5, or a 6?

# Mathematical Footnote

This sequence of numbers is very special.

1, 1, 2, 3, 5, 8, 13, ...

Perhaps you recognized this sequence as the Fibonacci numbers. Fibonacci, whose real name was Leonardo of Pisa, was an Italian mathematician.

Fibonacci numbers appear everywhere. For example, in an octave on the piano, there are 13 keys in all. Of these, 8 are white keys, 5 are black keys, 3 black keys are together, and 2 black keys are together.

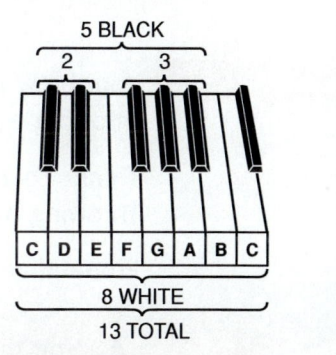

# LESSON 6-6

# Probability

**W**hy do teams toss a coin to determine who goes first in a competition? Since the coin is as likely to land heads up as tails up, most people feel that this is a fair way to decide.

When a coin is tossed, there are two **outcomes,** heads or tails. Each outcome is equally likely to happen. An **event** may be one or more possible outcomes. A list of all possible outcomes is called a **sample space.**

| Experiment | Sample Space |
|---|---|
| Tossing a coin | Head, Tail |
| Throwing a number cube | 1, 2, 3, 4, 5, 6 |
| Tossing two coins | (Head, Head), (Head, Tail), (Tail, Head), (Tail, Tail) |

If you throw a number cube, it will land so that one of these numbers is on top.

1  2  3  4  5  6

So the number cube can land in 6 different ways in all. The probability (or chances) that any particular number will be on top is 1 out of 6, or $\frac{1}{6}$.

**Probability** is a ratio between 0 and 1 that tells how likely it is that an event will happen. Probability is defined in this way.

$$\text{Probability} = \frac{\text{Number of successful outcomes}}{\text{Total number of outcomes}}$$

**EXAMPLE 1**  You throw a number cube once. What is the probability of throwing a 5?

**Solution**  **Think:** There are 6 possible outcomes (sample space). Throwing a 5 is the successful outcome in this case.

$$\text{Probability} = \frac{\text{Number of successful outcomes}}{\text{Total number of outcomes}} = \frac{1}{6}$$

**Try This**  **1.** You throw a number cube. What is the probability that it lands with an even number on top?

The probability in Example 1 is usually written like this.

$$P(5) = \frac{1}{6}$$

The closer a probability is to 1, the *more likely* it is that the event will happen.

**The probability of an event that is certain to happen is 1.**

For example, if you throw a number cube, you are sure that one of the numbers, 1, 2, 3, 4, 5, 6, will be on top.

$$P(\text{one of } 1, 2, 3, 4, 5, 6) = 1$$

The closer a probability is to 0, the *less likely* it is that the event will happen. For example, if you throw a number cube, you know that it cannot land with a 7 on top.

$$P(7) = 0$$

**The probability of an event that cannot happen is zero.**

**EXAMPLE 2** These cards are placed in a hat. One card is drawn at random. Find each probability.
a. Drawing a red card
b. Drawing a green card
c. Drawing a number less than 3

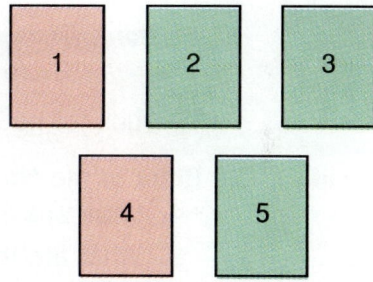

**Solutions** List the possible outcomes.

(red,1); (green,2); (green,3); (red,4); (green,5)

There are 5 outcomes in all.

a. Successful outcomes: (red,1); (red,4)

$P(\text{red card}) = \frac{2}{5}$

b. Successful outcomes: (green,2); (green,3); (green,5)

$P(\text{green card}) = \frac{3}{5}$

c. Successful outcomes: (red,1); (green,2)

$P(\text{number} < 2) = \frac{2}{5}$

**Try This** **Refer to the situation in Example 2 to find each probability.**
2. Drawing a number greater than 2
3. Drawing a number less than 5

The ratio of the successful outcomes to the unsuccessful outcomes of an event is called the **odds** in favor of the event.

Odds in favor = $\dfrac{\text{Number of successful outcomes}}{\text{Number of unsuccessful outcomes}}$

**EXAMPLE 3** Find the odds in favor of each event. Refer to the situation about drawing the cards in Example 2.
  a. What are the odds in favor of drawing a green card?
  b. What are the odds in favor of drawing a red 4?
  c. What are the odds that you will not draw an even number?

**Solutions**
  a. **Think:** There are 3 ways to draw a green card.
  There are 2 ways not to draw a green card.

  Odds in favor of green card = $\dfrac{3}{2}$, or 3 to 2, or **3 : 2**

  b. **Think:** There is 1 way to draw a red 4.
  There are 4 ways of not drawing a red 4.

  Odds in favor of red 4 = $\dfrac{1}{4}$, or 1 to 4, or **1 : 4**

  c. **Think:** There are 3 ways of not drawing an even number.
  There are 2 ways of drawing an even number.

  Odds for not drawing an even number = $\dfrac{3}{2}$, or 3 to 2, or **3 : 2**

**Try This** Refer to the situation about drawing the cards in Example 2.
  4. What are the odds in favor of not drawing an odd number?
  5. What are the odds in favor of drawing a red card?

# EXERCISES

**Objective:** To determine the probability of a simple event

## Check Your Understanding

**Skill Check** Write the letter of the correct answer.

1. Which best describes a sample space?
  a. An event
  b. A number of unsuccessful outcomes
  c. A set of all possible outcomes of an event
  d. A set of successful outcomes

2. In Example 2, what is the probability of drawing a green card having a number less than 3?
  a. $\dfrac{1}{5}$
  b. $\dfrac{2}{5}$
  c. $\dfrac{1}{3}$
  d. $\dfrac{2}{3}$

**3.** Which ratio describes the odds in favor of an event?

  a. $\dfrac{\text{Number of successful outcomes}}{\text{Total number of outcomes}}$

  b. $\dfrac{\text{Number of unsuccessful outcomes}}{\text{Total number of outcomes}}$

  c. $\dfrac{\text{Number of unsuccessful outcomes}}{\text{Number of successful outcomes}}$

  d. $\dfrac{\text{Number of successful outcomes}}{\text{Number of unsuccessful outcomes}}$

**4.** In Example 2, what are the odds in favor of drawing a green card with an even number on it?
  a. $4:1$    b. $1:5$
  c. $5:1$    d. $1:4$

**List the sample space for each probability experiment.**

**5.** Spin the spinner

**6.** Throw a number cube and toss a dime.

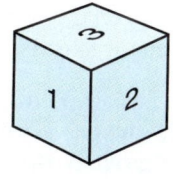

## Practice and Apply

**You spin the arrow once. Find each probability.**

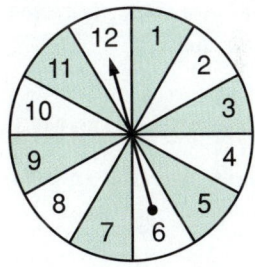

**7.** P(10)
**8.** P(multiple of 2)
**9.** P(factor of 6)
**10.** P(odd number)
**11.** P(number < 9)
**12.** P(prime number)
**13.** P(multiple of 12)
**14.** P(not an odd number)

**15.** You "win" if the arrow stops on an even number. What are the odds in favor of your winning?

**16.** You "lose" if the arrow stops on a number that begins with the letter "t". What are the odds that you will win?

**17.** There are 6 cards numbered 1 to 6. Joe must draw an even number to win. He draws a card at random. Are his chances of winning greater than $\frac{1}{4}$?

**18.** A box contains 15 chips. Three are white, 7 are blue, and 5 are green. You pick one at random. What is the probability that it is not green?

**19.** One of three identical boxes contains paper clips, one contains erasers, and a third contains stamps. The boxes are not labeled. Nick picks up one box. What are the chances that the box does not contain paper clips?

**20.** An experiment with mosquitoes showed that 8,000 out of 10,000 live for one week at most. Based on this experiment, what is the probability that a mosquito will live for more than one week?

A number is chosen at random from {4, 5, 6, 7, 8, 9, 10}.

**21.** What is the probability that it is a solution of $2x < 20$?

**22.** What is the probability that it is a solution of $2x - 3 = 15$?

## Connect and Extend

**23.** What is the sum of the probabilities of all the outcomes in a sample space? Give an example to explain your answer.

**Name some events in daily life that have the given probability of happening.**

**24.** Greater than $\frac{1}{2}$  **25.** Less than $\frac{1}{2}$  **26.** 1  **27.** 0

**28.** Draw a spinner having four different divisions, each having a different color, such that the chances of landing on a particular color are different from the chances of landing on the other colors.

**You win $10 in the Toss-A-Penny game at an amusement park if you toss a penny that lands in the blue area (see the figure at the right). You win $20 if you toss a penny that lands in the red area.**

**29.** Find P(landing in blue area).

**30.** Find P(landing in red area).

**31.** If you toss 200 pennies one after the other, in which area do you think more pennies will land? Give a reason for your answer.

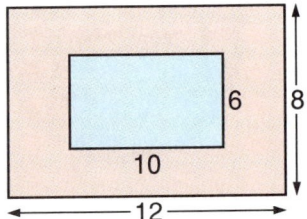

## Maintain Your Skills

**Express each fraction in lowest terms. (Pages 163–167)**

**32.** $\frac{5}{30}$   **33.** $\frac{24}{56}$

**34.** $\frac{8}{20}$   **35.** $\frac{12}{26}$

**Add. Write the answer in lowest terms. (Pages 194–205)**

**36.** $6\frac{5}{8} + 4\frac{2}{8}$   **37.** $\frac{19}{20} + \frac{11}{20}$

**38.** $\frac{4}{5} + \frac{2}{9}$

**39.** A 4 ounce glass of apple juice contains about 60 calories. How many calories are there in 24 ounces of apple juice? **(Pages 221–225)**

**40.** Ann gets $6 every 3 days for feeding a neighbor's dog. How many days does it take her to earn $14? **(Pages 221–225)**

# LESSON 6-7
# Independent and Dependent Events

**A** nickel and a quarter are tossed one after the other. What is the probability of tossing a head on the nickel and a tail on the quarter?

If two coins are tossed one after the other, the way one coin falls does not depend on how the other coin falls. That is, the tosses are independent, and the outcomes are **independent events.**

One way to find the probability that the first toss turns up heads and the second toss turns up tails is to list the sample space. In the sample space below, H represents turning up heads and T represents turning up tails.

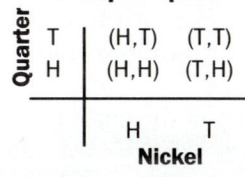

**Sample space:** (H,H), (T,H), (H,T), (T,T)

**Successful events:** (H on nickel, T on quarter), or (H,T). Then $P(H,T) = \frac{1}{4}$.

Another way to find P(H,T) is to multiply the separate probabilities. When you toss a coin, $P(T) = \frac{1}{2}$ for each toss and $P(H) = \frac{1}{2}$ for each toss.

$$P(H \text{ on nickel}) = \frac{1}{2} \qquad P(T \text{ on quarter}) = \frac{1}{2}$$

$$\begin{aligned} P(H,T) &= P(H) \cdot P(T) \\ &= \frac{1}{2} \cdot \frac{1}{2} \\ &= \frac{1}{4} \end{aligned}$$

This suggests the following rule for independent events $A$ and $B$.

> **Independent Events**
> For two independent events, $A$ and $B$,
> $P(A \text{ and } B) = P(A) \cdot P(B)$.

**EXAMPLE 1** Ten cards numbered from 0 to 9 are placed in a hat. One card is chosen at random and then returned to the hat. The cards are shuffled and a second card is picked.

What is the probability that the number on the first card is greater than 6 and the number on the second card is less than 2?

**Solution** Since the first card is replaced before the second is drawn, the events are independent.

The sample space is {0, 1, 2, 3, 4, 5, 6, 7, 8, 9}.

P(first event) = P(number > 6) = $\frac{3}{10}$

P(second event) = P(number < 2) = $\frac{2}{10}$, or $\frac{1}{5}$

P(number > 6, number < 2) = $\frac{3}{10} \cdot \frac{1}{5} = \frac{3}{50}$

**Try This** **Refer to the problem in Example 1 to find these probabilities.**
1. P(number > 8, even number)
2. P(odd number, odd number)

Think about what happens if the first card in Example 1 is not replaced before the second card is drawn. Since there is now one number less in the sample space, the second event is **dependent** on the first. When two events are dependent and one of the events happens, this affects the probability of the second event.

**EXAMPLE 2** There are 4 red marbles, 2 yellow marbles, and 2 blue marbles in a bowl. One marble is drawn at random and not replaced. Then a second marble is drawn. Find each probability.
a. P(blue, then yellow)
b. P(red, then red)

**Solutions**  Since the first marble is drawn and not replaced, the events are dependent.

a. **Sample space:** 4 red marbles, 2 yellow, 2 blue
Total number of marbles: 8
P(blue) = $\frac{2}{8}$, or $\frac{1}{4}$

**New sample space:** 4 red marbles, 2 yellow, 1 blue
Total number of marbles: 7
P(blue, then yellow) = $\frac{1}{4} \cdot \frac{2}{7}$
= $\frac{2}{28}$, or $\frac{1}{14}$

b. **Sample space:** 4 red marbles, 2 yellow, 2 blue
Total number of marbles: 8
P(red) = $\frac{4}{8}$, or $\frac{1}{2}$

**New sample space:** 3 red marbles, 2 yellow, 2 blue
Total number of marbles: 7
P(red, then red) = $\frac{1}{2} \cdot \frac{3}{7}$
= $\frac{3}{14}$

**Try This**  **Refer to the situation in Example 2 to find these probabilities.**

3. P(red, then blue)
4. P(yellow, then yellow)
5. P(blue, then red)
6. P(yellow, then blue)

> **Independent and Dependent Events**
>
> If two events are independent, the probability that both will happen equals the product of the probabilities that each event will happen.
>
> To find the probability of two dependent events, begin by finding the probability of the first event. Then, using the new sample space, find the probability of the second event. Multiply the probabilities.

# EXERCISES

**Objective:** To determine the probabilities of independent and dependent events

## Check Your Understanding

**Skill Check** Choose the letter of the correct answer.

1. A nickel and dime are tossed. What is P(T,T)?
   a. $\frac{1}{2}$
   b. 1
   c. $\frac{1}{8}$
   d. $\frac{1}{4}$

2. A number cube is thrown and a coin is tossed. Which statement is true?
   a. The events are dependent.
   b. The events are independent.
   c. There are exactly two outcomes in the sample space.
   d. P(6,T) > P(6,H)

**Refer to this information for Exercises 3–4.**
Six cards, three green, 1 red, and 2 black, are placed in a box.

3. One card is chosen at random and not replaced. Then a second card is chosen. What is P(green, then red)?
   a. $\frac{1}{2}$
   b. $\frac{1}{12}$
   c. $\frac{1}{7}$
   d. $\frac{1}{10}$

4. One card is chosen at random and then replaced. The cards are shuffled and another card is chosen. What is P(red, then green)?
   a. $\frac{1}{10}$
   b. $\frac{1}{12}$
   c. $\frac{1}{8}$
   d. $\frac{1}{4}$

5. Give an example of two independent events in daily life.
6. Give an example of two dependent events in daily life.

## Practice and Apply

A penny and a dime are tossed one after the other. Refer to the sample space at the right.

**Sample Space**

|  | Dime |  |
|---|---|---|
| T | (H,T) | (T,T) |
| H | (H,H) | (T,H) |
|  | H | T |
|  | Penny |  |

7. What is the probability of getting a head on the penny?
8. What is the probability of getting a tail on the dime?
9. What is the probability of getting a head on the penny and a tail on the dime?
10. What is the probability of getting (H,T) or (T,H)?
11. Explain why tossing (H,T) or (T,H) is more likely than tossing (H,H).

A box contains 2 green and 5 yellow marbles. Two marbles are drawn at random, one after the other. The first marble is replaced before the second one is drawn.

**12.** What is the probability that both marbles are green?

**13.** What is the probability that the first marble is green and the second one is yellow?

**14.** What is the probability that both marbles are yellow?

Suppose that the first marble drawn in Exercises 12–14 is not replaced before the second one is drawn.

**15.** What is the probability that both marbles are green?

**16.** What is the probability that both are yellow?

**17.** What is the probability that the first marble drawn is green and the second one is yellow?

Hannah always keeps her money in a sugar bowl. The bowl contains one $100 bill, three $20 bills, four $10 bills, and two $5 bills. Hannah reaches into the bowl and removes two bills at random, one after the other.

**18.** Are the two events dependent or independent? Explain.

**19.** What is the probability that both bills that Hannah drew are $20 bills?

**20.** What is the probability that the two bills total $10?

**21.** What is the probability that the two bills total $120?

**22.** If Hannah removes three bills, one after the other, what is the probability that they are all $10 bills?

## Connect and Extend

When it is impossible for two events to happen at the same time, they are called **mutually exclusive events.** For example, if you toss a number cube, you can toss a 5 *or* a 6, but you cannot toss a 5 *and* a 6. So tossing a 5 *and* tossing a 6 are mutually exclusive events.

**23.** Give an example of two mutually exclusive events.

**24.** Give an example of two events that are **not** mutually exclusive.

**25.** What is the probability that two mutually exclusive events can happen at the same time?

**A number cube is tossed.**

**26.** Find P(tossing a 5).

**27.** Find P(tossing a 6).

6-7 Independent and Dependent Events **245**

28. Find P(tossing a 5 *or* tossing a 6).
29. *Compare* your answer to Exercise 28 to the sum of your answers for Exercises 26 and 27.
30. Are tossing a 5 *and* tossing a 6 on one toss mutually exclusive events?
31. *Complete:* If $A$ and $B$ are mutually exclusive events, then P(A *or* B) = P(A) $\underline{?}$ P(B).

## Maintain Your Skills

**Evaluate each expression. (Pages 39–45, 150–153)**

32. $3(c - 10)$, for $c = -5$
33. $-6(y - x)$, for $x = -7$ and $y = 5$
34. $-151 - 2g^2h$, for $g = -4$ and $h = -5$
35. $t^5 \cdot t^3$, for $t = -2$

**Add or subtract. Write the answer in lowest terms. (Pages 194–209)**

36. $\frac{5}{6} + \frac{1}{8}$
37. $\frac{4}{9} + 1\frac{2}{3}$
38. $5\frac{2}{3} - 4\frac{7}{8}$
39. $\frac{9}{10} - \frac{1}{2}$

**Replace each $\underline{?}$ with the correct number. (Pages 226–230)**

40. 9 yd = $\underline{?}$ feet
41. $3\frac{2}{3}$ ft = $\underline{?}$ in
42. 82 fl oz = $\underline{?}$ c

## ★ Math Team Problems

43. Doris took some coins out of her purse and arranged the coins in four rows with four coins in each row. Each of the four rows, each of the four columns, and each of the two diagonals had exactly one penny, one nickel, and one quarter. Show how Doris could have arranged the coins.

44. Roberto needs to draw a line that is 6 inches long, but he does not have a ruler. He does have some sheets of notebook paper that are each $8\frac{1}{2}$ inches wide and 11 inches long. Describe how Roberto can use the notebook paper to measure six inches.

45. Grace has two minutes to get to the bus station before her bus departs. The bus station is two miles from her home. She drives at a speed of 30 miles per hour for the first mile traveled. Will she get to the station before her bus leaves? Explain.

## LESSON 6-8

# Making Choices

**M**eg has two jackets, one red and one blue. She also has three skirts, one red, one white and one blue. If Meg chooses a skirt and jacket to wear to school, how many different choices does she have?

You can draw a **tree diagram** to show the choices.

**TREE DIAGRAM**

| First Choice | Second Choice |
|---|---|
| Red jacket | Red skirt<br>White skirt<br>Blue skirt |
| Blue jacket | Red skirt<br>White skirt<br>Blue skirt |

The model shows that for each jacket, there are 3 different ways to choose a skirt.

**Number of Jacket Choices · Number of Skirt Choices**
$$2 \cdot 3$$

Total number of choices: $2 \cdot 3 = 6$

So there are 6 different ways for Meg to choose a jacket and a skirt.

This suggests the following rule.

> **Fundamental Counting Principle**
> If one choice can be made in $m$ ways, and a second choice can be made in $n$ ways, there are $m \cdot n$ ways of making the first choice followed by the second choice.

6-8  Making Choices

**EXAMPLE 1** When traveling from Midland to Houston, a tourist plans to make stops at Austin and San Antonio. There are 2 scenic routes from Midland to Austin. There are 4 scenic routes from Austin to San Antonio, and 2 scenic routes from San Antonio to Houston. How many different scenic routes can the tourist take?

**Solution** **Method 1:** Draw a tree diagram.

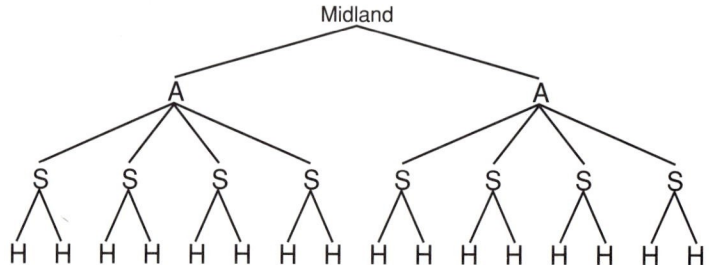

**Method 2:** Use the Fundamental Counting Principle.

| Number of Routes from | | |
|---|---|---|
| Midland to Austin · | Austin to San Antonio · | San Antonio to Houston |
| 2 | 4 | 2 |

There are 2 · 4 · 2, or **16** different scenic routes.

**Try This** 1. Tom has 5 shirts, 3 pairs of pants, and 2 jackets. How many different outfits can Tom make with these clothes?

The Fundamental Counting Principle can help you to find the number of events in the sample space of a probability problem.

**EXAMPLE 2** Ms. Sanchez' math quiz has 3 multiple choice questions. There are 4 possible answers for each of these questions. The quiz also has two True/False questions. What is the probability that you will get a perfect score if you guess the answer to each question?

**Solution** Use the Fundamental Counting Principle to find the total number of choices (sample space).

| Number of Choices | | | | |
|---|---|---|---|---|
| Question 1 · | Question 2 · | Question 3 · | Question 4 · | Question 5 |
| 4 | 4 | 4 | 2 | 2 |

There are 4 · 4 · 4 · 2 · 2, or 256 different ways for you to answer the questions. There is only one correct answer to each question, and 1 · 1 · 1 · 1 · 1, or 1 way to answer all 5 questions correctly.

P(perfect score by guessing) = $\frac{1}{256}$

**Try This**

2. Leona can take a one-week, two-week, or three-week vacation. She can go to the mountains to ski or to the shore to swim. How many choices does she have?

# EXERCISES

**Objective:** To find the total number of choices by using models and the Fundamental Counting Principle

## Check Your Understanding

**Skill Check** Write the letter of the correct answer.

A snack bar sells 5 different kinds of sandwiches, 3 different kinds of juice, and 4 different kinds of fruit.

1. How many choices are there for ordering a sandwich and one kind of juice?
   a. 8          b. 15
   c. 20         d. 12

2. How many choices are there for ordering a sandwich, one kind of juice, and one kind of fruit?
   a. 12   b. 15   c. 20   d. 60

3. If the snack bar runs out of two kinds of sandwiches, how many choices are there for ordering a sandwich and one kind of juice?
   a. 15         b. 5
   c. 6          d. 9

4. If the snack bar runs out of 2 kinds of fruit and 3 kinds of sandwiches, how many choices are there for ordering a sandwich, one kind of juice, and one kind of fruit?
   a. 12   b. 7   c. 8   d. 4

5. Redraw the tree diagram on the first page of this lesson so that the skirt choices come before the jacket choices.

## Practice and Apply

Refer to this information for Exercises 6–10.

The parts of the uniform that Martha wears to work must be chosen from the list at the right.

| Skirts | Blouses | Shoes |
|--------|---------|-------|
| white  | blue    | black |
| beige  | red     | brown |
|        | green   |       |

6. Martha always wears a beige skirt. Draw a tree diagram to show all of Martha's possible choices.

6-8  Making Choices   **249**

7. How many choices does Martha have to complete her uniform if she wears a beige skirt?

8. If Martha decides to wear brown shoes with her beige skirt, how many choices does she have to complete her uniform?

9. Do you think that Martha's choices of color for a blouse to wear with a beige skirt are all equally likely? Explain.

**Mr. Eagleton's math quiz has three multiple-choice questions. There are four possible answers for each question.**

10. How many different ways are there for you to answer the questions?

11. If you guess the answer to each question, what is the probability that you will get a perfect score?

**Suppose that, on each of the three questions, you can eliminate one choice as incorrect.**

12. If you guess when you choose the answer from the remaining three choices for each question, what is the probability that you will get a perfect score?

**In the final stage of a World Series, the winning team wins 4 out of 7 games. Suppose that Team A and B are evenly matched, but Team A won the first 3 games.**

13. Use the tree diagram to complete the list of possible outcomes, given that Team A wins the first 3 games.

| Possible Outcomes |
|---|
| W W W W |
| W W W L W |
| W W W _?_ _?_ _?_ |
| W W W _?_ _?_ _?_ _?_ |
| W W W _?_ _?_ _?_ _?_ |

14. How many possible outcomes are there, given that Team A has won the first three games of the series.

15. Given that Team A has won the first three games of the series, how many possible ways are there for Team A to win the series?

16. Given that Team A has won the first three games of the series, how many possible ways are there for Team B to win the series?

17. Which seems more likely, that Team A will win or that Team B will win?

## Connect and Extend

The diagram at the right shows the results of a random survey of 100 people in a small town. Each person was asked whether he or she heard the latest hit song on TV only, on the radio only, or on both TV and radio. Only 10 people said that they heard the song on *both* TV and radio.

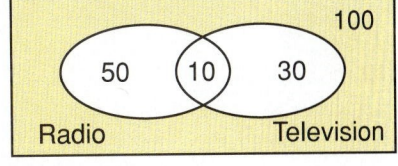

**18.** How many of the 100 people said they heard the song on TV?

**19.** How many of the 100 people said they heard the song on the radio?

**20.** How many said that they heard the song on TV only?

**21.** How many said that they heard the song on the radio only?

Suppose that a person from the town walks into the store that sponsored the survey.

**22.** What is the probability that the person heard the song on TV?

**23.** What is the probability that the person heard the song on the radio?

**24.** What is the probability that the person heard the song on TV *and* on the radio?

**25.** What is the probability that the person heard the song on TV *or* on the radio?

**26.** Why does P(song on TV *or* on radio) = P(song on TV) + P(song on radio) − P(song on TV *and* on radio)?

## Maintain Your Skills

**27.** Find the product of the first six negative integers. **(Pages 28–31)**

**Solve and check. (Pages 183–188)**

**28.** $\frac{1}{8}c = 1\frac{1}{4}$
**29.** $91 = \frac{7}{8}h$
**30.** $\frac{5}{12}q = 6\frac{1}{2}$
**31.** $\frac{5}{16} = \frac{3}{8}y$

Each of the 7 letters of the word "STUDENT" is written on a separate card. The cards are placed face down and shuffled. A card is chosen at random. Find each probability. **(Pages 236–240)**

**32.** The card chosen is the letter T

**33.** The card chosen is **not** a vowel.

**A box contains 18 eggs. Six of the eggs are cracked. (Pages 241–246)**

**34.** One egg is picked at random from the box and replaced. Then another egg is picked. Find P(cracked, then not cracked).

**35.** One egg is picked at random from the box and not replaced. Another egg is picked. Find P(cracked, then cracked).

## LESSON 6-9

# Permutations

**A**rt, Bob, and Chuck are running a race. How many different outcomes are possible?

Use a tree diagram to show the outcomes. In the tree diagram, A, B, and C represent the runners.

| First Place | Second Place | Third Place | Possible Outcome |
|---|---|---|---|
| A | B | C | ABC |
| A | C | B | ACB |
| B | A | C | BAC |
| B | C | A | BCA |
| C | A | B | CAB |
| C | B | A | CBA |

The ways in which the 3 runners can finish are listed below.

ABC     ACB     BAC     BCA     CAB     CBA

Each of these ways is an ordered arrangement of the same 3 letters. This means that ABC is not the same as ACB, or BAC, or BCA, and so on. An ordered arrangement is called a **permutation.**

In ordering the three letters A, B, and C, there are three choices for first place. For each first place, there are two choices for second place. For each second place, there is only one choice for third place. That is,

| Number of Choices for First Place | · | Number of Choices for Second Place | · | Number of Choices for Third Place | = | Total Outcomes |
|---|---|---|---|---|---|---|
| 3 | · | 2 | · | 1 | = | 6 |

So there are 6 permutations or ways in which the 3 runners can finish in first, second, and third places.

The number of ways for 3 runners, taken 3 at a time, to place can be written in permutation notation like this.

$$_3P_3$$

Number of runners     Taken 3 at a time

So $_3P_3 = 3 \cdot 2 \cdot 1 = 6$.

The product $3 \cdot 2 \cdot 1$ can be written in **factorial notation** as 3! (Read 3! as 3 factorial.)

Scientific calculators usually have a factorial key $\boxed{n!}$. To evaluate 5!, press 5 $\boxed{n!}$. On some calculators, you may have to use the second function key, $\boxed{\text{2nd F}}$. Press 5 $\boxed{\text{2nd F}}$ $\boxed{n!}$.

**EXAMPLE 1** Suppose that 4 runners enter a race. In how many different ways can the runners finish the race?

**Solution** **Think:** The number of ways in which 4 runners can finish is the same as $_4P_4$, the number of arrangements for 4 runners, taken 4 at a time.

$$\begin{aligned} _4P_4 &= 4! \\ &= 4 \cdot 3 \cdot 2 \cdot 1 \\ &= \mathbf{24} \text{ ways} \end{aligned}$$

6-9 Permutations

**Try This**

1. Suppose that 5 runners enter a race. In how many ways can they finish the race?

In a race with 4 runners, in how many different ways can they finish in first, second, and third places?

Use the Fundamental Counting Principle.

| First Place | · | Second Place | · | Third Place | = | Total Outcomes |
|---|---|---|---|---|---|---|
| 4 | · | 3 | · | 2 | = | 24 |

There are 4 · 3 · 2, or 24 ways in which 4 runners can finish in first, second, and third places. You can think of this as the number of permutations of 4 runners, taken 3 at a time.

So $_4P_3 = 4 \cdot 3 \cdot 2 = 24$.

**EXAMPLE 2**   There are 3 empty seats in the auditorium. There are 5 students to be seated.
a. Use the Fundamental Counting Principle to find the total number of possible ways for the 3 empty seats to be filled.
b. Express the situation in permutation notation.

**Solutions**   a. Use the Fundamental Counting Principle.

| First Chair | Second Chair | Third Chair | Total Outcomes |
|---|---|---|---|
| 5 · | 4 · | 3 = | **60** |

b. There are 60 ways, or ordered arrangements, in which the five students, taken three at a time, can be seated.

$$_5P_3 = 60 = \mathbf{5 \cdot 4 \cdot 3}$$

**Try This**   **Evaluate to find the number of permutations.**

2. $_5P_5$   3. $_5P_2$   4. $_5P_4$

## EXPLORE

Look at the pattern in these permutations.

$_4P_3 = 4 \cdot 3 \cdot 2$     $_5P_2 = 5 \cdot 4$     $_6P_4 = 6 \cdot 5 \cdot 4 \cdot 3$
$_5P_3 = \underbrace{5 \cdot 4 \cdot 3}_{\text{3 factors}}$     $_7P_2 = \underbrace{7 \cdot 6}_{\text{2 factors}}$     $_5P_4 = \underbrace{5 \cdot 4 \cdot 3 \cdot 2}_{\text{4 factors}}$

**Try to find the pattern by answering these questions.**

1. Look at the first factor in each product. With what number does it match in the permutation notation?
2. Count the number of factors in each product. With what number in the permutation notation can you match the number of factors?
3. *Complete:* $_9P_5 = \underline{?} \cdot \underline{?} \cdot \underline{?} \cdot \underline{?} \cdot \underline{?}$.
   Does your product begin with 9? Are there 5 factors?
4. *Complete:* To find $_nP_r$, multiply the first $\underline{?}$ factors of $n!$.

**EXAMPLE 3** An ice cream parlor sells 8 flavors of yogurt. How many ways are there to have a 2-dip cone of different flavors?

**Solution** **Think:** This is the same as the number of arrangements of 8 things, taken 2 at a time. Find $_8P_2$; that is, multiply the first two factors of 8!.

$$_8P_2 = 8 \cdot 7 = 56$$

There are **56** ways to have a 2-dip cone of different flavors.

**Try This** **Evaluate to find the number of permutations.**

5. $_{14}P_3$  6. $_9P_3$  7. $_8P_5$

# EXERCISES

**Objective:** To find the number of permutations in a given situation

## Check Your Understanding

**Skill Check** Write the letter of the correct answer.

**1.** Which is the same permutation as RST?
  a. TSR   b. SRT
  c. RST   d. RTS

**2.** Which is the number of ways in which 5 people can finish in a race?
  a. 5!   b. $_5P_1$
  c. 20   d. 60

**3.** Which is the number of ways in which 4 people can finish in first and second place in a race?
  a. $_4P_4$   b. $_4P_2$
  c. 24   d. 4

**4.** How many permutations of 5 objects taken 2 at a time are there?
  a. 25   b. 20
  c. 60   d. 120

6-9 Permutations  **255**

**5.** Use a factorial symbol to write the expression, $1 \cdot 2 \cdot 3 \cdot 4$.

**6.** *Complete:* $_7P_3$ represents a number of permutations of __?__ objects taken __?__ at a time.

## Practice and Apply

**Evaluate.**

**7.** $6!$  
**8.** $6! - 3!$  
**9.** $\dfrac{7!}{5!}$  
**10.** $\dfrac{4!}{6!}$  
**11.** $2! + 3!$  
**12.** $_7P_4$  
**13.** $_{20}P_2$  
**14.** $_3P_3 - {_2P_2}$  
**15.** $_{10}P_3$  
**16.** $\dfrac{_5P_3}{3!}$

**Express the possible number of ordered arrangements using permutation notation. You do not need to find the actual number of permutations.**

**17.** Seven objects taken 6 at a time

**18.** Choosing from 20 players to fill 9 positions on a baseball team

**19.** Filling 12 job openings from 15 applications

**20.** Twelve people standing in a line

**Find the number of permutations.**

**21.** How many different batting lineups are possible with 9 players?

**22.** In how many different ways can 15 different roller skaters finish in first, second, and third places in a race?

**23.** In how many different orders can 10 bands appear on stage to perform?

**24.** How many different obstacle courses can be set up using any 4 of 15 obstacles?

**A veterinarian is scheduling Friday appointments to treat 5 cats, Fluffy, Felix, Morris, Kitty, and Seymour.**

**25.** In how many different ways can the appointments be scheduled?

**26.** In how many different ways can the appointments be arranged if Kitty always has the first appointment of the day and Felix has the second?

**27.** List all the permutations of the letters, PERM. How can you be sure that you have a complete list?

**28.** A gas station is running short of gas and has only enough to fill up one truck, one bus, and one car. In how many ways can the three vehicles line up for gas?

## Connect and Extend

**Study the pattern shown below.**

$_3P_3 = 6$ and $_3P_2 = 6$     $_6P_6 = 720$ and $_6P_5 = 720$

$_5P_5 = 120$ and $_5P_4 = 120$     $_{10}P_{10} = 3{,}628{,}800$ and $_{10}P_9 = 3{,}628{,}800$

**29.** Which of these equations best describes the pattern above?
  **a.** $_nP_n = {_nP_n}$     **b.** $_nP_n = {_nP_{n-1}}$     **c.** $_nP_n = {_nP_{n-2}}$

**30.** Explain how the pattern works.

**Suppose that bicycle license plates in your town have 5 digits. A license number can start with any digit, including 0, and no digits can be repeated.**

**31.** How many different license plates can there be?

**32.** How many license plates will have only odd digits?

**33.** How many license plates can have 0 for the first digit?

**34.** If license plates are assigned at random, what is the probability of getting a plate having odd digits only?

**35.** If license plates are assigned at random, what is the probability of getting a plate that has the numbers 1, 2, 3, 4, 5 in that order?

## Maintain Your Skills

**Solve each equation. (Pages 105–109)**

**36.** $-(4 + 6f) = -16$     **37.** $2x + 5 = -1$     **38.** $-5y + 9 = 14$     **39.** $8(a + 10) = 32$

**Solve. (Pages 114–118)**

**40.** There are 48 students on a bus. There are 6 more boys than girls. Find the number of girls and boys on the bus.

**41.** Lyn worked twice as many hours as Stef. Together they worked 51 hours. How many hours did each person work?

**Solve and check each proportion. (Pages 216–220)**

**42.** $\frac{5}{6} = \frac{c}{42}$     **43.** $\frac{x}{18} = \frac{6}{3}$

**44.** $\frac{12}{15} = \frac{f}{45}$     **45.** $\frac{8}{e} = \frac{4}{13}$

**46.** How many different single dip, one topping, frozen yogurt sundaes can you make from 5 flavors of yogurt and 6 toppings? **(Pages 247–251)**

**47.** How many different outfits can Stan make from 2 jackets, 3 pairs of jeans, and 5 T-shirts? **(Pages 247–251)**

# SUMMARY

## Key Terms

Cross-products (p. 217)
Dependent events (p. 242)
Event (p. 236)
Extremes (p. 217)
Factorial (p. 253)
Independent events (p. 241)
Means (p. 217)
Mutually exclusive events (p. 245)
Odds (p. 238)

Outcome (p. 236)
Permutation (p. 253)
Probability (p. 236)
Proportion (p. 217)
Rate (p. 222)
Ratio (p. 216)
Sample space (p. 236)
Tree diagram (p. 247)
Unit rate (p. 222)

## Key Ideas

**A. Property of Proportions:** If $\frac{a}{b} = \frac{c}{d}$, $b \neq 0$, $d \neq 0$, then $ad = bc$.

**B.** Probability of an event $= \frac{\text{Number of successful outcomes}}{\text{Total number of outcomes}}$.

**C.** The probability of an event that is certain to happen is 1.

**D.** The probability of an event that cannot happen is 0.

**E.** Odds in favor of an event $= \frac{\text{Number of successful outcomes}}{\text{Number of unsuccessful outcomes}}$.

**F.** Probability of independent and dependent events: See page 243.

**G. Fundamental Counting Principle**
If one choice can be made in $m$ ways, and the second choice can be made in $n$ ways, there are $m \times n$ ways of making the first choice followed by the second choice.

# REVIEW

**Write each ratio as a fraction in lowest terms. (Pages 216–220)**

1. 12 out of 16
2. 36 to 15
3. 20 : 48
4. 18 wins to 4 losses

**Solve and check each proportion. (Pages 216–220)**

5. $\frac{c}{8} = \frac{18}{72}$
6. $\frac{18}{10} = \frac{e}{15}$
7. $\frac{18}{15} = \frac{7}{y}$
8. $\frac{d}{27} = \frac{14}{21}$

**Solve. (Pages 221–225)**

**9.** In a quilt, the ratio of patterned material to solid color material in the design is $3:5$. If 9 square feet of patterned material is used, how many square feet of solid color material is used?

**10.** The ratio of green jelly beans to yellow jelly beans in a candy dish is $5:6$. If there are 24 yellow jelly beans, how many green jelly beans are in the dish?

**Replace each ? with the correct number. (Pages 226–230)**

**11.** 5 yd = ? in   **12.** 9 c = ? pt   **13.** 28 qt = ? gal   **14.** $6\frac{1}{4}$ lb = ? oz

**On a map, $\frac{1}{4}$ inch represents 50 miles. (Pages 226–230)**

**15.** If the actual distance between two cities is 350 miles, what is the map distance?

**16.** If the map distance between two cities is $2\frac{1}{2}$ inches, what is the actual distance?

**You throw a number cube once. Find each probability. (Pages 236–240)**

**17.** P(multiple of 1)   **18.** P(factor of 8)   **19.** P(8)

**20.** A bag of marbles contains 4 red, 6 blue, and 2 yellow marbles. One marble is drawn. What are the odds for drawing a yellow marble? **(Pages 236–240)**

**A box contains 4 red and 3 yellow marbles. Two marbles are drawn, one after the other. (Pages 241–246)**

**21.** If the first marble is replaced before the second one is drawn, what is the probability that the first one is yellow and the second one is red?

**22.** If the first marble drawn is not replaced before the second one is drawn, what is the probability that both are red?

**For Exercises 23–24, refer to the pizza toppings listed below. (Pages 247–251)**

Meats: pepperoni, sausage, beef
Vegetables: peppers, onions, mushrooms, olives

**23.** You order a pizza with one meat item and one vegetable item. How many choices do you have?

**24.** How many fewer choices will you have if you have only 2 meat toppings and 3 vegetable toppings to choose from?

**Find the number of permutations. (Pages 252–257)**

**25.** In how many different ways can 12 swimmers finish in first, second, and third places in a race?

**26.** In how many different ways can 10 job openings be filled by 4 applicants?

# TEST

1. Write the ratio 40 out of 72 as a fraction in lowest terms.
2. Solve the proportion $\frac{24}{56} = \frac{x}{35}$.

**The ratio of apple trees to peach trees in an orchard is 7 to 9.**

3. If there are 45 peach trees, how many apple trees are in the orchard?
4. If there are 63 apple trees, how many peach trees are in the orchard?
5. If a construction crew paved 9680 yards of road, how many miles of road did they pave? (1760 yards = 1 mile)
6. On a map, $\frac{1}{2}$ inch represents 250 miles. If the map distance between two cities is $1\frac{1}{2}$ inches, what is the actual distance?

**There are 4 red pens, 3 blue pens, 4 black pens, and 1 green pen in a desk drawer. A pen is chosen at random. Find each probability.**

7. P(blue pen)
8. P(red pen)
9. P(not choosing a green pen)

**One number cube is thrown. Find each probability.**

10. P(multiple of 3)
11. P(number = 3!)

**A coin purse contains 3 pennies, a nickel, 3 dimes, and 2 quarters.**

12. If the first coin is not replaced before the second coin is drawn, what is the probability that the first coin is a dime and the second coin is a quarter?
13. If the first coin is replaced before the second coin is drawn, what is the probability that the first coin is a nickel and the second coin is a penny?

**Mr. Brown makes chairs. He uses leather, corduroy, vinyl, or plastic for the upholstery. The frames are made of oak, maple, or ash.**

14. How many choices does a customer have?
15. What is the probability that a choice is oak with leather upholstery?
16. In how many different ways can 8 cross country runners finish in first and second places in a race?

Replace each ? with <, >, or = to make a true statement.
Pages 7–10, 13–18, 23–25, 28–31)
  **1.** $-3 + 2$ ? $4(-1)$     **2.** $-3 - (-5)$ ? $-7 + 4$     **3.** $(-2)(5)$ ? $-2 - 7$

Replace each ? with the number that makes a true statement.
Write the name of the property shown. (Pages 54–58)
  **4.** $(-6 + 0) + 8 = -6 + $ ?     **5.** $(7b + 8) + 6c = 7b + ($ ? $+ 6c)$     **6.** $xyz + 4 = yzx + $ ?

Find the perimeter and area for each figure. (Pages 64–68)
  **7.** Rectangle: $l = 2\frac{1}{2}$ meters, $w = 1\frac{1}{2}$ meters     **8.** Square: each side $1\frac{1}{4}$ feet long

Find the solution set if the replacement set is $\{-3, -1, 0, 1, 3\}$. (Pages 80–84)
  **9.** $-4x + 3 < 3$     **10.** $3c + 5 = 2$     **11.** $-7g - 9 < 5$     **12.** $6a - 8 = -26$

Solve and check. (Pages 91–99)
  **13.** $f + 29 = -16$     **14.** $\frac{n}{-3} = -18$     **15.** $-48 = -12z$     **16.** $t - 47 = -29$

Tell whether the number is prime or composite. If the number has more than two factors, determine if the number is divisible by 2, 3, 4, 5, 6, 8, 9, or 10. (Pages 131–136, 139–141)
  **17.** 55     **18.** 17     **19.** 41     **20.** 93     **21.** 558

For each pair of fractions, tell whether the two are equivalent. For each pair of equivalent fractions, write a third equivalent fraction in lowest terms. (Pages 163–167)
  **22.** $\frac{15}{10}$ and $\frac{6}{9}$     **23.** $\frac{21}{35}$ and $\frac{27}{45}$     **24.** $\frac{26}{91}$ and $\frac{16}{56}$     **25.** $\frac{55}{90}$ and $\frac{35}{75}$

Solve. Write each answer in lowest terms.
(Pages 171–174, 221–225)
  **26.** Sandy spent $\frac{3}{4}$ of an hour exercising. She spent $\frac{1}{3}$ of that time jogging. What fractional part of an hour did she spend jogging?

  **27.** Tom's recipe requires $\frac{1}{2}$ cup of milk. He plans to make only $\frac{1}{2}$ of the recipe. How much milk will he need?

Solve. (Pages 216–220)
  **28.** If 20 pounds of dog food cost $11, how much will 35 pounds cost?
  **29.** A can of frozen orange juice concentrate sells at 24 ounces for 48¢. Fresh orange juices sells at 32 ounces for $1.60. How much more per ounce does the fresh orange juice cost?

Cumulative Review: Chapters 1–6

# CHAPTER 7
# DECIMALS

## LESSON 7-1
## Rational Expressions

**Y**ou can use algebra to write familiar formulas in another way. Do you recognize these formulas?

$$l = \frac{A}{w} \qquad \frac{d}{t} = r \qquad w = \frac{A}{l} \qquad h = \frac{2A}{b}$$

The expressions $\frac{A}{w}, \frac{d}{t}, \frac{A}{l},$ and $\frac{2A}{b}$ are ratios. An algebraic expression written as a ratio is called a **rational expression.** Here are some examples of rational expressions.

$$\frac{3}{c} \qquad \frac{5}{m+1} \qquad \frac{a+6}{4} \qquad \frac{a+b}{c} \qquad \frac{x^2-y^2}{x}$$

To evaluate the rational expression $\frac{a+b}{c}$ for $a = 2$, $b = -3$, and $c = 2$, substitute these values for the variables. Then simplify.

$$\frac{a+b}{c} = \frac{2+(-3)}{2}$$
$$= \frac{-1}{2}$$

The result, $\frac{-1}{2}$, is a rational number. A *rational number* can be written as the ratio of two integers, where the denominator does not equal zero.

> **Rational Number**
> A rational number is a number that can be expressed in the form $\frac{a}{b}$, where $a$ and $b$ are integers and $b$ does not equal zero.

Since any integer, fraction, or mixed number can be expressed in the form $\frac{a}{b}$, integers, fractions, and mixed numbers are rational numbers.

In Chapter 5, you used fractions to represent a part of a whole, as shown in the figure at the right.

$\frac{3}{4}$ of the figure is green.

In Chapter 6, you used fractions to compare. In the figure at the right, the ratio of green rectangles to white rectangles is

$$\frac{\text{green rectangles}}{\text{white rectangles}} = \frac{3}{4}.$$

- What is the location of the rational number, $-\frac{3}{4}$, on the number line?

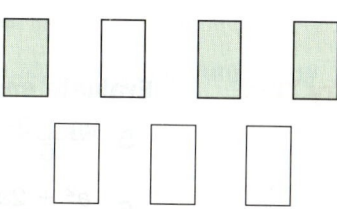

**EXAMPLE 1** Write each number in the form $\frac{a}{b}$, where $b \neq 0$.

a. $1\frac{3}{4}$  b. $-7$  c. $102$  d. $0$

**Solutions**
a. $1\frac{3}{4} = \frac{7}{4}$  b. $-7 = \frac{-7}{1}$
c. $102 = \frac{204}{2}$  d. $0 = \frac{0}{5}$

**Try This** Write each number in the form $\frac{a}{b}$, where $b \neq 0$.

1. $-9$  2. $1$  3. $3\frac{1}{8}$  4. $29$

Note that there are many ways to express each number in Example 1 in the form $\frac{a}{b}$. For example,

$$-7 = \frac{-7}{1} = \frac{-14}{2} = \frac{-21}{3} = \cdots \quad \text{and} \quad 0 = \frac{0}{1} = \frac{0}{15} = \frac{0}{99} = \cdots$$

7-1 Rational Expressions

## Practice and Apply

Replace each __?__ with <, >, or = to make a true statement.

9. $-\frac{2}{3}$ __?__ $-\frac{3}{5}$
10. $\frac{1}{12}$ __?__ $-\frac{6}{8}$
11. $\frac{5}{3}$ __?__ $\frac{12}{9}$
12. $-1\frac{3}{4}$ __?__ $\frac{-14}{8}$

Write in ascending order (least to greatest).

13. $-\frac{1}{5}, \frac{-2}{3}, \frac{3}{-2}, \frac{-1}{3}, \frac{-3}{4}$
14. $\frac{-1}{8}, \frac{3}{-8}, \frac{-2}{5}, -1, -\frac{4}{5}$

Simplify.

15. $\frac{2}{3} + \left(-\frac{4}{9}\right)$
16. $\frac{-3}{5} - \left(-\frac{13}{15}\right)$
17. $-\frac{2}{3} - \frac{3}{8}$
18. $-\frac{7}{8} \div \frac{5}{6}$
19. $-\frac{9}{4} \div \left(-\frac{3}{5}\right)$
20. $\frac{4}{5} \cdot \left(-\frac{6}{10}\right)$
21. $\frac{3}{2} - 1\frac{2}{3}$
22. $-3 \cdot \left(\frac{5}{3}\right)$

For Exercises 23–24, first write an equation to represent the problem. Solve the equation and answer the question.

23. If an elephant eats $5\frac{5}{12}$ bags of leaves, grasses, and bamboo shoots daily, about how many bags of vegetation would a herd of 18 elephants consume in a day?

24. The roasting time for a turkey is $\frac{1}{3}$ hours for each pound of turkey. About how much did the turkey weigh if it took $5\frac{1}{2}$ hours to cook?

## Connect and Extend

Simplify.

25. $\left(\frac{3}{4} - \frac{2}{3}\right)^2 \cdot \left(-\frac{4}{3}\right)$
26. $-\frac{4}{9} + \left[\frac{3}{5} - \left(-\frac{2}{5}\right)^2\right]$
27. $-\frac{3}{5}\left[-\frac{5}{4} + \left(-\frac{5}{3}\right)\right]$

28. Write a subtraction expression containing two rational numbers expressed as fractions with different denominators. The subtraction expression should equal $-\frac{3}{4}$.

29. Find a rational number between $-\frac{3}{5}$ and $-\frac{2}{5}$.

## Maintain Your Skills

Write each word expression as an algebraic expression. (Pages 39–43)

30. The number of hours in $x$ minutes
31. The cost of $y$ CD's at $9 each
32. The total cost for a $25 pair of shoes, a $5 bundle of socks, and tax, $t$

Write an equivalent fraction with a denominator of 100. (Pages 163–167)

33. $\frac{3}{4}$
34. $\frac{5}{8}$
35. $\frac{3}{8}$
36. $\frac{1}{2}$
37. $\frac{4}{5}$
38. $\frac{8}{125}$

> **Rational Number**
> A rational number is a number that can be expressed in the form $\frac{a}{b}$, where $a$ and $b$ are integers and $b$ does not equal zero.

Since any integer, fraction, or mixed number can be expressed in the form $\frac{a}{b}$, integers, fractions, and mixed numbers are rational numbers.

In Chapter 5, you used fractions to represent a part of a whole, as shown in the figure at the right.

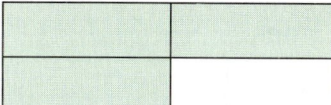

$\frac{3}{4}$ of the figure is green.

In Chapter 6, you used fractions to compare. In the figure at the right, the ratio of green rectangles to white rectangles is

$$\frac{\text{green rectangles}}{\text{white rectangles}} = \frac{3}{4}.$$

- What is the location of the rational number, $-\frac{3}{4}$, on the number line?

**EXAMPLE 1** Write each number in the form $\frac{a}{b}$, where $b \neq 0$.

a. $1\frac{3}{4}$    b. $-7$    c. $102$    d. $0$

**Solutions**  a. $1\frac{3}{4} = \frac{7}{4}$    b. $-7 = \frac{-7}{1}$

c. $102 = \frac{204}{2}$    d. $0 = \frac{0}{5}$

**Try This**  Write each number in the form $\frac{a}{b}$, where $b \neq 0$.

1. $-9$    2. $1$    3. $3\frac{1}{8}$    4. $29$

Note that there are many ways to express each number in Example 1 in the form $\frac{a}{b}$. For example,

$$-7 = \frac{-7}{1} = \frac{-14}{2} = \frac{-21}{3} = \cdots \quad \text{and} \quad 0 = \frac{0}{1} = \frac{0}{15} = \frac{0}{99} = \cdots$$

7-1 Rational Expressions

**EXAMPLE 2** Evaluate each rational expression.

a. $\dfrac{3b + 2}{c}$, for $b = 2$ and $c = -10$

b. $\dfrac{x^2 - y^2}{x + y}$, for $x = 3$ and $y = -2$

**Solutions** Substitute the given values for the variables and simplify.

a. $\dfrac{3b + 2}{c} = \dfrac{3(2) + 2}{-10}$

$= \dfrac{8}{-10} = \dfrac{4}{-5}$

b. $\dfrac{x^2 - y^2}{x + y} = \dfrac{(3)^2 - (-2)^2}{3 + (-2)}$

$= \dfrac{9 - 4}{1}$

$= \dfrac{5}{1}$, or **5**

**Try This** Evaluate each rational expression.

**5.** $\dfrac{m + 2k}{6}$, for $m = -1$ and $k = -4$

**6.** $\dfrac{a^2 + 2ab + b^2}{a + b}$, for $a = -2$ and $b = 5$

A number that is not rational is called an *irrational number*. Later in this textbook, you will study about irrational numbers such as $\pi$ and $\sqrt{2}$.

# EXERCISES

**Objectives:** To express rational numbers in the form $\dfrac{a}{b}$, where $b \neq 0$
To evaluate rational expressions

## Check Your Understanding

**Skill Check** Choose the letter of the correct answer.

**1.** Which is not a rational number?

a. $\dfrac{0}{1}$  b. 3

c. $-3$  d. $\pi$

**2.** Which is **not** expressed in the form $\dfrac{a}{b}$, where $a$ and $b$ are whole numbers?

a. $\dfrac{36}{6}$  b. 1

c. $\dfrac{0}{600}$  d. $\dfrac{1}{1000}$

3. What is the value of $\dfrac{2x - y}{8x}$ for $x = 3$ and $y = -4$?

   a. $\dfrac{1}{12}$   b. $\dfrac{5}{12}$   c. $\dfrac{-5}{12}$   d. $\dfrac{-5}{16}$

4. Evaluate $\dfrac{t^2 - 2rt + r^2}{t - r}$, for $t = 3$ and $r = 4$.

   a. 13   b. $-1$
   c. 1   d. $-13$

**True or False?** If a statement is false, give a reason.

5. Integers are rational numbers.
6. A rational number can be expressed as the ratio of two integers.
7. The ratio, $-\dfrac{3}{4}$, is not a rational number.
8. If $\dfrac{2a + b}{c}$ is a rational number, then $c$ cannot equal zero.

## Practice and Apply

Write each expression as a rational number in lowest terms.

9. $2 : 3$   10. $-4\dfrac{1}{3}$   11. $\dfrac{-5}{-10}$   12. $0 \div 3$   13. $8 \div 4$

14. Six out of 10
15. The quotient of $-4$ and 8

Write a rational number, in lowest terms, to represent each shaded area.

16.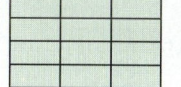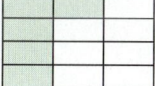

17.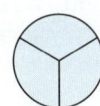

Evaluate each rational expression for $q = 2$, $r = -3$, and $s = 4$.

18. $\dfrac{2q}{r - s}$

19. $\dfrac{4r}{3s + 2rs}$

20. $\dfrac{sr + 6q}{3q - 2r}$

21. $\dfrac{r^2 - 2r}{-2(s - r) - 1}$

22. $\dfrac{s - q}{q - s}$

23. $\dfrac{2r - 3q}{3q - 2r}$

24. $\dfrac{s^2 - r^2}{s + r}$

25. $\dfrac{q^2 - 2qr + r^2}{q - r}$

26. $\dfrac{q - r}{r^2 - q^2}$

For Exercises 27–30, write each answer as a rational number in lowest terms.

27. In a class of 22 students, there are 3 under the age of 13. What fractional part of this class is under 13 years old?

28. If 2 cylinders on an 8-cylinder motor are not firing, what fractional part is firing?

29. Leon gave away 2 of his 25 baseball cards. What fractional part did he keep?

30. Allen bought $88 worth of camping equipment. If the tent cost $32, what fractional part of the total cost was this?

7-1  Rational Expressions   **265**

## Connect and Extend

**31.** Represent the rational number $2\frac{3}{5}$ with rectangles that are divided into 10 parts each.

**32.** Use the word "inverse" to explain why the denominator of a rational number in fractional form cannot be zero.

**33.** Use the term "Identity Property" to explain how a whole number can be written in fractional form.

**Determine the value of $x$ that will make the denominator equal zero.**

**34.** $\frac{5}{x}$  **35.** $\frac{x+3}{x-3}$  **36.** $\frac{5}{x+1}$  **37.** $\frac{x+10}{-5x}$

**38.** Use the guess-and-check strategy to determine the solution set for $x$.

$$\frac{3}{5+x} \geq 1$$

## Maintain Your Skills

**Write each group of integers in descending order. (Pages 7–10)**

**39.** $-6, 6, -7, 7, -5$   **40.** $-9, 8, -4, 2, -2$   **41.** $1, -5, 0, 6, -3$

**Find each product or quotient. Write the answers in lowest terms. (Pages 171–174, 189–193)**

**42.** $\frac{4}{9} \cdot \frac{35}{100}$   **43.** $\frac{1}{10} \div \frac{3}{12}$   **44.** $2\frac{1}{4} \div 6$   **45.** $\frac{16}{48} \cdot \frac{8}{22}$

**Solve. (Pages 200–205)**

**46.** Abdul plans the layout for the school newspaper. He uses $\frac{1}{4}$ of the sports page for advertisements and $\frac{2}{3}$ of the page for sports articles. How much of the page is left?

**47.** Eli worked $4\frac{3}{4}$ hours on Saturday and $6\frac{1}{2}$ hours on Sunday. How many hours did he work on Saturday and Sunday combined?

##  Math Team Problems

**48.** Measure out a length of string that is as long as you are tall. How many times will that length go around your head? Is the ratio of every person's head size to height the same?

**49.** Which would you prefer,—a stack of nickels equal to your height or a line of quarters laid out to equal your height? Explain.

# LESSON 7-2 Rational Numbers

You can use mental math to solve $n + \frac{1}{2} = 1$.

**Think:** Since $\frac{1}{2} + \frac{1}{2} = 1$, $n = \frac{1}{2}$.

What is the solution of $n + \frac{1}{2} = 0$?

To solve this problem, draw a number line that shows several fractions, each having 2 as a denominator. Each fraction to the right of 0 on the number line has an opposite to the left of 0, which is at the same distance from 0 on the number line.

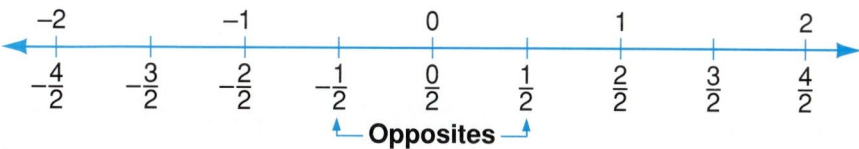

Since $\frac{1}{2}$ and $-\frac{1}{2}$ are opposites, their sum is 0. So the solution of $n + \frac{1}{2} = 0$ is $-\frac{1}{2}$.

The opposite of $\frac{1}{2}$ can be written in these three ways.

$$-\frac{1}{2} \qquad \frac{-1}{2} \qquad \frac{1}{-2}$$

Each of these ways represents the same number on the number line. Usually, the opposite of $\frac{1}{2}$ is written as $-\frac{1}{2}$.

You can use a number line to compare $-\frac{2}{3}$ and $-\frac{1}{3}$. Since $-\frac{2}{3}$ is to the left of $-\frac{1}{3}$ on the number line, $-\frac{2}{3} < -\frac{1}{3}$.

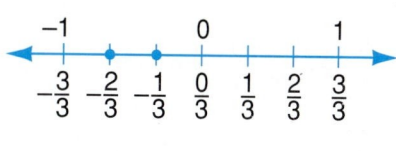

To compare rational numbers having like denominators, you can compare the numerators. When one or more of the rational numbers is negative, think of the numerator as having the negative sign.

$$-\frac{2}{3} < -\frac{1}{3} \text{ because } -2 < -1.$$

$$\frac{1}{5} > -\frac{1}{5} \text{ because } 1 > -1.$$

7-2 Rational Numbers **267**

To compare rational numbers having unlike denominators, rewrite the rational numbers as equivalent fractions having the same denominator. Then compare the numerators.

**EXAMPLE 1** Compare the two rational numbers. Write $<$ or $>$.

a. $-\frac{2}{3}$ and $-\frac{5}{6}$ 	b. $\frac{3}{4}$ and $\frac{7}{10}$

**Solutions**

a. The least common denominator of 3 and 6 is 6.

Write equivalent fractions. $\quad -\frac{2}{3} = -\frac{4}{6} \qquad -\frac{5}{6} = -\frac{5}{6}$

Compare the numerators. $\quad$ Since $-4 > -5$, $-\frac{4}{6} > -\frac{5}{6}$.

Compare $-\frac{2}{3}$ and $-\frac{5}{6}$. $\quad$ Since $-\frac{4}{6} > -\frac{5}{6}$, $-\frac{2}{3} > -\frac{5}{6}$.

b. The least common denominator of 4 and 10 is 20.

Write equivalent fractions. $\quad \frac{3}{4} = \frac{15}{20} \qquad \frac{7}{10} = \frac{14}{20}$

Compare the numerators. $\quad$ Since $14 < 15$, $\frac{14}{20} < \frac{15}{20}$.

Compare $\frac{3}{4}$ and $\frac{7}{10}$. $\quad$ Since $\frac{14}{20} < \frac{15}{20}$, $\frac{7}{10} < \frac{3}{4}$.

**Try This** Compare each pair of rational numbers. Write $<$ or $>$.

1. $-\frac{1}{2}$ and $-\frac{5}{8}$ 	2. $-\frac{3}{4}$ and $-\frac{4}{5}$

To add and subtract rational numbers, follow the rules for adding and subtracting fractions and integers.

**EXAMPLE 2** Simplify: a. $\frac{3}{4} + \left(-\frac{1}{2}\right)$ 	b. $-\frac{4}{5} - \left(-\frac{2}{3}\right)$

**Solutions**

a. $\frac{3}{4} + \left(-\frac{1}{2}\right)$

$\frac{3}{4} - \frac{1}{2}$

$\frac{3}{4} - \frac{2}{4}$

$\frac{3 - 2}{4}$

$\frac{1}{4}$

b. $-\frac{4}{5} - \left(-\frac{2}{3}\right)$

$-\frac{4}{5} + \frac{2}{3}$

$-\frac{12}{15} + \frac{10}{15}$

$\frac{-12 + 10}{15}$

$\frac{-2}{15}$, or $-\frac{2}{15}$

**Try This** Simplify: 3. $-\frac{1}{3} + \frac{5}{8}$ 	4. $\frac{1}{2} - \left(-\frac{1}{3}\right)$

## LESSON 7-3

# Decimals and Fractions

The mileage indicator, or odometer, on Reta's family car has recently turned 32,000.2 miles. She knows that the number at the extreme right indicates tenths of a mile.

What will be the next reading after 32,000.2 miles?

The **decimal** odometer reading, 32,000.3 is another way of expressing the rational number $32,000\frac{3}{10}$.

The ten basic symbols in our decimal system are 0, 1, 2, 3, 4, 5, 6, 7, 8, 9. Each of these symbols is called a **digit,** and the digits are used separately or in combination with other digits to write numbers. Each digit in a number has a **place value.**

| 2 | 6 | . | 5 | 7 | 3 |
|---|---|---|---|---|---|
| tens | ones | | tenths | hundredths | thousandths |

Notice that the decimal point separates the number into the whole number part and the fractional part.

You can use place values to write decimals in expanded form.

| Number | Expanded Form |
|---|---|
| 254 | $200 + 50 + 4$ |
| 106.5 | $100 + 0 + 6 + \frac{5}{10}$ |
| 9.83 | $9 + \frac{8}{10} + \frac{3}{100}$ |
| 0.306 | $\frac{3}{10} + \frac{0}{100} + \frac{6}{1000}$ |

**EXAMPLE 1** Write in expanded form: **a.** 520.6 **b.** 0.4302

**Solutions**
**a.** $520.6 = 500 + 20 + 0 + \frac{6}{10}$

**b.** $0.4302 = \frac{4}{10} + \frac{3}{100} + \frac{0}{1000} + \frac{2}{10,000}$

**Try This** Write in expanded form: **1.** 120.7 **2.** 0.0325

You can write a fraction for a decimal.

**EXAMPLE 2** Write a mixed number or a fraction for the decimals.
**a.** 3.65 **b.** −0.875

**Solutions** **a.** Write 3.65 as 3 + 0.65.
$$3.65 = 3 + 0.65$$
$$= 3 + \frac{65}{100}$$
$$= 3 + \frac{13}{20}, \text{ or } 3\frac{13}{20}$$

**b.** Since −0.875 is a negative decimal, the equivalent fraction is also negative.

$$-0.875 = -\frac{875}{1000}$$

Divide the numerator and denominator by 125.
$$= -\frac{875 \div 125}{1000 \div 125}$$
$$= -\frac{7}{8}$$

**Try This** Write a mixed number or a fraction for the decimal.
**3.** 8.225 **4.** −0.0625

# EXERCISES

**Objective:** To write a decimal in expanded form
To write a fraction for a decimal

## Check Your Understanding

**Skill Check** Write the letter of the correct answer.

**1.** What does the digit 9 in the number 1097.623 represent?
   **a.** tenths    **b.** hundreds
   **c.** tens      **d.** hundredths

**2.** Which fraction represents 0.48?
   **a.** $\frac{48}{10}$    **b.** $4\frac{4}{5}$
   **c.** $\frac{6}{125}$    **d.** $\frac{12}{25}$

**3.** Which is the decimal for $3 + \frac{0}{10} + \frac{1}{100} + \frac{5}{1000}$?
   **a.** 30.15    **b.** 301.5
   **c.** 3.15     **d.** 3.015

**4.** Which mixed number represents $-2.125$?
   **a.** $-2\frac{5}{8}$    **b.** $2\frac{5}{8}$
   **c.** $-2\frac{1}{8}$    **d.** $-2$

Write the decimal that represents the shaded part of the figure.

**5.** 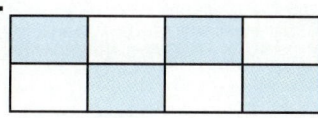    **6.** 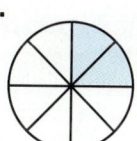    **7.**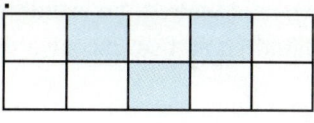

## Practice and Apply

Write each number in expanded form.

**8.** 351.042    **9.** 0.013    **10.** 70.021    **11.** 108.6374

Write a decimal for each of the following.

**12.** $900 + 2 + \frac{8}{10} + \frac{2}{100} + \frac{4}{1000}$    **13.** $400 + 70 + 9 + \frac{6}{100} + \frac{1}{1000}$

**14.** $7000 + 2 + \frac{5}{10} + \frac{9}{10,000}$    **15.** $50 + \frac{8}{1000}$

Write a mixed number or a fraction in lowest terms for the decimal.

**16.** 3.2       **17.** 47.032    **18.** $-8.48$    **19.** $-12.6$
**20.** 5.002    **21.** $-93.85$    **22.** 72.55     **23.** 0.004

Replace the ? with <, =, or > to obtain a true statement.

**24.** $\frac{1}{4}$ ? 0.25    **25.** $0 \cdot 1$ ? $\frac{1}{3}$    **26.** 0.70 ? $\frac{7}{10}$    **27.** 0.8 ? $\frac{7}{8}$

**28.** Use a dictionary to find the meaning of the prefix "deci-." Explain how this meaning is related to the word "decimal."

7-3  Decimals and Fractions   **273**

Use your answer to Exercise 28 to give the meaning of each of these words.

**29.** decimeter  **30.** decigram  **31.** deciliter

**32.** Explain this statement by using examples.

In our decimal system, the value of each decimal place is ten times the value of the place to its right.

## Connect and Extend

You can use exponents with base ten to write numbers in expanded form.

**Example:** $156.379 = 1 \times 10^2 + 5 \times 10^1 + 6 \times 10^0 + \frac{3}{10^1} + \frac{7}{10^2} + \frac{9}{10^3}$

Use exponents with base ten to write each number in expanded form.

**33.** 8.7025  **34.** 0.05003  **35.** 0.6203  **36.** 0.00001

Write each number as a mixed number or a fraction in lowest terms and as a decimal.

**37.** seventy-two and twenty-six hundredths

**38.** five and twenty-five thousandths

**39.** negative six hundred thousandths

**40.** three hundred and seventy-five thousandths

## Maintain Your Skills

Find each quotient. (Pages 32–35)

**41.** $\frac{-15}{3}$  **42.** $\frac{-72}{-9}$  **43.** $84 \div -7$  **44.** $\frac{-117}{13}$  **45.** $-162 \div -18$

Trudy is $t$ years old and Lori is $l$ years old. Write an algebraic expression to describe each situation. (Pages 110–113)

**46.** Stan is 7 years older than $\frac{1}{3}$ Trudy's age.

**47.** Virginia is 12 years younger than twice Lori's age.

A bag of marbles contains 4 red, 1 yellow, and 3 blue marbles. Find the probability of each of the following situations. (Pages 241–246)

**48.** Drawing a blue then a yellow marble, if the first marble drawn is *not* replaced.

**49.** Drawing two yellow marbles in a row, if the first marble drawn *is* replaced.

**50.** Drawing two red marbles in a row, if the first marble drawn is *not* replaced.

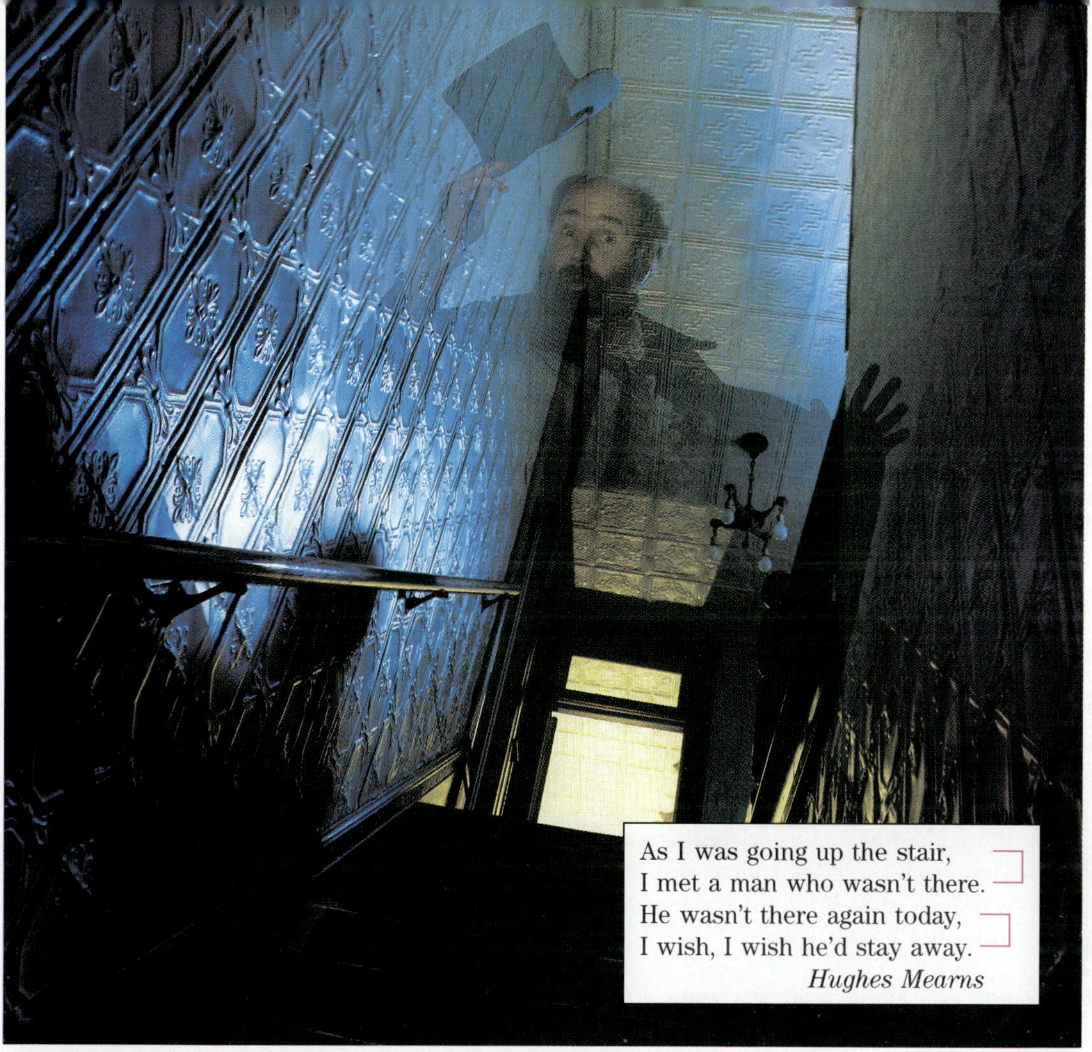

> As I was going up the stair,
> I met a man who wasn't there.
> He wasn't there again today,
> I wish, I wish he'd stay away.
> — *Hughes Mearns*

## LESSON 7-4  Problem Solving Strategies: Drawing a Diagram

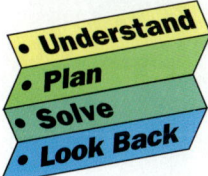

**N**otice that the lines of the poem that rhyme with each other are shown by drawing red connecting lines. A poem is **completely rhymed** when each line rhymes with at least one other line in the poem.

In how many ways can a four-line poem be completely rhymed?

Draw a diagram to show the possibilities.

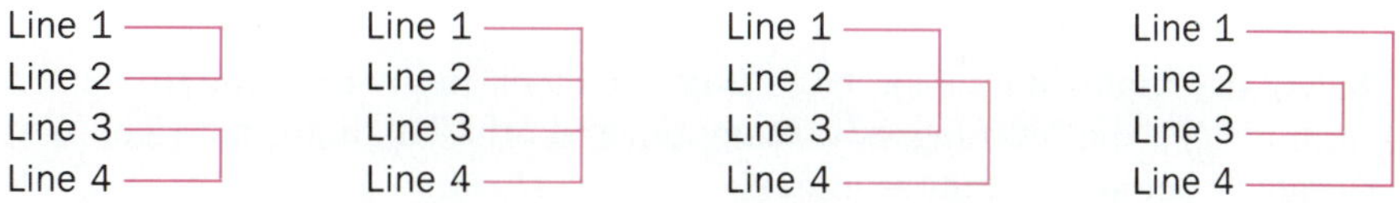

Line 1
Line 2
Line 3
Line 4

Line 1
Line 2
Line 3
Line 4

Line 1
Line 2
Line 3
Line 4

Line 1
Line 2
Line 3
Line 4

The diagram shows that there are 4 ways in which a four-line poem can be completely rhymed.

In how many ways can a five-line poem be completely rhymed?

The solution to this problem was given about 1000 years ago by Lady Murasaki of Japan. She showed that there are 11 ways for a five-line poem to be completely rhymed and 41 ways for a five-line poem to be incompletely rhymed.

A poem is **incompletely rhymed** when one or more lines have no other line to rhyme with them. In the 41 ways for a five-line poem to be incompletely rhymed, there are 52 different rhyme schemes!

# EXERCISES

**Objective:** To apply strategies in problem solving

1. If a poem has 3 lines, it can be completely rhymed in only one way. That is, all three lines must rhyme with each other. In how many ways can a three-line poem be incompletely rhymed? Draw a diagram to illustrate your answer.

**Refer to the diagram at the right to answer these questions.**

2. How long is the segment marked $t$?

3. How long is the segment marked $s$?

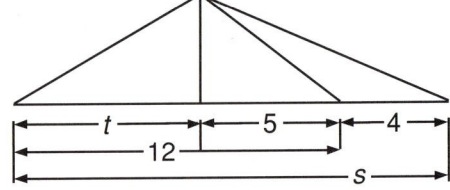

4. Maria has a square picture that is 8 inches on a side. She places it on a mat that forms a 2-inch border around the picture. What is the area of the matted picture?

5. Giorgio draws a square that is 9 inches on a side. He finds the midpoint of one side, and from that point he draws a line segment to each of the opposite corners of the square. What is the area of the largest triangle he has drawn?

6. As the hands of a clock move from 6 o'clock in the morning to 6 o'clock that evening, how many times do the hands form a right angle (an angle of 90 degrees)?

**276** CHAPTER 7

# Repeating Decimals

**A**ny rational number can be written as a *terminating* or as a *repeating* decimal. A number such as 0.75 is a **terminating** decimal because it terminates (ends) after the "5" in 0.75.

You know that a decimal is repeating when you see a repeating pattern. That is,

$$0.454545\cdots \text{ is a } \textbf{repeating decimal}$$

because the digits 4 and 5 form a pattern that repeats without end. You use a bar to indicate the digits that repeat.

$$0.454545\ldots = 0.\overline{45}$$

You can use a calculator to show that $\frac{5}{11} = 0.\overline{45}$.

$$5 \;\boxed{\div}\; 11 \;\boxed{=}$$

## EXPLORE

1. Copy and complete this table. Use a calculator. Use a bar to indicate the digits that repeat.

| Rational number | $\frac{4}{99}$ | $\frac{7}{99}$ | $\frac{11}{99}$ | $\frac{15}{99}$ | $\frac{25}{99}$ | $\frac{48}{99}$ |
|---|---|---|---|---|---|---|
| Decimal | $0.\overline{04}$ | ? | $0.\overline{1}$ | ? | ? | ? |

2. Guess the decimal for $\frac{66}{99}$. Use a calculator to check your answer. Explain why your guess was, or was not, correct.

3. Express $\frac{2}{3}$ as a decimal.

4. Why is the result the same as for $\frac{66}{99}$?

Since repeating decimals are unending, you can express them in two ways.

**a.** Use a bar to show the digits that repeat when you want to know the exact value of the decimal.

**b.** When an exact value is not needed, round the decimal to the nearest tenth, hundredth, thousandth, and so on, as required.

**EXAMPLE** Express $\frac{7}{11}$ as a repeating decimal and as a decimal rounded to the nearest thousandth.

**Solution** **Exact value:** $\frac{7}{11} = 0.\overline{63}$ (calculator) or $11\overline{)7.0000}^{\,0.6363...} = \mathbf{0.\overline{63}}$

**Rounded value:** Express the decimal to ten-thousandths, then round to the nearest thousandth.

$$\frac{7}{11} = 0.6363...$$

So, $\frac{7}{11} \approx \mathbf{0.636}.$

**Try This** Express each fraction as a repeating decimal and as a decimal rounded to the nearest thousandth.

1. $\frac{1}{3}$
2. $\frac{7}{6}$
3. $-\frac{8}{11}$
4. $\frac{2}{15}$

# EXPLORE

5. Copy and complete this table. Use a calculator or paper and pencil.

| Rational number | $\frac{3}{4}$ | $-\frac{5}{8}$ | $\frac{3}{16}$ | $\frac{8}{25}$ | $-\frac{9}{100}$ |
|---|---|---|---|---|---|
| Decimal | ? | ? | ? | ? | ? |

6. How are the decimals in the table alike?
7. If the denominator of a rational number factors into a product of 2's and 5's, can the rational number be expressed as a terminating or as a repeating decimal? Explain how you know.
8. If the denominator of a rational number in lowest terms has a prime factor that is not 2 or 5, can the rational number be expressed as a terminating decimal? Test your answers with some examples.
9. Write a summary that explains how the denominator of a rational number can be used to determine if the decimal form of the rational number repeats or terminates.

Decimals that terminate or repeat are called **rational numbers.** Some decimals, however, never terminate nor repeat. These are called **irrational numbers.** The number $\pi$ is an example of an irrational number.

# EXERCISES

**Objective:** To write rational numbers as decimals

## Check Your Understanding

**Skill Check** Write the letter of the correct answer.

1. Which is the repeating decimal for $\frac{5}{6}$?
   a. $0.8\overline{3}$
   b. $0.\overline{83}$
   c. $0.\overline{83}$
   d. $0.\overline{88}$

2. If $\frac{35}{63} = 0.\overline{5}$, which of these does **not** have the same value?
   a. $\frac{5}{9}$
   b. $\frac{1}{2}$
   c. $\frac{40}{72}$
   d. $\frac{55}{99}$

3. Which is the decimal for $\frac{3}{11}$ rounded to the nearest tenth?
   a. 0.3
   b. $0.\overline{27}$
   c. 0.2
   d. $0.\overline{2}$

4. Which is the exact value for $\frac{2}{25}$?
   a. $0.\overline{8}$
   b. 0.08
   c. $0.\overline{08}$
   d. 0.8

**True or false?** When a statement is false, explain why.

5. The exact value of $\frac{1}{3}$ expressed as a decimal is $1.\overline{3}$.
6. Any rational number can be written as a terminating or as a repeating decimal.
7. The exact value of $\frac{3}{5}$ expressed as a decimal is $0.\overline{6}$.

## Practice and Apply

**Tell whether the decimal form terminates or repeats.**

8. $\frac{1}{2}$
9. $-\frac{2}{3}$
10. $\frac{9}{4}$
11. $-\frac{3}{5}$
12. $\frac{11}{12}$
13. $-\frac{7}{10}$

**Round each decimal to the nearest tenth, hundredth, and thousandth.**

14. 431.6729
15. −38.02561
16. 0.7019
17. −1.9403

For Exercises 18–23, express each rational number as a decimal in two ways. First, give the exact value. Then give the value rounded to the nearest hundredth.

18. $\frac{2}{15}$
19. $-\frac{8}{9}$
20. $\frac{15}{18}$
21. $\frac{2}{27}$
22. $\frac{5}{33}$
23. $-\frac{1}{11}$

24. Write a repeating decimal that has 4 digits in the part that repeats.

25. Write a repeating decimal that has more than 5 digits in the part that repeats.

7-5  Repeating Decimals   **279**

## Connect and Extend

**26.** This exercise shows how to write a fraction for $0.\overline{1}$. Give a reason for each step.

**Statements**  
Let $n = 0.\overline{1}$.
a. Then $10n = 1.\overline{1}$.
b. $10n - n = 1.\overline{1} - 0.\overline{1}$
c. $9n = 1$
d. $n = \frac{1}{9}$

**Reasons**  
?  
?  
?  
?

**27.** This exercise shows how to write a fraction for $9.\overline{27}$. Give a reason for each step.

**Statements**  
Let $n = 9.\overline{27}$.
a. Then $100n = 927.\overline{27}$.
b. $100n - n = 927.\overline{27} - 9.\overline{27}$
c. $99n = 918$
d. $n = \frac{918}{99}$, or $\frac{102}{11}$

**Reasons**  
?  
?  
?  
?

**28.** Write a fraction in lowest terms for $0.\overline{3}$.

**29.** Write a fraction in lowest terms for $2.\overline{55}$.

## Maintain Your Skills

**Write the largest number and the smallest number in each set. (Pages 7–10)**

**30.** $\{-2, 0, -7, 6, 3\}$  **31.** $\{-12, 4, -3, 1, -15\}$  **32.** $\{-5, -12, -1, -7, -10\}$

**Copy the number. If the number is even, underline it. If it is divisible by four, circle it. If it is divisible by eight, draw a box around it. (Pages 131–136)**

**33.** 218  **34.** 91  **35.** 128  **36.** 812  **37.** 20

**Solve. (Pages 241–246, 252–257)**

**38.** A bag of marbles contains 4 blue, 2 yellow, and 3 red marbles. If one marble is drawn, what is the probability that it is a yellow or a red marble?

**39.** Diana wants to buy a toy, a cat bed, and a cat collar for her cat. The pet store carries 10 different cat toys, 3 different cat beds, and 12 different cat collars. In how many different ways can Diana choose a toy, a bed, and a collar for her cat?

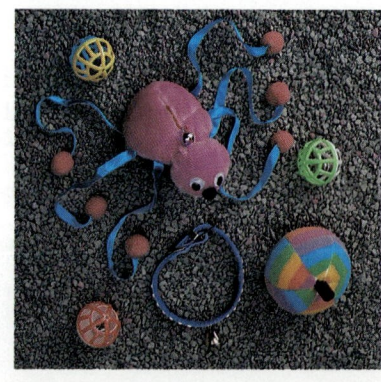

## MIDCHAPTER

**Write each expression as a rational number in lowest terms. (Pages 262–266)**

1. $-3\frac{7}{8}$
2. $10:12$
3. $15 \div (-3)$
4. $\frac{-5}{-30}$
5. Two out of 6

**Evaluate each rational expression for $r = 3$, $s = -3$, and $t = 2$. (Pages 262–266)**

6. $\frac{4s}{3t - 2rs}$
7. $\frac{s^2 - 3t}{-5(r - s)}$
8. $\frac{r^2 + s^2}{rt}$
9. $\frac{-7rs^2}{2t + 9s}$

**Simplify. (Pages 267–270)**

10. $-\frac{1}{2} + \frac{1}{5}$
11. $\left(-\frac{3}{8}\right)\left(-\frac{2}{3}\right)$
12. $\frac{2}{3} - \frac{1}{12}$
13. $-\frac{8}{16} + \frac{3}{4}$
14. $2\frac{1}{4} \div (-6)$
15. $-1\frac{1}{2} - \frac{2}{3}$
16. $1\frac{3}{4} \cdot \left(-\frac{4}{5}\right)$
17. $-2\frac{1}{2} \div \left(-1\frac{3}{4}\right)$

**For each decimal, write a mixed number or a fraction in lowest terms. (Pages 271–274)**

18. $-7.04$
19. $64.35$
20. $33.022$
21. $-0.0008$

**Express each rational number as a repeating decimal and as a decimal rounded to the nearest hundredth. (Pages 277–280)**

22. $\frac{2}{3}$
23. $\frac{7}{15}$
24. $\frac{1}{12}$
25. $\frac{10}{33}$
26. $\frac{6}{11}$
27. $\frac{7}{27}$

## Mathematical Footnote

The ancient Greek mathematicians thought that one, or unity, could not be divided. So the idea of a fraction made them uncomfortable.

The Egyptians could work only with fractions having a numerator of 1. To solve this problem found in the Rhind papyrus (around 1600 B.C.),

"Find a number such that the sum of it and one seventh of it together shall equal 19,"

the Egyptians wrote this expression.

$$16 + \frac{1}{2} + \frac{1}{8}.$$

Will the expression give the correct answer?

## LESSON 7-6  Estimating Sums and Differences

The bar graph above shows the number of inches of snow that fell in 5 days in White Valley. The bar graph can help you to compare the data.

- How can you tell from the bar graph that Friday had the greatest snowfall for the five days and Tuesday had the least?

**EXAMPLE 1**
a. Refer to the graph to compare the snowfall on Monday and Wednesday. Which is greater?
b. If the actual snowfall on Monday is 2.6 inches and the actual snowfall on Wednesday is 2.32 inches, estimate the difference.

**Solutions**

a. Compare the bars. Then make estimated readings from the graph to support your answer.

Mon: 2.6 in    Wed: 2.3 in
2.3 < 2.6
or
**2.6 > 2.3**

b. Round 2.32 to tenths, the same number of places as in 2.6.
2.6 − 2.3 = 0.3
So it snowed about **0.3** of an inch **more** on Monday.

282  CHAPTER 7

**Try This**

1. Estimate the difference in snowfall on Monday and on Thursday.
2. Estimate the difference in snowfall between the day it snowed the most and the day it snowed the least.

In many problem situations, you do not need to find an exact answer. An estimate may be close enough.

**EXAMPLE 2** The record snowfall for one week in White Valley is 14 inches. Did the snowfall for the week as recorded in the table below break the record?

**Solution** Estimate the snowfall for each day by rounding up to the next whole number. Then add the estimates.

|          | Mon | Tues | Wed  | Thur | Fri |
|----------|-----|------|------|------|-----|
| Actual   | 2.6 | 1.37 | 2.32 | 1.8  | 2.9 |
| Estimate | 3   | 2    | 3    | 2    | 3   |

**Total:** 3 + 2 + 3 + 2 + 3 = 13

Since the total snowfall is about 13 inches and this is less than the record of 14 inches, the record was **not** broken.

**Try This**

3. The record snowfall for three successive days in Snow Valley is 8 inches. Did the snowfall recorded in the table for Wednesday, Thursday, and Friday break the record?

**EXAMPLE 3** Filbert has $10. He wants to buy these items at the grocery store. Does he have enough money?

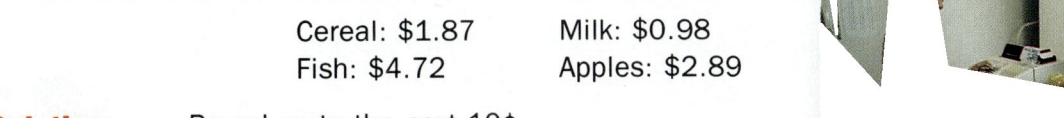

Cereal: $1.87   Milk: $0.98
Fish: $4.72     Apples: $2.89

**Solution** Round up to the next 10¢.

$1.87 ≈ $1.90     $0.98 ≈ $1.00
$4.72 ≈ $4.80     $2.89 ≈ $2.90

**Total estimate:** $1.90 + $1.00 + $4.80 + $2.90 = $10.60

You can also estimate by using the front digits.

Add the dollars (front digits): $1 + $4 + $2 = $7
Adjust the cents:   Think:      87¢ + 98¢ + 72¢ + 89¢ ≈ $4
Total estimate:                 $7 + $4 = $11

Both are good estimates, but rounding up to the next 10¢ gives the better estimate. Filbert does **not** have enough money.

**Try This**  Estimate the sum by using two different strategies.
4. $18.47 + $9.85 + $11.30

Sometimes it is useful to estimate before computing an exact answer.

**EXAMPLE 4**  Estimate this difference: 15.65 − 7.347. Then use paper and pencil or a calculator to find the exact answer and compare it with the estimate.

**Solution**  Use front digits or rounding to estimate.
**Front digits:** 15 − 7 = 8   and   0.65 > 0.347
The difference should be a little more than **8**.
Rounding: 15.65 ≈ 16   and   7.347 ≈ 7   16 − 7 = **9**
**Using a calculator:**   15.65   7.347 = **8.303**
Since the answer is close to the estimates, the answer is reasonable.

**Try This**  Estimate. Then find the exact answer. Compare your answer with the estimate.
5. 15.8 − 4.15
6. $92.87 − $44.38

# EXERCISES

**Objective:** To estimate sums and differences of decimals
To add and subtract decimals

## Check Your Understanding

**Skill Check**  Write the letter of the correct answer.

**1.** At Games Galore, Tim paid $2.79 for ping pong balls, $6.39 for dominoes, and $2.35 for a book of rules for the games. Estimate the total cost to the nearest 10¢.
   a. $11.00   b. $11.60   c. $11.70   d. $11.80

2. Estimate by rounding to the nearest whole number: 407.451 − 272.109
   a. 130   b. 135   c. 100   d. 125

3. For which situation would you need an exact answer?
   a. Reporting the number of onlookers at a parade
   b. Advertising the number of a newspaper's readers
   c. Recording the number of workers on a payroll
   d. Reporting the number of deer in a forest

4. Samantha rounded four prices up to the nearest dollar. None of the prices was an exact number of dollars. Which of these statements is true?
   a. The actual sum is greater than the estimated sum.
   b. The actual sum is less than the estimated sum.
   c. The actual sum equals the estimated sum.
   d. The estimated sum is twice the actual sum.

## Practice and Apply

This list shows the prices of 5 items at Breakfront's Grocery. Use < or > to compare the prices. Then estimate the difference in price by rounding to the nearest 10¢.

5. Bread and grapes
6. Detergent and light bulbs
7. Eggs and bread
8. Estimate the total cost of each pair in Exercises 5–7 by rounding up to the next dollar.
9. Estimate the total cost of each pair in Exercises 5–7 by rounding up to the next 10¢.

Grapes $1.89
Eggs $0.99
Light bulbs $2.79
Bread $1.34
Detergent $2.98

10. Find the actual total cost of each pair in Exercises 5–7. Tell which estimate is closer, the one in Exercise 8 or the one in Exercise 9.

**Estimate each sum or difference. Then find the exact answer. Compare your answer with the estimate.**

11. 30.02 + 7.304
12. 9.731 − 2.06
13. −5.92 − 7.843
14. 43.4 + (−6.23)
15. −9.41 + 10.306
16. 17.844 − 11.918
17. 25 − 6.772
18. 19 + 12.025
19. 7.81 − 12.005

**Solve each problem. Use estimation to check whether the answer is reasonable.**

20. Randolph took 3 buses to get across the city. He traveled 4.6 miles on the first bus, 8.2 miles on the second, and 11.05 miles on the third. How far did he travel?

21. Crystal bought a pair of shoes for $58.49, a blouse for $15.89, a pair of slacks for $32.58, and an umbrella for $14.99. The tax was $7.32. How much change will she get from $130?

7-6 Estimating Sums and Differences   **285**

22. Sheena wrote four checks for $12.34, $9.23, $15.41, and $22.05. If she has $75 in her checking account, is this enough to cover all the checks? If so, how much is left in the account? If not, how much is she short?

## Connect and Extend

23. Give an example of when you would not want to estimate an answer by rounding up.
24. Give an example of when you would not want to estimate an answer by rounding down.
25. Typo Card and Variety store sells pencils at 7 for $1.34. How much will Joan pay for one pencil? Explain.
26. Enter .01 [+] on a calculator. What number will show on the display after [=] is pressed 10 times? 65 times? 100 times?
27. Enter .1 [+] on a calculator. What number will show on the display after [=] is pressed 10 times? 43 times? 1000 times?
28. The figure at the right is a square. Find the value of s.
29. Which is greater, $\frac{17}{8}$ or 2.25? Explain.
30. Which is less, $\frac{1}{3}$ or 0.3? Explain.

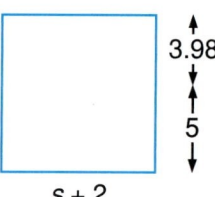

## Maintain Your Skills

**Use positive and negative numbers to represent each situation by an addition problem. Then find the sum. (Pages 2–6, 13–18)**

31. The Wildcats football team had a gain of 15 yards followed by a loss of 31 yards.
32. A boat sailed 18 kilometers due east. Then the boat turned and sailed 23 kilometers due west.
33. Cao has $44 in his savings account. He deposited $17.
34. Barbara has $25 in the bank. She withdrew $7.

**Write the answer only if it is an odd number. (Pages 131–136)**

35. 10,379 + 64,799      36. 60,842 ÷ 2      37. 52,438 − 21,979

**Write a proportion to find each amount. (Pages 226–230)**

38. 96 in = _?_ ft      39. 10 ft = _?_ yd      40. 24 oz = _?_ lb      41. $3\frac{1}{2}$ gal = _?_ qt

# LESSON 7-7 Problem Solving Exploration: Powers of Ten

**A.** Copy and complete the pattern at the right.

$10^1 = 10$
$10^2 = 100$
$10^3 = \underline{?}$
$10^4 = \underline{?}$
$10^5 = \underline{?}$

1. What is the relationship between the exponents and the number of zeros in the product?

2. Use your answer to Exercise 1 to write an argument that shows that $10^0 = 1$.

**B.** Copy and complete the pattern at the right.

$100 = 10^2$
$10 = 10^1$
$1 = 10^0$
$\frac{1}{10} = 10^?$
$\frac{1}{100} = 10^?$

3. Use the pattern to write an argument showing that $\frac{1}{10^2} = 10^{-2}$.

**C.** Copy and complete these patterns.

$10^4 \cdot 10^3 = (10 \cdot 10 \cdot 10 \cdot 10)(10 \cdot 10 \cdot 10)$
$\phantom{10^4 \cdot 10^3} = 10{,}000{,}000$
$\phantom{10^4 \cdot 10^3} = 10^7$

4. How many zeros are there in 10,000,000?

5. What is the exponent in $10^7$?

6. $10^5 \cdot 10^{-2} = (10 \cdot 10 \cdot 10 \cdot 10 \cdot 10)\left(\frac{1}{10} \cdot \frac{1}{10}\right)$
$\phantom{10^5 \cdot 10^{-2}} = 10 \cdot 10 \cdot 10$
$\phantom{10^5 \cdot 10^{-2}} = 10^?$

7. $10^3 \cdot 10^{-3} = (10 \cdot 10 \cdot 10)\left(\frac{1}{10} \cdot \frac{1}{10} \cdot \frac{1}{10}\right)$
$\phantom{10^3 \cdot 10^{-3}} = 1$
$\phantom{10^3 \cdot 10^{-3}} = 10^0$

8. $10^{-3} \cdot 10^{-2} = \left(\frac{1}{10} \cdot \frac{1}{10} \cdot \frac{1}{10}\right)\left(\frac{1}{10} \cdot \frac{1}{10}\right)$
$\phantom{10^{-3} \cdot 10^{-2}} = \frac{1}{100{,}000}$
$\phantom{10^{-3} \cdot 10^{-2}} = 10^?$

9. Refer to Exercises 6–8 to explain the relationship between the exponents in the factors and the power of 10 in the product.

10. Use the relationship in Exercise 9 to complete this generalization.

   If $a$ and $b$ are integers, then $10^a \cdot 10^b = 10^?$.

D. Copy and complete the pattern at the right.

11. Use the pattern to complete this statement.

   If $d$ is a whole number, then
   $$2^{-d} = \frac{1}{2^?}.$$

| | |
|---|---|
| $2^3$ | $= 8$ |
| $2^2$ | $= 4$ |
| $2^1$ | $= 2$ |
| $2^0$ | $= ?$ |
| $2^{-1}$ | $= \frac{1}{2}$ |
| $2^?$ | $= \frac{1}{4}$ |
| $2^?$ | $= \frac{1}{8}$ |

E. Copy and complete these patterns.

12. $2^3 \cdot 2^2 = (2 \cdot 2 \cdot 2)(2 \cdot 2)$
    $= 2 \cdot 2 \cdot 2 \cdot 2 \cdot 2$
    $= 2^?$

13. $2^3 \cdot 2^{-5} = (2 \cdot 2 \cdot 2)\left(\frac{1}{2} \cdot \frac{1}{2} \cdot \frac{1}{2} \cdot \frac{1}{2} \cdot \frac{1}{2}\right)$
    $= \frac{1}{2} \cdot \frac{1}{2}$
    $= \frac{1}{2^2} = 2^{-?}$

14. $2^3 \cdot 2^{-3} = (2 \cdot 2 \cdot 2)\left(\frac{1}{2} \cdot \frac{1}{2} \cdot \frac{1}{2}\right)$
    $= ?$
    $= 2^?$

15. Explain why this generalization is true.

   If $c$, $r$, and $s$ are integers and $c \neq 0$, then $c^r \cdot c^s = c^{r+s}$.

## ★ Math Team Problem

16. Rose and Jon Fisher gave some money to their three children. To the oldest, they gave half the money plus 50¢. To the middle child, they gave half of what was left plus 50¢. To the youngest, they gave half of what was left plus 50¢. After that, all the money was gone. How much did the parents give to their children in all?

# LESSON 7-8

## Scientific Notation

The approximate age of the earth is 4,500,000,000 years. Very large numbers such as this, as well as very small numbers, can be difficult to work with, because it is easy to leave out a digit and make a mistake. Another reason is that calculator displays do not accept such a large number of digits. So we use scientific notation to write and perform operations with such numbers.

To write a number in **scientific notation**, express it as a product of two factors.

**a.** Write the first factor as a number greater than or equal to 1 and less than 10.
**b.** Write the second factor as a power of 10.

You can express 4,500,000,000 in scientific notation in this way.

$$4.5 \times 10^9$$

Number greater than or equal to 1 and less than 10 ↗    ↖ Power of 10

Study the table on the next page which shows how to write a number in standard notation as a number in scientific notation.

| | Standard Notation | Scientific Notation | |
|---|---|---|---|
| **Numbers greater than 10** | 95672<br>9567.2<br>956.72<br>95.672 | $9.5672 \times 10^4$<br>$9.5672 \times 10^3$<br>$9.5672 \times 10^2$<br>$9.5672 \times 10^1$ | **Numbers greater than 10** |
| **Numbers between 1 and 10** | 9.5672 | $9.5672 \times 10^0$ | **Numbers between 1 and 10** |
| **Numbers less than 1** | 0.95672<br>0.095672<br>0.0095672<br>0.00095672 | $9.5672 \times 10^{-1}$<br>$9.5672 \times 10^{-2}$<br>$9.5672 \times 10^{-3}$<br>$9.5672 \times 10^{-4}$ | **Numbers less than 1** |

Notice the following.

When the number in standard notation is *greater than 10*, the power of 10 has a *positive exponent*.

When the given number is *between 1 and 10*, the power of 10 has a *zero exponent*.

When the given number is *less than 1*, the power of 10 has a *negative exponent*.

**EXAMPLE 1**   Write each number in scientific notation.
   **a.** 867,000,000,000     **b.** 0.00000305

**Solutions**   **a.** Think: The first factor must be greater than 1 and less than 10. Move the decimal point 11 *places* to the left.

867,000,000,000

Write as a product.     $8.67 \times 10^{11}$

**b.** Think: The first factor must be greater than or equal to 1 and less than 10. Move the decimal point 6 *places* to the right.

0.00000305

Write as a product.     $3.05 \times 10^{-6}$

**Try This**   Write each number in scientific notation.
   **1.** 47,500,000     **2.** 0.00009

To enter $8.67 \times 10^{11}$ into a scientific calculator that has an [EE] key, press 8.67 [EE] 11. The display will show $\boxed{8.67 \quad 11}$.

To change 123,000 to scientific notation, enter 123,000 and press [EE][=]. The display, $\boxed{1.23 \quad 05}$, tells you that $123{,}000 = 1.23 \times 10^5$.

To change back to standard notation, press [INV][EE].

To simplify an expression involving products or quotients of powers of 10, follow this rule.

> **Multiplying Powers of Ten**
> If $r$ and $s$ are integers, then
> $10^r \cdot 10^s = 10^{r+s}$.

**EXAMPLE 2**  Simplify each expression.
   **a.** $10^5 \cdot 10^2$             **b.** $10^6 \div 10^4$

**Solutions**  **a.** To multiply powers of 10, write the base, 10, and add the exponents.

$$10^5 \cdot 10^2 = 10^{5+2}$$
$$= \mathbf{10^7}$$

**b.** First, express $10^6 \div 10^4$ as a multiplication problem.

$$10^6 \div 10^4 = 10^6 \cdot \frac{1}{10^4}$$
$$= 10^6 \cdot 10^{-4}$$
$$= 10^{6 + (-4)}$$
$$= \mathbf{10^2}$$

**Try This**  **Simplify each expression.**
   **3.** $10^3 \cdot 10^6$       **4.** $10^5 \div 10^2$       **5.** $10^8 \cdot 10^{-7}$

You can use the rule for multiplying powers of ten to express products and quotients in scientific notation.

**EXAMPLE 3**  Express the product or quotient in scientific notation.
   **a.** $(2 \times 10^3)(8 \times 10^2)$       **b.** $\dfrac{1.5 \times 10^6}{0.3 \times 10^9}$

7-8  Scientific Notation

**Solutions**

a. Write the problem.  $(2 \times 10^3)(8 \times 10^2)$
   Regroup the factors.  $(2 \times 8)(10^3 \times 10^2)$
   Multiply.  $16 \times 10^5$
   Write in scientific notation.  **$1.6 \times 10^6$**

b. Write the problem.  $\dfrac{1.5 \times 10^6}{0.3 \times 10^9}$

   Regroup the factors and simplify.  $\dfrac{1.5}{0.3} \times \dfrac{10^6}{10^9}$

   Write in scientific notation.  **$5 \times 10^{-3}$**

**Try This**  Express the product or quotient in scientific notation.

6. $(4 \times 10^5)(7 \times 10^3)$

7. $\dfrac{6.4 \times 10^7}{0.8 \times 10^4}$

To use a scientific calculator to check the results of Example 3, press these keys.

Example 3a:  2 [EE] 3 [×] 8 [EE] 2 [=]

Example 3b:  1.5 [EE] 6 [÷] .3 [EE] 9 [=]

# EXERCISES

**Objective:** To write numbers in scientific notation
To compute products and quotients in scientific notation

## Check Your Understanding

**Skill Check**  Write the letter of the correct answer.

1. Which represents 0.000063 written in scientific notation.
   a. $63 \times 10^6$  b. $6.3 \times 10^5$
   c. $63 \times 10^{-6}$  d. $6.3 \times 10^{-5}$

2. Which represents 38,900,000 written in scientific notation.
   a. $3.89 \times 10^{-7}$  b. $3.89 \times 10^{-6}$
   c. $3.89 \times 10^7$  d. $38.9 \times 10^6$

3. Simplify: $10^3 \times 10^2$.
   a. $10^6$  b. $10^5$
   c. $100^5$  d. $30 \cdot 20$

4. Simplify: $10^8 \div 10^4$.
   a. $10^{12}$  b. $10^{-4}$
   c. $1^4$  d. $10^4$

5. What number is represented by this calculator display?

   | 4.2   -04 |

**True or False? When a statement is false, tell why.**

6. A number written in scientific notation is expressed as the product of two factors.
7. When a number in standard notation is less than 1, the power of 10 in scientific notation has a zero exponent.
8. $\frac{1}{a^2} = a^{-2}$

## Practice and Apply

**Write each number in scientific notation.**

9. 870
10. 0.325
11. 55.009
12. 694.003
13. 0.09
14. 24,000,000
15. 0.00057
16. 0.00000157

**Write each number in scientific notation.**

17. The approximate land area of planet Earth is 52,000,000 square miles.
18. The wave length of a light ray is 0.0000000072 centimeters.
19. In the 1790 census, there were about 4 million people in the United States. In the 1890 census, there were about 63 million people.
20. The volume of a droplet of water in a fog is about 0.00000008 cubic centimeters.
21. Special balances can weigh something as small as 0.00000001 gram.
22. The sun is about 150,000,000 kilometers from the planet Earth.

**Simplify each expression.**

23. $10^6 \cdot 10^3$
24. $10^9 \div 10^4$
25. $10^9 \div 10^9$
26. $10^8 \cdot 10^2$
27. $10^{12} \div 10$
28. $10^7 \div 10^3$
29. $10^{15} \cdot 10^4$
30. $10^7 \cdot 10^6$

**Express the product or quotient in scientific notation.**

31. $(7 \times 10^5)(8 \times 10^{-2})$
32. $(2.7 \times 10^{-3})(-3 \times 10^4)$
33. $\frac{3.6 \times 10^4}{6 \times 10^5}$
34. $\frac{7.2 \times 10^6}{9 \times 10^4}$
35. $(6 \times 10^{-1})(9 \times 10^2)$
36. $(9 \times 10^3)(5 \times 10^{-8})$
37. $\frac{2.1 \times 10^2}{0.7 \times 10^3}$
38. $\frac{1.44 \times 10^6}{12 \times 10^5}$

## Connect and Extend

39. $2^{100} \div 2^6 = \underline{\ ?\ }$
40. $3^{100} \div 81 = \underline{\ ?\ }$

For each exercise, write a story problem that can be solved by pressing the calculator keys shown. Then use a calculator to solve the problem. Express the answer in scientific notation.

**41.** 3.2 [EE] 3 [×] 4.5 [EE] 5 [=]    **42.** 6.3 [EE] 2 [÷] 1.5 [EE] 6 [=]

Write each number in standard notation.

**43.** The temperature of the sun is about $6 \times 10^3$ degrees Celsius.
**44.** The diameter of a helium atom is about $2.2 \times 10^{-8}$ centimeters.
**45.** The speed of a turtle is about $1.6 \times 10^{-2}$ kilometers per hour.
**46.** The number of seconds in a day is $8.64 \times 10^4$.

## Maintain Your Skills

Replace each __?__ with the correct number. (Pages 29–35)

**47.** $-5 \cdot \underline{\ ?\ } = 55$    **48.** $\underline{\ ?\ } \cdot -3 = -36$    **49.** $102 \div \underline{\ ?\ } = -17$    **50.** $\underline{\ ?\ } \div -8 = 15$

**51.** Angela bought $\frac{3}{4}$ yard of speaker wire to connect her stereo. She used $\frac{1}{3}$ of the speaker wire. How much did she use? **(Pages 171–174)**

**52.** Arthur has $\frac{5}{6}$ of a yard of cord on which to string beads. He cuts the cord into two equal parts. How long is each piece of cord? **(Pages 189–193)**

Write a mixed number or a fraction in lowest terms for each decimal. (Pages 271–274)

**53.** 6.4    **54.** $-26.075$    **55.** $-12.05$    **56.** 47.35

## Mathematical Footnote

Until about four hundred years ago, there was no standard way to write a fraction, such as $2\frac{3}{4}$, as a decimal. All of the following methods had been used to write the decimal 2.75.

$2^{⓪}7^{①}5^{②}$    2, 7′ 5″    2|75    2|75
$2^{75}$    2|75    2(75    $2.\overset{(1)}{7}\overset{(2)}{5}$    2:75

Even today different countries represent this number in various ways.

**United States:** 2.75    **England:** 2·75    **Europe:** 2,75

The decimal system is based on tens and tenths. Why do you think the number 10 was chosen rather than another number such as 5 or 12?

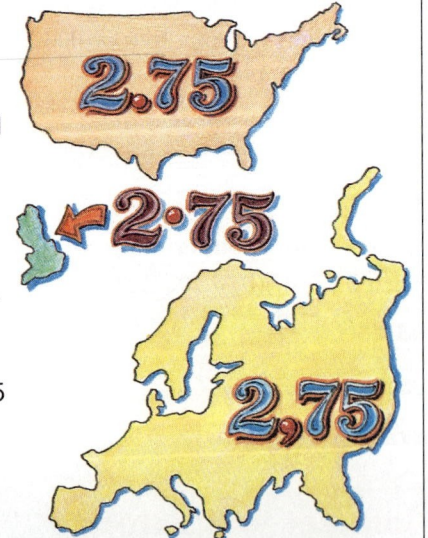

# LESSON 7-9
# Problem Solving Exploration
# The Metric System

**A.** Copy and complete the pattern in each column.

| Pattern 1 | Pattern 2 |
|---|---|
| 7.18 × 10 = 71.8 | 473.9 × 0.1 = 47.39 |
| 7.18 × 100 = ? | 473.9 × 0.01 = ? |
| 7.18 × 1000 = ? | 473.9 × 0.001 = ? |

1. When you multiply a whole number or a decimal by 10, 100, 1000, and so on, in which direction does the decimal point move?

2. When you multiply a whole number or a decimal by 0.1, 0.01, 0.001, and so on, in which direction does the decimal point move?

3. Write and test a rule you can use to multiply a number by 100.

4. Write and test a rule you can use to multiply a number by 0.001.

5. Multiply mentally.
   **a.** 0.38 × 100   **b.** 7.6 × 0.1   **c.** 8 × 0.0001

**B.** Copy and complete the pattern in each column.

| Pattern 1 | Pattern 2 |
|---|---|
| 19.8 ÷ 10 = 1.98 | 12.35 ÷ 0.1 = 123.5 |
| 19.8 ÷ 100 = ? | 12.35 ÷ 0.01 = ? |
| 19.8 ÷ 1000 = ? | 12.35 ÷ 0.001 = ? |

6. When you divide a whole number or a decimal by 10, 100, 1000, and so on, in which direction does the decimal point move?

7. When you divide a whole number or a decimal by 0.1, 0.01, 0.001, and so on, in which direction does the decimal point move?

8. Write and test a rule you can use to divide a number by 1000.

9. Write and test a rule you can use to divide a number by 0.01.

10. Divide mentally.
    **a.** 6.43 ÷ 10,000   **b.** 64.3 ÷ 0.01   **c.** 643 ÷ 0.1

C. Nearly every country in the world today uses the **metric system**. One reason for this is that it is a decimal system of measurement.

These are the basic units of measurement in the metric system.

**Length:** meter  **Mass:** gram  **Capacity:** liter

Prefixes are used with basic units. This table shows how each prefix is related to a decimal place value.

11. Copy and complete the table.

| kilo- (k) | hecto- (h) | deka- (da) | Basic Unit | deci- (d) | centi- (c) | milli- (m) |
|---|---|---|---|---|---|---|
| 1000 thousands | 100 ? | 10 ? | ? | 0.1 ($\frac{1}{10}$) tenths | 0.01 ($\frac{1}{100}$) ? | 0.001 ($\frac{1}{1000}$) ? |

When a prefix is attached to one of the basic units, a new unit is created.

12. Is 10 times a gram a dekagram or a decigram?
13. Is one-thousandth of a liter a kiloliter or a milliliter?
14. Is one thousand times a meter a kilometer or a millimeter?
15. Is one tenth of a gram a decigram or a centigram?
16. *Complete:* Each unit of measure is ? times the next smaller unit.
17. *Complete:* Each unit of measure is ? of the next larger unit.
18. Which unit is larger, the kilometer or the meter? To change meters to kilometers, do you multiply, or divide, by 1000?
19. Which unit is larger, the gram or the centigram? To change grams to centigrams, do you multiply, or divide, by 100?
20. Write a short paragraph that explains how multiplying and dividing by 10, 100, 1000, 0.1, 0.01, and 0.001 can be compared to converting (changing) metric units.

## LESSON 7-10 Estimating Products and Quotients of Decimals

**W**hen the Jordan family moved from Florida to Indiana, the driver of the moving van said that his truck, which holds 8.9 tons when fully loaded, had only $\frac{3}{4}$ of a load. How many tons did the Jordan's furnishings weigh?

You can estimate the weight by rounding 0.75 (which equals $\frac{3}{4}$) to the nearest tenth and 8.5 to the nearest whole number.

**Think:** 0.75 is about 0.8    8.9 is about 9.
So $0.8 \times 9 = 7.2$.

**Estimate:** Since both numbers were rounded up, you know that the furnishings weigh a little less than 7.2 tons.

You can use a calculator to compute a closer answer.
$$0.75 \: \boxed{\times} \: 8.9 \: \boxed{=} \: \boxed{6.675}.$$

- What is the difference, in pounds, between the estimated answer and the calculated answer?

**EXAMPLE 1**   Estimate each product.
  a. 0.875 × 7.4    b. 0.5249 × 0.64

**Solutions**  a. **Think:** 0.875 is a little less than 1.
  7.4 is a little more than 7.
  So 0.875 × 7.4 ≈ 1 × 7, or 7.
  **Estimate:** The product will be a **little less than 7.**

  b. **Think:** 0.5249 is a little more than $\frac{1}{2}$.
  So, 0.5249 × 0.64 is about $\frac{1}{2}$ × 0.64, or 0.32.
  **Estimate:** The product will be a **little more than 0.32.**

**Try This**   **Estimate each product.**
  **1.** 19.36 × 6.41    **2.** 0.249 × 0.84    **3.** 1.31 × 12.63

In the move from Florida to Indiana, the Jordans drive their own car. One day, they drive 318.25 miles on 11.7 gallons of gasoline. How can the Jordans compute their car's fuel economy?

The **fuel economy** of a car refers to how many miles it can travel on one gallon of gasoline. To find a car's fuel economy, you can use this formula.

  **Fuel Economy = Number of Miles ÷ Number of Gallons**

First, the Jordan's used the formula to estimate their car's fuel economy.

**Think:** 11.7 is about 12.
318.25 is close to 324 which is divisible by 12.

**Estimate:** 324 ÷ 12 = 27

**Using a calculator:** 318.25 ÷ 11.7 = 27.2008547.

Since the estimate and the computed answer are close, an answer of about 27 miles per gallon is reasonable.

**EXAMPLE 2**   Estimate each quotient.
  **a.** 9.8 ÷ 0.24    **b.** 68.19 ÷ 2.1

**Solutions:**  **a. Think:** 9.8 is about 10.  0.24 is about $\frac{1}{4}$.
  **Estimate:** 10 ÷ $\frac{1}{4}$ = 10 × 4 = **40**

**298**   CHAPTER 7

**b. Think:** 68.19 is about 70.  2.1 is about 2.
   **Estimate:** 70 ÷ 2 = **35**

**Try This**    Estimate each quotient.
   **4.** 8.3 ÷ 0.09   **5.** 47.36 ÷ 2.5

# EXERCISES

**Objectives:** To estimate products of decimals
To estimate quotients of decimals

## Check Your Understanding

**Skill Check**   Write the letter of the correct answer.

1. Which best describes the product of 2.39 and 0.07?
   a. Less than 0
   b. Between 0 and 1
   c. Between 1 and 10
   d. Greater than 10

2. Which best describes the product of 15.1 and 0.38?
   a. Between 0 and 1
   b. Between 1 and 5
   c. Between 5 and 7
   d. Greater than 7

3. Which best describes the result of dividing 2.08 by 2.1?
   a. Less than 0
   b. Between 0 and 1
   c. Between 1 and 5
   d. Greater than 5

4. Which best describes the result of dividing 14.8 by 3.01?
   a. Between 0 and 1
   b. Closer to 5 than to 6
   c. Closer to 6 than to 7
   d. Close to 50

5. If 23 × 58 = 1334, what is 2.3 and 0.58?
6. If 29 × 17 = 493, what is 2.9 × 1.7?
7. If 732 ÷ 2 = 366, what is 7.32 ÷ 0.2?
8. If 1311 ÷ 23 = 57, what is 1.311 ÷ 0.23?

## Practice and Apply

Estimate each product. Show the factors you used in your estimate.
Then tell whether it is an underestimate, U, or an overestimate, O.

9. 2.321 × 254.786   10. 89.02 × 6.3201   11. 9.375 × 67.4102
12. 28.37 × 35.24    13. 49 × 0.02       14. 7.98 × 5.55

Estimate the quotient by rounding.

15. 62.94 ÷ 7.04    16. 302.78 ÷ 2.6   17. 80.8 ÷ 9.37
18. 907.01 ÷ 9.67   19. 135.2 ÷ 8.89   20. 1207.2 ÷ 2.02

7-10   Estimating Products and Quotients of Decimals

**Estimate the answer to each problem. Then find the exact answer and compare it with the estimate to check for reasonableness.**

21. A car travels 725 miles on 24 gallons of gasoline. What is its fuel economy?

22. A car travels 2186 miles on 95 gallons of gasoline. Is its fuel economy greater than, or less than, the fuel economy of the car in Exercise 21?

23. A car rental agency charges $31.49 per day plus $0.30 per mile. What is the charge for renting a car for 7 days and driving it 507 miles?

24. An electric generator uses 0.75 of a gallon of gasoline per hour to keep its motor running. If the tank holds 12.75 gallons of gas, about how many hours will it run on a full tank?

## Connect and Extend

25. Write a paragraph that explains the difference in the way in which decimal points are treated in addition, subtraction, multiplication, and division.

26. What happens when you divide a number by smaller and smaller numbers?

27. Will the product of $(-3.5)^5$ and $(-2.8)^{32}$ be positive or negative? Explain.

28. Explain what is shown below.

$$0.36\overline{)28.8} = \frac{28.8}{0.36} = \frac{28.8 \times 100}{0.36 \times 100} = \frac{2800}{36} = 36\overline{)2800}^{\,80}$$

## Maintain Your Skills

29. A football field is 360 feet long and 160 feet wide. Find the area of the football field. (Pages 64–68)

30. The base of the triangle is 3 meters and the height is 4 meters. Find the area. (Pages 69–74)

**Solve and check. (Pages 101–109)**

31. $\frac{c}{3} - 5 = 8$  32. $54 = 3(4x - 2)$  33. $2f + 5 = 21$  34. $8(g - 2) + 3g = 61$

**Write each fraction in lowest terms. (Pages 163–167)**

35. $\frac{2}{16}$  36. $\frac{6}{45}$  37. $\frac{16}{28}$  38. $\frac{24}{64}$  39. $\frac{57}{90}$

## ★ Math Team Problem

40. A train is one mile long and is traveling at a rate of one mile per minute. How long will it take the train to pass through a tunnel one mile long?

# SUMMARY

## KEY TERMS

Decimal (p. 271)
Digit (p. 271)
Expanded form of a decimal (p. 272)
Irrational number (p. 278)
Metric system (p. 296)

Place value (p. 271)
Rational expression (p. 262)
Rational number (p. 263)
Repeating decimal (p. 277)
Scientific notation (p. 289)
Terminating decimal (p. 277)

## KEY IDEAS

**A.** To compare rational numbers having like denominators, compare the numerators. If one or more of the rational numbers is negative, think of the numerator as having the negative sign.

**B.** To compare rational numbers having unlike denominators, rewrite the rational numbers as equivalent fractions having the same denominator. Then compare the numerators.

**C.** Repeating decimals are written with a bar over the digits that repeat.

**D.** To write a number in scientific notation, express it as a product of two factors.
   **1.** Write the first factor as a number greater than or equal to 1 and less than 10.
   **2.** Write the second factor as a power of 10.

**E.** When a number in standard notation is:
   **a.** greater than 10, the power of 10 has a positive exponent.
   **b.** between 1 and 10, the power of 10 has a zero exponent.
   **c.** less than 1, the power of 10 has a negative exponent.

**F. Multiplying Powers of Ten**
   If $r$ and $s$ are integers, then $10^r \cdot 10^s = 10^{r+s}$

# REVIEW

**Write each expression as a rational number in simplest form. (Pages 262–266)**

1. $6:8$
2. $0$
3. $-3\frac{2}{5}$
4. 20 out of 28
5. $\frac{-15}{-27}$

**Evaluate each rational expression for $a = 2$, $b = -2$, and $c = -3$. (Pages 262–266)**

6. $\frac{5c}{b-a}$
7. $\frac{b-a}{a-b}$
8. $\frac{3a+4ac}{a-b}$

**Replace each ? with <, >, or =. (Pages 267–270)**

9. $-\frac{3}{5}$ ? $\frac{4}{7}$
10. $\frac{7}{8}$ ? $\frac{5}{6}$
11. $-\frac{10}{4}$ ? $-2\frac{1}{2}$
12. $-4\frac{2}{9}$ ? $-\frac{13}{3}$

**Simplify. (Pages 267–270)**

13. $\left(-\frac{3}{7}\right) + \frac{4}{5}$
14. $\frac{2}{3} \cdot \left(-\frac{3}{7}\right)$
15. $-\frac{3}{4} \div \left(-\frac{9}{16}\right)$
16. $-\frac{9}{16} - \left(-\frac{5}{12}\right)$

**Write each number in expanded form. (Pages 271–274)**

17. 32.1
18. 0.048
19. 464.051
20. 197.23

**For each decimal, write a mixed number or a fraction in lowest terms. (Pages 271–274)**

21. $-7.4$
22. 12.09
23. $-0.025$
24. 40.36

25. Gina wishes to cut a board into 7 equal pieces. How many cuts must she make to do this? (Pages **275–276**)

**Express each rational number as a decimal in two ways. First give the exact value. Then give the value rounded to the nearest thousandth. (Pages 277–280)**

26. $-\frac{4}{9}$
27. $\frac{7}{18}$
28. $\frac{10}{11}$
29. $-\frac{13}{27}$
30. $\frac{7}{12}$
31. $-\frac{22}{27}$

**Estimate each sum or difference. Then find the exact answer. Compare the answer with the estimate. (Pages 282–286)**

32. $-8.9 + (-5.013)$
33. $8.633 - 11.06$
34. $-5.07 + 0.98$
35. $38.2 + 7.351$
36. $3.098 - 2.1243$
37. $-0.048 + 9$

**Write each number in scientific notation. (Pages 289–294)**

38. 28.001
39. 0.677
40. 0.0007205
41. 732,000

**Estimate each product. (Pages 297–300)**

42. $8.24 \times 790.002$
43. $12.0437 \times 0.004$
44. $7.85 \times 4.71$

**Estimate each quotient by rounding. (Pages 297–300)**

45. $38.204 \div 1.7$
46. $224.21 \div 3.93$
47. $708.12 \div 9.65$

# TEST

**1.** Write the ratio, 48 : 36, as a rational number in simplest form.

Evaluate each rational expression for $d = -2$, $e = 4$, and $f = -6$.

**2.** $\dfrac{3e - df}{de + f}$  **3.** $\dfrac{f - f^2}{-2(d - e) - 5}$  **4.** $\dfrac{f^2 - f^2}{f - e}$  **5.** $\dfrac{3e - 4d}{4d - 3e}$

Replace each __?__ with <, >, or = to make a true statement.

**6.** $\dfrac{-3}{8}$ __?__ $\dfrac{-2}{5}$  **7.** $-\dfrac{20}{6}$ __?__ $-3\dfrac{1}{3}$  **8.** $2\dfrac{1}{3}$ __?__ $3\dfrac{1}{2}$  **9.** $-1\dfrac{4}{9}$ __?__ $-\dfrac{4}{3}$

Simplify.

**10.** $-\dfrac{3}{4} + \dfrac{3}{10}$  **11.** $-\dfrac{5}{9} - \dfrac{7}{12}$  **12.** $-\dfrac{4}{10} + \dfrac{6}{25}$  **13.** $-9 \cdot \left(-\dfrac{4}{21}\right)$

Write each number in expanded form.
**14.** 21.036    **15.** 365.505    **16.** 0.0075    **17.** 82.2848

For each decimal, write a mixed number or a fraction in lowest terms.
**18.** $-4.4$    **19.** 15.76    **20.** 82.18    **21.** $-0.005$

Express each rational number as a decimal in two ways. First give the exact value. Then give the value rounded to the nearest hundredth.

**22.** $\dfrac{8}{9}$   **23.** $\dfrac{7}{11}$   **24.** $\dfrac{22}{27}$   **25.** $\dfrac{11}{12}$   **26.** $\dfrac{5}{18}$   **27.** $\dfrac{4}{33}$

Estimate each sum or difference. Then find the exact answer. Compare the answer with the estimate.
**28.** $-7.93 + 0.099$   **29.** $-5 - 2.007$   **30.** $-0.48 + 8.032$   **31.** $18.04 - 6.904$

Write each number in scientific notation.
**32.** 0.0889    **33.** 1907.34    **34.** 16,800,000    **35.** 0.0004092

Express the product or quotient in scientific notation.
**36.** $(6.4 \times 10^{-3})(2 \times 10^5)$   **37.** $\dfrac{3.6 \times 10^{-4}}{1.2 \times 10^3}$   **38.** $\dfrac{8.5 \times 10^5}{0.5 \times 10^9}$

Estimate each product or quotient.
**39.** $10.27 \times 681.81$   **40.** $227.61 \div 4.009$   **41.** $82.098 \times 5.32$

# CUMULATIVE REVIEW: CHAPTERS 1-7

1. The lowest temperatures ever recorded in Florida, Indiana, Kentucky, Michigan, and Texas are $-2, -35, -34, -51,$ and $-23$ respectively. What is the average low temperature for the six states? **(Pages 64–68)**

**Use the diagram at the right to find the area of each figure. (Pages 69–74)**

2. Parallelogram ACDF
3. Triangle AEF
4. Trapezoid ACDE

**Solve and check. (Pages 101–109)**

5. $-7 + 5n = -102$
6. $\frac{e}{8} - 2 = 7$
7. $-6(f + 3) = 36$
8. $-67 = 5(g - 3) - g$

**True or False? Explain your answer. Let $a = 2^2 \cdot 3^2 \cdot 5$ and $b = 2^2 \cdot 5^2 \cdot 7$. (Pages 154–158)**

9. $2^2 \cdot 5$ is a common factor of $a$ and $b$.
10. $2^2 \cdot 3^2 \cdot 5 \cdot 7$ is the GCF of $a$ and $b$.
11. $2^2 \cdot 3^2 \cdot 5^2 \cdot 7^2$ is the LCM of $a$ and $b$.

**Solve. (Pages 183–193)**

12. Norman spent $\frac{4}{5}$ of his money on clothes. If he spent $40, how much money did he have originally?
13. Mr. Chan has a board that is 15 feet long. How many lengths of $1\frac{1}{2}$ feet can he cut from the larger board?

**Solve and check. (Pages 210–220)**

14. $\frac{8}{70} = \frac{e}{35}$
15. $\frac{c}{10} = \frac{8}{25}$
16. $\frac{7}{z} = \frac{8}{15}$
17. $\frac{4}{5} = \frac{10}{p}$

**Solve. (Pages 226–230)**

18. A container holds 15 cups. How many pints is this?
19. A board is 4 yards long. How many inches is this?
20. One inch on a road map represents 25 miles on land. How many miles on land does $2\frac{1}{6}$ inches represent?
21. One inch on a road map represents 35 miles on land. How many inches on a map represent 175 miles on land?

**A drawer contains 8 brown socks, 6 black socks, and 4 white socks. One sock is chosen at random. Find each probability. (Pages 236–240)**

22. P(black sock)
23. P(brown sock)
24. P(not a brown sock)

**Write each answer in lowest terms. (Pages 262–266)**

25. Stacy gave away 6 of 20 football cards. What fractional part did she give away?
26. Alan has 14 fish. Two of his fish are goldfish. What fractional part of his total number of fish are not goldfish?

# CHAPTER 8 / PERCENT

## LESSON 8-1  The Meaning of Percent

**O**ne hundred students were asked this question.

Which sport do you like most to watch on TV, basketball, soccer, baseball, or football?

Out of 100 students, 45 chose football. You can write a ratio to compare the number of students who chose football to the total number of students in the survey.

$$\frac{\text{Number who chose football}}{\text{Total number surveyed}} = \frac{45}{100}$$

Notice that the denominator of the ratio is 100. A ratio that compares a number to 100 is called a **percent**.

So, $\frac{45}{100}$ = 45 percent, or 45%.

When the second term of a ratio is not 100, you can write a proportion to find the percent equal to that ratio.

8-1  The Meaning of Percent  **305**

**EXAMPLE 1** Write the percent for $\frac{17}{20}$.

**Solution** Think: $\frac{17}{20} = \frac{?}{100}$

Let $p$ = the first term (numerator) of the percent ratio.

Write a proportion. Use $\frac{17}{20}$ and $\frac{p}{100}$. $\qquad \frac{17}{20} = \frac{p}{100}$

Solve for $p$. $\qquad 17 \cdot 100 = 20p$

Apply the Division Property for Equality. $\qquad \frac{1700}{20} = \frac{20p}{20}$

$\qquad 85 = p$

Since $p = 85$, $\frac{17}{20} = \frac{85}{100} =$ **85%**.

**Try This** Write the percent for each ratio.

1. $\frac{1}{4}$  2. $\frac{3}{5}$  3. $\frac{7}{10}$  4. $\frac{9}{20}$  5. $\frac{41}{50}$

**EXAMPLE 2** Write the percent for $\frac{1}{6}$.

**Solution** Think: $\frac{1}{6} = \frac{?}{100}$

Let $m$ = the first term (numerator) of the percent ratio.

Write a proportion. $\qquad \frac{1}{6} = \frac{m}{100}$

Solve for $m$. $\qquad 6 \cdot m = 100$

$\qquad \frac{6m}{6} = \frac{100}{6}$

$\qquad m = 16\frac{4}{6} = 16\frac{2}{3}$

Since $m = 16\frac{2}{3}$, $\frac{1}{6} = \frac{16\frac{2}{3}}{100} =$ **$16\frac{2}{3}$%**.

**Try This** Write the ratio as a percent.

6. $\frac{1}{3}$  7. $\frac{5}{6}$  8. $\frac{7}{8}$  9. $\frac{5}{12}$  10. $\frac{9}{16}$

If the first term (numerator) and second term (denominator) of a ratio are equal, then the ratio equals 100%.

$\qquad \frac{50}{50} = 100\% \qquad \frac{7.3}{7.3} = 100\% \qquad \frac{2\frac{1}{2}}{2\frac{1}{2}} = 100\%$

If the first term of a ratio is greater than the second term, then the ratio is greater than 100%.

**EXAMPLE 3** Write the percent for $\frac{11}{8}$.

**Solution** Think: $\frac{11}{8} = \frac{?}{100}$

Let $t =$ the first term of the percent ratio.

Write a proportion. $\qquad \frac{11}{8} = \frac{t}{100}$

Solve for $t$. $\qquad 8 \cdot t = 11 \cdot 100$

$$\frac{8t}{8} = \frac{1100}{8}$$

$$t = 137\frac{4}{8} = 137\frac{1}{2}$$

Since $t = 137\frac{1}{2}$, $\frac{11}{8} = \frac{137\frac{1}{2}}{100} = \mathbf{137\frac{1}{2}\%}$.

**Try This** Write the percent for each ratio.

**11.** $\frac{100}{100}$  **12.** $\frac{1.5}{1.5}$  **13.** $\frac{7}{2}$  **14.** $\frac{25}{4}$  **15.** $\frac{39}{12}$

# EXERCISES

**Objective:** To write a fraction as a percent

## Check Your Understanding

**Skill Check** Write the letter of the correct answer.

**1.** Which best describes a ratio having a denominator less than its numerator?
  **a.** Equal to 100%
  **b.** Greater than 100%
  **c.** Less than 100%
  **d.** Less than 1%

**2.** Which best describes a ratio having a denominator equal to its numerator?
  **a.** Equal to 100%
  **b.** Greater than 100%
  **c.** Less than 100%
  **d.** Equal to 1%

**3.** Which is the percent for $\frac{5}{4}$?
  **a.** 80%
  **b.** 120%
  **c.** 125%
  **d.** 150%

**4.** Which is the percent for $\frac{3}{8}$?
  **a.** 30%
  **b.** $37\frac{1}{2}\%$
  **c.** $12\frac{1}{2}\%$
  **d.** 300%

## Practice and Apply

Write the proportion you would use in writing a percent for each ratio. Use $n$ for the variable.

**5.** $\frac{3}{10}$  **6.** $\frac{7}{20}$  **7.** $\frac{2}{5}$  **8.** $\frac{5}{12}$  **9.** $\frac{20}{19}$

**Write the ratio as a percent.**

10. $\frac{1}{4}$  11. $\frac{1}{20}$  12. $\frac{2}{25}$  13. $\frac{3}{50}$  14. $\frac{5}{8}$

15. $\frac{7}{40}$  16. $\frac{5}{9}$  17. $\frac{21}{16}$  18. $\frac{17}{6}$  19. $\frac{9}{5}$

20. $\frac{3}{16}$  21. $\frac{7}{11}$  22. $\frac{19}{10}$  23. $\frac{11}{5}$  24. $\frac{3}{2}$

**Write the ratio as a percent.**

25. In 1990, about $\frac{19}{50}$ of the population of the United States was 40 years old or older.

26. In 1990, about $\frac{8}{25}$ of the population of the United States was younger than 21 years old.

27. Refer to your answer to Exercise 25 to determine what percent of the population was under age 40 in 1990.

## Connect and Extend

28. Write a mixed number for 343%.

29. A worker receives time and a half for overtime. What percent of the worker's usual pay is this?

30. When the percent for $\frac{1}{8}$ is known, what is a shortcut for finding the percent for $\frac{3}{8}$? Show that your answer is correct.

31. The percent for the ratio $\frac{g}{100}$ is $g$%. Write a percent for $\frac{g}{25}$.

32. The percent for the ratio $\frac{x}{100}$ is $x$%. Write a percent for $\frac{x}{300}$.

## Maintain Your Skills

**Write an equivalent fraction with a denominator of 100. (Pages 163–167)**

33. $\frac{3}{4}$  34. $\frac{7}{10}$  35. $\frac{6}{25}$  36. $\frac{19}{20}$  37. $\frac{2}{40}$

**Express each fraction in lowest terms. (Pages 163–167)**

38. $\frac{45}{80}$  39. $\frac{20}{36}$  40. $\frac{39}{51}$  41. $\frac{16}{96}$  42. $\frac{70}{154}$

43. Mrs. Klein has $\frac{2}{3}$ of a quart of orange juice to distribute equally among her 4 grandchildren. How much will each child get? **(Pages 189–193)**

## LESSON 8-2

# Decimals and Percents

The weather forecaster on Channel 10 made this prediction: "There is a 40% chance of rain tomorrow."

When weather forecasters predict a 40% chance of rain, they mean that, for a specific designated area such as a county or a large city, chances are that 40% of the area will have rain and 60% of the area will not. Notice that the forecast does not say which part or parts of the area will have rain and which will not.

Although a probability is usually written as a percent or as a fraction, it can also be written as a decimal. So a probability of 40% can be written in these ways.

$$40\% \qquad \frac{40}{100} \qquad 0.40 \qquad 0.4$$

Since percent means *per hundred* or *hundredths*, you can write a percent as a decimal.

$$23\% = \frac{23}{100} = 0.23$$

**EXAMPLE 1** Write each percent as a decimal.
  a. 15.5%      b. 106%      c. 0.5%

**Solutions**  Write the percent with a denominator of 100. Then divide by 100.

  a. $15.5\% = \frac{15.5}{100} =$ **0.155**      b. $106\% = \frac{106}{100} =$ **1.06**

  c. $0.5\% = \frac{0.5}{100} =$ **0.005**

8-2  Decimals and Percents  **309**

**Try This**  **Write each percent as a decimal.**
1. 8%   2. 125%   3. 12.5%   4. 0.2%

Here is a shortcut for writing a percent as a decimal. Move the decimal point two places to the left and drop the percent symbol.

$0.5\% = 000.5 = 0.005 \qquad 125\% = 125 = 1.25 \qquad 90\% = 90 = 0.90$

Sometimes you need to write a decimal as a percent.

**EXAMPLE 2** Write each decimal as a percent.
 a. 0.32   b. 0.8   c. 4   d. 0.352

**Solutions** Write a ratio with a denominator of 100 for each decimal. Then write the percent.

a. $0.32 = \frac{32}{100} =$ **32%**

b. $0.8 = \frac{8}{10} = \frac{8 \times 10}{10 \times 10} = \frac{80}{100} =$ **80%**

c. $4 = \frac{4 \times 100}{1 \times 100} = \frac{400}{100} =$ **400%**

d. $0.352 = \frac{352}{1000} = \frac{352 \div 10}{1000 \div 10} = \frac{35.2}{100} =$ **35.2%**

**Try This**  **Write each decimal as a percent.**
5. 0.26   6. 0.9   7. 0.135   8. 9

Here is a shortcut method for writing a decimal as a percent. Move the decimal point two places to the right and write the percent symbol.

$$0.26 = 0.26 = 26\%$$

Look for the pattern among these percents.

$50\% = \frac{50}{100} = \frac{1}{2} \qquad 5\% = \frac{5}{100} = \frac{1}{20} \qquad 0.5\% = \frac{0.5}{100} = \frac{1}{200}$

Be careful when you work with percents that are less than 1%. For example, $\frac{1}{2}\%$ is $\frac{1}{2}$ of 1%. That is,

$$\frac{1}{2}\% = \frac{\frac{1}{2}}{100} = \frac{1}{2} \div 100 = \frac{1}{2} \cdot \frac{1}{100} = \frac{1}{200}.$$

Since $0.5\% = \frac{1}{200}$ and $\frac{1}{2}\% = \frac{1}{200}$, $0.5\% = \frac{1}{2}\%$.

**EXAMPLE 3**  Write $\frac{1}{5}$% as a fraction in lowest terms.

**Solution**  Method 1: Change $\frac{1}{5}$ to a decimal first.

$$\frac{1}{5}\% = 0.2\% = \frac{0.2}{100} = 0.002 = \frac{2}{1000} = \frac{1}{500}$$

Method 2: Think of $\frac{1}{5}$% as $\frac{\frac{1}{5}}{100}$.

$$\frac{1}{5} = \frac{\frac{1}{5}}{100} = \frac{1}{5} \div 100 = \frac{1}{5} \cdot \frac{1}{100} = \frac{1}{500}$$

**Try This**  Write each percent as a fraction in lowest terms.

9. $\frac{1}{8}$%    10. 0.3%    11. $\frac{1}{3}$%    12. $\frac{2}{5}$%

# EXERCISES

**Objectives:** To write a decimal as a percent
To use percents less than 1%

## Check Your Understanding

**Skill Check**  Write the letter of the correct answer.

1. Which equals $\frac{\frac{1}{3}}{100}$?
   a. $\frac{1}{30}$    b. $\frac{100}{3}$
   c. $\frac{1}{3}$    d. $\frac{1}{300}$

2. *Complete:* If 40% = $\frac{40}{100}$ = $\frac{2}{5}$, then 4% = $\underline{\ ?\ }$.
   a. $\frac{1}{25}$    b. $\frac{2}{500}$
   c. $\frac{2}{25}$    d. $\frac{4}{25}$

3. Which percent is equivalent to 0.001?
   a. 1%    b. 0.1%
   c. 0.01%    d. 10%

4. Which decimal is equivalent to 4.5%?
   a. 4.5    b. 450
   c. 0.45    d. 0.045

Complete.

5. $0.3 = \frac{?}{10} = \frac{?}{100} = \underline{\ ?\ }\%$

6. $0.07 = \frac{?}{100} = \underline{\ ?\ }\%$

7. $0.60 = \frac{?}{100} = \underline{\ ?\ }\%$

8. $0.008 = \frac{?}{1000} = \frac{?}{100} = \underline{\ ?\ }\%$

## Practice and Apply

Write each decimal as a fraction with a denominator of 100.

9. 0.41    10. 0.7    11. 0.98    12. 7.9    13. 0.223

8-2  Decimals and Percents

**Write each percent as a decimal.**
14. 17.8%   15. 507%   16. 0.9%   17. 66.1%   18. 5.8%
19. 4.5%    20. 0.1%   21. 91%    22. 0.66%   23. 0.25%

**Write each decimal as a percent.**
24. 0.4     25. 3.6    26. 8.6    27. 0.04    28. 0.8
29. 0.26    30. 5.1    31. 0.003  32. 0.207   33. 9.4

**Write each percent as a fraction in lowest terms.**
34. $\frac{4}{5}$%   35. $\frac{7}{10}$%   36. $\frac{3}{4}$%   37. $\frac{2}{3}$%   38. $\frac{7}{8}$%

**Write a fraction, a decimal, and a percent for each shaded region.**

39. 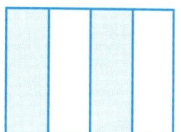   40. 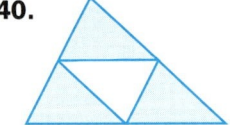   41.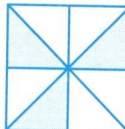

42. 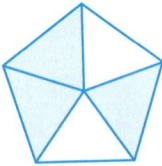   43. 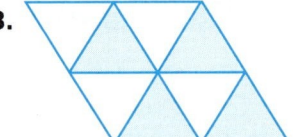   44.

## Connect and Extend

45. If $\frac{a}{5} = 20\%$, find $a$.
46. If $\frac{t}{10} = 30\%$, find $t$.
47. If $\frac{r}{8} = 62\frac{1}{2}\%$, find $r$.
48. If $\frac{w}{3} = 66\frac{2}{3}\%$, find $w$.
49. What is 50% of $62.80?
50. What is 200% of $3.25?

## Maintain Your Skills

**Write the answer if it is positive. (Pages 13–18, 23–25)**
51. $-56{,}632 + 56{,}633$   52. $17{,}438 - (-21{,}291)$   53. $44{,}009 - 44{,}010$

**Find each product. Write answers in lowest terms. (Pages 171–174)**
54. $\frac{5}{12} \cdot \frac{3}{5}$   55. $\frac{5}{9} \cdot 24$   56. $\frac{3}{4} \cdot \frac{10}{21}$   57. $20 \cdot \frac{3}{8}$

**Solve. (Pages 171–174, 252–257)**

58. If a recipe calls for $\frac{1}{3}$ cup of sugar, how much sugar will be needed to make $\frac{1}{2}$ of the recipe?

59. A yogurt store sells 7 flavors of frozen yogurt. How many ways are there to have a 3-dip cone of different flavors?

## LESSON 8-3
# Estimating the Percent of a Number

The Smith family's yearly budget and their state's yearly budget are shown below.

        Smith Family's Budget:        $48,000
        State Budget:        $480 million

If both have spent 78% of their budget by the month of July, how much money has each spent?

Since 480 million is a multiple of 48 thousand, you can work with the smaller numbers in the Smith family's budget first. The amount spent from the state budget will be a multiple of the amount spent from the family budget.

Use fractions to estimate the amount the Smiths spent. Think of fractional equivalents close to 78% that make it easy to compute mentally.

**Think:** 78% is between 75% and 80%.

$$75\% = \frac{75}{100} = \frac{3}{4} \qquad\qquad 80\% = \frac{80}{100} = \frac{4}{5}$$

8-3   Estimating the Percent of a Number   **313**

Find $\frac{3}{4}$ of 48,000.   Find $\frac{4}{5}$ of 48,000.

$\frac{3}{4} \cdot 48{,}000 = 36{,}000$   $\frac{4}{5} \cdot 48{,}000 = 38{,}400$

So 78% of $48,000 is between $36,000 and $38,400.

To get a closer estimate, think of 78% as halfway between 75% and 80%. Find $\frac{1}{2}(38{,}400 - 36{,}000)$.

Find the difference between 38,400 and 36,000.   $38{,}400 - 36{,}000 = 2400$

Find $\frac{1}{2}$ of 2400.   $\frac{1}{2} \cdot 2400 = 1200$

Add 1200 to 36,000.   $36{,}000 + 1200 = 37{,}200$

So 78% of the Smith's budget is about $37,200.

Now consider the state budget of 480 million. This is 10,000, or $10^4$, times the Smith's yearly budget. So, by July, the state has spent about $(37{,}200 \cdot 10{,}000)$, or $372 million.

Memorizing this table will help you to estimate mentally.

**Equivalent Percents and Fractions**

| | | | |
|---|---|---|---|
| $\frac{1}{4} = 25\%$ | $\frac{1}{2} = 50\%$ | $\frac{3}{4} = 75\%$ | |
| $\frac{1}{5} = 20\%$ | $\frac{2}{5} = 40\%$ | $\frac{3}{5} = 60\%$ | $\frac{4}{5} = 80\%$ |
| $\frac{1}{6} = 16\frac{2}{3}\%$ | $\frac{1}{3} = 33\frac{1}{3}\%$ | $\frac{2}{3} = 66\frac{2}{3}\%$ | $\frac{5}{6} = 83\frac{1}{3}\%$ |
| $\frac{1}{8} = 12\frac{1}{2}\%$ | $\frac{3}{8} = 37\frac{1}{2}\%$ | $\frac{5}{8} = 62\frac{1}{2}\%$ | $\frac{7}{8} = 87\frac{1}{2}\%$ |

**EXAMPLE 1** Use fractions to estimate the percent of each number.
a. 23% of 127    b. 83% of 145

**Solutions**  a. Think: 23% is about $\frac{1}{4}$. 127 is about 128 (divisible by 4).

Estimate: $\frac{1}{4}$ of $128 = \frac{1}{4} \cdot 128$

$= 32$

So 23% of 127 is about **32**.

**b. Think:** 83% is about $83\frac{1}{3}\%$ and $83\frac{1}{3}\% = \frac{5}{6}$.

145 is about 144, which is divisible by 6.

Estimate: $\frac{5}{6}$ of $144 = \frac{5}{\cancel{6}_1} \cdot \cancel{144}^{24}$

$= 120$

So 83% of 145 is about **120**.

**Try This**    Use fractions to estimate the percent of each number.
1. 19% of 95     2. 33% of 48     3. 88% of 65

**EXAMPLE 2**   The Chess Club needs $84.50 to buy new chess sets. The club is 32% short of its goal. About how much more money is needed?

**Solution**    Estimate 32% of $84.50.

**Think:** 32% is about $33\frac{1}{3}\%$ and $33\frac{1}{3}\% = \frac{1}{3}$.

84.50 is about $84, which is divisible by 3.

Estimate: $\frac{1}{3}$ of $84 = \frac{1}{\cancel{3}_1} \cdot \cancel{84}^{28}$

$= 28$

So the Chess Club needs about **$28.**

**Try This**
4. Dominic earns a bonus of 12% of his salary. In January, Dominic earned $965 (salary plus bonus). Estimate the amount of the bonus.

# EXERCISES

**Objective:** To find the percent of a number using estimation

## Check Your Understanding

**Skill Check**   Write the letter of the correct answer.

1. Which will give the best estimate for 26% of $125?
   a. $\frac{1}{3} \cdot 120$    b. $\frac{1}{5} \cdot 125$
   c. $\frac{1}{4} \cdot 128$    d. $\frac{3}{10} \cdot 130$

2. Which will give the best estimate for 62% of 496?
   a. $\frac{3}{5} \cdot 500$    b. $\frac{3}{4} \cdot 496$
   c. $\frac{2}{3} \cdot 498$    d. $\frac{5}{8} \cdot 400$

8-3   Estimating the Percent of a Number

**3.** Which will give the best estimate for 66% of 910?

a. $\frac{2}{3} \cdot 912$  b. $\frac{4}{5} \cdot 910$
c. $\frac{3}{4} \cdot 908$  d. $\frac{9}{10} \cdot 910$

**4.** Which will give the best estimate for $29\frac{1}{2}\%$ of 498?

a. $\frac{1}{3} \cdot 498$  b. $\frac{1}{4} \cdot 496$
c. $\frac{2}{5} \cdot 500$  d. $\frac{3}{10} \cdot 500$

## Practice and Apply

Use fractions to estimate the percent of each number.

**5.** 21% of 149
**6.** 9.5% of 231
**7.** 26% of 608
**8.** 33% of 98
**9.** 12% of 26
**10.** 19% of 49
**11.** 37% of 115
**12.** 88% of 249
**13.** 13% of 95
**14.** 98% of 47
**15.** 61% of 91
**16.** 31% of 59

Solve by using estimation.

**17.** Penrose High School has 925 students. Last fall, 88% of the students voted in the student council elections. About how many students was this?

**18.** Marie spent $49.50 on clothes. She paid a sales tax of $8\frac{1}{2}\%$. Did she pay more than, or less than, $4.95 for the tax?

**19.** After 4 years, a car which cost $7925 lost 62% of its value. Estimate the re-sale value of the car at the end of the 4 years.

**20.** A survey at a school found that about 68% of its 1150 students hold part-time jobs. About how many students hold part-time jobs?

## Connect and Extend

Solve by using estimation. Explain your reasoning.

**21.** Which is more, 81% of $9000 or 89% of $7000?

**22.** Which is more, 33% of $600 or 31% of $700?

**23.** Which is less, 76% of $2000 or 79% of $1500?

**24.** Which is less, 21% of $7000 or 26% of $6000?

**25.** Which will equal $1.00, $\frac{1}{2}\%$ of $200 or $\frac{1}{3}\%$ of $300?

## Maintain Your Skills

Estimate to place the decimal point in the product. (Pages 297–300)

**26.** $3.15 \times 21.4 = 6741$
**27.** $37 \times 2.8 = 1036$
**28.** $6.45 \times 2.16 = 13932$
**29.** $23.5 \times 6.4 = 15040$

Correct any wrong answers. (Pages 309–312)

**30.** 2.5% = 0.25
**31.** 25% = 0.25
**32.** 0.25% = 0.25
**33.** 250% = 0.25

Write each percent as a fraction in lowest terms. (Pages 309–312)

**34.** $\frac{3}{8}\%$
**35.** 90%
**36.** 48%
**37.** $1\frac{1}{4}\%$
**38.** $4\frac{1}{2}\%$

| Use |
|---|
| Average daily use per person |
| Taking a shower |
| Brushing teeth (with water running) |
| Shaving (with water running) |
| Washing dishes by hand |
| Running a dishwasher |

## LESSON 8-4
# Problem Solving Strategies
# Deciding on Estimates

- Understand
- Plan
- Solve
- Look Back

The table above gives figures from a recent World Almanac for water use in the United States.

In the table, the number of gallons for each use is an *average*, or estimate. To make the data even easier to compare, you can use the averages to draw a bar graph. A bar graph will help you to see relative sizes more easily and to compare the various amounts of water used more quickly.

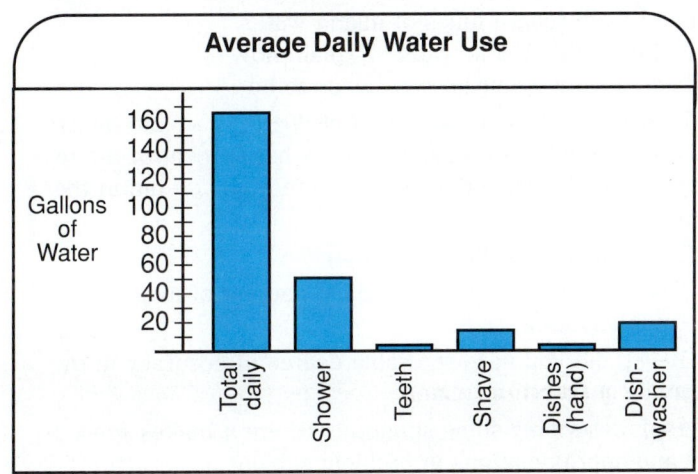

8-4 Problem Solving Strategies **317**

- Which daily activity uses about 30% of the total daily water usage?
- Estimate the percent of total daily water usage consumed by daily brushing of teeth and by shaving.

## EXERCISES

**Objective:** To apply strategies in problem solving

Write *Approximate* or *Exact* to tell which kind of answer would be best in the situation. Give a reason for your answer.

1. Measuring a dose of medicine
2. Buying refreshments for a class party
3. Making a monthly budget
4. Keeping the money accounts for a class project
5. Measuring the amount of trash produced by one family in a week

**For Exercises 6–7, refer to the graph on page 317.**

6. If you changed from using a dishwasher to washing dishes by hand, about how many gallons of water will be saved each time?
7. What other changes might a person make to save water, based on the information given?
8. The area of the state of Texas consists of 262,017 square miles of land and 4,790 square miles of inland water. What percent of the total area is water? Explain how you decided how accurate your answer needs to be.
9. The area of the state of Florida consists of 58,664 total square miles of which 4,511 square miles are inland water. What percent of the total area is water? Use the same degree of accuracy as you did in the previous exercise.
10. Which state, Texas or Florida, has more water?
11. Which state, Texas or Florida, has the highest percent of inland water compared to total land area?
12. In Exercises 10–11, did you need the same degree of accuracy in the data to answer the question? Explain.
13. Write a paragraph describing some situations where it makes sense to use estimates and approximations in problem solving.

# Finding the Percent of a Number

A city plans to build a monorail from a residential area to a new industrial area. The city planners estimate that 38.9% of the 2,800 industrial workers will use the monorail regularly. How many workers are expected to use the monorail regularly?

Often, an estimated answer is close enough for practical use. To estimate 38.9% of 2,800, think:

38.9% is about 40% and 40% = $\frac{40}{100}$ = $\frac{2}{5}$.

$\frac{2}{5}$ of 2800 = $\frac{2}{5} \cdot \overset{560}{\cancel{2800}}$
$\phantom{\frac{2}{5} \text{ of } 2800 } = 1120$

So about 1120 workers can be expected to use the monorail.

An exact answer can be found by using paper and pencil or by using a calculator.

### USING A CALCULATOR
**Think:** 38.9% of 2800 = 2800 × 38.9%

2800 [×] 38.9 [%] [1089.2]

The actual number is 1089.2 to the nearest tenth.

### USING PAPER AND PENCIL
Write the percent as a decimal.   $38.9\% = \frac{38.9}{100}$
$\phantom{Write the percent as a decimal.\ \ \ 38.9\%} = 0.389$

Multiply.
$$\begin{array}{r} 2800 \\ \times\ 0.389 \\ \hline 1089.2 \end{array}$$

In problems such as this, the given percent may be an approximation that has been rounded to the nearest tenth. Therefore the actual percent may lie between (38.9 + 0.05)% and (38.9 − 0.05)%. That is, the actual percent may lie between 38.95% and 38.85%.

- *Complete:* If 41.3% is an approximation, then the actual percent may lie between __?__ % and __?__ %.

8-5  Finding the Percent of a Number   **319**

**EXAMPLE 1** What is $87\frac{1}{2}\%$ of 1600?

**Solution** **Method 1:** Using fractions

Think: $87\frac{1}{2}\% = \frac{7}{8}$

$$\frac{7}{\underset{1}{8}} \cdot \overset{200}{\underset{}{1600}} = 1400$$

**Method 2:** Using decimals

Think: $87\frac{1}{2} = 87.5\% = \frac{87.5}{100} = 0.875$

Multiply.
$$\begin{array}{r} 1600 \\ \times\ 0.875 \\ \hline 1400.000 \end{array}$$

**Method 3:** Using a calculator.

1600  87.5  1400

**Try This**  1. What is 25% of 172?   2. What is $66\frac{2}{3}\%$ of 96?

**EXAMPLE 2** In addition to her regular salary at Central Auto Sales, Louise Wong receives a commission of 4.5% of total sales. Her total sales last week amounted to $32,000. How much commission did she receive?

**Solution** Find 4.5% of $32,000.

**Write 4.5% as a decimal.**  $4.5\% = \frac{4.5}{100} = 0.045$

**Multiply.** $0.045 \times 32,000 = 1440$

Louise's commission was **$1440.**

**Try This** 3. A real estate agent receives a commission of $3\frac{1}{4}\%$ of sales. What is the agent's commission on the sale of a house for $128,000?

You can also use mental math to find the commission in Example 2.

**Think:** 1% of 32,000 = $\frac{1}{100} \cdot 32,000 = 320$

$\frac{1}{2}\%$ of 32,000 = $\frac{1}{2}$(1% of 32,000) = 160

4% of 32,000 = 4(1% of 32,000) = 1280

So $4\frac{1}{2}\%$ of $32,000 = $1280 + $160 = $1440.

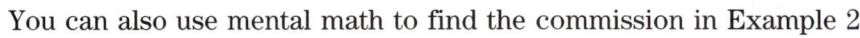

Estimation and mental math are also useful in computing the amount of a tip.

**EXAMPLE 3**  Your family has a dinner bill of $46.85. You want to leave a 15% tip. How much will you leave?

**Solution**  Think: 10% of 46 is 4.60.   5% of 46 is $\frac{1}{2}$(10% of 46), or 2.30.
Estimate: 4.60 + 2.30 = 6.90   You will probably leave **$7.00.**

**Try This**   4. A restaurant bill totals $29.60. You want to leave a 15% tip. How much will you leave?

# EXERCISES

**Objective:** To find the percent of a number using paper and pencil, using a calculator, or using mental computation

## Check Your Understanding

**Skill Check**  Write the letter of the correct answer.

1. Which expression **cannot** be used to find 7% of 31?
   a. $\frac{7}{10} \times 31$
   b. $\frac{7}{100} \times 31$
   c. $0.07 \times 31$
   d. $31 \; \boxed{\times} \; 7 \; \boxed{\%}$

2. A real estate agent receives a commission of $6\frac{1}{2}$% on a sale of $78,000. Find the commission.
   a. $4,680
   b. $4,719
   c. $507
   d. $5,070

3. Which is the best estimate for a 15% tip on a restaurant bill of $38.90?
   a. $4.00
   b. $6.00
   c. $5.00
   d. $8.00

4. John spent $36.90 for bowling shoes. The sales tax was $5\frac{1}{2}$%. How much was the sales tax?
   a. $20.29
   b. $2.03
   c. $1.85
   d. $2.95

## Practice and Apply

**Find the percent of the number.**

5. $33\frac{1}{3}$% of 96
6. 15% of 144
7. 105% of 60
8. $1\frac{1}{2}$% of 30

**Solve.**

9. What is 83% of 150?
10. $83\frac{1}{3}$% of 144 is what number?
11. Find 240% of 580.
12. What is 12% of 900?
13. 4% of 7.8 is $n$. Find $n$.
14. 30% of 102 is $r$. Find $r$.
15. What is 71% of 490?
16. 98% of 116 is what number?

17. What is $16\frac{2}{3}$% of 12?
18. Find $62\frac{1}{2}$% of $7,750.
19. What is 0.6% of 18?
20. 0.2% of 90 is what number?
21. After a restaurant meal, Eileen gave the waiter a 15% tip on her bill of $18.60. About how much was the tip?
22. A basketball team played 25 games this year. They won 60% of the games played. How many games did they lose?

For Exercises 23–25, find the answer to each pair of problems.

23. **a.** 28% of 50
    **b.** 50% of 28
24. **a.** 16% of 12.5
    **b.** 12.5% of 16
25. **a.** 32% of 75
    **b.** 75% of 32

26. Write a generalization for the pattern you observed in Exercises 23–25.

Use the generalization you wrote in Exercise 26 to solve these problems.

27. 16% of 25
28. 32% of $87\frac{1}{2}$
29. 60% of $16\frac{2}{3}$

## Connect and Extend

30. The city of Palomino plans to build a shopping mall in a rectangular lot measuring 1500 feet by 2000 feet. That city estimates that 59% of this space will be needed for parking. About how many square feet will the shopping area cover?

31. The shaded area in the figure at the right covers 30% of the area of the large triangle. What is the area of the large triangle?

32. If you throw a dart at random, what is the probability that it will land in the small triangle?

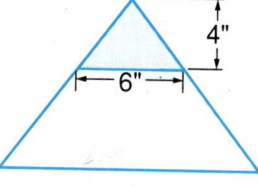

33. What is the probability that a dart thrown at random will land inside the larger triangle but outside the smaller triangle?

## Maintain Your Skills

34. Mr. Williams is installing carpet in his living room. The room is $15\frac{1}{3}$ feet long and $12\frac{3}{4}$ feet wide. How many square feet of carpeting are needed? **(Pages 64–68, 175–179)**

Evaluate for $p = -4$, $d = 2$, and $e = -3$. (Pages 262–266)

35. $\dfrac{e^2}{pd - 1}$
36. $\dfrac{d - e}{e^2 - d^2}$
37. $p(e - d)^2$

Write each percent as a decimal. (Pages 309–312)

38. 18.7%
39. 4.4%
40. $2\frac{1}{4}$%
41. 0.3%
42. $22\frac{1}{2}$%

Write each number as a percent. (Pages 305–308)

**1.** $\frac{3}{10}$   **2.** $\frac{13}{25}$   **3.** $\frac{11}{12}$   **4.** 7.3   **5.** 0.02   **6.** 0.614

Write each percent as a decimal. (Pages 309–312)

**7.** 26.5%   **8.** 0.7%   **9.** 8.1%   **10.** 903%   **11.** 0.41%

Write each percent as a fraction in lowest terms. (Pages 309–312)

**12.** $\frac{3}{5}$%   **13.** $\frac{5}{8}$%   **14.** $\frac{9}{10}$%   **15.** $\frac{5}{6}$%   **16.** $\frac{1}{2}$%

Use fractions to estimate the percent of each number. (Pages 313–316)

**17.** 48% of 131   **18.** 16% of 154   **19.** 89% of 108

Solve. (Pages 319–322)

**20.** Miss White receives a commission of 5.5% on sales of $4570. How much commission did she receive?

**21.** Your restaurant bill totals $25.40. You want to leave a 15% tip. How much will you leave?

# Mathematical Footnote

Consider this headline.

Does this mean that an individual worker will receive a pay raise of more dollars than a manager? Not necessarily!

Suppose that a worker is earning $12,000 a year. How much does a 5% pay raise amount to? You might think: "10% of 12,000 is 1200 and half of that is 600." So, for a worker, a pay raise of 5% amounts to $600.

Now suppose that a manager earns $30,000 a year. What does the manager's 3% raise amount to? Here you might think: "1% of 30,000 is 300 and three times that is 900." So the manager's pay raise amounts to $900.

Because the manager's base salary is greater, the smaller percent (3%) raise amounts to more dollars than the 5% raise. This is the reason why many people try to negotiate for the highest possible salary when starting a new job.

## LESSON 8-6

# Interest

**Interest** is money paid for the use of money. Banks pay their customers interest on their savings. Customers pay the bank interest on money they borrow from the bank.

When customers borrow money, the amount borrowed is called the **principal**. The **rate of interest** is the percent of the principal charged for borrowing the principal. The rate of interest is given as a yearly percent. However, the amount of interest can be calculated for any period of time.

This is the formula for computing simple interest.

Interest = Principal · Rate per Year · Time in Years
$$I = p \cdot r \cdot t \quad \text{or} \quad I = prt$$

**EXAMPLE 1** Find the interest on a loan of $2400 that is borrowed for 5 months at a rate of 9.5%.

**Solution** Write the time in years and write the rate as a decimal.

Time ($t$): 5 mo = $\frac{5}{12}$ yr         Rate ($r$): 9.5% = $\frac{9.5}{100}$ = 0.095

$I = p \cdot r \cdot t$

$= (\overset{200}{\cancel{2400}})(0.095)\left(\frac{5}{\underset{1}{\cancel{12}}}\right) = 200(0.095)(5) = 95$

The interest is **$95.00**.

**Try This**  Find the interest on each loan.
  1. $1600 for 3 years at 9%
  2. $8500 for $2\frac{1}{2}$ years at $8\frac{1}{2}$%

The **amount** of money paid back by the borrower over the time of a loan is the sum of the principal and the interest.

$$\begin{array}{ccccc} \text{Amount} & = & \text{Principal} & + & \text{Interest} \\ A & = & P & + & I \end{array}$$

In Example 1, the amount is $2400 + $95, or $2495.

Interest paid on savings accounts is often computed using compound interest. **Compound interest** is computed on the principal plus the interest previously earned. The compound interest formula shown below is used to compute the amount of money accumulated (principal plus interest).

$$A = P(1 + r)^n$$

$A$: Amount of money accumulated
$P$: Principal
$r$: yearly interest rate
$n$: number of years

**EXAMPLE 2**  Four years ago, Erica Kanter put $400 in a savings account. The account earns 8% interest compounded yearly. Find the amount accumulated.

**Solution**  In the formula, replace $P$ with 400, $r$ with 0.08, and $n$ with 4.

$A = P(1 + r)^n$
$A = 400(1 + 0.08)^4$
$A = 400(1.08)^4$

You can use a calculator with a $\boxed{y^x}$ key to find the amount.

400 $\boxed{\times}$ 1.08 $\boxed{y^x}$ 4 $\boxed{=}$ $\boxed{544.19558}$

If your calculator does not have a $\boxed{y^x}$ key, press these keys.

400 $\boxed{\times}$ 1.08 $\boxed{=}$ $\boxed{=}$ $\boxed{=}$ $\boxed{=}$ $\boxed{544.19558}$

The amount accumulated after 4 years is **$544.19,** rounded down to the nearest cent.

**Try This**   Use the compound interest formula to find the amount accumulated. The interest is compounded yearly. Round down to the nearest cent.

3. $900 at 7% for 2 years
4. $12,000 at $8\frac{1}{2}$% for 3 years

- Why do banks have higher interest rates for the money they loan than for the money they pay depositors on savings accounts?

# EXERCISES

**Objectives:** To compute simple interest
To compute compound interest

## Check Your Understanding

*Skill Check*   Write the letter of the correct answer.

1. Which expression can be used to find the simple interest on $1200 at $7\frac{1}{2}$% for 6 months?
   a. 1200 × 0.75 × 6
   b. 1200 × 0.075 × 6
   c. 1200 × 0.075 × $\frac{1}{2}$
   d. 1200 × 0.75 × $\frac{1}{2}$

2. Which expression can be used to find the compound interest on $900 compounded yearly at 8% for 3 years?
   a. 900 × 0.08 × 3
   b. $900(0.08)^3$
   c. $900(1.8)^3$
   d. $900(1.08)^3$

3. Find the simple interest on a loan of $500 borrowed for 8 months at 12%.
   a. $40      b. $480
   c. $400     d. $30

4. Find the amount accumulated to the nearest cent on savings of $720 compounded yearly for 4 years at $8\frac{1}{2}$%.
   a. $997.00      b. $244.80
   c. $979.55      d. $997.82

True or False? When a statement is false, explain why.

5. Compound interest is computed on interest previously earned.
6. In banking, the word principal can refer to an amount borrowed or to an amount invested.
7. In the simple interest formula and in the compound interest formula, the letter $r$ represents a monthly interest rate.
8. In the formula $A = P(1 + r)^n$, the value of $A$ will always be greater than the value of $P$.

## Practice and Apply

**Find the simple interest.**

9. On $600 borrowed for one year at 10%
10. On $900 borrowed for 3 months at 8%
11. On $1800 borrowed for 4 months at 12.5%
12. On $720 borrowed for 2 months at 13.5%
13. On $900 borrowed for 4 months at $11\frac{1}{2}$%
14. On $3200 borrowed for 6 months at $13\frac{1}{2}$%

**Use the compound interest formula to find the accumulated amount. The interest is compounded yearly. Round answers down to the nearest cent.**

15. On $900 at 7% for 2 years
16. On $8000 at 9% for 3 years
17. On $750 at 8.5% for 3 years
18. On $4000 at 7.8% for 5 years
19. On $12,000 at $11\frac{1}{2}$% for 4 years
20. On $250 at $8\frac{1}{4}$% for 10 years

**Solve.**

21. George and Clara Lightfoot borrow $20,000 to start a business. The yearly interest rate is $11\frac{1}{2}$%. How much interest is due on the loan in 3 months?

22. Stephanie invests $5,000 in an account paying 9% interest compounded yearly. How much will she have in the account after 5 years?

Jasper borrowed $6500.00 from his bank for home improvements. The bank charged a yearly interest rate of 13%.

23. How much simple interest will Jasper owe at the end of one year?
24. What is the total amount Jasper will owe the bank at the end of one year?

Carrie invests $100 in a savings account at a yearly interest rate of 5%. Bob invests $100 in a savings account at a yearly interest rate of 10%. Both leave the money in the bank for 1 year.

25. How much simple interest will each person's account earn in one year?
26. Does Bob's account earn exactly twice the interest that Carrie's does?

## LESSON 8-7

# Discount

**P**hil McGuire works as a salesperson in Central City Department Store. As an employee, he receives a 20% discount on all items he purchases.

A **discount** is the amount that is subtracted from the **regular price**, or **list price**, of an item. The **rate of discount** is used to find the amount of the discount and the sale price.

**EXAMPLE 1** Phil decides to buy a bicycle from Central City Department Store. The regular price is $117.50 and his employee discount rate is 20%. How much is the discount?

**Solution**   Think: The discount is 20% of $117.50.

$$20\% \text{ of } 117.50 = \frac{1}{\cancel{5}} \times \cancel{117.50}^{23.50}$$

$$= 23.50$$

The amount of discount is **$23.50.**

**Try This**   1. An item with a list price of $50 is on sale at a discount of 25%. What is the amount of discount?

**330** CHAPTER 8

The price of an item after the discount is subtracted from the regular price is called the **sale price** or **net price**. To find the sale price when the regular price and the rate of discount are given, you must first answer the question:

What is the amount of discount?

Think of this as a hidden question in the problem. Then use the amount of discount to find the sale price.

**EXAMPLE 2** A calculator with a list price of $26.50 is on sale at a 40% discount. Find the sale price.

**Solution** First, find the amount of discount. Then find the sale price.

Find the discount.   40% of 26.50 = $\frac{2}{5} \times 26.50$

$$= 10.60$$

Find the sale price.   Regular price − Discount = Sale Price
26.50 − 10.60 = 15.90

The sale price is **$15.90**.

**Try This** Find the sale price.
2. List price: $52.90; discount: 20%
3. List price: $17.32; discount: 25%

# EXERCISES

**Objectives:** To compute the amount of discount
To find the sale price

## Check Your Understanding

**Skill Check** Write the letter of the correct answer.

1. A camera is on sale for 20% off the regular price of $165. What is the sale price?
   a. $53
   b. $33
   c. $132
   d. $165

2. The sale price of a cassette is $4.80. This list price is $7.20. What is the amount of discount?
   a. $9.60
   b. $2.40
   c. $12.00
   d. $3.60

3. The list price of a baseball glove is $24. It is on sale at a discount of 30%. What is the amount of discount?
   a. $7.20   b. $8.00
   c. $16.00  d. $16.80

4. How much will you pay for a dress with a regular price of $78 on sale at a 35% discount?
   a. $27.30  b. $52.00
   c. $50.70  d. $26.00

**Complete each statement.**

5. The price of an item before it is marked down at a discount is called the  ?  .

6. The amount a customer saves by buying items on sale is called a  ?  .

7. Amount of Discount =  ?  × Rate of Discount

8. Sale Price = List Price −  ?

## Practice and Apply

**Find the amount of discount on each item.**

9. A theater ticket purchased for $25 is sold at a 20% discount.

10. A sailboat with a list price of $3300 is offered for sale at a $33\frac{1}{3}$% discount.

11. A health club is offering a 15% discount on new memberships that regularly cost $230.

12. A sweater with a regular price of $36.40 is on sale at 40% off.

**Find the sale price of each item.**

13. Running shoes listed at $49.50 are on sale at a 30% discount.

14. A motorbike regularly priced at $576 is on sale at a $12\frac{1}{2}$% discount.

15. Slacks on sale at 35% off have a list price of $32.

16. Skateboards advertised at a list price of $40 are on sale at 15% off.

**Use mental math to determine which of the two discounts in each pair will have the lower sale price.**

17. A $33\frac{1}{3}$% discount on a regular price of $69 or a 30% discount on a regular price of $90.

18. A 10% discount on a list price of $99 or a 15% discount on a list price of $105.

19. A 5% discount on a regular price of $58 or a 10% discount on a regular price of $91.

20. A 25% discount on a regular price of $87 or a 30% discount on a list price of $90.

You can find the sale price of an item in one step by thinking of a discount in this way.

If the rate of discount is 20% of the regular price, then the amount you pay, or the sale price, is 80% of the regular price.

**21.** What percent of the sale price do you pay if the discount is 45%?

**22.** What percent of the sale price do you pay if the discount is $33\frac{1}{3}$%?

**23.** What percent of the sale price do you pay if the discount is 40%?

**24.** What percent of the sale price do you pay if the discount is $37\frac{1}{2}$%?

**25.** A $24 pair of slacks is on sale for 15% off. Find the sale price in one step.

**26.** Craig bought 3 exercise suits at 25% off. If the regular price for each suit is $24, find the sale price of a suit.

## Connect and Extend

Write *True* or *False*. If you write *True*, give an example to show your reasoning. If you write *False*, give a counterexample.

**27.** Conrad wants to buy a suit that is on sale at 20% off. One week later, the store advertises an additional discount of 10%. Conrad thinks: "The sale price would be the same if the store had given a 30% discount at the first sale."

**28.** Lottie plans to buy some luggage with a list price of $340 at a 30% discount. Her state has a 5% sales tax. She thinks: If there is a 30% discount and then I pay a 5% tax, this will be the same as a discount of (30 − 5) or 25%.

## Maintain Your Skills

**Solve. (Pages 114–118)**

**29.** Kimiko worked twice as many hours as Roscoe. Together they worked 57 hours. How many hours did each person work?

**30.** A cat weighs $\frac{1}{3}$ as much as a dog. Together the cat and dog weigh 44 pounds. How much does each animal weigh?

**Solve each proportion. (Pages 216–220)**

**31.** $\frac{6}{9} = \frac{8}{a}$

**32.** $\frac{4}{12} = \frac{c}{15}$

**33.** $\frac{12}{f} = \frac{8}{10}$

**34.** $\frac{z}{4} = \frac{180}{30}$

**Solve. Use fractions to estimate the percent. (Pages 313–316)**

**35.** The Spanish Club needs $46.80 to buy refreshments for a fiesta. The club is 24% short of its goal. About how much more money does the club need?

**36.** Carmen saves $9\frac{1}{2}$% of her monthly earnings. If her monthly earnings amount to $912, about how much does she save each month?

## LESSON 8-8: Solving Percent Equations and Proportions

Sue bought a $20 blouse for $15. What was the percent discount?

Gerald's 10% commission of a sale amounted to $12. What was the amount of the sale?

John had 85% of the 20 questions on a test correct. How many were correct?

Each of these problems is a different type of percent problem. These are the three general types.
1. Finding a percent of a number
2. Finding what percent one number is of another
3. Finding the original number when a percent of it is known

Each type of percent problem can be solved either by writing an equation or by writing a proportion.

**EXAMPLE 1** Adam's test score was 80%. There were 120 test questions. How many of Adam's answers were correct?

**Solution** Think: Let $n$ = the number of correct answers.
Then 80% of 120 = $n$.

**USING AN EQUATION**

80% of 120 = $n$

$\frac{4}{5} \times \overset{24}{\cancel{120}} = n$

96 = $n$

**USING A PROPORTION**

$\frac{80}{100} = \frac{n}{120}$

$100n = 80 \times 120$

$\frac{100n}{100} = \frac{80 \times 120}{100}$

$n = 96$

Adam answered **96** questions correctly.

**Try This**
1. What is 125% of 68?
2. What is 0.5% of 220?

- Which of the three problems at the beginning of the lesson can be solved in the same way as the problem in Example 1?
- What is the first step in writing a proportion to solve a percent problem?

**EXAMPLE 2** A bird seed mixture contains 2 pounds of sunflower seeds for every 10 pounds of millet seeds. What percent of the mixture of the two kinds of seeds is the sunflower seeds?

**Solution** **Think:** The mixture contains a total of 2 + 10, or 12 pounds of seeds. Of these 12 pounds, 2 pounds are sunflower seeds.

Let $p$ = the percent of sunflower seeds.
Then 2 is $p$ percent of 12.

**USING AN EQUATION**

$$2 = p\% \text{ of } 12$$
$$2 = \frac{p}{100} \times 12$$
$$2 \times 100 = 12p$$
$$\frac{200}{12} = \frac{12p}{12}$$
$$16\tfrac{2}{3} = p$$

**USING A PROPORTION**

$$\frac{2}{12} = \frac{p}{100}$$
$$12p = 200$$
$$\frac{12p}{12} = \frac{200}{12}$$
$$p = 16\tfrac{2}{3}$$

The sunflower seed is **$16\tfrac{2}{3}$%** of the mixture.

**Try This**
3. 36 is what percent of 240?
4. 45 is what percent of 12?

**EXAMPLE 3** The Turner family budgets 6.5% of monthly income for savings. If they save $117 monthly, what is their monthly income?

**Solution** **Think:** Let $c$ = the Turner's monthly income.
Then 117 = 6.5% of $c$.

**USING AN EQUATION**

$$117 = 6.5\% \times c$$
$$117 = \frac{6.5}{100} \times c$$
$$11700 = 6.5c$$
$$\frac{11700}{6.5} = \frac{6.5c}{6.5}$$
$$1800 = c$$

**USING A PROPORTION**

$$\frac{117}{c} = \frac{6.5}{100}$$
$$6.5 \times c = 117 \times 100$$
$$\frac{6.5c}{6.5} = \frac{11700}{6.5}$$
$$c = 1800$$

The Turner family's monthly income is **$1800.**

**Try This**
5. 4.5 is 2.4% of what number?
6. 14 is 125% of what number?

8-8 Solving Percent Equations and Proportions

# EXERCISES

**Objective:** To solve percent problems by using equations or proportions

## Check Your Understanding

*Skill Check* Write the letter of the correct answer.

**1.** Which equation could you use to solve this problem?
$$3 \text{ is what percent of } 10?$$
 **a.** $10 = \frac{n}{100} \times 3$   **b.** $3 = \frac{n}{100} \times 10$
 **c.** $\frac{3}{100} = 10n$   **d.** $3 = 10n$

**2.** Which proportion could you use to solve this problem?
$$14 \text{ is } 5\% \text{ of what number?}$$
 **a.** $\frac{14}{r} = \frac{5}{100}$   **b.** $\frac{r}{14} = \frac{5}{100}$
 **c.** $\frac{5}{r} = \frac{14}{100}$   **d.** $\frac{14}{r} = \frac{5}{1}$

**3.** Which proportion could you use to solve this problem?
$$\text{What is } 72\% \text{ of } 18?$$
 **a.** $\frac{72}{100} = \frac{n}{18}$   **b.** $\frac{72}{100} = \frac{18}{n}$
 **c.** $\frac{72}{n} = \frac{18}{100}$   **d.** $\frac{72}{n} = \frac{18}{1}$

**4.** Which equation could you use to solve this problem?
$$45 \text{ is } 30\% \text{ of what number?}$$
 **a.** $45 \times \frac{30}{100} = t$   **b.** $30 = \frac{45}{100} \cdot t$
 **c.** $45 = \frac{30}{100} \cdot t$   **d.** $45 = 30t$

## Practice and Apply

**Write the equation you would use to solve the problem. Do *not* solve the problem.**

**5.** What is 8% of 21?

**6.** 4 is 20% of what number?

**7.** 17 is what percent of 85?

**8.** 6 is 15% of what number?

**Solve by using an equation.**

**9.** What percent of 92 is 69?

**10.** What is 8% of 75?

**11.** 13% of 651 is what number?

**12.** What percent of 65 is 39?

**13.** 7 is 16% of what number?

**14.** What percent of 52 is 78?

**Write the proportion you could use to solve each problem. Do *not* solve the problem.**

**15.** What is 8% of 90?

**16.** What percent of 36 is 9?

**17.** 102 is 30% of what number?

**18.** 28 is 5% of what number?

**Solve by using a proportion.**

**19.** What percent of 30 is 27?

**20.** 25 is what percent of 60?

**21.** 42 is what percent of 30?

**22.** What is 25% of 28?

**23.** 45.5 is 91% of what number?

**24.** What is 3% of 825?

**Solve. Use the method you prefer.**

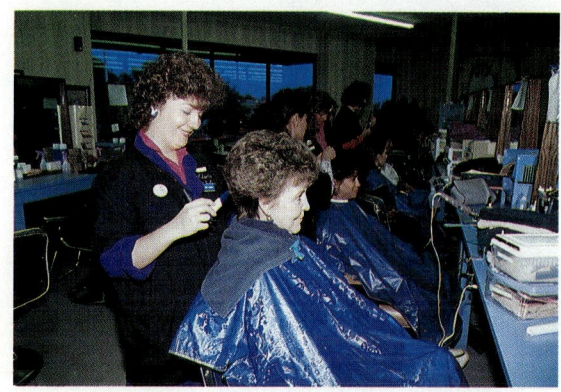

**25.** Warren spelled 30 words correctly on a 40-word spelling test. What percent of the words did he spell correctly?

**26.** Lisa earns a 5% commission on sales. Last week, her commission was $35.50. What were her sales?

**27.** The training program at a beautician school lasts 6 months. The tuition costs $1,900. In addition, the student must buy a personal set of tools for $600. What percent of the total cost of tuition and tools is the cost of the tools?

## Connect and Extend

**28.** In the figure at the right, there are 18 small squares in D. If D is 75% of C + D, how many small squares of the same size as in rectangle D can you put in the rectangle C + D?

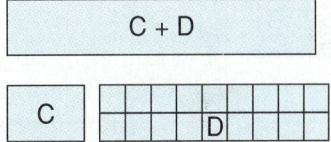

**29.** *Complete:* If 18 is 75% of some number, then the number is  ?  .

**30.** Rheta owns a boutique. When an item arrives, Rheta marks it up 60% from what she paid for it. If the item doesn't sell after 9 weeks, she takes 25% off its price. If the item still is in the shop after 13 weeks, she takes another 30% off. If she sells the item at this price, does she earn or lose money on it?

**31.** What percent of the price that Rheta originally paid for this item is her earnings or loss?

## Maintain Your Skills

**Solve. (Pages 105–109)**

**32.** $b + 9 = 2(b - 3)$  **33.** $-5(r + 2) = 30$  **34.** $25 - 5t = 3(2t + 1)$

**Solve by estimating. (Pages 282–286)**

**35.** Tim has $79 in his checking account. He wrote two checks for $29.17 and $36.64. Does he have enough money in his account to write another check for $13.20?

**36.** Debbie has $86. She needs to buy two new tires which cost $31.45 each. She also wants to have the front wheels aligned for $22.98. Does she have enough money?

**Write each ratio as a percent. (Pages 305–308)**

**37.** $\frac{29}{50}$  **38.** $\frac{3\frac{1}{2}}{3\frac{1}{2}}$  **39.** $\frac{1}{12}$  **40.** $\frac{5}{16}$  **41.** $\frac{27}{18}$

## LESSON 8-9: Percent Increase and Decrease

You can save on the cost of energy during the warmer months of the year by keeping the thermometer set at 80°F.

The figure at the right shows the approximate percent increase in energy costs when the thermostat is set at 79°F, 78°F, 77°F, 76°F, and 75°F.

The **percent increase** is the ratio of the amount of increase to the original amount multiplied by 100.

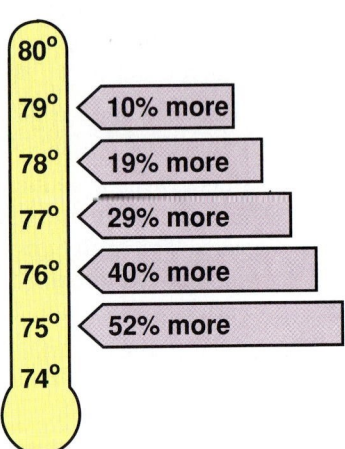

**EXAMPLE 1** The Bassett family paid $120 in cooling costs for the month of June. During August, they lowered the thermostat setting by 1°. Cooling costs for August amounted to $132. Find the percent increase.

**Solution**

Find the amount of increase.    $132 - 120 = 12$

Find the ratio: $\dfrac{\text{Amount of increase}}{\text{Original amount}}$    $\dfrac{12}{120} = \dfrac{1}{10}$

Write a percent for $\dfrac{1}{10}$.    $\dfrac{1}{10} = 10\%$

There is a **10%** increase in cooling costs from June to August.

**Try This**

1. The Wilson's paid $140 in cooling costs last month. Their bill increased to $196 this month. Find the percent increase.

You can save on heating costs during the cooler months of the year by lowering the thermostat setting to about 65°F. The **percent decrease** is the ratio of the amount of decrease to the original amount multiplied by 100.

**EXAMPLE 2** The Gadbois family paid $90 in heating costs last month. This month's bill was reduced by lowering the thermostat setting 3°. Heating costs for this month were $82. Find the percent decrease (nearest whole percent).

**Solution**

Find the amount of decrease.   $90 - 82 = 8$

Write the ratio: $\dfrac{\text{Amount of decrease}}{\text{Original amount}}$   $\dfrac{8}{90} = \dfrac{4}{45}$

Write a percent for $\dfrac{4}{45}$.   Let $p$ = the first term of the percent ratio.

Use a proportion.
$$\dfrac{p}{100} = \dfrac{4}{45}$$
$$45p = 400$$
$$p = 8.8$$

There is a **9%** decrease (nearest whole percent).

**Try This**

2. The Lopez family paid $123 in heating costs during December. Their bill for January was $105. Find the percent decrease (nearest whole percent).

# EXERCISES

**Objectives:** To apply percent increase
To apply percent decrease

## Check Your Understanding

**Skill Check** Write the letter of the correct answer.

1. Which ratio can be used to find the percent decrease?
   a. $\dfrac{\text{Amount of decrease}}{\text{Amount of increase}}$
   b. $\dfrac{\text{Amount of increase}}{\text{Amount of decrease}}$
   c. $\dfrac{\text{Amount of decrease}}{\text{Original amount}}$
   d. $\dfrac{\text{Amount of increase}}{\text{Original amount}}$

2. Which proportion can be used to find the percent increase from 125 to 170?
   a. $\dfrac{p}{100} = \dfrac{45}{170}$
   b. $\dfrac{p}{170} = \dfrac{45}{100}$
   c. $\dfrac{p}{125} = \dfrac{45}{100}$
   d. $\dfrac{p}{100} = \dfrac{45}{125}$

**3.** What is the percent increase from 72 to 108?

   **a.** 50%    **b.** $66\frac{2}{3}$%
   **c.** $33\frac{1}{3}$%   **d.** 36%

**4.** What is the percent decrease from 420 to 140?

   **a.** 200%    **b.** $66\frac{2}{3}$%
   **c.** 50%    **d.** 300%

Find the percent increase or decrease.

**5.** From 50 to 10    **6.** From 15 to 45    **7.** From 25 to 15

## Practice and Apply

Write the ratio you could use to find the percent increase or decrease. Then find the percent increase or decrease.

| | Item | Original Cost | Present Cost |
|---|---|---|---|
| **8.** | Rent | $320/mo | $368/mo |
| **9.** | Car | $4650 | $3720 |
| **10.** | Shirt | $25 | $21.25 |

| | Item | Original Cost | Present Cost |
|---|---|---|---|
| **11.** | Groceries | $24.20 | $25.41 |
| **12.** | Airline ticket | $383 | $421.30 |
| **13.** | Electric bill | $76/mo | $53.20/mo |

Solve. Round answers to the nearest whole percent.

**14.** During the day, teenagers watch about 3 hours of TV. During the evening, they watch 4 hours of TV. What is the percent increase?

**15.** During the football strike in 1987, attendance at some stadiums fell from about 60 thousand to about 8 thousand. What was the percent decrease?

**16.** The Stafford Company paid $330 in heating costs one month. The next month, after lowering the thermostat, their bill decreased to $290. What was the percent decrease?

**17.** Dan paid $95 in cooling costs this month. Last month his bill was $85. Find the percent increase.

This table shows the number of games or events won by several of Pearson High's clubs or teams in the last 2 years. Use an estimate to name the club or team whose wins increased or decreased by each percent.

| Number of Wins | | |
|---|---|---|
| Team or Club | Last Year | This Year |
| Chess Club | 12 | 16 |
| Debating Team | 13 | 10 |
| Math Team | 16 | 19 |
| Quiz Team | 29 | 23 |
| Trivia Club | 25 | 38 |

**18.** About 50%

**19.** About 25%

**20.** About $33\frac{1}{3}$%

**21.** About 20%

**22.** If a change from 40 to 48 is a 20% increase, is a change from 48 to 40 a 20% decrease? Explain.

## Connect and Extend

23. A record player is on sale at 20% off. After 2 weeks, the record players are marked back up to the original price. What is the percent increase? Explain your answers.

**There were 30 students in an exercise class last year. Use this information for Exercises 24–27.**

24. If the number of students increased 50% since last year, how many students are in the class this year?
25. If the number of students increased 10% since last year, how many students are in the class this year?
26. If the number of students increased 200% since last year, how many students are in the class this year?
27. If the number of students in the class this year is 36, and this number has increased 200% since last year, how many students were in the class last year?

## Maintain Your Skills

**Solve. (Pages 114–118)**

28. Find three consecutive numbers whose sum is $-93$.
29. Find three consecutive multiples of 4 whose sum is 144.
30. Write the prime factorization of 2,772. (Pages 143–146)

**Solve. (Pages 183–188, 206–209)**

31. Earl spent $\frac{4}{5}$ of his money on CDs. If he spent $32, how much did he have originally?
32. Tommy had 7 feet of rope. He used $3\frac{5}{6}$ feet of it. How much rope did he have left?

**Express the product or quotient in scientific notation. (Pages 289–294)**

33. $\dfrac{1.69 \times 10^8}{13 \times 10^6}$
34. $(-3.4 \; 10^{-7})(5 \times 10^3)$
35. $\dfrac{5.1 \times 10^4}{17 \times 10^5}$

**Solve. (Pages 324–328, 330–333)**

36. Find the simple interest on a loan of $800 that is borrowed for 3 months at a rate of 16.5%.
37. A camera with a list price of $69.90 is on sale at a 30% discount. Find the sale price.

# SUMMARY

## KEY TERMS

Amount (p. 325)
Commission (p. 320)
Compound interest (p. 325)
Discount (p. 330)
Interest (p. 324)
List price (p. 330)
Net price (p. 331)
Percent (p. 305)

Percent decrease (p. 339)
Percent increase (p. 338)
Principal (p. 324)
Rate of discount (p. 330)
Rate of interest (p. 324)
Regular price (p. 330)
Sale price (p. 331)
Simple interest (p. 324)

## KEY IDEAS

**A.** To determine whether a ratio is less than 100%, equal to 100%, or greater than 100%, look at the first term.
  1. If it is less than the second term, then the ratio is less than 100%.
  2. If it is equal to the second term, then the ratio equals 100%.
  3. If it is greater than the second term, then the ratio is greater than 100%.

**B.** To write a percent as a decimal, move the decimal point two places to the left. Omit the percent symbol.

**C.** To write a decimal as a percent, move the decimal point two places to the right. Write the percent symbol.

**D.** To find the percent of a number, first change the percent to a decimal. Then multiply.

**E. Formula for simple interest:** $i = prt$ where $i =$ interest, $p =$ principal, $r =$ yearly interest rate, and $t =$ the number of years.

**F. Formula for compound interest:** $A = p(1 + r)^n$ where $A =$ the accumulated amount, $p =$ principal, $r =$ yearly interest rate, and $n =$ the number of years.

**G.** To find the sale price when the regular price and the rate of discount are given, first find the amount of discount. Then subtract the discount from the regular price.

# REVIEW

**Write each ratio as a percent. (Pages 305–308)**

1. $\frac{8}{25}$
2. $\frac{23}{10}$
3. $\frac{17}{20}$
4. $\frac{5}{12}$
5. $\frac{23}{15}$
6. $\frac{7}{18}$

**Write each decimal as a percent. (Pages 309–312)**

7. 4.8
8. 0.5
9. 0.34
10. 0.07
11. 0.621

**Write each percent as a fraction in lowest terms. (Pages 309–312)**

12. $\frac{5}{6}$%
13. $\frac{3}{4}$%
14. $\frac{1}{2}$%
15. $\frac{3}{10}$%
16. $\frac{7}{20}$%

**Use fractions to estimate the percent of each number. (Pages 313–316)**

17. 38% of 50
18. 15% of 161
19. 81% of 213

**Write *Approximate* or *Exact* to tell which kind of answer would be best in the situation. Give a reason for your answer. (Pages 317–318)**

20. Buying tickets for a rock concert
21. Measuring ingredients for a recipe
22. Ordering packages of nuts to sell as a fund raising activity

**Find the percent of the number. (Pages 319–322)**

23. What is 135% of 80?
24. $2\frac{1}{2}$% of 40 is what number?

**Solve. (Pages 324–328, 330–341)**

25. A baseball team played 160 games this season. They lost 30% of the games played. How many games did they win?
26. Terri has a lunch bill of $10.60. She wants to leave a 15% tip. How much will she leave?
27. Find the simple interest on a loan of $800 borrowed for 9 months at $6\frac{1}{2}$%.
28. Find the amount accumulated to the nearest cent on $340 compounded yearly for 3 years at 7%.
29. A sweater with a list price of $38.90 is on sale at a 30% discount. Find the sale price.
30. 42 is 7% of what number?
31. What percent of 40 is 15?
32. Elizabeth was earning $5.50 an hour before her raise to $5.83 an hour. What is the percent increase?
33. Mr. Jones bought a used car for $5700 and sold it for $4674. What was his percent loss?

Chapter 8 Review  **343**

# TEST

**Write each ratio as a percent.**

1. $\frac{2}{5}$  2. $\frac{3}{4}$  3. $\frac{5}{2}$  4. $\frac{17}{20}$  5. $\frac{42}{33}$  6. $\frac{1}{3}$

**Write each decimal as a percent.**

7. 0.79  8. 0.06  9. 5.3  10. 0.418  11. 0.2

**Write each percent as a fraction in lowest terms.**

12. $\frac{1}{4}$%  13. $\frac{1}{2}$%  14. $\frac{7}{10}$%  15. $\frac{1}{6}$%  16. $\frac{2}{3}$%

**Use fractions to estimate the percent of each number.**

17. 11% of 133  18. 42% of 626  19. 84% of 171

**Find the percent of the number.**

20. $12\frac{1}{2}$% of 56 is what number?
21. What is 125% of 68?

**Solve.**

22. 3 is what percent of 20?
23. 56% of what number is 28?
24. 12 is $4\frac{1}{2}$% of what number?
25. What percent of 50 is 39?
26. Find the simple interest on a loan of $500 borrowed for 10 months at 18%.
27. Find the amount accumulated to the nearest cent on $150 compounded yearly for 2 years at 5%.
28. A home computer with a list price of $1078 is on sale at a discount of 20%. Find the sale price.
29. A cassette tape with a list price of $9.85 is on sale at a 4% discount. What is the amount of discount?
30. When Mr. Smith went to the movies several years ago, his admission ticket cost $4.00. When he went to the movies last week, his ticket cost him $6.00. Find the percent increase.
31. Last year a certain television cost $483. This year the same television costs $345. Find the percent decrease.

**Use an integer to represent the opposite of each amount. (Pages 2–6)**

1. 17 degrees above zero
2. 8 years ago
3. A weight loss of 5 pounds

**Solve. (Pages 32–35)**

4. When an integer is divided by −8, the quotient is 7. What is the integer?
5. The total change in the temperature over a four-week period was −28°C. What was the average change per week?

**Use the numbers, 2, −5, and 8, to illustrate each property. (Pages 54–58)**

6. Associative Property of Multiplication
7. Commutative Property of Addition

**Solve. (Pages 126–130, 200–209, 236–240, 252–257)**

8. Some of the factors of 96 are 1, 2, 3, 6, 12, 24, 48, and 96. Find the remaining factors.
9. Find the third and the seventh nonzero multiple of 31.
10. Susan made a dry fruit mix with $2\frac{1}{2}$ cups of raisins, $2\frac{1}{3}$ cups of dried apples, and $2\frac{1}{4}$ cups of dried apricots. How much fruit did she use in the mix?
11. On Monday, Tim jogged $3\frac{1}{2}$ miles. On Wednesday, he jogged $4\frac{1}{4}$ miles. How much further did he jog on Wednesday than on Monday?
12. You throw a number cube. What is the probability of throwing a factor of 6?

13. Raymond has 20 different books on a bookshelf. In how many ways can he pick 3 books?

**Simplify. (Pages 267–270)**

14. $-\frac{1}{5} + \frac{3}{4}$
15. $1\frac{3}{4} \cdot \left(-\frac{4}{5}\right)$
16. $-\frac{1}{10} \div -\frac{3}{12}$
17. $-\frac{11}{12} - \left(-\frac{5}{3}\right)$

**Write the decimal as a mixed number or a fraction in lowest terms. (Pages 277–280)**

18. −14.4
19. 7.75
20. −22.008
21. −9.16

**Write the ratio as a percent. (Pages 305–308)**

22. In Saturday's football game, 6 out of 15 passes were incomplete.
23. Four out of 32 students were absent on Friday.

Cumulative Review: Chapters 1–8

# UNIT 4

### PROJECT 1
You are on the food committee that is planning a BBQ picnic at a local park for a group of about 120 exchange students. You want to show them what an all-American picnic is like. Work as a team to plan the menu, determine the amount of food, drink, and other supplies needed, and estimate the total cost.

### PROJECT 3
How much TV do teenagers watch? Why do teenagers watch TV? What types of shows do teenagers watch? These and many other questions are asked by research companies to learn about the viewing habits, likes, and dislikes of teenage viewers. You and your team will decide on some issues about TV and teenagers you would like to investigate, conduct a survey of about 50 teenagers, then organize and present your data to the class.

# Using Graphs

## PROJECT 2

Do you or does someone in your family regularly make long-distance calls? Can you estimate the cost of each call before you receive the telephone bill? Telephone companies provide a schedule of rates for long-distance calls. For this project, you will investigate rate charges of different long-distance services. Then you will compare the rates and costs of the different services and consider their advantages and disadvantages.

## CONTENTS

### 9 Analyzing Data
Misleading Graphs
Using Data From Graphs and Tables
Organizing and Presenting Data
Measures of Central Tendency
Problem Solving Strategies: Analyzing Sample Data
Stem and Leaf Plots
Box and Whisker Plots

### 10 The Number Line
The Set of Real Numbers
The Addition Property of Inequality
The Multiplication Property of Inequality
Problem Solving Strategies: Using Generalizations
Solving Inequalities
Conjunctions
Disjunctions

### 11 The Coordinate Plane
Coordinate Graphs
Graphing Linear Equations
The Standard Form of a Linear Equation
The Slope of a Line
Problem Solving Strategies: Revising the Solution
Graphing Equations and Inequalities
Problem Solving: Using Two Variables
Translations

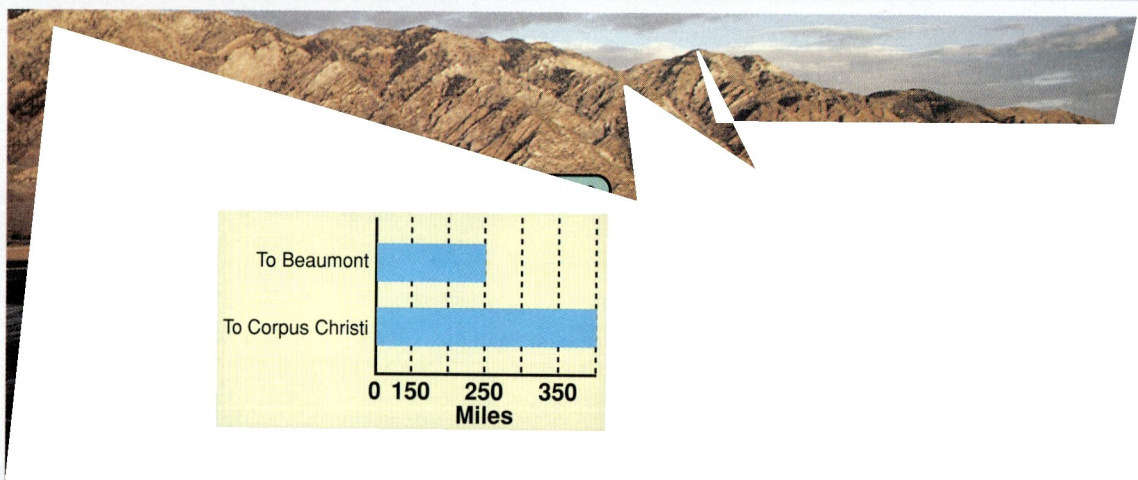

# LESSON 9-1

## Misleading Graphs

Sometimes the information shown in graphs can be misinterpreted because of the way in which the information is displayed. Special attention should be paid to the horizontal and vertical scales used in the graphs.

**EXAMPLE** The graph above shows the approximate driving mileages from Dallas to Beaumont and Dallas to Corpus Christi.

   **a.** How many miles is it from Dallas to Beaumont?
   **b.** How many miles is it from Dallas to Corpus Christi?
   **c.** Is the distance from Dallas to Corpus Christi twice the distance from Dallas to Beaumont?
   **d.** Measure the length of each bar on the graph. Why does the length of each bar make the mileage between Dallas and each city misleading?

**Solutions**   **a. 250 miles**    **b. 400 miles**    **c. No**

   **d.** The length of the bottom bar is twice as long as the top bar, but the actual distance from Dallas to Corpus Christi (400) is not twice the distance from Dallas to Beaumont (250).

**Try This** Explain how you would redraw the graph so it is not misleading.

# EXERCISES

**Objective:** To discover the ways in which graphs can be misleading

## Check Your Understanding

The graph at the right compares the sales of Super Flakes cereal with those of its leading competitor.

1. How does the height of the Super Flakes box compare with the height of the Chux box?

2. Do the heights of the boxes accurately represent the fact that the sales of Super Flakes are 3 times the sales of Chux?

3. How do the widths of the cereal boxes compare?

4. How do the areas of the cereal boxes compare?

5. Do the areas of the cereal boxes accurately compare the sales of the two cereals?

6. What information is missing from the graph?

## Practice and Apply

The two bar graphs below show the number of books sold by two book stores during a four-month period.

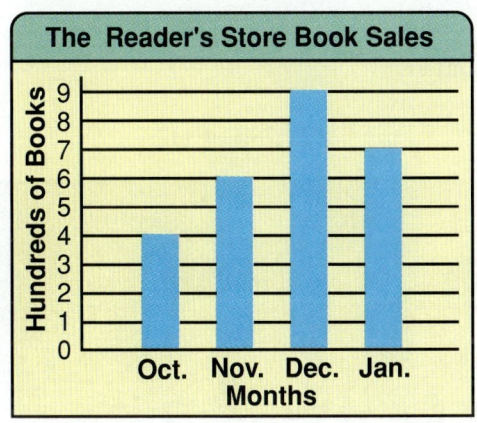
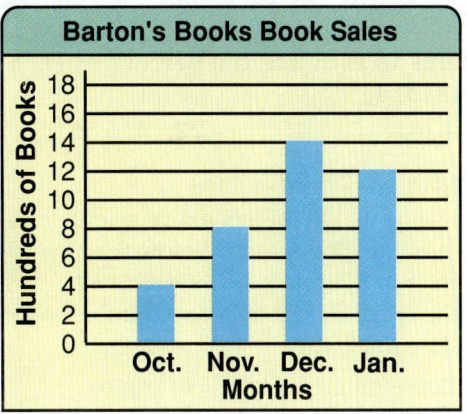

Compare the graphs. Then answer these questions.

7. Look at the bars on the two graphs. Which store appears to have sold more books during October? during December?

9-1 Misleading Graphs 349

8. Looking at the bars on the two graphs, which store appears to have sold more books during the 4-month period?

9. How many books did the Reader's Store actually sell over the 4 months?

10. How many books did Barton's Books actually sell over the 4 months?

11. Does placing the graphs side by side make it appear that the Reader's Store sold more than Barton's Books over the 4-month period? Explain.

12. Suppose that the two graphs were not placed side by side. Would it still appear that the Reader's Store sold more books than Barton's Books during the 4-month period? Explain.

## Connect and Extend

13. Describe how bar graphs used in advertisements can be misleading.
14. Describe how line graphs used in advertisements can be misleading.

This table shows the price of a share of stock for the XYZ Ticker Tape Corporation from 1987 to 1991.

| XYZ Ticker Tape Corporation | | | | | |
|---|---|---|---|---|---|
| End of Year | 1987 | 1988 | 1989 | 1990 | 1991 |
| Price | 95 | $104\frac{1}{2}$ | 102 | $100\frac{3}{8}$ | 105 |

15. Make a line graph that suggests that the prices stayed about the same over the 5 years.
16. Make a line graph that suggests that the prices varied widely over the 5 years.

## Maintain Your Skills

**Multiply. Write answers in lowest terms. (Pages 171–174)**

17. $\frac{5}{8} \cdot \frac{4}{15}$  
18. $\frac{3}{18} \cdot 12$  
19. $6 \cdot \frac{3}{20}$  
20. $\frac{8}{10} \cdot \frac{5}{14}$

**Use fractions to estimate the percent of each number. (Pages 313–316)**

21. 17% of 100
22. 62% of 105
23. 78% of 117

**Solve. (Pages 23–25, 282–286)**

24. A helicopter is 140 meters above sea level. A submarine directly below the helicopter is 34 meters below sea level. How far apart are the helicopter and the submarine?

25. The admission price for a field trip to the planetarium was $2.95 for adults and $1.75 for students. Estimate to determine whether the total cost will be less than, or greater than, $70 for 3 adults and 28 students.

350  CHAPTER 9

## LESSON 9-2

# Using Data from Graphs and Tables

**N**ewspapers and magazines contain data organized in tables or graphs. This makes the information easier to read and to use. When using such data it is useful to have a calculator available.

**EXAMPLE 1**  Joann's father bought a new car last year. This year Joann's older sister, Ashley, plans to buy the same model.

While reading the paper, Ashley saw this graph, which shows the inflation rate for several consumer items over the past year. Inflation refers to increase in costs.

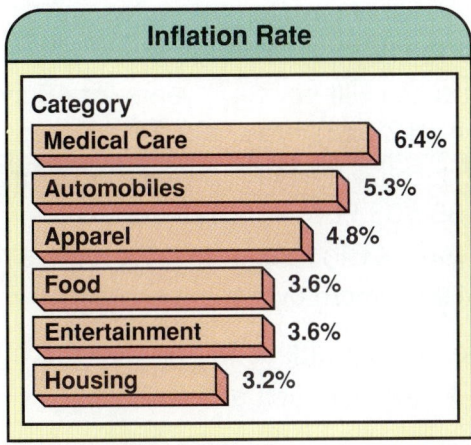

**Inflation Rate**

| Category | |
|---|---|
| Medical Care | 6.4% |
| Automobiles | 5.3% |
| Apparel | 4.8% |
| Food | 3.6% |
| Entertainment | 3.6% |
| Housing | 3.2% |

How can Ashley estimate the cost of a car this year if her father paid $11,800 for last year's model?

**Solution**  Read the graph to find the inflation rate: 5.3%

**Think:** 5.3% is about 5% and 5% = $\frac{1}{20}$.   $11,800 \approx \$12,000$

**Multiply to find the amount of increase.**   $\frac{1}{20} \times 12,000 = 600$

**Estimate.**   $12,000 + 600 = 12,600$

The car will cost Ashley about **$12,600**.

**Try This**

1. A family spent about $100 per month on entertainment last year. Estimate how much they can expect to spend this year if the inflation rate is 3.1%.

**EXAMPLE 2** Jacob is borrowing money to buy a used car. He plans to borrow $1000 at an interest rate of 14%. Use the table below to find how much more he will pay if he borrows the money for 4 years instead of for 3 years.

| Monthly Payments for a Loan of $1000 at Rates Indicated | | | | |
|---|---|---|---|---|
| Years | 14% | $14\frac{1}{8}$% | $14\frac{1}{4}$% | $14\frac{3}{8}$% |
| 1 | 89.60 | 89.65 | 89.71 | 89.76 |
| 2 | 47.82 | 47.88 | 47.93 | 47.99 |
| 3 | 33.98 | 34.04 | 34.10 | 34.15 |
| 4 | 27.13 | 27.18 | 27.24 | 27.30 |
| 5 | 23.06 | 23.12 | 23.18 | 23.24 |

**Solution** The table shows monthly payments for a loan of $1000 at four different interest rates.

Find the amount Jacob will pay for 4 years, or 48 months.
48($27.13) = $1,302.24

Find the amount Jacob will pay for 3 years, or 36 months.
36($33.98) = $1,223.28

Find the difference: $1302.24 − $1,223.28 = $78.96
Jacob will pay **$78.96** more over the four years.

**Try This** Suppose that the interest rate is $14\frac{3}{8}$%. How much more will Jacob pay for a 4-year loan than a 3-year loan?

# EXERCISES

**Objective:** To use information presented in graphs and tables

## Check Your Understanding

**Skill Check** Write the letter of the correct answer.
For Exercises 1–2, refer to the graph in Example 1.

**1.** Last year, a family spent about $150 per week for food. About how much can they expect to spend per week this year?
 a. $145  b. $160
 c. $156  d. $210

**2.** Bobbie's clothing budget for last year was $1600. If she adjusts her budget for inflation, about how much will it be this year?
 a. $1520  b. $1664
 c. $1920  d. $1680

**For Exercises 3–4, refer to the table in Example 2.**

**3.** Phil Harper borrows $1000 at $14\frac{1}{4}$% for 3 years. Find the amount he will pay for the loan.
 a. $102.30   b. $1,227.60
 c. $307.52   d. $1,307.52

**4.** Gloria Kay borrows $1000 at $14\frac{3}{8}$% for 4 years. How much less will she pay if she borrows the same amount for 4 years at $14\frac{1}{8}$%?
 a. $0.48   b. $2.88
 c. $3.96   d. $5.76

## Practice and Apply

This graph shows braking distances for cars on wet and dry asphalt roads while driving at speeds from 10 miles per hour to 60 miles per hour. *Braking distance* is the distance a car travels after the brakes are applied and until it stops completely.

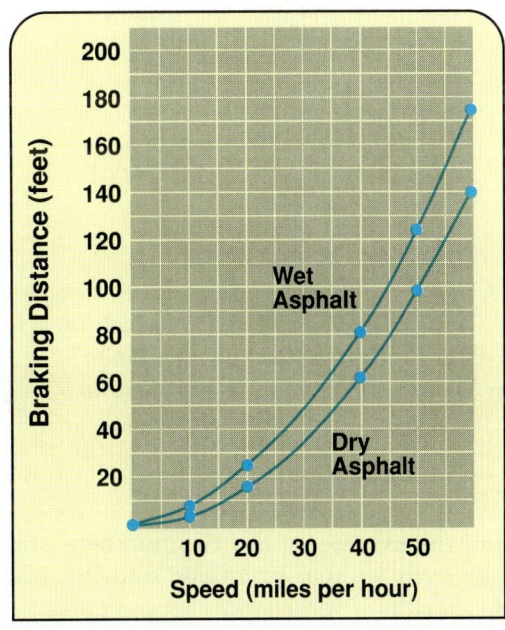

**5.** What is the braking distance for a car traveling at 50 miles per hour on dry asphalt?

**6.** What is the braking distance for a car traveling at 30 miles per hour on wet asphalt?

**7.** Estimate the difference in stopping distance for a car traveling on wet asphalt at 35 miles per hour and for a car traveling on dry asphalt at the same rate.

**For Exercises 8–11, use the graph to estimate the speed of the car.**

**8.** Braking distance: 50 feet
 Driving surface: dry asphalt

**9.** Braking distance: 100 feet
 Driving surface: wet asphalt

**10.** Braking distance: 130 feet
 Driving surface: wet asphalt

**11.** Braking distance: 10 feet
 Driving surface: dry asphalt

**12.** Karla was traveling 45 miles per hour on wet asphalt when she saw a branch lying in her path. She was 90 feet from the branch when she applied the brakes. Could she stop in time?

**13.** Refer to the graph to determine whether this generalization is true. Explain your answer.

 As the rate of speed increases, braking distances on wet asphalt increase faster than those on dry asphalt.

**14.** What conclusion can be drawn from Exercise 13 for persons driving cars in rainy weather?

Cora Martinez, the manager of the shoe department for Dillcris, Inc., recorded the number of customers in the department from 9:00 A.M. to 6:00 P.M. on a typical day.

| Hours | Customers |
|---|---|
| 9:00–10:00 | 11 |
| 10:00–11:00 | 24 |
| 11:00–12:00 | 26 |
| 12:00–1:00 | 48 |
| 1:00–2:00 | 40 |
| 2:00–3:00 | 31 |
| 3:00–4:00 | 24 |
| 4:00–5:00 | 27 |
| 5:00–6:00 | 52 |

**15.** During what hour did the shoe department have the most customers?

**16.** Over what 2-hour period did the shoe department have the most customers?

**17.** Cora decided to hire part-time help for the 12:00–2:00 and 5:00–6:00 periods. Why is this a good decision?

**18.** Cora has 3 full time employees on duty from 10:00 A.M. to 6:00 P.M. each day. If she wants the ratio of employees on duty to customers to be 1 to 10, how many part-time employees should she hire to work from 12:00–1:00? from 5:00–6:00?

## Connect and Extend

**19.** What are the advantages of presenting data on a graph instead of in a table? Give an example in which presenting data on a graph would be more effective than using a table.

**20.** When is it more useful to show data in a table instead of on a graph? Give an example to explain your answer.

## Maintain Your Skills

**Find the average of the two numbers and state whether the average is an even number or an odd number. (Pages 131–136)**

**21.** 6 and 14    **22.** −3 and 1    **23.** 37 and 43    **24.** 8 and −1

**Solve each proportion. (Pages 216–220)**

**25.** $\frac{3}{4} = \frac{e}{15}$    **26.** $\frac{7}{5} = \frac{6}{n}$    **27.** $\frac{f}{13} = \frac{1}{5}$    **28.** $\frac{14}{g} = \frac{4}{10}$

**Solve. (Pages 334–337)**

**29.** 19 is 50% of what number?

**30.** What percent of 30 is 75?

## ★ Math Team Problem

**31.** A man is shipwrecked on an island and a crate of apples is washed ashore with him. He calculates that if he eats one apple per day, the apples will last 14 days longer than if he eats two apples per day for two days and no apples on the third day. How many apples are there in the crate?

# LESSON 9-3
# Organizing and Presenting Data

A telephone company wanted an estimate of how long it should take to install telephone service in private residences. A survey of 50 telephone installers was taken and the average time in minutes that it took each installer was recorded.

| 15 | 12 | 20 | 7  | 9  | 12 | 10 | 12 | 15 | 12 | 8  | 21 | 13 | 15 | 9  |
|----|----|----|----|----|----|----|----|----|----|----|----|----|----|----|
| 10 | 11 | 8  | 16 | 12 | 15 | 9  | 11 | 10 | 14 | 10 | 13 | 9  | 15 | 17 |
| 12 | 13 | 11 | 20 | 9  | 14 | 8  | 10 | 15 | 11 | 22 | 9  | 10 | 11 | 13 |
| 14 | 18 | 17 | 19 | 12 |    |    |    |    |    |    |    |    |    |    |

One way to organize data is to use a **frequency table.** Then you can use the frequency table to construct a histogram. A **histogram** is a kind of bar graph that shows the frequency of intervals of data.

**Frequency Table**

| Interval | Tally | Frequency |
|----------|-------|-----------|
| 7–9      | ||||| ||||| | 10 |
| 10–12    | ||||| ||||| ||||| ||| | 18 |
| 13–15    | ||||| ||||| ||| | 13 |
| 16–18    | |||| | 4 |
| 19–21    | |||| | 4 |
| 22–24    | | | 1 |

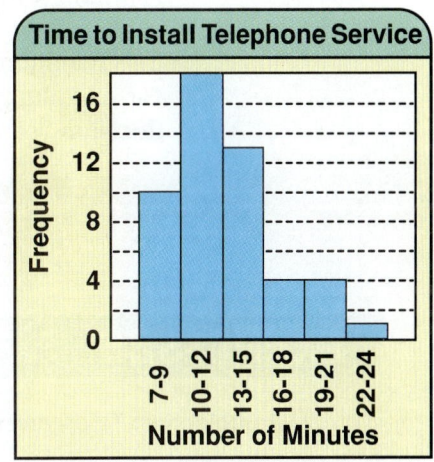

### EXAMPLE

Use the frequency table to answer these questions.
a. How many installers took 10–12 minutes to install telephone service?
b. What percent of the installers took 10–12 minutes to install telephone service?

**Solutions**
a. Since the frequency for 10–12 minutes is **18**, there were 18 installers who completed the job in 10–12 minutes.

9-3 Organizing and Presenting Data **355**

**b. Think:** There were 50 installers in all in the survey. What percent of 50 is 18?

Let $n$ = the percent. Write an equation.

$$18 = \frac{n}{100} \cdot 50$$
$$\frac{18}{50} = \frac{n}{100}$$
$$50n = 1800$$
$$n = 36$$

So **36%** of the installers took 10–12 minutes to complete the job.

**Try This** What percent of the installers took more than 18 minutes to install telephone service?

- The telephone company decided that a reasonable length of time for installing telephone service was 10–15 minutes. Was the decision reasonable? Why or why not?

# EXERCISES

**Objective:** To organize data as a frequency table and as a histogram

## Check Your Understanding

*Skill Check* Write the letter of the correct answer.

**Refer to the frequency table on page 355.**

**1.** How many installers took 15 minutes or less to complete this job?
   a. 13    b. 31
   c. 41    d. 28

**2.** What percent of the installers took less than 13 minutes to complete this job?
   a. 82%   b. 56%
   c. 26%   d. 18%

**3.** How many installers took from 16 to 21 minutes to complete the job?
   a. 4     b. 9
   c. 21    d. 8

**4.** Which interval gives the best average time for installing telephone service?
   a. 22–24   b. 19–24
   c. 7–12    d. 10–15

**5.** Explain how you could use the histogram on page 355 to answer Exercise 1.

## Practice and Apply

Pre-Algebra midterms test scores for a class of 30 students are listed at the right.

**Midterm Test Scores**

| 70 | 86 | 70 | 74 | 77 |
|---|---|---|---|---|
| 82 | 62 | 69 | 79 | 71 |
| 87 | 68 | 72 | 72 | 91 |
| 98 | 73 | 64 | 81 | 77 |
| 99 | 76 | 68 | 95 | 85 |
| 95 | 80 | 87 | 73 | 80 |

6. Make a frequency table that shows the scores. Use the intervals 60–64, 65–69, 70–74, and so on.

7. Make a histogram from your frequency table.

8. About what percent of the students had grades of 85 or better?

9. About what percent of the students had grades of 75 or better?

The table at the right shows the results of a survey of 30 families who were asked how many magazines they read per month.

**Number of Magazines Read in One Month by 30 Families**

| 9 | 2 | 7 | 12 | 4 |
|---|---|---|---|---|
| 10 | 8 | 14 | 1 | 8 |
| 7 | 3 | 5 | 17 | 3 |
| 9 | 15 | 20 | 2 | 5 |
| 6 | 11 | 8 | 16 | 1 |
| 3 | 18 | 6 | 9 | 11 |

10. Make a frequency table to show the information. Use the intervals 1–3, 4–6, 7–9, 10–12, 13–15, 16–18, and 19–21.

11. Make a histogram from the frequency table.

12. About what percent of the families read from 1–3 magazines per month?

13. Estimate the average number of magazines read per month by the families.

## Connect and Extend

14. In the histogram on page 355, what if intervals of 5 had been chosen. Which bar would be the highest?

15. What are the advantages of grouping data in a frequency table? What are the disadvantages?

## Maintain Your Skills

For Exercises 16–19, write the answer only if it is negative. (Pages 13–18, 23–25, 28–35)

16. $-26 + (-23)$
17. $52 \div (-4)$
18. $(-12)(-35)$
19. $-30 - (-42)$

Solve. (Pages 64–68)

20. The average age of five employees at Johnson's Hardware is 29. What is the age of the fifth employee if the ages of the other employees are 24, 30, 46, and 26?

Evaluate each expression. Write answers in lowest terms. (Pages 39–43, 46–50, 194–199)

21. $s - \frac{9}{14}$, for $s = \frac{13}{14}$

22. $c + \frac{27}{2}$, for $c = 6\frac{1}{2}$

23. $u - t$ for $u = \frac{19}{24}$, and $t = \frac{5}{24}$

# COMPUTER EXPLORATION

## Organizing and Presenting Data

**Objective:** To use a computer to explore data

1. **a.** With the Holt *Pre-Algebra* disk in drive A (or drive 1), run the program, *Organizing and Presenting Data*, as follows.
   **IBM:** Type **a:f**, and press **ENTER**.
   **Apple:** Turn on the computer. When the menu appears on the screen, type **F**.
   **b.** Read the opening messages and press **RETURN** (or **ENTER**).

2. **a.** Enter the following test scores of 12 students one at a time. Press **RETURN** after each entry.
   83, 91, 88, 92, 91, 88, 80, 95, 78, 67, 88, 92 END
   **b.** Does the frequency table list the values in ascending order or descending order?
   **c.** Which grade occurs most often?
   **d.** What percent (nearest whole percent) of the values in the table occur exactly once?

3. Press any key to continue.
   **a.** Does the horizontal axis show intervals of grades? If not, what does it show?
   **b.** What do the units on the vertical axis represent?
   **c.** Press any key. Press **Y** to (Y/N?) to enter another list of data.

4. **a.** Enter this list of data which represents net gains and losses on the New York Stock Exchange over a 15-day period. You do not need to enter the "+" when entering positive values. Press **RETURN** after each entry.
   3, −1, 2, 0, −1, 2, 3, −2, 5, 4, −3, 1, 0, 0, 2, END
   **b.** What is the range (highest and lowest values) of the data?
   **c.** Which two values occur most often?
   **d.** Press any key to obtain the histogram. Use the histogram to compute the net loss or gain.
   **e.** What is the advantage of using a histogram that shows the frequency of each item in the data?
   **f.** What is the advantage of using a histogram that shows the frequency of data in intervals?

## LESSON 9-4
# Measures of Central Tendency

A group of Grade 9 students are lined up from tallest to shortest as shown above. You can find the most common height, the average or mean height, and the height of the student in the middle.

These numbers are called the *mode*, the *mean*, and the *median*. They are measures of central tendency. A **measure of central tendency** is a statistic, usually a number, that is representative or typical in some way of a set of data.

The **mode** of a set of data is the number that occurs most often.

The mode of the student heights is 5 feet.

A set of data may have more than one mode, or it may have no mode.

The **mean** or **average** of a set of data having $n$ numbers is the sum of the numbers divided by $n$.

Mean of student heights: $\dfrac{6 + 5\frac{1}{2} + 5 + 5 + 4\frac{1}{2}}{5} = \dfrac{26}{5}$
$= 5\frac{1}{5}$

The mean of the student heights is $5\frac{1}{5}$ feet.

9-4  Measures of Central Tendency  **359**

The **median** of a set of data is the middle number in the set when the numbers are arranged in descending order or in ascending order.

The median of the student heights is 5 feet.

When there is an even number of numbers in a set, the median is the mean of the two middle numbers.

**EXAMPLE** The Weedless Wonder company has developed a new spray for destroying weeds. A test of the spray on 15 plots of land sown with weeds showed these results.

**Percent of Weeds Destroyed:** 98 98 91 89 97 99 90 85
99 99 71 75 71 71 89

a. Find the mean, median, and mode of the data.

b. If you were the company president, which of these measures would you use in an advertising campaign. Why?

**Solutions** Arrange the numbers in descending order.
99, 99, 99, 98, 98, 97, 91, 90, 89, 89, 85, 75, 71, 71, 71

a. Mean: $\frac{\text{Sum of numbers}}{15} = \frac{1322}{15} = 88.1$

Median: **90**     Modes: **71** and **99**

b. The president would use the **mode,** 99, because it suggests the best results.

# EXERCISES

**Objective:** To determine the mean, median, and mode of a set of data and to decide which measure is best to use in a given situation

## Check Your Understanding

**Skill Check** Write the letter of the correct answer.

**1.** What is the median of this set of numbers?

25  29  51  40

a. 29         b. 40
c. 34.5       d. 39.5

**2.** What is the mode of this set of numbers?

1, 1, 2, 2, 2, 3, 3, 3, 3, 4, 4, 4, 4

a. 4             b. 4 and 3
c. 1, 2, and 3   d. 1, 2, 3, and 4

**3.** If the mean of three numbers is 60, and two of the numbers are 56 and 59, which statement about the third number is true?
   a. It is less than 60.
   b. It is greater than 60.
   c. It equals 60.
   d. It equals $\frac{56 + 59}{2}$.

**4.** Which statement is **not** true?
   a. The mean and the median are always equal.
   b. Every set of data has a mean.
   c. Every set of data has exactly one median.
   d. The mean of a set of numbers is the same as the average.

## Practice and Apply

Find the mean, median, and mode of each set of data.

**5.**

| Number of Fiction Books Checked Out During Six Days ||
|---|---|
| Day | Number |
| Monday | 30 |
| Tuesday | 45 |
| Wednesday | 19 |
| Thursday | 30 |
| Friday | 36 |
| Saturday | 50 |

**6.**

| Heights in Feet of Six Mountains ||
|---|---|
| Mountain | Height |
| Blackburn | 16,390 |
| Churchill | 15,638 |
| McKinley | 20,320 |
| Pike's Peak | 14,110 |
| Rainier | 14,410 |
| San Luis | 14,010 |

**7.** Alexis is trying to convince her parents to raise her monthly allowance from $30 to $40. She surveyed some of her friends at school and made a table showing their monthly allowances. Then she told her parents: "The average is $40.20."

Do you think that this gives a true picture of how much allowance her friends are receiving? Explain.

| Allowance Amounts ||||
|---|---|---|---|
| $150 | $40 | $40 | |
| $35 | $35 | $35 | $35 |
| $33 | $30 | $30 | $30 |
| $30 | $30 | $25 | $25 |

Mary and Bill own a small business. Of the twelve employees, one earns $21,500, one earns $19,000, two earn $14,500 each, five earn $13,100 each, and three earn $11,000 each. Mary and Bill pay themselves $25,000 each.

**8.** Find the mean of the fourteen salaries.

**9.** Find the median and the mode of the fourteen salaries.

**10.** Which of these measures, the mean, median, or mode, is the highest *average* salary?

**11.** Which of these measures, the mean, median, or mode, is the lowest *average* salary?

**12.** Which of these measures would you use to argue for an increase in salary? Explain.

**13.** Which of these measures would you use to argue that "most people" earn more than you do? Explain.

## Connect and Extend

Pierre showed his classmates how to use the "assumed mean" to find the average of a set of numbers.

**Example:** Find the average of 52, 47, 40, 38, 29, and 24.

**Solution**
1. Estimate the mean. Choose one of the given numbers.
   Assumed mean: 38
2. Subtract each number in the given set from the estimate.
   $38 - 52 = -14$    $38 - 47 = -9$    $38 - 40 = -2$
   $38 - 38 = 0$    $38 - 29 = 9$    $38 - 24 = 14$
3. Find the sum of the differences in Step 2.
   $-14 + (-9) + (-2) + 0 + 9 + 14 = -2$
4. Find the average difference.    $-2 \div 6 = -0.3$
5. Since the average difference is negative, the assumed mean is lower than the true mean. So add 0.3 to the assumed mean.
   $38 + 0.3 = 38.3$

   The mean is **38.3**.

**14.** If the result of Step 4 is a positive number, the assumed mean is too high. Will you add or subtract the average difference to the assumed mean?

**Use Pierre's method to find the mean (nearest tenth).**

**15.** 53, 32, 49, 24, 62      **16.** 159, 167, 171, 162      **17.** 100, 125, 145, 132, 150

## Maintain Your Skills

**Solve and check. (Pages 183–188)**

**18.** $\frac{1}{6}e = \frac{5}{12}$      **19.** $1\frac{3}{8}f = \frac{3}{4}$      **20.** $\frac{1}{12}a = 2\frac{3}{10}$      **21.** $12 = 3\frac{1}{3}x$

**Solve. (Pages 221–225, 236–240)**

**22.** Mrs. Harvey drove 186 miles in 3 hours. If she drives at the same speed, how many miles can she drive in 4 hours?

**23.** Frank bought 15 gallons of gasoline for $21. If he pays the same price per gallon, how much will 12 gallons cost?

**24.** A drawer contains 4 black socks and 6 white socks. Without looking, Tom pulls one sock out of the drawer. What are the odds that the sock is not white? Write the answer in lowest terms.

**Write each ratio as a percent. (Pages 305–308)**

**25.** $\frac{3}{4}$      **26.** $\frac{2}{5}$      **27.** $\frac{9}{10}$      **28.** $\frac{5}{8}$      **29.** $\frac{9}{4}$

The graph on the right shows the cost per ounce for five brands of cereal. (Pages 348–350)

1. Suppose you compare the cost for Cereal A and Cereal B. How are the heights of the bars misleading?
2. Explain how you can change the graph so that it is not misleading.

The table on the right shows the monthly rainfall totals in inches for a southwest city for the first six months of 1987–1990. (Pages 351–354)

| Month | 1987 | 1988 | 1989 | 1990 |
|---|---|---|---|---|
| January | 0.92 | 0.27 | 3.79 | 1.28 |
| February | 2.87 | 0.32 | 0.85 | 3.55 |
| March | 1.36 | 2.66 | 2.12 | 2.08 |
| April | 0.45 | 2.02 | 2.43 | 3.12 |
| May | 6.75 | 3.33 | 6.90 | 3.65 |
| June | 10.85 | 2.60 | 3.10 | 1.55 |

3. During which month and year did the city receive the greatest amount of rain? the least amount of rain?
4. Mrs. Long always wants to have an umbrella with her when the chances of rain are the most likely. During which month will Mrs. Long always carry her umbrella?

The average class size in twenty schools is listed at the right. (Pages 355–362)

| 22 | 25 | 20 | 23 | 31 |
| 37 | 24 | 19 | 29 | 32 |
| 39 | 35 | 18 | 32 | 34 |
| 38 | 25 | 38 | 26 | 33 |

5. Make a frequency table that shows the class sizes. Use the intervals 16–20, 21–25, 26–30, 31–35, and 36–40.
6. Make a histogram from your frequency table.

# Mathematical Footnote

How much do you know about the mathematics of peanut butter? Do you know that 80% of all Americans have a jar of peanut butter in the home? Make an informal survey of your classmates to find how many have a jar of peanut butter at home. Compare the percent to the national average.

George Washington Carver, an African American scientist of international renown, made over 300 products from peanuts, including soap, printers ink, and face powder.

## LESSON 9-5 Problem Solving Strategies: Analyzing Sample Data

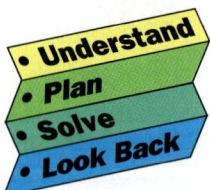

**M**ost salmon spend part of their life in the salt water of the ocean, thousands of miles from the freshwater stream where they were born. Yet scientists know that salmon return to the same freshwater stream to spawn.

How do scientists know about the habits of salmon? One way of knowing is a result of tagging some of the fish, a procedure used to study the habits and migrations of other fish, birds, and animals.

Scientists cannot tag every fish, so they capture, tag, and then release a **sample** of the total salmon population. By tracing the sample, they can discover patterns in the behavior of the whole population and make predictions about the size of the population.

### EXAMPLE

A fisheries ranger took a sample of 100 fish from Clearwater Lake, tagged the fish, and put them back. Sometime later, the ranger took another sample of 250 fish. Of these 250, 6 had tags. How could the ranger estimate the size of the population of fish in the lake?

**Solution**

Write a proportion. Use this formula.

$$\frac{\text{Fish tagged in Sample 1}}{\text{Total fish in lake}} = \frac{\text{Tagged fish found in Sample 2}}{\text{Total fish in Sample 2}}$$

Let $x$ = the number of fish in the lake.

Substitute the known values.   $\dfrac{100}{x} = \dfrac{6}{250}$

Cross-multiply.   $6 \cdot x = 100 \cdot 250$

Simplify.   $6x = 25{,}000$

$x = 4166.6$

A good estimate of the number of fish in the lake would be about 4,170.

# EXERCISES

**Objective:** To apply strategies in problem solving

1. A scientist marked out one square yard in a field, and counted 17 cicada larvae within that square. Use this sample to estimate the total number of larvae for a field with an area of 1000 square yards.

2. At Clearwater Lake the following year, the ranger again took a sample of 100 fish and marked them with a new tag. A second sample of 175 fish taken that same year showed that 4 had the new tags. Do you think the fish population in Clearwater Lake increased or decreased from the previous year? Explain.

3. Jules told Sara, "Three-fourths of the people I talked to said they liked bananas better than apples." Based on this information, should Sara bring more bananas than apples to their 9th grade class picnic?

4. Suppose that the sample Jules talked about in Exercise 3 had only 4 people. Should Sara act on the information?

5. Suppose that the sample Jules talked about in Exercise 3 was made up of 15 five-year olds. Should Sara act on the information?

6. Work with a team to make a list of some of the characteristics of a *reliable sample*, that is, a sample that will result in information you can trust.

Problem Solving Strategies   **365**

# LESSON 9-6

## Stem and Leaf Plots

**T**his table shows the lifetime home-run record of Henry (Hank) L. Aaron.

| Year | Home Runs | Year | Home Runs | Year | Home Runs | Year | Home Runs |
|------|-----------|------|-----------|------|-----------|------|-----------|
| 1954 | 13 | 1960 | 40 | 1966 | 44 | 1972 | 34 |
| 1955 | 27 | 1961 | 34 | 1967 | 39 | 1973 | 40 |
| 1956 | 26 | 1962 | 45 | 1968 | 29 | 1974 | 20 |
| 1957 | 44 | 1963 | 44 | 1969 | 44 | 1975 | 12 |
| 1958 | 30 | 1964 | 24 | 1970 | 38 | 1976 | 10 |
| 1959 | 39 | 1965 | 32 | 1971 | 47 | | |

Ms. Tallchief's class decided to use a **stem-leaf plot** to display the information on Hank Aaron's home run record.

To draw a stem-leaf plot, first find the highest and lowest scores. Use the tens digits as the **stems.** List the stems in a vertical line as shown at the right.

Stems $\begin{cases} 1 \\ 2 \\ 3 \\ 4 \end{cases}$

366   CHAPTER 9

Now read and list the data on the number of home runs. The first number, 13, has a stem of 1. The units digit, 3, is called a **leaf**.

```
1 | 3, 2, 0
2 | 7, 6, 4, 9, 0
3 | 0, 9, 4, 2, 9, 8, 4
4 | 4, 0, 5, 4, 4, 4, 7, 0
```

Write the leaf at the right of the vertical line and next to the stem.

The second number, 27, has a stem of 2 and a leaf of 7. Write the 7 on the right side of the vertical line and next to the 2.

If you continue with the rest of the data, you will have the plot shown.

Now rearrange the leaf for each stem in order from smallest to largest.

```
1 | 0, 2, 3
2 | 0, 4, 6, 7, 9
3 | 0, 2, 4, 4, 8, 9, 9
4 | 0, 0, 4, 4, 4, 4, 5, 7
```

Notice that leaves that appear more than once in the data are listed the same number of times in the stem-leaf plot.

Notice that the highest score is 47 and the lowest is 10. The interval from 10 to 47 is called the **range** of the scores. The range can also refer to the difference of the highest and lowest scores: $47 - 10 = 37$.

**EXAMPLE** Refer to the stem-leaf plot above to answer these questions.
a. What is the greatest number of home runs Hank Aaron scored in a season?
b. How many times did Hank Aaron score 39 home runs in a season?
c. What is the number of home runs Hank Aaron scored most often in a season? What measure of central tendency is this?
d. What is the median of the data?
e. In what interval would you estimate that the mean number of home runs will fall?

**Solutions** Read the information from the stem-leaf plot.
a. **47**
b. **Twice** (There are two 39s.)
c. **44** (There are four 44s.) This is the mode of the data.
d. There are 23 leaves. The median is the twelfth leaf, **34**.
e. **30–39**

**Try This**

1. For how many seasons did Hank Aaron score 44 home runs?
2. For how many seasons did Hank Aaron score more than 35 home runs?

# EXERCISES

**Objective:** To construct a stem and leaf plot

## Check Your Understanding

**Skill Check** Write the letter of the correct answer. Refer to the stem and leaf plot at the right. The plot shows students' scores on a math test.

```
2 | 8
3 | 8
4 | 3, 4, 8, 9
5 | 1, 3, 4, 4, 7, 9
6 | 8, 8
7 | 3, 6, 7, 7, 9
8 | 1, 2, 2, 2, 2, 5, 7
9 | 0, 1, 6
```

1. What is the range of the student's math scores?
   **a.** 0–9   **b.** 28–96   **c.** 28–90   **d.** 29

2. How many student's scores are reported in the stem and leaf plot?
   **a.** 37   **b.** 62   **c.** 22   **d.** 29

3. What was the median score on the test?
   **a.** 68   **b.** 76
   **c.** 73   **d.** 71

4. How many students had scores higher than 75 on the test?
   **a.** 13   **b.** 15   **c.** 11   **d.** 14

5. Estimate the percent of students who scored higher than 75 on the math test.
6. About what percent of the students had a score lower than 50 on the math test?

## Practice and Apply

The data in this chart represents the number of movies rented from Video Special over a two-week period.

| 2 | 85 | 32 | 98 | . . | |

7. Construct a stem and leaf plot for the data.
8. Find the mode and median for the data.
9. On what percent of the days were there more than 90 rentals? Round the percent to the nearest tenth.
10. Why do certain days have a greater number of rentals than others?

This table shows U.S. presidents' ages at their inauguration.

| Age at Inauguration | | | | | |
|---|---|---|---|---|---|
| 57 | 61 | 57 | 57 | 58 | 57 | 61 |
| 54 | 68 | 51 | 49 | 64 | 50 | 48 |
| 65 | 52 | 66 | 46 | 54 | 49 | 50 |
| 47 | 55 | 55 | 54 | 42 | 51 | 56 |
| 55 | 51 | 54 | 51 | 60 | 62 | 43 |
| 55 | 56 | 61 | 52 | 69 | 64 | |

11. Make a stem and leaf plot for the data.
12. What was the youngest inaugural age? What was the oldest?
13. How many presidents were in their fifties at inauguration?
14. What is the most common age at inauguration?
15. What is the median age at inauguration?
16. Estimate the interval for the average inaugural age.

## Connect and Extend

This stem-leaf plot compares the writing and math scores of 20 students. The stems, 2, 3, 4, 5, 6, and 7 are in the center. The leaves for the writing scores are on the left and the leaves for the math scores are on the right.

```
       Writing |   | Math
             9 | 2 |
       8, 7, 4, 2 | 3 | 3, 4, 7, 8, 9
    8, 7, 6, 5, 4, 2, 0 | 4 | 0, 0, 0, 0, 3, 4, 4, 8, 9
          7, 7, 1, 0 | 5 | 3, 8
             5, 0, 0 | 6 | 4, 5, 8, 9
                   3 | 7 |
```

17. What is the mode of the math scores?
18. What is the mode of the writing scores?
19. What is the median writing score?
20. What is the median math score?
21. What percent of the writing scores were higher than 59?
22. What percent of the math scores were higher than 59?
23. How could you use the stem and leaf plot for the writing scores to make a histogram?

## Maintain Your Skills

**Write two inequalities for each pair of numbers. (Pages 7–10)**

24. 0 and 6
25. 2 and $-12$
26. $-4$ and 0
27. $-5$ and $-7$

**Use the order of operations to simplify each expression. (Pages 46–50)**

28. $16 \div 4 \cdot 3 - 6$
29. $9 + [(3 + 7) \div 2]$
30. $36 \div 3 \div 6 \div 2$

**Solve and check. (Pages 105–109, 183–193)**

31. $13 = 10 - 3x$
32. $6y - 4 = -28$
33. $11 + 4x = -11$
34. $0.3n - 4.2 = 2.7$
35. $3.2x + 7.1 = 0.7$
36. $8.6 = 2.1 - 1.3y$
37. $\frac{4}{7}t + 3 = 15$
38. $-\frac{2}{3}a + 3 = 11$
39. $-\frac{4}{5}r - 6 = -30$

# LESSON 9-7

# Box and Whisker Plots

After data is collected and organized, it is ready to be analyzed. One way of doing this is to draw a box and whisker plot. A **box and whisker plot** provides a picture of the central tendency of the data and is useful in comparisons.

**EXAMPLE** The table at the right shows a 19-year record of the number of hours of sunshine in a northern city. Make a box plot for the data.

| Hours of Sunshine | | | |
|---|---|---|---|
| 272 | 260 | 248 | 245 |
| 240 | 239 | 238 | 238 |
| 238 | 235 | 234 | 230 |
| 230 | 228 | 226 | 225 |
| 212 | 190 | 185 | |

**Solution**

1. Find the greatest and least values (extremes).
   Greatest value: **272**   Least value: **185**

2. Find the median of the data.   Median: **235**

3. Find the median of the 9 scores above 235. This is called the **upper quartile.**
   Upper quartile: **240**
   Find the median of the 9 scores below 235. This is called the **lower quartile.**
   Lower quartile: **226**

4. Draw a number line that includes the extreme points of the data. Choose and mark an appropriate scale. On a straight line above the line, mark the extreme values, the median, the lower quartile, and the upper quartile as shown below.

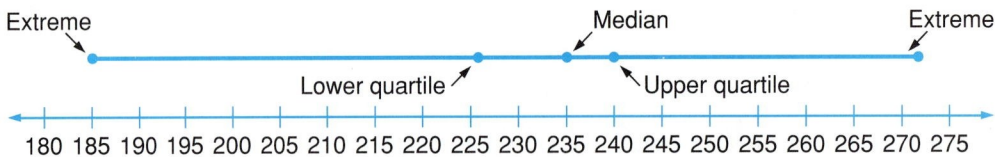

5. Draw a box around the median values.

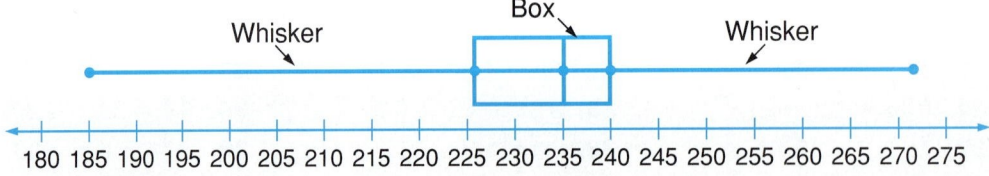

**Try This**  Refer to the Example on page 369 to answer these questions.
1. Find this difference: Lower quartile − Least value
2. Find this difference: Greatest value − Upper quartile
3. How do the answers to the previous two questions help to explain why the whiskers have different lengths?

Notice that the five numbers, 185, 226, 235, 240, and 272 separate the box and whisker plot into 4 sections. Each section represents about 25% of the data.

• What percent of the data is in the box?

# EXERCISES

**Objective:** To analyze data using a box and whisker plot

## Check Your Understanding

**Skill Check**  Write the letter of the correct answer. Refer to the box and whisker plot at the right.

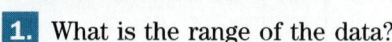
17   42   51   68   88

1. What is the range of the data?
   a. 42 to 68     b. 51 to 68
   c. 17 to 68     d. 17 to 88
2. What is the median of the data?
   a. 42     b. 51
   c. 88     d. 55
3. What is the median of the lower half of the data?
   a. 17 to 42     b. 42
   c. 51           d. 46.5
4. What fraction of the data is in the whiskers?
   a. $\frac{1}{4}$    b. $\frac{1}{10}$    c. $\frac{3}{4}$    d. $\frac{1}{2}$

**Find the extremes for each set of data.**
5. 10, 27, 12, 19, 13, 7, 5, 21
6. 24, 31, 3, 10, 17

## Practice and Apply

**Find the median, the upper quartile, and the lower quartile of each set of numbers.**
7. 21, 26, 23, 27, 23, 22, 32, 24, 25, 20, 23, 20, 21
8. 5, 13, 62, 17, 49, 20, 40, 62, 47, 60, 75

**Construct a box and whisker plot for the data in the stem and leaf plot.**

9. 1 | 2, 8
   2 | 3, 5, 6
   3 | 0, 3, 3, 4, 7, 9
   4 | 2, 3
   5 | 1, 4

10. 1 | 8
    2 | 3, 6
    3 | 3, 4, 5
    4 | 0, 2, 5, 7, 8
    5 | 2
    6 | 5

9-7   Box and Whisker Plots   **371**

The box plot below shows the cost in dollars of twenty different hair dryers.

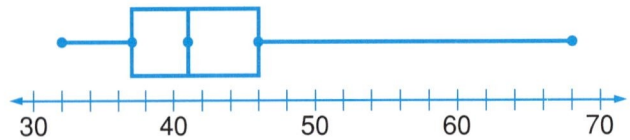

11. About how much does the most expensive hair dryer cost?
12. About how much does the least expensive hair dryer cost?
13. What is the median price of a hair dryer?
14. What percent of the hair dryers cost more than $46.50?
15. How many of the hair dryers cost less than $41?
16. What percent of the hair dryers cost less than $41?
17. How many hair dryers could you choose from if you had $41 to spend?

## Connect and Extend

The data at the right shows test scores for two Pre-Algebra classes.

| First Pre-Algebra Class | Stems | Second Pre-Algebra Class |
|---|---|---|
| 8 7 0 | 9 | 3 4 6 |
| 9 7 7 6 | 8 | 0 1 1 1 2 7 |
| 9 8 8 7 5 5 5 3 2 | 7 | 0 2 3 3 4 6 9 |
| 7 4 2 2 2 0 | 6 | 4 7 8 |
|  | 5 |  |

18. Find the mean, median, and mode for each class.
19. Draw a box and whisker plot for each class.
20. Combine both sets of data in a single stem and leaf plot.
21. Find the mean, median, and mode for the stem and leaf plot in Exercise 20.
22. Draw a box and whisker plot for the stem and leaf plot in Exercise 20.

## Maintain Your Skills

**Find the solution set if the replacement set R = {−2, −1, 0, 1, 2, 3}. (Pages 80–84)**

23. $3f + 5 = 2$
24. $4c - 7 > 1$
25. $9x - 41 = -41$
26. $6r - 19 = -7$

**Estimate each product or quotient. Then find the exact answer. How close are the two answers? (Pages 297–300)**

27. $4.3 \times 50.1$
28. $55.8 \div 7.2$
29. $8.8 \times 45.1$

**Solve. (Pages 324–328, 330–333)**

30. Find the simple interest on a loan of $3200 that is borrowed for $1\frac{1}{2}$ years at a rate of $9\frac{1}{2}\%$.
31. A VCR with a list price of $279 is on sale at a 15% discount. Find the sale price.

# SUMMARY

## KEY TERMS

Box and whisker plot (p. 370)
Frequency table (p. 355)
Histogram (p. 355)
Leaf (p. 367)
Lower quartile (p. 370)
Mean (p. 359)

Median (p. 360)
Mode (p. 359)
Range (p. 367)
Stem (p. 366)
Stem-leaf plot (p. 366)
Upper quartile (p. 370)

## KEY IDEAS

**A.** The mode of a set of data is the number that occurs most often. A set of data may have more than one mode, or no mode.

**B.** The mean of a set of data having $n$ numbers is the sum of the numbers divided by $n$.

**C.** The median of a set of data is the middle number in the set when the numbers are arranged in descending order or in ascending order. When there is an even number of numbers in a set, the median is the mean of the two middle numbers.

**D.** A stem-leaf plot is useful because it shows the median, the mode, and the range for a set of data.

**E.** A box and whisker plot is useful in analyzing data because it illustrates the median values and the range of a set of data.

# REVIEW

**Michael walked at a rate of 3 kilometers per hour. The line graph at the right shows the distance in kilometers Michael walked in 5 hours. (Pages 348–350)**

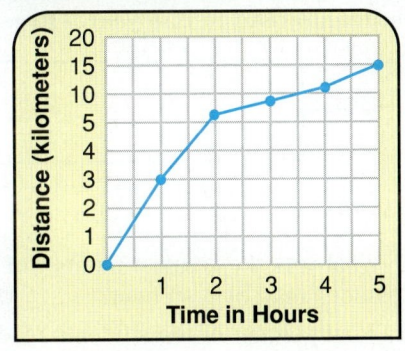

1. Is the distance Michael walked the first hour farther than the distance he walked the fifth hour? Explain your answer.
2. How is the steepness of the graph misleading?
3. Explain how you can change the graph so that it is not misleading.

The number of records sold at Music Warehouse each day for one month is listed at the right. (Pages 351–357)

**Number of Records Sold at Music Warehouse Each Day**

| 48 | 53 | 41 | 69 | 69 | 51 |
| 59 | 57 | 64 | 63 | 59 | 42 |
| 54 | 50 | 62 | 68 | 44 | 46 |
| 63 | 57 | 59 | 54 | 49 | 47 |
| 72 | 60 | 57 | 58 | 64 | 48 |

4. Make a frequency table that shows the number of records sold. Use the intervals 40–44, 45–49, 50–54, and so on.
5. Make a histogram from your frequency table.
6. About what percent of the days did the store sell 49 records or fewer?

The table at the right shows the yearly salaries for 7 employees at a company. (Pages 359–362)

**Yearly Salaries**

| Manager | $18,000 | Clerk | $12,500 |
| Assistant | $15,500 | Clerk | $10,000 |
| Clerk | $12,000 | Clerk | $11,500 |
| Clerk | $11,500 | | |

7. Find the mean of the seven salaries.
8. Find the median and the mode of the seven salaries.
9. Which of these measures, the median or mode, would you use to argue for an increase in salary? Explain.
10. A ranger took a sample of 85 trout from a lake, tagged the fish, and put them back. Later, the ranger took another sample of 108 trout. Of these, 3 had tags. Use this sample to estimate the size of the population of trout in the lake. (Pages 364–365)

This table shows the number of games won during a season by different major league baseball teams. (Pages 366–369)

**Number of Wins**

| 90 | 73 | 84 | 63 | 77 | 84 |
| 87 | 70 | 71 | 71 | 90 | 99 |
| 81 | 81 | 73 | 74 | 72 | 80 |
| 70 | 88 | 75 | 83 | 71 | 64 |

11. Make a stem and leaf plot for the data.
12. What is the median number of wins?
13. Find the mode for the set of data.

The box and whisker plot below shows the number of books a class of students read in one year. (Pages 370–372)

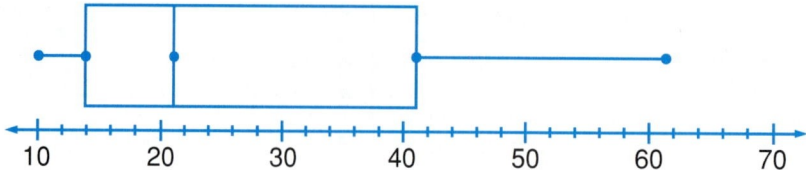

14. What was the greatest number of books read by a student?
15. What was the median number of books read by the students?
16. What percent of the students read more than 41 books?

# TEST

The bar graph at the right shows the number of people who attended an airshow on Saturday and on Sunday.

1. How many people attended the airshow on Saturday? on Sunday?
2. Measure the length of each bar on the graph. Why does the length of each bar make the graph misleading?
3. Explain how you would redraw the graph so it is not misleading.

**Airshow Attendance**

| | |
|---|---|
| Saturday | |
| Sunday | |

0  30,000  50,000  70,000
**Number of People**

Carlos Arroyo is a sales clerk at United Shoe Store. He uses a table similar to the one at the right to determine the amount of sales tax.

4. How much is the sales tax on a purchase of $9.32?
5. What is the total cost, including tax, for a purchase of $9.99?

| Amount of Sale | Tax |
|---|---|
| 8.92–9.08 | 0.54 |
| 9.09–9.24 | 0.55 |
| 9.25–9.41 | 0.56 |
| 9.42–9.58 | 0.57 |
| 9.59–9.74 | 0.58 |
| 9.75–9.91 | 0.59 |
| 9.92–10.08 | 0.60 |

The scores on a Pre-Algebra test for a class of 25 students are listed at the right.

6. Make a frequency table that shows the scores. Use the intervals 65–69, 70–74, and so on.
7. Determine the median and mode of the data.

| 73 | 88 | 84 | 79 | 68 |
|---|---|---|---|---|
| 76 | 66 | 73 | 66 | 80 |
| 73 | 88 | 90 | 84 | 73 |
| 90 | 84 | 72 | 90 | 84 |
| 96 | 73 | 70 | 84 | 96 |

The table at the right shows the number of students on 14 school buses.

8. Make a stem and leaf plot to show the data.
9. What percent of the buses had more than 30 but fewer than 40 students?

| The Number of Students on Each School Bus |||||||
|---|---|---|---|---|---|---|
| 27 | 25 | 36 | 31 | 40 | 32 | 31 |
| 40 | 33 | 35 | 28 | 32 | 41 | 27 |

The box and whisker plot below shows the cost in dollars of 22 different clock radios.

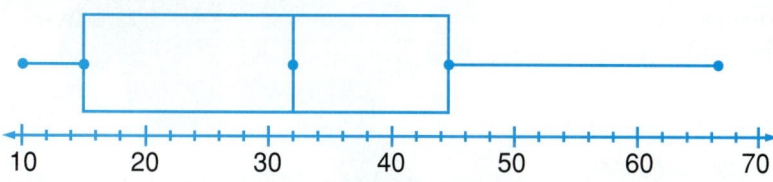

10. What is the median price of a clock radio?
11. About how much does the most expensive clock radio cost?
12. How many of the clock radios cost less than $32?

Chapter 9 Test  **375**

**Write two inequalities for each pair of numbers. (Pages 7–10)**

**1.** $-6$ and $-1$   **2.** $10$ and $-10$   **3.** $4$ and $0$   **4.** $-21$ and $-12$

**Find the area of each figure. (Pages 69–74)**

**5.** 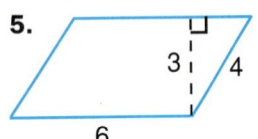   **6.** 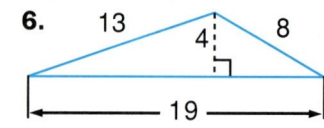   **7.**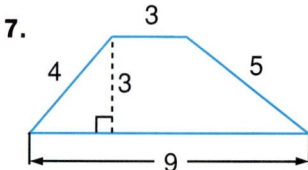

**Solve. (Pages 114–118)**

**8.** Angelo's present salary is $22,000. This is $360 more than twice what he was making when he began the job. What was his starting salary?

**9.** Cynthia earns $6.00 an hour at her job. Last week, she got a bonus of $55. If her total earnings for the week were $229, how many hours did she work?

**Use the following list of numbers for Exercises 10–11. (Pages 131–136)**

124   218   263   816   342   7254   43,248

**10.** Which numbers are divisible by 4?

**11.** Which numbers are divisible by 6?

**Solve and check. (Pages 183–188)**

**12.** $\frac{3}{4}z = 1\frac{1}{2}$   **13.** $\frac{5}{6}c = \frac{7}{12}$   **14.** $4\frac{3}{10} = 2\frac{3}{5}f$   **15.** $12 = 3\frac{1}{3}g$

**A bag contains 5 blue, 2 red, and 3 green marbles. Two marbles are drawn, one after the other. (Pages 241–246)**

**16.** If the first marble is replaced before the second one is drawn, what is the probability that the first one is blue and the second one is green?

**17.** If the first marble is not replaced before the second one is drawn, what is the probability that both are red?

**Solve. (Pages 282–286, 334–337)**

**18.** At Debbie's Variety Store, Chris paid $18.73 for a watch, $2.69 for 35 mm film, and $3.49 for a package of batteries. Estimate the total cost to the nearest 10¢.

**19.** Mrs. Jones budgets $7\frac{1}{2}$% of her monthly income for savings. If she saves $90 monthly, what is her monthly income?

# CHAPTER 10 THE NUMBER LINE

| Toss | 1 | 2 | 3 | 4 | 5 |
|---|---|---|---|---|---|
| Record | 0.1 | 0.10 | 0.100 | 0.1001 | 0.10010 |

## LESSON 10-1

# The Set of Real Numbers

**N**ancy tosses a coin and then writes a decimal to record the result. When the coin shows a head, she writes a 1. When the coin shows a tail, she writes a 0. After each toss, she writes a 0 or 1 to the right of the digit that shows the result of the previous toss. So, after the fifth toss, Nancy's record, written as a decimal, looks like this.

$$0.10010$$

So far this year, you have worked with decimals that either terminate or repeat a pattern forever. In Chapter 7, you learned that any terminating or repeating decimal can be written in the form of a fraction. The decimal that Nancy is writing will be a *nonrepeating, infinite decimal* because the ones and zeros will not terminate, nor will they form a repeating pattern. That is, the ones and zeros will occur randomly.

Nancy's number is called an **irrational number.** Taken together, the set of all rational and irrational numbers is called the set of **real numbers.**

10-1  The Set of Real Numbers  **377**

For every real number, there is exactly one point on the number line that corresponds to it. For every point on the number line, there is exactly one real number that corresponds to it. This means that there is a **one-to-one correspondence** between the set of real numbers and the points on the number line.

Notice that the graph of $y = 4$ separates the number line into three different sets of points.

1. The point representing 4.
2. The set of points to the right of 4, that is, the set of points greater than 4.
3. The set of points to the left of 4, that is, the set of points less than 4.

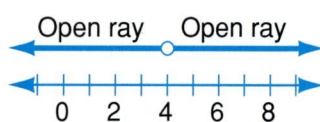

So each point on the number line makes one of these sentences true.

$$y = 4 \qquad y > 4 \qquad y < 4$$

The graph of $y > 4$ is an **open ray** because the number 4 is not included in the solution set. To indicate this, the end point of the ray is shown as an open circle. This tells you that 4 is not included in the solution set.

When a graph also includes the endpoint, the endpoint is shown as a closed circle. In this case, the graph is a **closed ray**. The inequality symbol, ≤ or ≥, is used to indicate this.

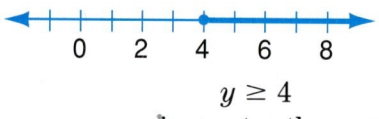 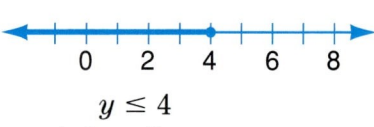

$y \geq 4$ 
$y$ is greater than or equal to 4.

$y \leq 4$ 
$y$ is less than or equal to 4.

**EXAMPLE 1** Write a description of the graph.

a.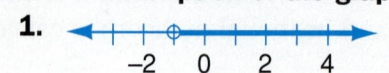

b.

**Solutions**  a. {All real numbers greater than or equal to $-5$}
b. {All real numbers less than $\frac{1}{2}$}

**Try This** Write a description of the graph.

1. 

2. 

**378**  CHAPTER 10

- A number, $x$, does not equal 5 and it is not greater than 5. What conclusion can you draw?

When the replacement set of an inequality is finite, such as $\{-1, 0, 1, 2, 3\}$, the solution set can be found by testing each replacement. The solution set can be represented as a graph or by listing the solutions. For example, the solution set of $x \leq 1$ with replacement set $\{-1, 0, 1, 2, 3\}$ can be shown in these ways.

$\{-1, 0, 1\}$

However, when the replacement set is the set of all real numbers, the solution set is usually infinite and must be graphed or described.

| Inequality | Graph | Description |
|---|---|---|
| $y < 5$ | | {All real numbers less than 5} |
| $c > 1$ | | {All real numbers greater than 1} |
| $r \geq 3$ | | {All real numbers greater than or equal to 3} |

**EXAMPLE 2** Graph each inequality.

**a.** $m > -5$   **b.** $c \leq -3$

**Solutions**

**a.** The graph of $m > -5$ includes all points on the number line to the right of $-5$.

**b.** The graph of $c \leq -3$ includes all points on the number line to the left of, and including, $-3$.

**Try This** Graph each inequality.

**3.** $t < -4$   **4.** $c \geq \frac{1}{2}$

- What point on the graph of $x \geq 1$ is *not* on the graph of $x > 1$?

10-1 The Set of Real Numbers **379**

# EXERCISES

**Objectives:** To describe the solution set of an inequality
To graph an inequality

## Check Your Understanding

**Skill Check**  Write the letter of the correct answer.

1. Which inequality has an open ray for its graph?
   **a.** $t = 9$   **b.** $c < -3$   **c.** $m \geq 0$   **d.** $q \leq 1$

2. Which describes the solution set for $r \geq -1$?
   **a.** {All real numbers greater than $-1$}
   **b.** {All real numbers less than $-1$}
   **c.** {All real numbers greater than or equal to $-1$}
   **d.** {All real numbers less than or equal to $-1$}

3. Which is the graph of $d \leq 4$?

4. Which graph is described by {all real numbers greater than 10}?

## Practice and Apply

**Write a description of the graph.**

5.

6.

7.

8.

9.

10.

11.

12.

**380**   CHAPTER 10

Graph the equation or inequality.

13. $a < 2$
14. $b \leq -3$
15. $c < 7$
16. $d \geq 4$
17. $e \leq 0$
18. $f \geq -1$
19. $g > -2$
20. $f < -5$
21. $s = 4$
22. $j \leq -5$
23. $k < 7$
24. $r < 9$
25. $m \geq 8$
26. $f < 6$
27. $t = -7$
28. $p < 10$

Write a description of the solution set.

29. $t \geq 3$
30. $r < -6$
31. $p \geq 9$
32. $g \leq 0$

33. When you compare two numbers, there are three possible comparison statements you can write. What are they?

## Connect and Extend

For Exercises 34–35, the replacement set is $\{-1, 0, 1, 2, 3\}$.

34. If one number is selected at random, what is the probability it will be a solution of $x < 2$?

35. If two numbers are selected at random, what is the probability that each will be a solution of $y \geq 1$?

For Exercises 36–39, the replacement set is {all real numbers}. Write True or False for each statement. When a statement is false, explain why.

36. The graph of $x \leq a$ is an open ray.
37. The graph of $x \leq a$ is always pointed to the left.
38. As $a$ increases, the endpoint of $x \leq a$ moves to the right.
39. Suppose $x$ is decreased by 1 and $a$ is also decreased by 1. The endpoint of $x \leq a$ will then move 1 unit to the left.

## Maintain Your Skills

Find the solution set of each inequality if the replacement set is $\{-7, -5, 0, 5, 7\}$. (Pages 80–84)

40. $-12 \geq 12s$
41. $5t + 2 \geq 12$
42. $10 - 4f > 30$
43. $-3t + 1 > -20$

Write an equation for each problem. Then solve the problem. (Pages 114–118)

44. A coat costs $45.50 more than a pair of pants. The total cost of both items is $108.48. How much does each item cost?

45. A kitten is $\frac{2}{3}$ as tall as a puppy. The difference between their heights is 2 inches. How tall is the kitten?

Estimate the sum or difference. (Pages 282–286)

46. $37.012 + 9.96$
47. $19.81 - 4.009$

# LESSON 10-2
# The Addition Property of Inequality

After Tad paid his lunch bill of $9, he had less than $2.25 left. How much money could he have had to start with?

**UNDERSTAND THE PROBLEM**
What is the question?  How much money did Tad have before he paid the bill?
What is given?  Tad spent $9. He has less than $2.25 left.

**DEVELOP A PLAN**
**Strategy:** Write an inequality. Use guess and check or logical reasoning to solve the problem.

**CARRY OUT THE PLAN**
Represent the unknown. Let $x$ represent the original amount.
Write an inequality.

**Think:** Original amount − Amount spent < $2.25
**Translate:**     $x$     −     9     < $2.25
**Solve:** You know that Tad had at least $9.00 because he paid the bill. If he has less than $2.25 left, he must have had less than $11.25 to start with.

**LOOK BACK**
**Check:** Is $11.24 a solution of $x - 9 < 2.25$?   Yes ✓

Therefore, Tad had less than $11.25 and more than $9.00 to start with.

Another way to solve the inequality $x - 9 < 2.25$ is to use the *Addition Property of Inequality*. This property is similar to the Addition Property of Equality which you have already studied. That is, you can add the same number to each side of an inequality. The result is an equivalent inequality.

> **Addition Property of Inequality**
> If $a \leq b$, then $a + c \leq b + c$.
> If $a \geq b$, then $a + c \geq b + c$.

To solve the inequality, $x - 9 < 2.25$, think about how to solve the equation, $x - 9 = 2.25$.

| | |
|---|---|
| Write the inequality. | $x - 9 < 2.25$ |
| Add 9 to each side. | $x - 9 + 9 < 2.25 + 9$ |
| Simplify. | $x < 11.25$ |

**EXAMPLE 1** Solve and graph: $x - 3 \geq 2$

**Solution**

| | |
|---|---|
| Write the inequality. | $x - 3 \geq 2$ |
| Add 3 to each side. | $x - 3 + 3 \geq 2 + 3$ |
| Simplify. | $x \geq 5$ |
| Graph the solution set. | |

The solution set is {all real numbers greater than or equal to 5}.

**Try This** Solve and graph.

1. $x + 1 < -5$
2. $s - 4 \geq 1\frac{1}{2}$

**EXAMPLE 2** Solve and graph: $2\frac{1}{2} + p \geq -1\frac{1}{4}$

**Solution**

| | |
|---|---|
| Rewrite $2\frac{1}{2} + p$ as $p + 2\frac{1}{2}$. | $p + 2\frac{1}{2} \geq -1\frac{1}{4}$ |
| Add $\left(-2\frac{1}{2}\right)$ to each side. | $p + 2\frac{1}{2} + \left(-2\frac{1}{2}\right) \geq -1\frac{1}{4} + \left(-2\frac{1}{2}\right)$ |
| Simplify. | $p \geq -3\frac{3}{4}$ |
| Graph the solution set. | |

The solution set is {all real numbers greater than or equal to $-3\frac{3}{4}$}.

**Try This** Solve and graph.

3. $b - \frac{1}{2} > 3$
4. $m + 2\frac{1}{3} \leq 1\frac{3}{4}$

- Can you name the greatest number in the solution set for $x > 0$ if the replacement set is {all real numbers}?

- What is the smallest number in the solution set for $x > 0$ if the replacement set is {all real numbers}?

# EXERCISES

**Objective:** To solve and graph an inequality that involves addition or subtraction

## Check Your Understanding

**Skill Check** Write the letter of the correct answer.

**1.** This graph shows the solution set of which inequality?

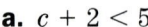

**a.** $c + 2 < 5$    **b.** $c - 2 < 5$
**c.** $c + 5 < 2$    **d.** $c - 5 < 2$

**2.** This graph shows the solution set of which inequality?

**a.** $m - 1 \geq 2$    **b.** $m + 1 \geq 2$
**c.** $m + 1 \geq 1$    **d.** $m - 2 \geq 1$

**3.** Which inequality is equivalent to $d + \frac{1}{2} \leq 1$?

**a.** $d > 1\frac{1}{2}$    **b.** $d \geq \frac{1}{2}$    **c.** $d \leq 1\frac{1}{2}$    **d.** $d \leq \frac{1}{2}$

**4.** Which is the first step in solving $r - 2.5 > -1$?
   **a.** Add 1 to each side.    **b.** Subtract 1 from each side.
   **c.** Add 2.5 to each side.    **d.** Subtract 2.5 from each side.

## Practice and Apply

**Solve and graph the inequality.**

**5.** $x + 5 \leq 9$     **6.** $y - 2 \leq 3$     **7.** $s - 4 \geq -1$
**8.** $6 + m < 4$     **9.** $5 + r > -1$     **10.** $y - 1 \geq 2$
**11.** $p - 4 < -4$     **12.** $d - 1 \leq 5$     **13.** $k + 2 < -2$
**14.** $s - \frac{1}{2} \geq 5\frac{1}{2}$     **15.** $\frac{1}{4} + t \geq 1$     **16.** $b + \frac{3}{4} \leq -\frac{1}{4}$

**Write the letter of the inequality that represents each statement.**

**17.** The cost, $c$, is *no more than* $50.
   **a.** $c < 50$     **b.** $c \leq 50$     **c.** $c > 50$     **d.** $c \geq 50$

**18.** The amount, $a$, is *at least* $75.
   **a.** $a = 75$     **b.** $a \leq 75$     **c.** $a \geq 75$     **d.** $a > 75$

**19.** The savings, $s$, is *at most* $21.
   **a.** $s > 21$     **b.** $s \geq 21$     **c.** $s < 21$     **d.** $s \leq 21$

**20.** The distance, $d$, is *more than* 20 feet.
   **a.** $d = 20$     **b.** $d < 20$     **c.** $d > 20$     **d.** $d \geq 20$

Write an inequality to represent the situation.
Then solve the inequality.

21. Tandi had some money in her savings account. After she deposited $50, there was more than $180 in her account. What is the least amount that could have been in the account originally?

22. The cost of a VCR is no more than $280. Rod has saved $90. What is the greatest amount Rod has to save in order to buy a VCR?

## Connect and Extend

Evaluate the expression, $b^2 - 4ac$, for the given values of $a$, $b$, and $c$. Then determine whether $b^2 - 4ac \leq 0$.

23. $a = 2, b = 5, c = 3$
24. $a = 1, b = 2, c = -1$
25. $a = 3, b = 7, c = 4$

26. If $a < 0$, which is greater, $a$ or $a + 1$?
27. If $a > 0$, which is greater, $a$ or $a + 1$?
28. Find a counterexample for the following: If $a > b$, and $c > d$, then $a - c > b - d$.

## Maintain Your Skills

Determine whether each group of numbers is written in ascending order, descending order, or neither. Explain your answer. (Pages 7–10)

29. $5, 1, 0, -1, -2$
30. $-4, -6, 0, 4, 6$
31. $-7, -3, -2\frac{1}{2}, -1\frac{1}{2}, -\frac{3}{4}$

Solve. (Pages 64–68, 95–99)

32. A rectangular drawing board is 36 inches long and 29 inches wide. What is the area of the drawing board?

33. When Tom solved the equation, $3x = 6$, he divided both sides by 6. Did he make an error? Explain your answer.

Replace each ? with the correct number. (Pages 175–179)

34. $4\frac{3}{4} = \frac{?}{4}$
35. $5\frac{5}{8} = \frac{?}{8}$
36. $7\frac{5}{6} = \frac{?}{6}$
37. $8\frac{4}{9} = \frac{?}{9}$

##  Math Team Problems

38. The figure at the right shows how two squares can be drawn by forming just 7 lines. Show how two squares can be formed by drawing 6 lines.

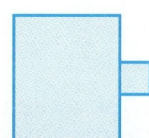

39. A football field is about 53 yards wide and 100 yards long. If a balloon floats down and lands on the field, what is the probability that it will land between the two 40-yard lines? Write the answer as a fraction, a decimal, and a percent.

# LESSON 10-3
# The Multiplication Property of Inequality

The numbers in Row A are in ascending order.

**Row A:**   $-3$   $-2$   $-1$   $0$   $1$   $2$   $3$

If each number in Row A is multiplied by a positive number such as 5, the resulting products in Row B are also in ascending order.

$\phantom{\text{Row B: }}$ $-3 \cdot 5$   $-2 \cdot 5$   $-1 \cdot 5$   $0 \cdot 5$   $1 \cdot 5$   $2 \cdot 5$   $3 \cdot 5$
**Row B:**   $-15$   $-10$   $-5$   $0$   $5$   $10$   $15$

What happens when the numbers in Row A are multiplied by a negative number, such as $-5$? The resulting products in Row C are now in descending order.

$\phantom{\text{Row C: }}$ $-3(-5)$   $-2(-5)$   $-1(-5)$   $0(-5)$   $1(-5)$   $2(-5)$   $3(-5)$
**Row C:**   $15$   $10$   $5$   $0$   $-5$   $-10$   $-15$

This suggests that the Multiplication Property of Equality has two parts. Part 1 refers to multiplying each side of an inequality by a positive number.

> **Multiplication Property of Inequality I**
> If $a \leq b$, and $c \geq 0$, then $ca \leq cb$.
> If $a \geq b$, and $c \geq 0$, then $ca \geq cb$.

**EXAMPLE 1** Solve and graph: $3x > 15$.

**Solution**

Write the inequality.  $\phantom{xxxx}$ $3x > 15$

Multiply each side by $\frac{1}{3}$.  $\phantom{xx}$ $\frac{1}{3}(3x) > \frac{1}{3}(15)$

Simplify.  $\phantom{xxxxxxxxxxxx}$ $x > 5$

Graph the solution set.

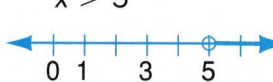

The solution set is {all real numbers greater than 5}.

**Try This**   Solve and graph:   **1.** $4x > 16$   **2.** $5b < -1\frac{1}{4}$

Part 2 of the Multiplication Property of Inequality refers to multiplying each side of an inequality by a negative number.

> **Multiplication Property of Inequality II**
> If $a \leq b$, and $c \leq 0$, then $ca \geq cb$.
> If $a \geq b$, and $c \leq 0$, then $ca \leq cb$.

Notice that when you multiply each side of an inequality by a negative number, the order of the inequality is reversed.

**EXAMPLE 2** Solve and graph: $-2\frac{1}{2}c > 15$

**Solution**

Write $-2\frac{1}{2}$ as $-\frac{5}{2}$. $\qquad -\frac{5}{2}c > 15$

Multiply each side by $-\frac{2}{5}$. $\qquad \left(-\frac{2}{5}\right)\left(-\frac{5}{2}c\right) < \left(-\frac{2}{5}\right)15$
Reverse the inequality symbol.

Simplify. $\qquad c < -6$

Graph the solution set.

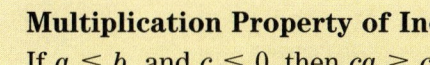

The solution set is {**all real numbers less than $-6$**}.

**Try This** Solve and graph: **3.** $-4t < 2\frac{1}{4}$     **4.** $-1\frac{1}{3}n > -2$

To check whether you have solved an inequality correctly, choose a number from the solution set and substitute the value in the inequality. To check Example 2, chose $-10$ from the solution set.

$$-2\frac{1}{2}c > 15$$

Replace $c$ with $-10$. $\qquad -\frac{5}{2}(-10) \stackrel{?}{>} 15$

$$25 > 15 \checkmark$$

**EXAMPLE 3** The area of a rectangle is less than or equal to 200 square centimeters. If the length is 20 centimeters, what can the width be?

**Solution** Let $w$ represent the width.

$20w \leq 200$

$\frac{1}{2}(20w) \leq \frac{1}{20}(200)$

$w \leq 10$

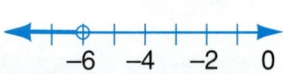

$A \leq 200$    $w$

$l = 20$

The width can be **10 centimeters or less.**

**Try This**

5. Cheri can spend $250 for a party. She estimates that the cost per person will be more than $15. What is the greatest number of people Cheri can invite to the party?

# EXERCISES

**Objective:** To solve and graph an inequality that involves multiplication

## Check Your Understanding

**Skill Check** Write the letter of the correct answer.

1. Which inequality is equivalent to $5x < 25$?
   a. $x < -5$  b. $x > -5$
   c. $x > 5$  d. $x < 5$

2. Which inequality is equivalent to $-2x > 7$?
   a. $x < 3\frac{1}{2}$  b. $x > 3\frac{1}{2}$
   c. $x > -3\frac{1}{2}$  d. $x < -3\frac{1}{2}$

3. Which is the graph of $4x \leq 12$?
   a.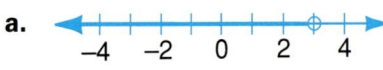
   b.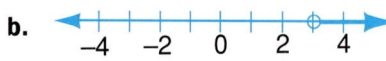
   c.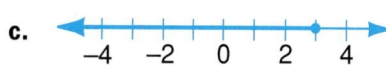
   d.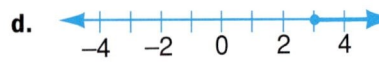

4. Which is the graph of $-3x \leq 15$?
   a.
   b.
   c.
   d.

**Complete each statement.**

5. If $12 > 3$, $\frac{1}{3}(12)$ _?_ $\frac{1}{3}(3)$.
6. If $-1 < 5$, $10(-1)$ _?_ $10(5)$.
7. If $\frac{1}{5} < 10$, $(-5)(\frac{1}{5})$ _?_ $(-5)(10)$.
8. If $-3t \leq 27$, then $(-3)(-3t)$ _?_ $(-3)(27)$.

## Practice and Apply

**Solve and graph each inequality.**

9. $2a \leq 16$
10. $-3b < 9$
11. $-2c \geq 7$
12. $-5t < 30$
13. $-4c \geq -10$
14. $-9f > -45$
15. $3r \leq -18$
16. $7k \geq 54$
17. $-5r \leq -30$
18. $-3p \leq -18$
19. $-2s \geq -19$
20. $-9r > 36$

**21.** Carey and Bert together earned at least $91. If they divide the money equally, what is the least amount each boy will receive?

**22.** The area of a rectangle is at least 20 square inches. The length of one side is at least 8 inches. What is the least possible width of the rectangle?

**23.** Art spent three times more money than Ron. If Ron spent less than $50, what is the greatest amount that Art could have spent?

## Connect and Extend

**Find the values that make the inequality true. The replacement set is $\{-2, -1, 0, 1, 2, 3\}$.**

**24.** $x < 0$  **25.** $x^2 < 0$  **26.** $x^2 \geq x$

**27.** $-x < 1$  **28.** $x^3 \leq -1$  **29.** $-x \leq -2$

**30.** Write a paragraph that compares the Multiplication Property of Inequality with the Multiplication Property of Equality. How are they alike? How do they differ?

## Maintain Your Skills

**Solve. (Pages 64–75)**

**31.** The grades on four math tests can range from 0 to 100. Lee scored 79, 83, and 88 on 3 of the tests. What is the highest Lee can average on the 4 tests?

**32.** The mainsail on Anna's catamaran has the shape of a triangle. The base of the sail is 7 feet and the height is 20 feet. Find the area of the sail.

**Solve and check each equation. (Pages 101–104)**

**33.** $7 + 6y = -29$  **34.** $\frac{c}{4} - 14 = 5$  **35.** $17 = \frac{e}{-9} + 4$  **36.** $-140 = -8 + 11n$

**Express the product in scientific notation. (Pages 289–294)**

**37.** $(9 \times 10^{-6})(7 \times 10^8)$   **38.** $(4.2 \times 10^{-7})(-6 \times 10^{-2})$

**39.** Richard spent $27.37 on a school jacket. If the same jacket sold for $19.55 last year, what is the percent increase? **(Pages 338–341)**

## Mathematical Footnote

When mathematician Mary Ellen Rudin was asked, "How do you actually do mathematics?", she replied: "I lie on the sofa in the living room with my paper and pencil and try this thing and that thing. Actually, I'm very geometric in my thinking."

What seems to be Ms. Rudin's favorite strategy for solving problems?

# LESSON 10-4
## Problem Solving Strategies Using Generalizations

Aaron claims that 12 and 35 are magic numbers because any integer can be formed just by adding and subtracting the two numbers. Here are some examples that Aaron gave.

$$35 - 12 - 12 - 12 = -1$$
$$35 - 35 = 0$$
$$-(35 - 12 - 12 - 12) = 1$$
$$-(35 - 12 - 12 - 12) - (35 - 12 - 12 - 12) = 2$$

- How can you use Aaron's generalization to form a 3? a 4? a −2?

- Do you think Aaron's generalization is true? Why or why not?

- Do you think that 35 and 12 are "special" numbers or it is possible that any two integers will work? How can you find out?

# EXERCISES

**Objective:** To apply strategies in problem solving

Write a sentence that describes the rule or generalization for each pattern.

1. triangle   square   pentagon   ?   ?
2. *a*   *bb*   *c*   *ddd*   *e*   *ffff*   ?   ?
3. ⊐  ⊔  ⊏   ?   ?

4. Use your generalizations to write or draw the next two entries in the patterns for Exercises 1–3.

Write a rule that generalizes each pattern.

5. 7, 15   9, 19   11, 23   13, 27   ?   ?
6. 0, 1, 4, 9, 16, ?, ?
7. Use the rules you wrote for Exercises 5 and 6 to write the next two entries in each pattern. Are your rules correct?

Solve these problems. You may wish to use a generalization.

8. How many eggs will it take to serve scrambled eggs for 10 people if you can serve 6 people with 11 eggs, 7 people with 13 eggs, and 8 people with 15 eggs?

9. Jillian earned $16 in simple interest in a year from her bank account of $200. At the same rate of interest, how much will she earn in a year on $300? on $475?

Work with a team to write and test a generalization that you can use to solve these problems. You may need to do some investigating outside of class.

10. With how many of your classmates can you share an orange if you give each person one section? Is the answer the same if you share a tangerine under the same conditions? A grapefruit?

11. Can you predict the length of a person's foot if you know the length from wrist to elbow of that person?

Write a description of each graph. (Pages 377–381)

1.
2.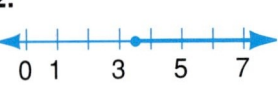
3.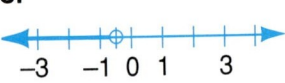

Graph each inequality. (Pages 377–381)
4. $g \geq -4$   5. $z < 0$   6. $r \leq 2$   7. $f < -1$

Solve and graph each inequality. (Pages 382–385)
8. $q - 2 \leq 5$   9. $s + 6 > 4$   10. $r - \frac{1}{5} > \frac{4}{5}$   11. $3\frac{1}{4} + p > 5\frac{1}{2}$

Solve and graph each inequality. (Pages 386–389)
12. $5c > -25$   13. $-4t > -24$   14. $-7e < 56$   15. $12f < 72$

# Mathematical Footnote

A psychiatrist studies the brain and thoughts and feelings; an oceanographer studies currents and life in the sea. Recently, a psychiatrist, an oceanographer, and a mathematician found an area where their interests come together. Mathematics provided a way to describe both currents in the ocean and the movement of impulses along nerve cells.

Often a mathematician will develop a way to think and talk about numbers that is purely abstract. If you ask "What is this abstract idea good for? What can I do with it?", the answer is "Nothing." Later, however, someone else discovers that this "useless" mathematics provides a helpful way to describe and study some very practical field such as how to communicate with a spacecraft, how to make an artificial limb, or how to save an endangered species.

# LESSON 10-5

# Solving Inequalities

**H**annah needs an average math test score of at least 85 to qualify for a scholarship. Her scores on the first two tests were 78 and 86. There is one more test to take. What is the lowest Hannah can score on the third test and still qualify for the scholarship?

**UNDERSTAND THE PROBLEM**

What is the question?   What is the lowest score Hannah can get on the third test?

What is given?   Hannah scored 78 and 86 in the first two tests. She needs an average of 85.

**DEVELOP A PLAN**

**Strategy:** Use the formula for finding an average. Write an inequality.

**CARRY OUT THE PLAN**

Represent the unknown.   Let $x$ represent Hannah's score on the third test.

**Think:** An average of at least 85 means an average that is greater than or equal to 85.

Write an inequality.    $\dfrac{78 + 86 + x}{3} \geq 85$

Add 78 and 86.    $\dfrac{164 + x}{3} \geq 85$

Multiply each side by 3.    $\overset{1}{\cancel{3}}\left(\dfrac{164 + x}{\underset{1}{\cancel{3}}}\right) \geq 3(85)$

$164 + x \geq 255$

Add ($-164$) to each side.    $-164 + 164 + x \geq -164 + 255$
Simplify.    $x \geq 91$

**LOOK BACK**

**Check**   Is the average of 78, 76, and 91 $\geq$ 85? Yes, since 85 $\geq$ 85.

So Hannah must score at least 91 on the third test. You can use the inequality properties to solve problems. It is usually easier to apply the Addition Property of Inequality first.

**EXAMPLE 1** Solve and graph: $-2x + 3 < 7$.

**Solution**

| | |
|---|---|
| Write the inequality. | $-2x + 3 < 7$ |
| Add $(-3)$ to each side. | $-2x + 3 + (-3) < 7 + (-3)$ |
| Simplify. | $-2x < 4$ |
| Multiply each side by $\left(-\frac{1}{2}\right)$. Remember to reverse the inequality symbol. | $-\frac{1}{2}(-2x) > -\frac{1}{2}(4)$ $x > -2$ |

The solution set is {all real numbers greater than $-2$}.

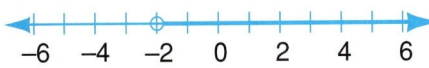

**Try This** Solve and graph.
1. $5x - 3 \leq 2$
2. $-4x - 1 > 7$

**EXAMPLE 2** Solve and graph: $2 - \frac{1}{3}p \leq 4$.

**Solution**

| | |
|---|---|
| Write the inequality. | $2 - \frac{1}{3}p \leq 4$ |
| Add $(-2)$ to each side. | $-2 + 2 - \frac{1}{3}p \leq -2 + 4$ |
| Simplify. | $-\frac{1}{3}p \leq 2$ |
| Multiply each side by $-3$. Reverse the inequality. | $-3\left(\frac{1}{3}p\right) \geq (-3)2$ |
| Simplify. | $p \geq -6$ |

The solution set is {all real numbers greater than or equal to $-6$}.

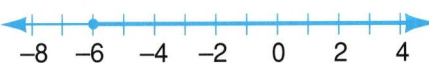

**Try This** Solve and graph.
3. $-5x + 1 > -9$
4. $7 - \frac{3}{4}d \leq 4$

Some word problems involving inequalities may have more than one answer.

**EXAMPLE 3** The sum of three consecutive whole numbers is less than or equal to 12. What integers are possible?

**Solution**  Let x = the smallest integer.
Then x + 1 = the next integer, and x + 2 = the greatest integer.

| | |
|---|---|
| Write an inequality. | x + x + 1 + x + 2 ≤ 12 |
| Combine like terms. | 3x + 3 ≤ 12 |
| Add (−3) to each side. | 3x ≤ 9 |
| Multiply each side by $\frac{1}{3}$. | x ≤ 3 |

Since x represents the smallest whole number, and the sum of the numbers must be less than or equal to 12, there are four possibilities.

| Consecutive Whole Numbers | Check | |
|---|---|---|
| 0, 1, and 2 | 0 + 1 + 2 ≤ 12? | Yes ✓ |
| 1, 2, and 3 | 1 + 2 + 3 ≤ 12? | Yes ✓ |
| 2, 3, and 4 | 2 + 3 + 4 ≤ 12? | Yes ✓ |
| 3, 4, and 5 | 3 + 4 + 5 ≤ 12? | Yes ✓ |

**Try This**  5. The sum of two consecutive whole numbers is less than 46. Find the pair with the greatest sum.

# EXERCISES

**Objective:** To solve inequalities using a multi-step process

## Check Your Understanding

**Skill Check**  Write the letter of the correct answer.

1. Which inequality is equivalent to $-3x + 1 < 7$?
   a. $x > -2$  b. $x < -2$
   c. $x > 2$   d. $x < 2$

2. Which inequality is equivalent to $5x - 3 \leq -8$?
   a. $x \geq 1$   b. $x \leq 1$
   c. $x \geq -1$  d. $x \leq -1$

3. Which inequality is equivalent to $1 - 4x > -3$?
   a. $x < 1$   b. $x < -1$
   c. $x > 1$   d. $x > -1$

4. Which inequality is equivalent to $3x - 4x \geq 2$?
   a. $x \geq 2$   b. $x \geq -2$
   c. $x \leq 2$   d. $x \leq -2$

5. What is the first step in solving the inequality $-2x + 5 > 10$?
6. What is the first step in solving the inequality $-3a - 4a > -21$?

## Practice and Apply

**Solve and graph the inequality.**

7. $-5x + 2 \geq 27$    8. $2x - 6 < 8$    9. $4 + 3x \leq -5$

10. $-3x - 5 < -23$
11. $6 - 3x \leq 9$
12. $12 + 8x < -12$
13. $6x + 4 \geq 4$
14. $10 - 3x \leq -20$
15. $9 + 4x > 11$
16. $\frac{1}{2}x - 7 \leq -3$
17. $5 - \frac{1}{4}x \geq 3$
18. $\frac{2}{3}x + 4 > -6$
19. $3x + 2 + 2x \leq 7$
20. $2(x + 3) \geq -9$
21. $-3(x - 1) \leq 9$

22. Nikki's bowling scores for three games are 100, 121, and 125. To make the next round of the tournament, Nikki must have an average of at least 116 in four games. What is the lowest score Nikki can bowl in the fourth game and make the average of 116?

23. The sum of four consecutive whole numbers is less than 66. Find the numbers with the greatest sum.

24. Find all of the values of $x$ that give this rectangle a perimeter greater than 51.

25. Find the values of $x$ that give the rectangle a perimeter less than 51.

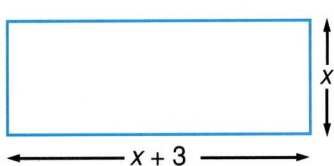

26. What is the value of $x$ that gives a perimeter of exactly 51?

## Connect and Extend

**This program computes and prints the solution of any inequality of the form, $ax + b < c$.**

```
100 PRINT "ENTER VALUES FOR A, B, C"
110 PRINT "(IN THAT ORDER, SEPARATED BY COMMAS)";
120 INPUT A, B, C
130 IF A = 0 THEN 200
140 LET X = (C - B)/A
150 IF A < 0 THEN 180
160 PRINT "X < "; X
170 GOTO 240
180 PRINT "X > "; X
190 GOTO 240
200 IF B < C THEN 230
210 PRINT "NO SOLUTION"
220 GOTO 240
230 PRINT "ALL REAL NUMBERS ARE SOLUTIONS."
240 PRINT "ANY MORE INEQUALITIES TO SOLVE
     (1 = YES, 0 = NO)";
250 INPUT Z
260 IF Z = 1 THEN 120
270 END
```

Run the program for the following values of $a$, $b$, and $c$.

**27.** $a = 2; b = 5; c = 9$
**28.** $a = 3; b = 4; c = 10$
**29.** $a = 2; b = -6; c = 0$
**30.** $a = -2; b = 5; c = 9$
**31.** $a = 0; b = 7; c = 8$
**32.** $a = 0; b = 8; c = 7$

## Maintain Your Skills

**Solve. (Pages 334–337)**

**33.** During one 50-hour week, Martha spent 25% of her work time in staff meetings. How many hours did she spend in meetings?

**34.** A teacher's assistant earns 68% of the salary of a teacher. If a teacher earns $460 per week, how much does a teacher's assistant earn?

The table at the right shows the number of books checked out of the school library each day for five days. (Pages 359–362)

**35.** What is the median for the data?
**36.** What is the mode for the data?
**37.** What was the mean number of books checked out each day?

## ★ Math Team Problems

The cards below are made in such a way that the number in the lower right corner of each card is obtained by doing something with the other three numbers. The cards in each row follow the same pattern. For each row, find the pattern and determine what number should replace the question mark to fit the pattern.

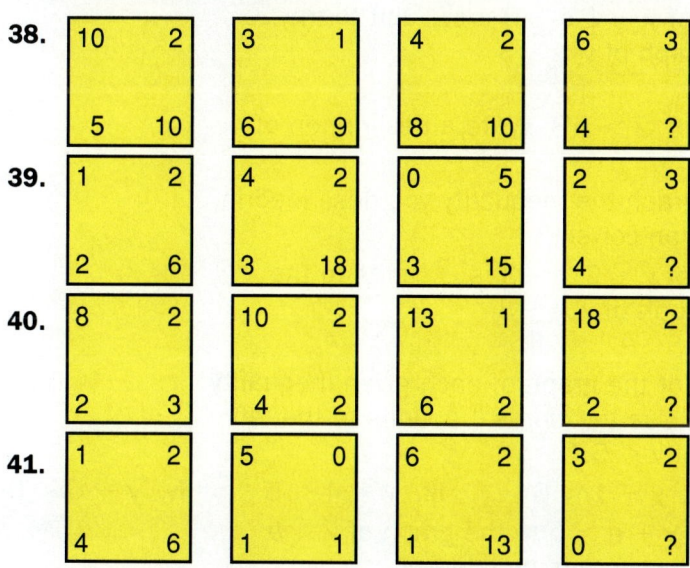

# COMPUTER EXPLORATION

## Graphing Inequalities

**Objective:** To explore the graphs of inequalities using a computer

1. a. With the Holt *Pre-Algebra* disk in drive A (or drive 1), run the program, *Graphing Inequalities*, as follows.
   **IBM:** Type **a:g,** and press **ENTER**.
   **Apple:** Turn on the computer. When the menu appears on the screen, press **G**.
   b. Read the opening messages and press **RETURN**.

2. a. To **"PICK A NUMBER ON THE LINE,"** type $-3$ and press **RETURN**.
   b. Write a description of the graph of $y \leq -3$.
   c. Press **3** to have the computer graph $y \leq -3$. Was your description correct?
   d. Repeat steps a, b, and c for the graphs of $y > -3$ and $y \geq -3$.

3. a. Solve the inequality $y - 1 \leq -3$. Write a description of the graph.
   b. Use the computer to graph the inequality you solved in 3a. Was your description correct?
   c. *Complete:* The graph of $y - 1 \leq -3$ is __?__ unit to the __?__ (left or right) of the graph of $y \leq -3$.

4. a. Solve the inequality $y + 2 > -3$. Write a description of the graph.
   b. Use the computer to graph the inequality you described in 4a. Was your description correct?
   c. *Complete:* The graph of $y + 2 > -3$ is __?__ units to the __?__ (left or right) of the graph of $y > -3$.

5. a. Compare the position of the graph of each given inequality to the graph of $y < 5$. Use the words "__?__ units to the left or right of the graph of $y < 5$."
       **i.** $y - 3 < 5$     **ii.** $y + 1 < 5$     **iii.** $y + 4 < 5$     **iv.** $y - 6 < 5$
   b. Compare the graph of $y + a > b$ to the graph of $y > b$ when $a$ is positive and when $a$ is negative.

# LESSON 10-6

## Conjunctions

This year, the city math and volleyball championships will be held on the same day. The math competition is in the morning. The volleyball competition is in the afternoon. Which students will compete in the morning *and* in the afternoon?

Both Maria and Jamal are on the math team *and* on the volleyball team. So Maria and Jamal are the only students who will compete in the morning and in the afternoon.

In mathematics, the word *and* has a special meaning.

A statement formed by joining two sentences with the word *and* is called a **conjunction.** A conjunction is true only if both sentences are true.

**EXAMPLE 1** Graph the solution set of $t \geq -1$ and $t \leq 4$.

**Solution** The solution set of the conjunction is the set of solutions that make $t \geq -1$ true *and* those that make $t \leq 4$ true.

Graph the solution set of
$t \geq -1$.

Graph the solution set of
$t \leq 4$.

**Think:** What values of $t$ make both inequalities true?

Graph $t \geq -1$ and $t \leq 4$.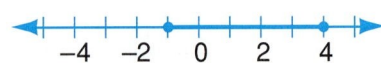

**Try This**   **Graph the solution set.**
1. $m \geq -4$ and $m \leq 2$
2. $c \geq -3$ and $c \leq 0$

The graph of the conjunction $t \geq -1$ and $t \leq 4$ can be described as a **closed interval** between, and including, 1 and 4. The conjunction can be written in a more compact way like this.

$$-1 \leq t \leq 4$$

The interval, $-1 \leq t \leq 4$, is also called the **intersection** of $t \geq -1$ and $t \leq 4$ because it contains all of the numbers belonging to both solution sets.

**EXAMPLE 2**   Graph the solution set for $1 < a < 3$.

**Solution**   **Think:** $1 < a < 3$ means $1 < a$ and $a < 3$.
What values of $a$ will make both inequalities true?

Graph $1 < a < 3$.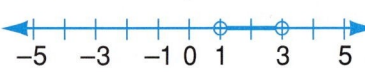

**Try This**   **Graph the solution set.**
3. $-1 < y < 3.5$
4. $0 < q < 5$

The graph of the conjunction $1 < a < 3$ can be described as an **open interval**. Why do you think it is called this?

Sometimes the graph of a conjunction is a ray. It is also possible that the solution set of a conjunction is the empty set.

**EXAMPLE 3**   Graph the solution set.
a. $t \geq -1$ and $t \geq 3$
b. $w \leq 0$ and $w \geq 1$

**Solutions**   a. Graph $t \geq -1$: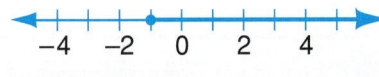

Graph $t \geq 3$: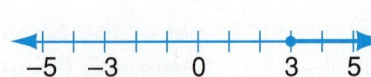

**Think:** What values of $t$ will make both inequalities true? Graph the points common to both inequalities.

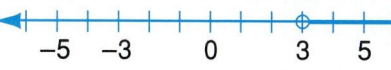    $t \geq -1$ and $t > 3$

**b.** Graph $w \leq 0$:

Graph $w \geq 1$:

Since the graphs have no points in common, the solution set is the **empty set,** written as { }.

**Try This**   **Graph the solution set.**
5. $a \geq 2$ and $a < -1$     6. $d > 5$ and $d < -1$

# EXERCISES

**Objective:** To solve and graph a conjunction

## Check Your Understanding

**Skill Check**   Write the letter of the correct answer.

1. Which best describes the solution set of $x \geq -1$ and $x \leq 3$?
   a. An open interval
   b. A closed interval
   c. A closed ray
   d. { }

2. Which best describes the solution set of $x < 3$ and $x > -2$?
   a. An open interval
   b. An open ray
   c. All real numbers
   d. { }

3. Which conjunction best describes this graph?

   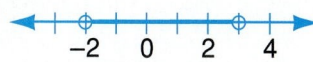

   a. $-1 < x < 3$
   b. $3 < x < -2$
   c. $x < -1$ and $x < 3$
   d. $x > -2$ and $x < 3$

4. Which graph is the solution set of $x \leq 1$ and $x \geq -1$?

   a.

   b.

   c.

   d.

10-6   Conjunctions   **401**

## Practice and Apply

**Write each statement as an inequality without the word *and*.**

**5.** $t < 4$ and $t > -1$  
**6.** $a > -1$ and $a < 0$  
**7.** $-3 < x$ and $x < 0$

**Graph the solution set. Write { } when the solution is the empty set.**

**8.** $x < -1$ and $x < 3$  
**9.** $t < -2$ and $t < 0$  
**10.** $y \le 4$ and $y < 1$  
**11.** $w \le 6$ and $w < 2$  
**12.** $-1 \le z \le 4$  
**13.** $a < 5$ and $a \ge 3$  
**14.** $r < 4$ and $r \le 4$  
**15.** $-6 \le m \le 2$  
**16.** $x > 5$ and $x \le -1$  
**17.** $-3 \le y \le 0$  
**18.** $s \le 4$ and $s < 2$  
**19.** $2 \le v < 7$

**Write an inequality with *and* to represent each situation.**

**20.** The weight of each boy on a football team is greater than 130 pounds *and* less than 205 pounds.

**21.** In one wrestling weight class, wrestlers must weigh more than 167 pounds but no more than 185 pounds.

## Connect and Extend

**Graph the solution set.**

**22.** $3x + 1 < 7$ and $2x - 5 \ge 1$  
**23.** $-5x - 1 > -11$ and $4 - 3x \le 7$

**Suppose that Set *A* and Set *B* contain these numbers.**
Set $A = \{1, 2, 3, 4, 5\}$   Set $B = \{4, 5, 6, 7\}$

The diagram at the right shows that the intersection of Set *A* and Set *B* (the numbers that are in both sets) is $\{4, 5\}$. Draw a diagram to show the intersection of Sets A and B for each given pair.

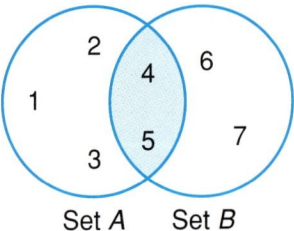

**24.** $A = \{1, 3, 4, 7\}$; $B = \{2, 5, 7\}$  
**25.** $A = \{1, 3, 5, 7\}$; $B = \{2, 4, 6, 8\}$  
**26.** $A = \{1, 3, 5, 7\}$; $B = \{3, 5\}$  
**27.** $A = \{1, 5, 10, 15\}$; $B = \{0, 5, 10\}$

## Maintain Your Skills

**Simplify each expression. (Pages 46–50)**

**28.** $18 + 7 \cdot (32 - 6)$  
**29.** $(10 + 15) \div 5$  
**30.** $7 \cdot (3 - 3) \div 3$

**Solve. (Pages 126–130)**

**31.** Find all the factors of 96.  
**32.** List the first five nonzero multiples of 13.

**A bag contains 4 pennies, 6 quarters, 7 dimes, and 3 nickels. One coin is drawn at random. Find each probability. (Pages 236–240)**

**33.** P(drawing a quarter)  
**34.** P(not drawing a penny or a quarter)

## LESSON 10-7 Disjunctions

Eartha took a poll of the 4th and 5th period math classes to determine their favorite one-digit number.

Which numbers are the favorites of the 4th *or* 5th period classes?

Favorite of *both* classes: {1, 2, 3, 5, 7, 9}
Note that 7 is listed only once.

| Favorite One-Digit Number | |
|---|---|
| 4th Period | 5th Period |
| 1, 3, 7 | 2, 5, 7, 9 |

The word *or* has a special meaning in mathematics. A statement formed by joining two sentences with the word *or* is called a **disjunction.** A disjunction is true if either sentence is true. A disjunction is also true if each of its sentences is true.

When you graph an inequality containing ≤, you are graphing a disjunction. That is, the inequality $x \leq 2$ means $x < 2$ or $x = 2$.

**EXAMPLE 1** Graph the solution set of $y \leq 3$ or $y \geq 6$.

**Solution** Graph $y \leq 3$.

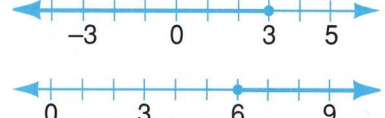

Graph $y \geq 6$.

**Think:** The solution set is the solutions that make *either* inequality, or *both* inequalities, true.

Graph $y \leq 3$ or $y \geq 6$:

**Try This** Graph the solution set.
1. $k \leq -2$ or $k \geq 1$
2. $m > -1$ or $m < -3$

Notice that the solution set contains all the solutions of $y \leq 3$ and all the solutions of $y \geq 6$. That is, the graph of an inequality involving *or* is the **union** of the graphs of the two inequalities.

10-7 Disjunctions **403**

**EXAMPLE 2** Solve and graph: $x > -2$ or $2x + 1 \geq 3$.

**Solution**  Graph $x > -2$.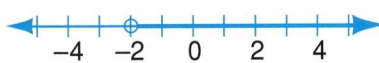

Solve $2x + 1 \geq 3$ for $x$.
$$2x + 1 \geq 3$$
$$2x \geq 2$$
$$x \geq 1$$

Graph $x \geq 1$.

Graph $x > -2$ or $x \geq 1$.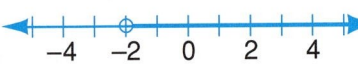

The solution set is {all real numbers greater than $-2$}.

**Try This**  Solve and graph.
3. $3x + 1 > -5$ or $x \geq 2$
4. $-b + 6 \leq 3$ or $2b - 1 \geq -1$

# EXERCISES

**Objective:** To solve and graph a disjunction

## Check Your Understanding

**Skill Check**  Write the letter of the correct answer.

1. Which is the graph of $x < -1$ or $x > 2$?

   a.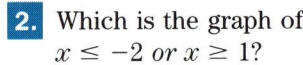
   b.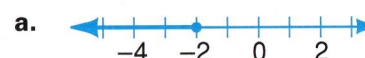
   c. 
   d. 

2. Which is the graph of $x \leq -2$ or $x \geq 1$?

   a. 
   b. 
   c. 
   d. 

3. Which is the graph of $2x + 1 \leq -1$ or $x \geq 1$?

   a. 
   b. 
   c. 
   d. 

404   CHAPTER 10

**4.** Which statement best describes this graph?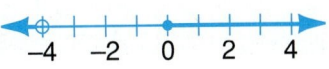

  a. $t < -4$ and $t > 0$  
  b. $t > 0$ or $t < -4$  
  c. $t \leq -4$ or $t \geq 0$  
  d. $t < -4$ or $t \geq 0$

## Practice and Apply

**Write each statement without *or*.**

**5.** $x > 5$ or $x = 5$  
**6.** $x < 0$ or $x = 0$  
**7.** $x < -3$ or $x = -3$

**Solve and graph.**

**8.** $x \leq 3$ or $x < 1$  
**9.** $x > -1$ or $x > 3$  
**10.** $x \geq 1$ or $2x < -4$  
**11.** $x \leq -3$ or $-3x < -12$  
**12.** $4x \leq -12$ or $6x > 18$  
**13.** $-5x > -10$ or $2x > -8$  
**14.** $3x < -9$ or $2x + 5 > 7$  
**15.** $3x - 1 \leq -5$ or $7x > 21$

## Connect and Extend

For Exercises 16–17, the replacement set is $\{-3, -2, -1, 0, 1, 2, 3\}$. If one number is chosen at random from the replacement set, find the probability that it will be in the solution set of the inequality.

**16.** $a \leq 2$ or $a \geq 1$  
**17.** $c \leq 2$ and $c > -1$

**18.** Are the words "and" and "or" always used in English sentences in the same way as you have been using them in conjunctions and disjunctions? Explain.

**19.** At Skyline High School, 20% of the students received an A in math, 35% of the students received an A in English, and 15% of the students received an A in both math and English. What percent of the students received an A in math *or* in English?

## Maintain Your Skills

**Solve each equation. (Pages 105–109)**

**20.** $3(a + 5) = 60$  
**21.** $2(y - 8) = -44$  
**22.** $-3(d + 4) - 2(9 + d) = 0$

**Evaluate each expression for $e = -2, f = -1, g = 3,$ and $h = 4$. (Pages 150–153)**

**23.** $-3h^3$  
**24.** $(2f^3)^5$  
**25.** $(-3efg)^5$  
**26.** $(e - f + h)^2$

**Solve. (Pages 171–174, 200–205, 282–286)**

**27.** Elizabeth had $\frac{5}{7}$ of a tank of gas. She used up $\frac{1}{5}$ of this amount during her trip. How much gas remains in her tank?

**28.** Richard has $15. He wants to buy the following items at the grocery store. Does he have enough money?

Cereal: $2.79  Juice: $3.38  
Fish: $5.63  Cheese: $3.18

# SUMMARY

## KEY TERMS

Closed interval (p. 400)
Conjunction (p. 399)
Disjunction (p. 403)
Empty set (p. 401)
Intersection (p. 400)
Irrational number (p. 377)

One-to-one correspondence (p. 378)
Open interval (p. 400)
Real number (p. 377)
Union (p. 403)

## KEY IDEAS

**A.** When the graph of an inequality does not include the endpoint in the solution set, the endpoint of the ray is shown as an open circle.

**B.** When the graph of an inequality includes the endpoint in the solution set, the endpoint of the ray is shown as a closed circle.

**C. Addition Property of Inequality:** Adding the same number to each side of an inequality results in an equivalent inequality.

**D. Multiplication Property of Inequality**
   **a.** Multiplying each side of an inequality by a positive number does not change the order of the inequality symbol.
   **b.** Multiplying each side of an inequality by a negative number reverses the order of the inequality symbol.

**E.** A conjunction is true only if both sentences are true.

**F.** A disjunction is true if either sentence is true or if both sentences are true.

# REVIEW

**Graph each equation or inequality. (Pages 377–381)**
1. $f \geq 2$
2. $g \geq -6$
3. $h = -3$
4. $k \leq 5$

**Write a description of each graph. (Pages 377–381)**

5. [number line with closed dot at 2, arrow left; marks −6 −4 −2 0 2]
6. [number line with open dot at 0, arrow left; marks −4 −2 0 2 4]
7. [number line with closed dot at 2, arrow right; marks −4 −2 0 2 4]
8. [number line with open dot at −4, arrow right; marks −8 −6 −4 −2 0]

**Solve and graph each inequality. (Pages 382–385)**
9. $x + 7 < 10$
10. $w - 1 \leq -4$
11. $f + 5 \geq -1$
12. $z - 4 \leq -7$
13. $y - 2 \geq -3\frac{1}{2}$
14. $s - 5\frac{1}{2} > -3\frac{1}{3}$

**Solve and graph each inequality. (Pages 386–389)**
15. $3c > -12$
16. $-6y > 24$
17. $-4t \leq -18$
18. $-6z \leq 21$
19. $9g > -45$
20. $-5e \geq 0$

21. If Pierre sells more than $100 worth of tickets, he will win a radio. Each ticket sells for $3.75. How many tickets must he sell to win the radio?

22. Gina saved $82 toward a new bike. This is less than $\frac{1}{2}$ of the cost of the bike. What is the least amount the bike could cost?

**Find the next two entries in each pattern. (Pages 390–391)**
23. a    ab    abc    abcd    abcde    ?    ?
24.     ?    ?
25. 38    383    3838    ?    ?

**Solve and graph each inequality. (Pages 393–397)**
26. $2c + 3 < 5$
27. $9r - (-7) \geq 25$
28. $-8f - 4 < 28$
29. $15d + 6 \geq -24$
30. $-2v - 1 \leq -6$
31. $-y + 8 \leq 2\frac{1}{2}$

**Solve. (Pages 393–397)**
32. Rafael's scores on four Pre-Algebra exams are 88, 74, 76, and 82. He wants his mean score on five tests to be greater than 80. What is the lowest score he can get on the fifth test?

**Graph the solution set. (Pages 399–405)**
33. $f \geq -3$ and $f < -1$
34. $d \geq -1$ and $d \leq 2$
35. $-3 < t \leq 6$
36. $q < 2$ or $q < -1$
37. $f \leq 1$ or $f \geq 3$
38. $5g \leq -10$ or $6g > 36$

# CHAPTER TEST

**Write a description of each graph.**

1. [number line with open circle at -2, shaded right]
2. [number line with closed circle at -3, shaded left]
3. [number line with closed circle at -5, shaded left]
4. [number line with open circle at 5, shaded left]
5. [number line with open circle at 1, shaded left]
6. [number line with open circle at 0, shaded right]
7. [number line with closed circle at -2, shaded left]
8. [number line with open circle at -3, shaded right]

**Graph each inequality.**

9. $k \geq -4$
10. $g \leq 8$
11. $x \geq 6$
12. $y < -5$
13. $t < -3$
14. $f \leq -12$
15. $g > -\frac{5}{2}$
16. $d \leq -\frac{1}{2}$

**Solve and graph each inequality.**

17. $9y \leq 63$
18. $e - 7 > -12$
19. $52 \leq -8w$
20. $-\frac{1}{5} + q \geq \frac{3}{10}$
21. $31 > 10z - 9$
22. $7 - 4x > -1$

**Solve.**

23. The cost of a radio that George would like to buy is greater than $60. George has saved $28. What is the least amount that he has to save before he can buy the radio?

24. Alix paid less than $8.70 for three pounds of meat. What was the greatest possible cost per pound?

25. The sum of two consecutive whole numbers is less than 88. Find the pair with the greatest sum.

26. Ava scored 12 points in one basketball game and 18 points in another. She wants her average for the three games to be greater than 18 points. What is the least number of points she must score in the next game?

**Graph.**

27. $y \leq 1$ and $y > -1$
28. $2 \leq t \leq 5$
29. $x \leq 2$ or $x > 4$
30. $x < -1$ or $2x - 1 > 3$

## CHAPTERS 1-10

**Solve. (Pages 69–75, 114–118, 139–141, 163–167)**

1. The red pennants flown for gale warnings are triangular in shape. The base of the triangle is 8 feet and the height is 15 feet. Find the area of the pennant.

2. Angela spent a total of $204 on rock concert tickets. If each ticket cost $12, how many did she buy?

3. List all the numbers between 20 and 50 that have exactly two different factors.

4. Write three fractions that are equivalent to $\frac{70}{112}$.

**For Exercises 5–8, refer to the spinner at the right. (Pages 236–240)**

5. List the possible outcomes of a spin.
6. What is the probability of the pointer's stopping on an even number?
7. What is the probability of the pointer's stopping on a number greater than 3?
8. What is the probability of the pointer's stopping on a prime number?

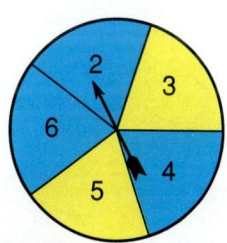

**For Exercises 9–14, express each rational number as a decimal in two ways. First, give the exact value. Then give the value rounded to the nearest tenth. (Pages 277–280)**

9. $\frac{2}{9}$
10. $-\frac{4}{15}$
11. $-\frac{6}{11}$
12. $\frac{7}{33}$
13. $-\frac{13}{22}$
14. $\frac{11}{27}$

**Write each decimal as a percent. (Pages 309–312)**

15. 0.07
16. 0.68
17. 4.3
18. 0.9
19. 0.502

**Tony borrowed $4500 from the bank for a car. The bank charged a yearly interest rate of 12%. (Pages 324–328)**

20. How much simple interest will Tony owe at the end of one year?
21. What is the total amount Tony will owe the bank at the end of one year?

**Solve. (Pages 334–337)**

22. The Harris family spent $234 one month for food. This was 26% of their income. What was their monthly income?

23. On a final exam with 75 questions, Rosa got 66 questions correct. What percent of the questions was this?

Cumulative Review: Chapters 1–10 **409**

## LESSON 11-1

# Coordinate Graphs

**Y**ou can find the location of a city on a map by using latitude and longitude. On the map above, the city of Austin, Texas is close to where the lines showing a latitude of +30° and a longitude of +98° meet. So the ordered pair, (30, 98) gives the approximate coordinates of Austin and helps you locate it on the map. The map is an example of a **coordinate** graph.

In a *rectangular coordinate system,* two perpendicular lines, called **axes,** meet at a point called the **origin.** The horizontal axis is called the *x*-**axis** and the vertical axis is called the *y*-**axis.** This rectangular coordinate system allows you to graph any ordered pair with **coordinates** $(x, y)$. The *x*-**coordinate** gives the distance to the left or right of the *y*-axis. The *y*-**coordinate** gives the distance of this point above or below the *x*-axis.

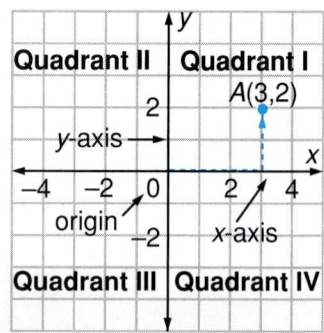

410  CHAPTER 11

The $x$-coordinates of points to the right of the $y$-axis are assigned positive values; those to the left are assigned negative values. The $y$-coordinates of points above the $x$-axis are assigned positive numbers; those below the $x$-axis are assigned negative values.

The coordinates of the origin are $(0, 0)$. The point $A(3, 2)$ is 3 units to the right of the $y$-axis and 2 units above the $x$-axis. The axes separate the coordinate plane into four regions called **quadrants.**

**EXAMPLE 1** Name the coordinates of points $P$, $Q$, and $R$. Then name the axis or quadrant in which each point lies.

**Solutions** Point $P$ is located 2 units to the right of the $y$-axis and 3 units above the $x$-axis.
Coordinates of $P$: **(2, 3)**
$P(2, 3)$ is in **Quadrant 1.**

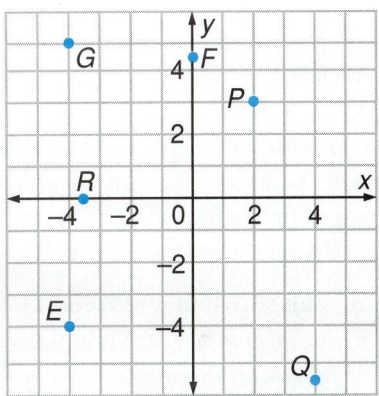

Point $Q$ is located 4 units to the right of the $y$-axis and 5.5 units below the $x$-axis.
Coordinates of $Q$: **(4, −5.5)**
$Q(4, −5.5)$ is in the **Quadrant IV.**

Point $R$ is $3\frac{1}{2}$ units to the left of the $y$-axis. It is on the $x$-axis.
Coordinates of $R$: $\left(-3\frac{1}{2}, 0\right)$. $R\left(-3\frac{1}{2}, 0\right)$ is **on the $x$-axis.**

**Try This** Refer to the graph in Example 1 to locate each point. Name the quadrant or axis where each point lies.
1. $E$  
2. $F$  
3. $G$

**EXAMPLE 2** Graph these points: **a.** $A(-2.5, -4)$  **b.** $B\left(-3\frac{1}{2}, 0\right)$

**Solutions** 
a. Start at the origin. Move 2.5 units to the left. Then move 4 units down. Draw a dot and label the point as the ordered pair $A(-2.5, -4)$.

b. Start at the origin. Move $3\frac{1}{2}$ units to the left. Draw a dot. Label the point $B\left(-3\frac{1}{2}, 0\right)$.

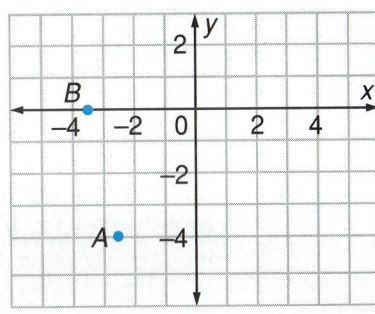

11-1 Coordinate Graphs

# EXERCISES

**Objectives:** To find the coordinates of points
To graph points on the coordinate plane

## Check Your Understanding

**Skill Check** Write the letter of the correct answer.
Refer to the graph at the right below.

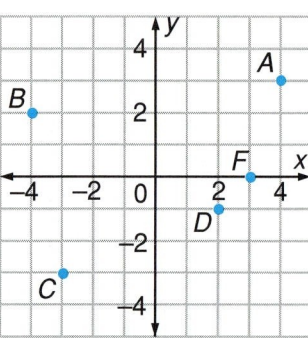

1. What are the coordinates of point *A*?
   **a.** (3, 4)   **b.** (4, 3)   **c.** (−4, 3)   **d.** (4, −3)
2. What are the coordinates of point *D*?
   **a.** (−1, 2)   **b.** (2, 1)   **c.** (−2, 1)   **d.** (2, −1)
3. Which best describes the location of point *B*?
   **a.** It is in Quadrant I.   **b.** It is in Quadrant II.
   **c.** It is on the *x*-axis.   **d.** It is on the *y*-axis.
4. Which best describes the location of point *F*?
   **a.** It is in Quadrant III.   **b.** It is in Quadrant IV.
   **c.** It is on the *x*-axis.   **d.** It is on the *y*-axis.

**True or False?** When a statement is false, explain why.
5. $P(-5, -6)$ is the same point as $Q(-6, -5)$.
6. Any point on the *y*-axis has coordinates $(0, y)$.
7. Any point on the *x*-axis has coordinates $(0, x)$.

## Practice and Apply

For Exercises 8–9, name the coordinates of points *A*, *B*, *C*, and *D*.

8.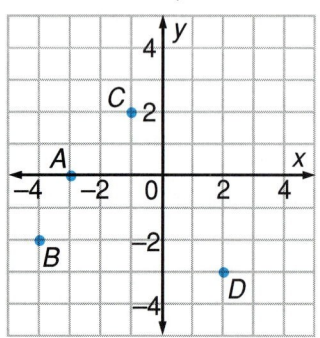

9.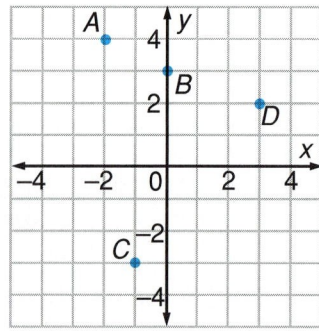

Name the quadrant or axis where the point is located.
10. $A(-3, -5)$    11. $B(0, 4)$    12. $C(-2, 1)$    13. $D(-3, 0)$
14. $E(4, -1)$    15. $F(-1, -1)$    16. $G(4, 4)$    17. $H(-2, -3)$

**412**    CHAPTER 11

**Graph the point for the ordered pair. Label the point.**

18. $A(2, -4)$
19. $B(3, 0)$
20. $C(-1, -1)$
21. $D(-3, 2)$
22. $R(-3, -2)$
23. $J(-4, 3)$
24. $K(3, -4)$
25. $S(2, 2)$
26. $M\left(1\frac{1}{2}, 4\right)$
27. $N\left(4, -3\frac{1}{2}\right)$
28. $P(-2.5, 0)$
29. $Q(-1.5, -3.5)$

30. Suppose that you put the tip of your pencil on point $S$ in Exercise 25. Then you move 4 units to the right and 3 units down. Name the coordinates of your new location.

## Connect and Extend

For Exercises 31–32:
   a. Graph the given points.
   b. Draw lines connecting the points.
   c. Find the perimeter and the area of the figure you just drew.

31. $A(3, 2); B(3, 6); C(4, 2), D(4, 6)$
32. $E(2, 3); F(2, -2); G(-4, -2); H(-4, 3)$

Each of Exercises 33 and 34 lists three vertices (corners) of a rectangle. Graph the vertices. Then draw the rectangles and give the coordinates of the fourth vertex.

33. $P(-6, 1), Q(-1, 5), R(-6, 5)$
34. $S(-5, -1), T(-8, -4), V(-6, -6)$

## Maintain Your Skills

**Find the solution set if the replacement set is $\{-3, -1, 0, 1, 3\}$. (Pages 80–84)**

35. $19d - 1 = 56$
36. $-12a + 6.2 = 42.2$
37. $-17(f + 3) = -51$

**Solve and check each equation. (Pages 101–104)**

38. $9r + 13 = -41$
39. $\frac{a}{2} + 8 = -10$
40. $-7 = \frac{x}{4} - 14$
41. $53 = 6n - 19$

**Write an equation to represent the situation. Then solve. (Pages 114–118)**

42. The distance from Phoenix to San Diego is 180 miles more than the distance from San Diego to Los Angeles. The sum of the two distances is 402 miles. Find each distance.

43. The width of a rectangle is 10 inches less than twice the length. The perimeter is 22 inches. What is the length and width of the rectangle?

44. A large glass of juice at Healthy Foods holds 275 milliliters, the medium glass holds 175 milliliters, and the small glass holds 125 milliliters. If Mr. Brown buys 1 large, 2 medium, and 4 small glasses for his family, how many liters of juice did he buy? Round the answer to the nearest tenth. **(Pages 295–296)**

# LESSON 11-2
## Graphing Linear Equations

To mow lawns in his neighborhood, Julio charges $8.00 per hour plus an additional fee of $5.00 per lawn. The table and the graph below show his earnings, $y$, for the number of work hours, $x$.

| x | 1 | 2 | 3 | 4 | 5 | 6 | 7 | ... |
|---|---|---|---|---|---|---|---|-----|
| y | 13 | 21 | 29 | 37 | 45 | 53 | 61 | ... |

The table represents a *function* that pairs the number of hours Julio works with his earnings. The function can be described in words.

**Function:** Multiply 8 and the number of hours worked. Then add 5.

**Figure 1** (graph showing earnings vs. number of hours)

If you let $x$ represent the number of hours worked and $y$ represent the amount of earnings, you can write this equation for the function.

$$y = 8x + 5$$

Each ordered pair in the table is a solution of the equation.

If the replacement set for $x$ and $y$ is the set of real numbers, there is an infinite number of solutions for this equation.

You can graph the solution set by drawing a line through at least 3 of the points graphed in Figure 1. Since the graph is a straight line, the function $y = 8x + 5$ is a linear function.

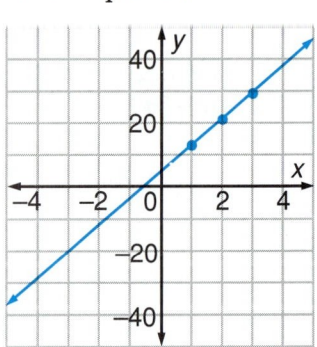

**EXAMPLE 1** Which ordered pairs are solutions of $y = 2x - 1$?
  a. (1, 1)    b. (0, 1)    c. (−1, −3)

**Solutions** Substitute the given values for x and y in the equation.

a. $y = 2x - 1$
$1 \stackrel{?}{=} 2(1) - 1$
$1 = 1$ ✓
**Solution**

b. $y = 2x - 1$
$1 \stackrel{?}{=} 2(0) - 1$
$1 \neq -1$ ✓
**Not a solution**

c. $y = 2x - 1$
$-3 \stackrel{?}{=} 2(-1) - 1$
$-3 = -3$ ✓
**Solution**

**Try This**  Which ordered pairs are solutions of $y = -3x + 4$?
1. $(-1, 7)$  2. $(0, 4)$  3. $(-2, 10)$

To find a solution of a linear equation, choose a value for $x$. Then substitute the chosen value in the equation and solve for $y$.

**EXAMPLE 2**  Make a table showing three solutions of $y = -2x + 3$.

**Solution**  Choose 3 different values for $x$.

Let $x = 0$.  
$y = -2x + 3$  
$y = -2(0) + 3$  
$y = 0 + 3$  
$y = 3$

Let $x = 1$.  
$y = -2x + 3$  
$y = -2(1) + 3$  
$y = -2 + 3$  
$y = 1$

Let $x = 2$.  
$y = -2x + 3$  
$y = -2(2) + 3$  
$y = -4 + 3$  
$y = -1$

Make a table to show the solutions.

| x | 0 | 1 | 2 |
|---|---|---|---|
| y | 3 | 1 | -1 |

**Try This**  Make a table to show three solutions of each equation.
4. $y = 5x - 3$  5. $y = 3x - 4$

Since the solution set of a linear equation is infinite, the solution set is usually represented as a graph.

**EXAMPLE 3**  Graph the equation $y = 3x - 2$.

**Solution**  Make a table that lists three solutions of $y = 3x - 2$.

If $x = 0$, $y = 3(0) - 2 = -2$.
If $x = 1$, $y = 3(1) - 2 = 1$.
If $x = 2$, $y = 3(2) - 2 = 4$.

| x | 0 | 1 | 2 |
|---|---|---|---|
| y | -2 | 1 | 4 |

Use the ordered pairs in the table to draw the graph.

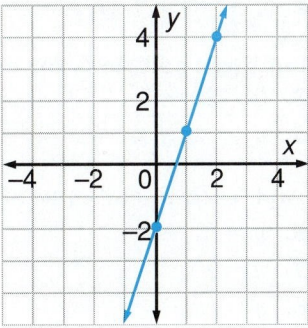

**Try This**  Graph each equation.
6. $y = 2x + 1$  7. $y = 3x + 2$

• At which point on the graph in Example 3 does $x = -1$?

# EXERCISES

**Objectives:** To determine whether an ordered pair is a solution of an equation
To graph linear equations

## Check Your Understanding

*Skill Check* Write the letter of the correct answer.

1. Which ordered pair is a solution of $y = 2x + 5$?
   a. $(7, 1)$   b. $(-7, 1)$
   c. $(1, 7)$   d. $(-1, 7)$

2. Which ordered pair is **not** a solution of $y = 2x + 5$?
   a. $(0, 5)$   b. $(1, 3)$
   c. $(-4, -3)$   d. $(-1, 3)$

3. The points $(-1, 6)$ and $(0, 4)$ lie in the graph of a line. Which is the equation of the line?
   a. $y = 2x + 4$   b. $y = 2x - 4$
   c. $y = -2x + 4$   d. $y = -2x - 4$

4. Which is the equation of the line that passes through the points $(3, 12)$ and $(4, 17)$?
   a. $y = 5x + 3$   b. $y = -5x + 3$
   c. $y = -5x - 3$   d. $y = 5x - 3$

## Practice and Apply

Determine whether the given ordered pair is a solution of the given equation. Answer Yes or No.

5. $y = 3x + 1$; $(2, 5)$
6. $y = 2x + 4$; $(-1, 2)$
7. $y = 2x + -1$; $(0, 5)$
8. $y = 3x - 1$; $(-2, -4)$
9. $y = 5x - 3$; $(-1, -8)$
10. $y = 2x + 4$, $(2, 0)$
11. $y = -x + 1$; $(1, 1)$
12. $y = -4x - 3$; $(-2, 5)$

Find three solutions for each equation.

13. $y = 2x - 5$
14. $y = x + 4$
15. $y = 3x + 4$
16. $y = -x - 4$
17. $y = 4x + 3$
18. $y = 2x - 6$

Graph each linear equation in the coordinate plane.

19. $y = 3x - 4$
20. $y = 2x + 7$
21. $y = 3x - 1$
22. $y = x$
23. $y = -x$
24. $y = 4x + -6$
25. $y = 5x - 1$
26. $y = 7 - 2x$
27. $y = -2x$

A home repair service charges $10 per hour plus a base fee of $12 per call.

28. Find the cost for three hours of work.
29. Write an equation that can be used to find the cost, $C$, for $w$ hours of work by the repair service.
30. Graph the equation you wrote in Exercise 29.

A supermarket charges $3 per hour plus a base fee of $5 for customers who wish to rent a rug cleaning machine.

**31.** Write an equation that can be used to find the cost, $C$, of renting the machine for $h$ hours.

**32.** If a customer paid $29 to rent the machine, how many hours was it rented?

## Connect and Extend

**33.** When does $(x, y) = (y, x)$? Describe the graph.

**Find the points where the graph of the equation crosses the $x$-axis and the $y$-axis.**

**34.** $y = 2x - 4$     **35.** $y = 5 - x$     **36.** $y = x$

**Make a table of three ordered pairs that fit each description. Draw the graph. Write an equation that describes the graph.**

**37.** Each $x$-coordinate is the opposite of its $y$-coordinate.

**38.** Each $x$-coordinate is twice its $y$-coordinate.

## Maintain Your Skills

**Solve each equation. (Pages 105–109)**

**39.** $3(4 + x) - 4x = 16$     **40.** $-23 = 9f - (5f + 7)$     **41.** $7h - 9(6 + h) = -61$

**Solve. (Pages 206–209, 330–333)**

**42.** The price of a stereo is reduced by $\frac{2}{7}$ of the original cost. What fraction of the original cost is the sale price?

**43.** Hugh had $8\frac{1}{4}$ gallons of gasoline in his gas tank. He used $3\frac{1}{2}$ gallons. How much gas does he have left?

**44.** The Johnson family paid $80 in heating costs last month. This month, their heating costs were $76. Find the percent decrease.

**Solve and graph each inequality. (Pages 393–397)**

**45.** $4z - 6 \leq 6$     **46.** $-3c + 5 \geq -4$     **47.** $5k + 9 \geq 4$     **48.** $-2f - 10 \leq -13$

## ★ Math Team Problem

**49.** Maria leaves at 10 A.M. on her bicycle for her friend, LaShana's home, which is 20 miles away. She arrives at 1 P.M. After spending the night, she leaves at 10 A.M. and arrives home at 1 P.M.

Draw a graph to show a point on the road home that Maria passes at exactly the same time as she did the day before. Assume that she travels at the same speed on both days.

# LESSON 11-3
# The Standard Form of a Linear Equation

The equation $y = 3x - 2$ is a linear equation and its graph is a straight line. You can use the Properties of Equality to write this equation in **standard form.**

> **Standard Form of a Linear Equation**
> A linear equation expressed in the form
> $$Ax + By = C$$
> is said to be in standard form. In the standard form, $A$, $B$, and $C$ represent integers and $A$ and $C$ cannot both equal 0.

| | |
|---|---|
| Write the equation. | $y = 3x - 2$ |
| Add $(-3x)$ to each side. | $-3x + y = -3x + 3x - 2$ |
| Simplify. | $-3x + y = -2$ |
| Multiply each side by $(-1)$. | $3x - y = 2$ |

For the equation $3x - y = 2$, $A = 3$, $B = -1$, and $C = 2$.

When a linear equation is in standard form, you can draw its graph by finding where the graph crosses the $x$- and $y$-axes. These points are called the **intercepts** of the line.

The **$x$-intercept** is the $x$-coordinate of the point where the graph crosses the $x$-axis. At this point, $y = 0$. The **$y$-intercept** is the $y$-coordinate of the point where the graph crosses the $y$-axis. At this point, $x = 0$.

**EXAMPLE 1**  Find the x-intercept and the y-intercept of $3x - 2y = 6$. Use the intercepts to draw the graph.

**Solution**

**x-intercept:** Replace y with 0 and solve.

$$3x - 2y = 6$$
$$3x - 2(0) = 6$$
$$3x = 6$$
$$x = 2$$

**y-intercept:** Replace x with 0 and solve.

$$3x - 2y = 6$$
$$3(0) - 2y = 6$$
$$-2y = 6$$
$$y = -3$$

The x-intercept is **2** and the y-intercept is **−3.**

Graph the points (2, 0) and (0, −3).

Draw a line passing through the points.

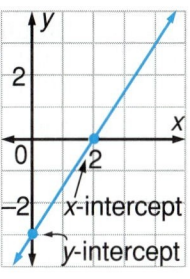

**Try This**

Find the x- and y-intercepts. Use the intercepts to graph the equation.

1. $2x - 5y = 10$
2. $4x - y = -8$

- Can more than one line have the same x-intercept?
- Can more than one line have the same y-intercept?
- Can more than one line have the same x-intercept *and* the same y-intercept? Explain.

When $A = 0$, the equation $Ax + By = C$ becomes $By = C$. Equations of the form $By = C$ look like this.

$$y = 2 \qquad y = \tfrac{1}{2} \qquad y = -6 \qquad 2y = -3$$

**EXAMPLE 2** Graph $y = 2$.

**Solution** Write $y = 2$ in standard form. $\qquad 0 \cdot x + y = 2$

**Think:** For any value of x you choose, y will always equal 2.

The graph is a line 2 units above the x-axis and parallel to it. The y-intercept is 2.

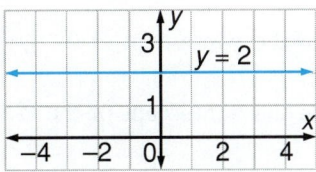

**Try This**

Describe the graph of each equation.

3. $y = -2$
4. $y = 0$

When $B = 0$, the equation $Ax + By = C$ becomes $Ax = C$. Equations in the form $Ax = C$ look like this.

$$x = -1 \qquad x = -\tfrac{1}{4} \qquad x = -3 \qquad 4x = 8$$

**EXAMPLE 3** Graph $x = -1$.

**Solution** Write $x = -1$ in standard form.  $x + 0 \cdot y = -1$

**Think:** For any value of $y$ you choose, $x$ will always equal $-1$.

The graph is a line one unit to the left of the y-axis and parallel to it. The x-intercept is $-1$.

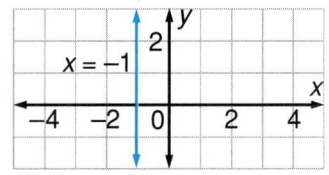

**Try This** Describe the graph of each equation.

5. $x = 2$
6. $x = 0$

# EXERCISES

**Objective:** To graph a linear equation using intercepts

## Check Your Understanding

**Skill Check** Write the letter of the correct answer.

1. Which is the standard form for $y = 3x + 1$?
   a. $3x - y = 1$
   b. $3x - y = -1$
   c. $y - 3x = 1$
   d. $3x + y = 1$

2. Which is the x-intercept of $3x - 2y = -6$?
   a. $-2$
   b. $2$
   c. $-3$
   d. $3$

3. Which is the y-intercept of $5x - y = 10$?
   a. $-2$
   b. $2$
   c. $10$
   d. $-10$

4. Which is the equation of a line parallel to the y-axis?
   a. $y = 1$
   b. $x = 3$
   c. $y = x$
   d. $x + y = 0$

5. *Complete:* You can find the y-intercept of the graph of a line by replacing $x$ with __?__.

6. *Complete:* You can find the x-intercept of the graph of a line by replacing $y$ with __?__.

## Practice and Apply

Rewrite the equation in standard form. Give the x- and y-intercepts.

7. $y = 4x + 8$
8. $3y = 2x - 6$
9. $y = 3x - 9$
10. $2(x - 3y) = -12$
11. $x = 3y + 9$
12. $4x + 6 = y$

13. Write three equations of the form $By = C$.
14. Write three equations of the form $Ax = C$.

**Find the $x$- and $y$-intercepts. Use the $x$- and $y$-intercepts to draw the graph.**

**15.** $3x + 2y = 12$  **16.** $4x - 3y = 12$  **17.** $2x - y = 4$
**18.** $y = -5$  **19.** $-2x + 4y = 8$  **20.** $5x - 2y = -10$
**21.** $-x + 5y = -5$  **22.** $x = 3$  **23.** $3x + 3y = -6$
**24.** $4x - 5y = -20$  **25.** $-5x - 3y = 15$  **26.** $x - y = 0$
**27.** $2x + y = -6$  **28.** $3x - \frac{1}{2}y = 3$  **29.** $9x - 6y = -18$

## Connect and Extend

**30.** Write an equation of the line that is parallel to the $x$-axis and has a $y$-intercept of $-5$.

**31.** Write an equation of the line that is parallel to the $y$-axis and has an $x$-intercept of 3.

**32.** Write the equation $y = Sx + T$ in standard form.

**33.** Graph these three equations in the same coordinate plane.
  **A:** $2x + 3y = 6$  **B:** $2x + 3y = 12$  **C:** $2x + 3y = -6$

**34.** Refer to the graphs you drew for Exercise 33 to write a paragraph that explains what happens to the graph of $2x + 3y = C$ under these two conditions.
  **a.** $C$ increases
  **b.** $C$ decreases

## Maintain Your Skills

**In a veterinarian's office, 14 of the 18 dogs and cats boarded are dogs. (Pages 216–220)**

**35.** Write the ratio, in lowest terms, of dogs to cats.

**36.** Write the ratio, in lowest terms, of cats to the total number of animals boarded.

**Evaluate. (Pages 252–257)**

**37.** $_8P_5$  **38.** $_{22}P_2$  **39.** $_9P_4$  **40.** $_5P_5 - {}_4P_4$  **41.** $\dfrac{_6P_2}{4!}$

**Name the quadrant or axis where the point is located. (Pages 410–413)**

**42.** $N(-4, 4)$  **43.** $T(5, -2)$  **44.** $C(-7, 0)$  **45.** $E(-5, -2)$

## ★ Math Team Problem

**46.** Copy the figure at the right. Then place 8 dots on the figure so that there are two dots on each of the three line segments and two dots on each of the four circles.

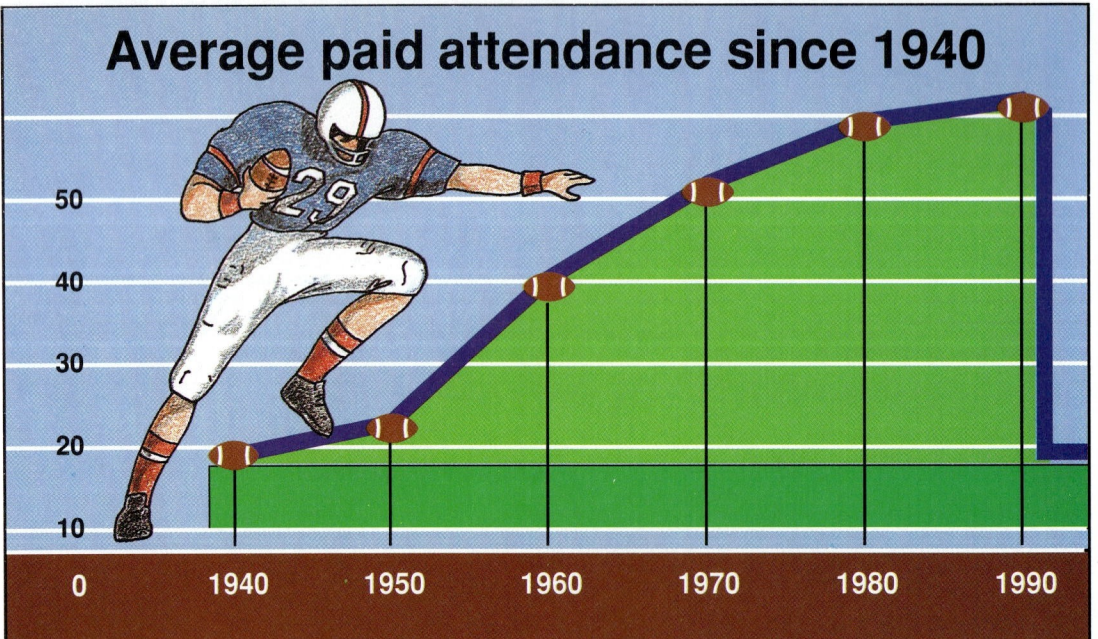

## LESSON 11-4

# The Slope of a Line

The graph above shows the average paid attendance at professional football games from 1940 to 1989. Notice how the steepness of the line shows that the rate of change in paid attendance was greatest from 1950 to 1960. The nearly flat line from 1980 to 1989 shows that the rate of change in paid attendance was least during these years.

The word used in mathematics to describe the steepness of a line is *slope*. The slope of a line is measured by a ratio that compares the change in vertical units to the change in horizontal units for two points on the line.

The ordered pairs in the table below were used to graph the line at the right.

| x | 0 | 2 | 4 |
|---|---|---|---|
| y | 1 | 2 | 3 |

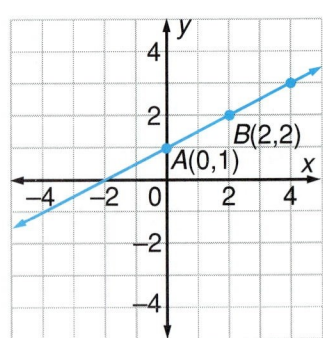

Notice that as $x$ increases by 2, $y$ increases by 1. Notice also that as $y$ decreases by 1, $x$ decreases by 2. In either case, the ratio of the change in the $y$-coordinates to the change in the corresponding $x$-coordinates is $\frac{1}{2}$.

That is, $\text{slope} = \dfrac{\text{change in } y\text{-coordinates}}{\text{change in } x\text{-coordinates}}$.

**EXAMPLE 1** Draw a line having a slope of $-\frac{2}{3}$.

**Solution** Since slope is a ratio, you can write the following.

$\dfrac{\text{change in } y\text{-coordinates}}{\text{change in } x\text{-coordinates}} = -\frac{2}{3}$, or $\frac{-2}{3}$

1. Choose a point on the x-axis. Label the point A.
2. Starting at point A, move 2 units down and 3 units to the right. Mark and label the new point B.
3. Draw a line through points A and B.

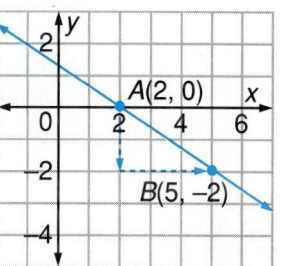

**Try This** Draw a line with the given slope.

1. $\frac{1}{2}$  2. $-\frac{1}{2}$  3. $\frac{2}{1}$  4. $-\frac{2}{1}$

- One line has a slope of $-\frac{2}{3}$, another line has a slope of $\frac{-2}{3}$, and another has a slope of $\frac{2}{-3}$. Do they have the same slope? Explain.

When you know the coordinates of two points on a line, you can subtract the y-coordinates and the x-coordinates to find its slope.

**EXAMPLE 2** Find the slope of the line that passes through the points $A(-1, 3)$ and $B(2, 4)$.

**Solution**
1. Subtract the y-coordinates.
$3 - 4 = -1$
2. Subtract the x-coordinates in the same order.
$-1 - 2 = -3$
3. Write the slope ratio.

$\text{slope} = \dfrac{\text{change in } y\text{-coordinates}}{\text{change in } x\text{-coordinates}} = \dfrac{-1}{-3} = \dfrac{1}{3}$

The slope of the line is $\frac{1}{3}$.

**Try This** Find the slope of the line passing through each pair of points.

5. $A(-1, 3)$ and $B(-3, 4)$
6. $C(0, -3)$ and $D(4, 0)$

11-4 The Slope of a Line

What is the slope of a line, such as $y = 3$, which is parallel to the $x$-axis? Since the change is the $y$-coordinates is always 0,

$$\frac{\text{change in } y\text{-coordinates}}{\text{change in } x\text{-coordinates}} = \frac{0}{\text{change in } x\text{-coordintes}} = 0.$$

So the slope of any line parallel to the $x$-axis is 0.

What is the slope of a line, such as $x = 3$, which is parallel to the $y$-axis? Since the change in the $x$-coordinates is always 0, the denominator of the slope ratio is zero. But division by zero is impossible! We say that the slope of any line parallel to the $y$-axis is undefined.

**EXAMPLE 3** Find the slope of $y = \frac{2}{3}x - 1$.

**Solution**
1. Find the coordinates of two points on the graph of $y = \frac{2}{3}x - 1$.
   
   Let $x = 0$.  $y = \frac{2}{3}(0) - 1 = -1$  Ordered pair: $(0, -1)$
   
   Let $x = 3$.  $y = \frac{2}{3}(3) - 1 = 1$  Ordered pair: $(3, 1)$

2. Find the slope.

   $$\text{slope} = \frac{\text{change in } y\text{-coordinates}}{\text{change in } x\text{-coordinates}} = \frac{-1 - 1}{0 - 3} = \frac{-2}{-3}, \text{ or } \frac{2}{3}$$

The slope of $y = \frac{2}{3}x - 1$ is $\frac{2}{3}$.

**Try This** Find the slope of the line.

7. $y = \frac{3}{4}x - 1$  8. $y = -3x + 4$

The results of Example 3 and Try This Exercises 7 and 8 show this pattern.

| Equation | Slope | Coefficient of $x$ |
|---|---|---|
| $y = \frac{2}{3}x - 1$ | $\frac{2}{3}$ | $\frac{2}{3}$ |
| $y = \frac{3}{4}x - 1$ | $\frac{3}{4}$ | $\frac{3}{4}$ |
| $y = -3x + 4$ | $-3$, or $\frac{-3}{1}$ | $-3$, or $-\frac{3}{1}$ |

The pattern can be generalized by this statement.

> **Slope of a Linear Equation**
> The slope of a linear equation expressed in the form $y = mx + b$ is $m$.

**EXAMPLE 4** Find the slope of the line whose equation is $3x + 2y = 6$.

**Solution** Rewrite the equation in the form $y = mx + b$.

$$3x + 2y = 6$$

Add $(-3x)$ to each side. $\qquad 2y = -3x + 6$

Multiply each side by $\frac{1}{2}$. $\qquad y = -\frac{3}{2}x + 3$

The slope is $-\frac{3}{2}$.

**Try This** Find the slope of the line having the given equation.

9. $2x + y = 4$ 　　　　　　10. $5x - 2y = 10$

# EXERCISES

**Objectives:** To graph a line, given its slope
To find the slope of a line

## Check Your Understanding

**Skill Check** Write the letter of the correct answer.

1. Which is the slope of the line through $A(1, 3)$ and $B(-3, 5)$?
   a. 2　　　b. $-2$
   c. $\frac{1}{2}$　　　d. $-\frac{1}{2}$

2. Which is the slope of $x = -4$?
   a. $-4$　　　b. 0
   c. Undefined　　d. 1

3. Which is the slope of $y = \frac{3}{4}x + 6$?
   a. $\frac{3}{4}$　　　b. $-\frac{3}{4}$
   c. 6　　　d. $-6$

4. Which is the slope of $3x + 3y = 8$?
   a. $-1$　　　b. 1
   c. $\frac{8}{3}$　　　d. $-\frac{8}{3}$

**True or False?** When a statement is false, explain why.

5. The slope of a line is a ratio.
6. The slope of any line parallel to the $x$-axis is always zero.
7. The slope of the $x$-axis is undefined.
8. The slope of the $y$-axis is zero.
9. The slope of a line can be positive, negative, or equal to zero.

## Practice and Apply

**Draw a line having the given slope.**

10. $\frac{1}{5}$　　　11. 3　　　12. $-4$　　　13. $\frac{3}{4}$　　　14. $-\frac{1}{2}$

**Find the slope of the line passing through the two given points. Then graph the line.**

**15.** (3, 1) and (2, 4)  **16.** (2, 0) and (0, 2)  **17.** (−1, −1) and (1, 1)
**18.** (−2, 5) and (5, 2)  **19.** (5, 1) and (−2, 1)  **20.** (0, −3) and (2, 1)
**21.** (3, 4) and (1, 2)  **22.** (−3, 5) and (2, 5)  **23.** (1, −4) and (−1, 2)

**Find the slope of the graph of each equation.**

**24.** $5x + 2y = 10$  **25.** $2x - 3y = 0$  **26.** $y = -3x + 5$
**27.** $4y = 2x - 10$  **28.** $y = \frac{3}{5}x - 9$  **29.** $-3x = 5$
**30.** $-3x + 2y = 12$  **31.** $x = y$  **32.** $4x + 4y = 7$

## Connect and Extend

**Graph each pair of equations in the same coordinate plane.**

**33.** $y = 2x + 1$ and $y = -\frac{1}{2}x + 1$  **34.** $y = \frac{2}{3}x + 3$ and $y = -\frac{3}{2}x + 3$

**35.** What is the slope of $y = 2x + 1$? Of $y = -\frac{1}{2}x + 1$?

**36.** What is the slope of $y = \frac{2}{3}x + 3$? Of $y = -\frac{3}{2}x + 3$?

**37.** Compare the graphs of the pairs of equations. Do the graphs in each pair appear to be perpendicular to each other?

**38.** Write a short summary about the relationship between the graphs of each pair of equations and the slopes of the graphs.

## Maintain Your Skills

**Solve. (Pages 114–118, 131–136)**

**39.** Find three consecutive numbers whose sum is −18.

**40.** Find three consecutive even numbers whose sum is 102.

**41.** Find three numbers between 100 and 130 that are divisible by 8.

**42.** Find three numbers between 220 and 265 that are divisible by 9.

**Write an inequality to represent the situation. Then solve the inequality. (Pages 382–385, 386–389)**

**43.** Rammel and Deven must take in more than $250 to make a profit on the T-shirts they are selling. Each T-shirt sells for $7.50. How many must they sell to make a profit?

**44.** If Sabrena gains 5 pounds, she will weigh more than 117 pounds. How much does Sabrena weigh now?

**Name the coordinates of each point. Then name the axis or quadrant where each point is located. (Pages 410–413)**

1. Point M
2. Point N
3. Point O
4. Point P
5. Point Q
6. Point R

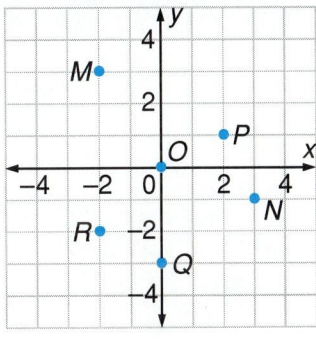

**Graph the point for each ordered pair. Label the point. (Pages 410–413)**

7. $G(-4, 6)$
8. $H(-5, -1)$
9. $S(6, -3)$
10. $J(2, 0)$

**Graph each linear equation in the coordinate plane. (Pages 414–417)**

11. $y = 4x - 5$
12. $y = 1 - 3x$
13. $y = -2x$

**Find the $x$- and $y$-intercepts. Use the $x$- and $y$-intercepts to draw each graph. (Pages 418–421)**

14. $6x - 2y = 18$
15. $10x + 5y = -20$
16. $-4x - y = -8$

**Find the slope of the line passing through the two given points. (Pages 422–426)**

17. $(3, 2)$ and $(6, 8)$
18. $(-4, 3)$ and $(-8, 6)$
19. $(0, 4)$ and $(-6, 4)$

**Find the slope of the graph of each equation. (Pages 422–426)**

20. $9y = 6x - 3$
21. $7x + 2y = 14$
22. $8 = -4x$

# Mathematical Footnote

Benjamin Banneker (1731–1806) was an astronomer, mathematician, and surveyor. As a student of astronomy, he accurately predicted a solar eclipse in 1789. Later he published a series of almanacs for which he performed all the intricate mathematical calculations himself. Banneker, an accomplished surveyor, was appointed by George Washington to be a member of the commission which surveyed and laid out the streets for the District of Columbia.

# LESSON 11-5
## Problem Solving Strategies
## Revising the Solution

- Understand
- Plan
- Solve
- Look Back

**J**oann wants to know how long it will take her to complete her paper route if she adds 10 new customers. It now takes her about 30 minutes to deliver papers to 25 customers.

Joann thought: "It now takes me a little more than a minute to deliver the paper to each customer. So I should be able to deliver papers to 35 customers in about 30 + 10, or 40 minutes."

When Joann actually added the new route to her old one, she discovered that it took her 75 minutes! She knew that she would have to revise her route so that the total distance she traveled in delivering the papers would be less.

She drew the diagram above showing the route in red for the 25 original customers and the location of the homes of the 10 new customers in blue. Joann's starting and ending point, her house, is marked with a J.

- Can you see why Joann's estimate was too low?

# EXERCISES

**Objective:** To apply strategies in problem solving

1. Find a new route for Joann that will include all 35 houses and require fewer steps. Remember that the route starts and stops at Joann's house and that you can travel only on the horizontal and vertical lines.

2. Compare your solution to Exercise 1 with that of a classmate. Is there more than one route that Joann can use?

3. This is a map of the paths in a campground. The paths are all one-way as shown by the arrows. How many different routes are there from the registration center at A to the baseball park at T?

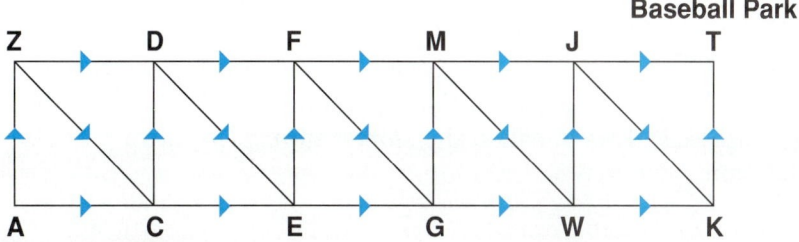

**Plan at least two ways to solve each problem. Solve both ways and compare. Which method was easier? Which method was more efficient? If necessary, revise your solution.**

4. How many raisins are there in a pound?

5. How much does the average pair of shoes weigh?

6. What is the second most commonly-used letter in the English language?

7. Can every even number after 2 be written as the sum of two prime numbers?

11-5 Problem Solving Strategies **429**

# LESSON 11-6 Graphing Equations and Inequalities

The figure at the right shows the graphs of three equations on the same coordinate plane. The slopes and $y$-intercepts of these equations are compared in the chart below.

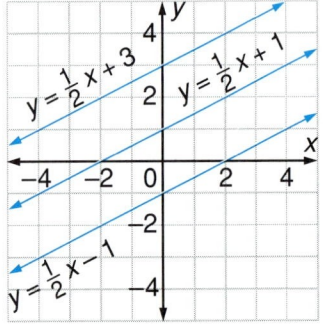

| Equation | Slope | $y$-intercept |
|---|---|---|
| $y = \frac{1}{2}x + 3$ | $\frac{1}{2}$ | 3 |
| $y = \frac{1}{2}x + 1$ | $\frac{1}{2}$ | 1 |
| $y = \frac{1}{2}x - 1$ | $\frac{1}{2}$ | -1 |

For each equation, the slope is the coefficient of $x$ and the $y$-intercept is the last term. Each equation is written in the form $y = mx + b$. This is called the *slope-intercept form* of a linear equation.

> **Slope-Intercept Form**
>
> The equation of a line having slope $m$ and $y$-intercept $b$ is $y = mx + b$.

**EXAMPLE 1** Find the slope and $y$-intercept of the graph of $5x + 3y = 6$.

**Solution** Rewrite the equation in slope-intercept form; that is, solve the equation for $y$.

| | |
|---|---|
| **Rewrite the equation.** | $5x + 3y = 6$ |
| **Add $(-5x)$ to each side.** | $-5x + 5x + 3y = -5x + 6$ |
| **Simplify.** | $3y = -5x + 6$ |
| **Divide each side by 3.** | $y = -\frac{5}{3}x + 2$ |

$y = -\frac{5}{3}x + 2$ — slope is $-\frac{5}{3}$, $y$-intercept is 2

The slope is $-\frac{5}{3}$ and the $y$-intercept is **2**.

**Try This**   Find the slope and y-intercept of the graph of each equation.
  1. $5x + 3y = 12$
  2. $2x - 3y = 12$
  3. $3x - 2y = 12$

You can use the slope and y-intercept of a line to draw its graph.

**EXAMPLE 2**   Graph $2x - 3y = 9$ using the slope and y-intercept.

**Solution**
  1. Rewrite the equation in slope-intercept form.
    $$2x - 3y = 9$$
    $$-3y = -2x + 9$$
    $$y = \tfrac{2}{3}x - 3$$
    The slope is $\tfrac{2}{3}$; the y-intercept is $-3$.

  2. Graph the equation. Since the y-intercept is $-3$, graph $P(0, -3)$. Then, starting at $P$, move 3 units to the right and 2 units up. Label this point $Q$.

  3. Draw a straight line through $P$ and $Q$.

**Try This**   Graph each equation. Use the slope and y-intercept.
  4. $y = 3x + 2$
  5. $-2y = 3x + 4$

The graph of a linear equation separates the coordinate plane into three sets of points.
  1. The points on one side of the line
  2. The points on the line
  3. The points on the other side of the line

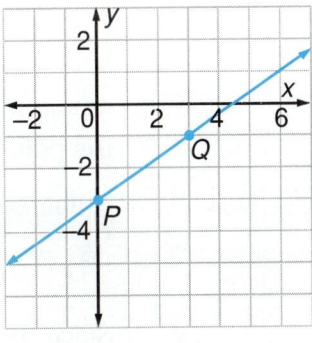
Half-plane | Half-plane

The equation of the line in the graph at the right is $y = \tfrac{3}{2}x - 3$. So the coordinates of every point in the coordinate plane make one of these equations true.

$$y = \tfrac{3}{2}x - 3 \qquad y > \tfrac{3}{2}x - 3 \qquad y < \tfrac{3}{2}x - 3$$

11-6   Graphing Equations and Inequalities

The graph of an inequality that contains the symbol > or < does not include the line that is the graph of the corresponding equation. To indicate this, use a *dashed line* for the boundary of the half-plane.

**EXAMPLE 3** Graph $y > 2x - 1$.

**Solution** Draw the graph of $y = 2x - 1$ as a **dashed** line. To identify the plane for which $y > 2x - 1$, test a point in each half-plane.

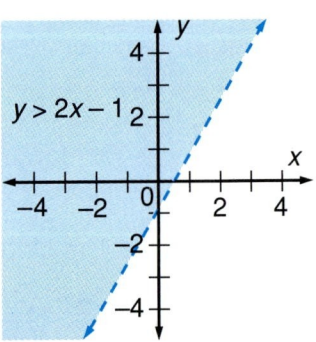

Try (0, 1).         Try (3, 1).
Is $1 > 2(0) - 1$?  Is $1 > 2(3) - 1$?
$1 > -1$?  Yes ✓   $1 > 5$?  No

Shade the half-plane that contains the point (0, 1).

**Try This** Graph each inequality.

6. $y < x + 3$         7. $y > -\frac{3}{4}x - 1$

Since the graph in Example 3 does not include the graph of $y = 2x - 1$, the graph is called an **open half-plane.**

**EXAMPLE 4** Graph $3x - 4y \leq 12$.

**Solution** Draw the graph of $3x - 4y = 12$ as a **solid** line. To identify the half-plane for which $3x - 4y < 12$, test a point in each half-plane.

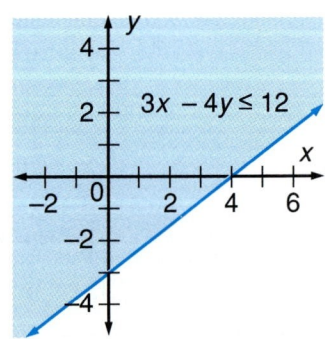

Try (0, 0).
Is $3(0) - 4(0) \leq 12$?
$0 \leq 12$?    Yes ✓

Try (3, −2).
Is $3(3) - 4(-2) \leq 12$?
$17 \leq 12$?  No

Shade the half-plane containing the point (0, 0).

**Try This** Graph each inequality: 8. $2x + 3y \geq 6$    9. $4x - 2y \leq 6$

The graph of an inequality that contains ≤ or ≥ includes the line that is the boundary of a half-plane. The graph is called a **closed half-plane.**

# EXERCISES

**Objective:** To graph equations and inequalities using the slope-intercept form

## Check Your Understanding

*Skill Check* Write the letter of the correct answer.

1. For the equation $y = 5x + 3$, which statement is true?
   a. $m = 5$; $y$-intercept $= 3$
   b. $m = -5$; $y$-intercept $= 3$
   c. $m = 5$; $y$-intercept $= -3$
   d. $m = -5$; $y$-intercept $= -3$

2. Which statement about the graph of $x - 3y \leq 6$ is **not** true?
   a. $(3, 1)$ is in the solution set.
   b. The graph is an open half-plane.
   c. $(0, 2)$ is in the solution set.
   d. The graph is a closed half-plane.

3. Which is a solution of $y < 3x + 2$?
   a. $(1, 5)$    b. $(0, 3)$    c. $\left(\frac{1}{3}, 1\right)$    d. $(-1, 0)$

4. For the equation $3x - 2y = -18$, which statement is true?
   a. $m = \frac{3}{2}$; $y$-intercept $= 9$
   b. $m = -\frac{3}{2}$; $y$-intercept $= 9$
   c. $m = \frac{3}{2}$; $y$-intercept $= -9$
   d. $m = -\frac{3}{2}$; $y$-intercept $= -9$

5. How many points in the coordinate plane are solutions for $y = 3x - 4$?

6. How many points in the coordinate plane are solutions for $y > 3x - 4$?

7. Does the solution set of $y \leq 3x - 4$ contain all the solutions of $y = 3x - 4$?

## Practice and Apply

Find the slope and *y*-intercept of the graph of each equation.

8. $5x + y = -3$
9. $y - 4x = 1$
10. $6 + 2y = 8x$
11. $2y + 8 = 0$
12. $9 - 3y = 3x$
13. $4x - 2y = 12$

Graph each equation, using the slope and *y*-intercept.

14. $-2x + y = 9$
15. $4y + 8x = 4$
16. $3x - 2y = -6$
17. $3x - y = -7$
18. $6y + 3x = 6$
19. $x + y = 0$
20. $2y - 8x = 6$
21. $y = 7$
22. $3x + 3y - 6 = 0$

**Graph each inequality.**

23. $y \geq -2x + 3$
24. $5y \leq 10x - 5$
25. $x > 0$
26. $y \leq 5$
27. $2y - x \geq -10$
28. $3x - y > -4$
29. $y - x \geq 0$
30. $3y + 6 \leq x$
31. $-2(x + y) \geq 6$

## Connect and Extend

**Describe the graph of each inequality. Use one of the descriptions below.**

**a.** A straight line   **b.** An open half-plane   **c.** A closed half-plane

32. $Ax + By = C$
33. $Ax + By \leq C$
34. $Ax + By > C$
35. $Ax = C$
36. $Ax \leq C$
37. $By > C$
38. Draw the graph of $x \geq 1$ *or* $y \leq 2$.
39. Draw the graph of $x \geq 1$ *and* $y \leq 2$.

## Maintain Your Skills

**Solve by the guess-and-check method. (Pages 85–88)**

40. $93 + r = 64$
41. $-81 - s = 106$
42. $\dfrac{e}{-7} = 43$
43. $24c = 552$

**Evaluate each expression. (Pages 150–153)**

44. $(3t)^2$, for $t = 2$
45. $(s - t)^3$, for $s = 5$, $t = 3$
46. $-56a^2b^2$, for $a = -1$, $b = 3$
47. $6^2 - d^2$, for $d = -2$

**Write an inequality to represent each word expression. (Pages 393–397)**

48. Half of Cory's age is less than 18 years.
49. The sum of two consecutive positive integers is greater than 15.
50. Two less than 3 times the amount Sara earned is at most $60.
51. The sum of a number and 5 times the number is at least 72.

## ★ Math Team Problem

52. Use the Holt *Pre-Algebra* disk (see page 398) and the program, *Slope and Y-Intercept*, to explore the graphs of straight lines. Type **s** to select the program. Write a summary of the results.

53. Kijo has two jars of equal size. One jar is full of nickels. The other is half full of dimes. Which amount of money will be greater, the amount contained in the jar of nickels or the amount contained in the jar of dimes? Explain.

# LESSON 11-7
## Problem Solving Using Two Variables

**M**any equations can be solved by using two variables.

 Donya and Cleon rode a total of 15 miles on a bike hike. Donya rode 3 miles farther than Cleon. How far did each biker ride?

**Solution**

**Think:** There are two conditions.
The total of the two rides is 15 miles.
Donya rode 3 miles farther than Cleon.

**Strategy:** Write an equation for each condition.
Use guess and check to solve the equation.

Let $x$ represent the number of miles Donya rode.
Let $y$ represent the number of miles Cleon rode.

Write an equation for the total distance.  $\quad x + y = 15$
Write an equation for the difference
between Donya's and Cleon's distances.  $\quad x - y = 3$

**Think:** Which two numbers have a sum of 15 and a difference of 3? Try 9 and 6.

**Check:** Does $9 + 6 = 15$? Yes ✓
Does $9 - 6 = 3$? Yes ✓

Donya rode **9 miles** and Cleon rode **6 miles.**

**Try This**

**Solve by using two variables.**

1. Marty and Tom together have saved a total of $177. Marty has saved twice as much as Tom. How much has each person saved?

2. While on vacation, Charlie and Gail took turns driving. The distance that Charlie drove was 115 miles less than the distance that Gail drove. Together, they drove 235 miles. How far did each drive?

Another strategy you can use to solve the problem in Example 1 is to graph the equations in the same coordinate plane. If the graphs of the lines meet at a point (intersect), the coordinates of the point is a solution of both equations.

Find the $x$- and $y$-intercepts of the equations and draw the graphs.

$x + y = 15$     $x$-intercept: 15
                 $y$-intercept: 15

$x - y = 3$      $x$-intercept: 3
                 $y$-intercept: $-3$

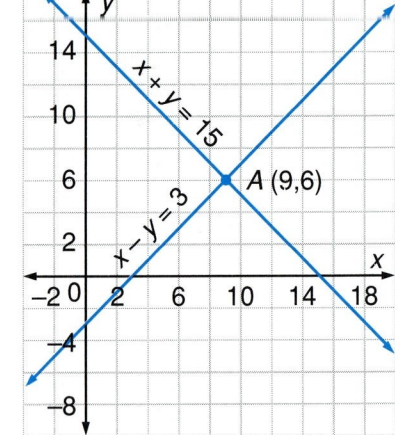

The graph shows that the point of intersection is (9, 6). You know from the Check in Example 1 that (9, 6) is a solution of both equations.

Two equations that use the same two variables form a **system** of equations. The system may be written in these ways.

$$x = 3 \text{ and } y = 2$$

or

$$x = 3$$
$$y = 2$$

For two lines that intersect at a point, the coordinates of that point are the **solution** of the system.

**EXAMPLE 2** Solve by graphing: $2x + y = 9$ and $x - 2y = 2$

**Solution** Rewrite each equation in slope-intercept form. Then graph both equations in the same coordinate plane.

$2x + y = 9 \longrightarrow y = -2x + 9$
Slope: $-2$
y-intercept: $9$

$x - 2y = 2 \longrightarrow y = \frac{1}{2}x - 1$
Slope: $\frac{1}{2}$   y-intercept: $-1$
The solution is **(4, 1)**.

Check in the original equations.
Does $2(4) + 1 = 9$?   Yes ✓
Does $4 - 2(1) = 2$?   Yes ✓

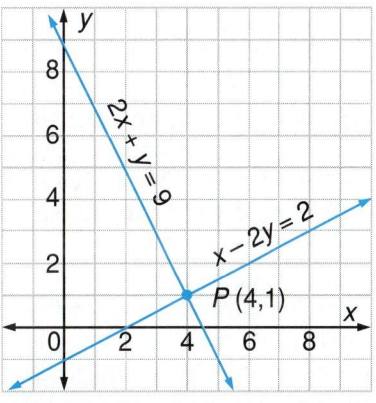

**Try This** **Solve by graphing:** 3. $2x - y = 3$ and $4x + y = 9$

# EXERCISES

**Objective:** To solve a system of linear equations in two variables by graphing

## Check Your Understanding

**Skill Check**   Write the letter of the correct answer.

1. Which is the solution of $x = 1$ and $y = 4$?
   **a.** $(-1, 4)$   **b.** $(4, -1)$
   **c.** $(1, 4)$   **d.** $(-4, 1)$

2. Which is the solution of $x = 4$ and $2y + x = 0$?
   **a.** $(4, 2)$   **b.** $(4, 0)$
   **c.** $(4, -2)$   **d.** $(4, -4)$

3. Which is the solution of $x + y = 4$ and $x - y = -2$?
   **a.** $(1, -3)$   **b.** $(1, 3)$
   **c.** $(-3, -1)$   **d.** $(3, 1)$

4. Which is the solution of $2x - y = 1$ and $-x + 2y = 4$?
   **a.** $(3, 2)$   **b.** $(-3, 2)$
   **c.** $(-2, 3)$   **d.** $(2, 3)$

5. Without graphing, find the solution of the system $x = 4$ and $y = 6$.
6. What is the solution, if any, of the system $x = 1$ and $x = -5$?
7. What is the solution, if any, of the system $y = -8$ and $y = 3$?
8. What is the solution, if any, of the system $x = 0$ and $y = 0$?

11-7   Problem Solving: Using Two Variables

# Practice and Apply

**Read the solution of each system of equations from the graph. Check whether your answer is correct.**

9. $x + y = 10$
   $x - y = 4$

10. $2x - y = 3$
    $3x + y = 7$

11. $x + 2y = 4$
    $2x - y = 8$

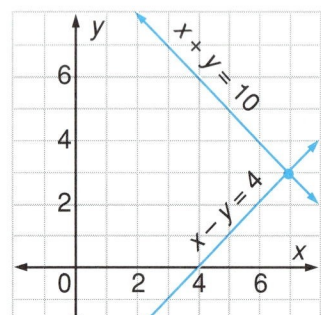 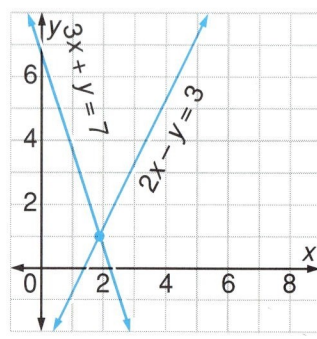 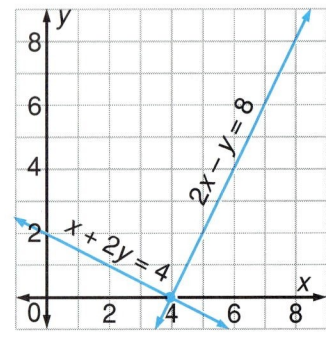

**Solve each system by graphing. Check your solutions.**

12. $y = 6 - x$
    $y = x - 4$

13. $y = 2x - 3$
    $y = x - 2$

14. $y = 2x + 3$
    $y = x - 1$

15. $x - y = -2$
    $x + y = 4$

16. $x - y = 8$
    $x + y = 4$

17. $y = 2x - 3$
    $y = -x$

18. $x + 4y = -5$
    $3x + y = -4$

19. $x + y = 2$
    $2y - x = 10$

20. $2x - y = 10$
    $x + 2y = -5$

21. $-x + 2y = 4$
    $x = -2y$

22. $x = y + 4$
    $2x - 5y = 2$

23. $3x + y = 6$
    $x - 2y = 2$

**Write two equations in two variables to solve each problem. Then solve by guess-and-check or by graphing.**

24. Carlos saved $30 less than his brother for a vacation. Together they saved $90. How much did each person save?

25. Together, Wonda and Jackson sold 68 tickets for a school play. Jackson sold 26 more tickets than Wonda. How many tickets did Jackson sell?

26. The length of a poster is 6 inches more than the width. The perimeter is 56 inches. Find the length and the width.

27. The difference between two positive numbers is 10. One number is 3 times the other. What are the numbers?

## Connect and Extend

A supermarket offers two plans for renting cleaning equipment. Under Plan A, the charge is $5 per hour. Under Plan B, there is a base fee of $6 and an additional charge of $2 per hour.

| Rent-a-Cleaner | |
|---|---|
| **Plan A** | **Plan B** |
| $5 / hr | Fee; $6 + $2 / hr |

28. Write an equation in two variables to represent each plan. Let $y$ = the total rental cost, and let $x$ = the number of hours the equipment was rented.
29. Draw a graph of each equation in the same coordinate plane. The graph should show rental costs from 0 to 6 hours.
30. How many hours will it take for both costs to be the same?
31. For how long will it cost less to use Plan A?
32. Which graph has the steeper slope? What does this mean in terms of rental costs?

## Maintain Your Skills

**Estimate each sum or difference. Then find the exact answer. (Pages 282–286)**

33. $-29.06 + 11.281$
34. $-5.37 - 18.032$
35. $42 - 27.009$

**Solve. (Pages 330–333, 338–341)**

36. A sweater with a list price of $32 is on sale at a 20% discount. Find the sale price.
37. Last year, a certain tape player cost $30. This year, the same tape player costs $42. Find the percent increase.

**Graph the point for each ordered pair. Label the point. (Pages 410–413)**

38. $Q(-1, 4)$
39. $R(-3, -2)$
40. $S(6, 0)$
41. $T(4, -1)$

## Mathematical Footnote

When is a pattern not a pattern?

These are all prime numbers. That is, they have no factors other than one and themselves.

$$7 \quad 37 \quad 337 \quad 3{,}337 \quad 33{,}337 \quad 333{,}337$$

Would you conjecture that 3,333,337 is also a prime number? Try dividing it by 7 and by 31.

A pattern is a pattern only if you can *prove* that it must continue to repeat itself in the same way.

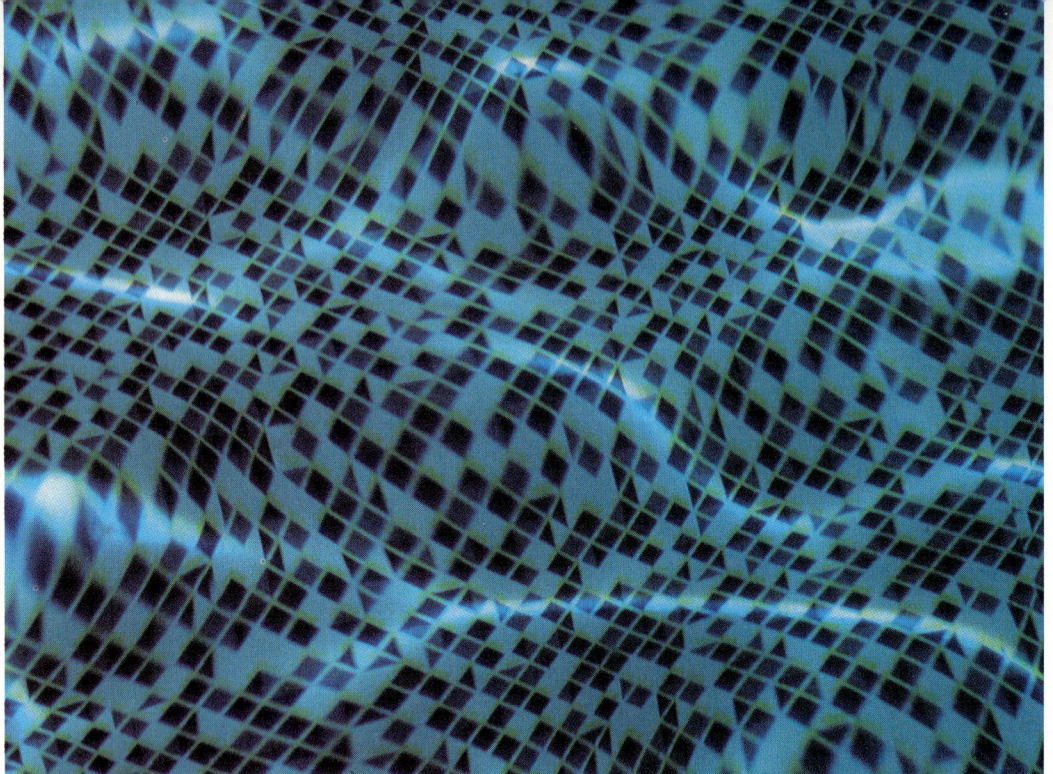

# LESSON 11-8
## Translations

**G**raphic artists use computers to create designs. Designs for floor tiles, for example, may repeat one basic pattern many times. The overall design is then made by sliding, or **translating**, the basic pattern in various directions.

The graph at the right shows the point $A(2, 3)$. By moving the $x$-coordinate one unit to the right and the $y$-coordinate 4 units down, the point $A$ can be moved, or translated, to $A'$ (read as $A$ prime), the **image point** of point $A$.

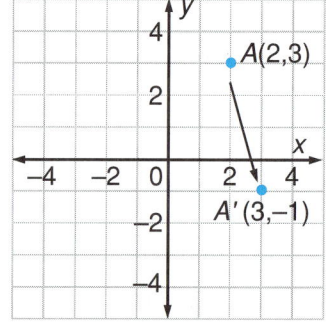

A translation of a point to the right or upward is considered positive. A translation to the left or downward is considered negative. So the translation of point $A(2, 3)$ one unit to the right and 4 units down can be shown like this.

$$A(2, 3) \rightarrow A'(2 + 1, 3 + (-4))$$
$$A(2, 3) \rightarrow A'(2 + 1, 3 - 4)$$
$$A(2, 3) \rightarrow A'(3, -1)$$

In words, you say that the image of A(2, 3) under the translation
$$(x, y) \rightarrow (x + 1, y - 4)$$
is A'(3, −1).

**EXAMPLE 1** Given the translation $(x, y) \rightarrow (x - 3, y + 2)$ and points S(1, −4) and T(−1, 1), find and graph the image points S' and T'.

**Solution** **Think:** The translation $(x, y) \rightarrow (x - 3, y + 2)$ moves each x-coordinate 3 units to the left and each y-coordinate 2 units up.

Find S', the image point of S.
S(1, −4) → S'(1 − 3, −4 + 2)
S(1, −4) → **S'(−2, −2)**

Find T', the image point of T.
T(−1, 1) → T'(−1 − 3, 1 + 2)
T(−1, 1) → **T'(−4, 3)**

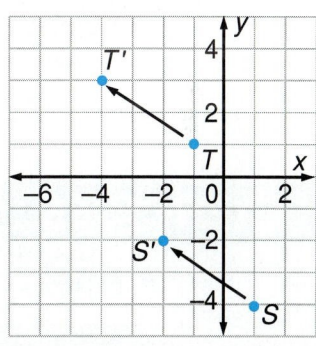

**Try This**
1. Given the translation $(x, y) \rightarrow (x + 2.5, y + 3.5)$ and the points R(−1.5, 2) and Q(3, −2.5), find and graph the image points, R' and Q'.

You can also use translations to slide a geometric figure, such as a triangle, from one position to another.

**EXAMPLE 2** Graph the points A(2, 3), B(5, 6) and C(7, 1). Connect the points to form a triangle. Find and graph the image of triangle ABC under the translation $(x, y) \rightarrow (x - 5, y - 3)$.

**Solution** **Think:** The translation moves each x-coordinate 5 units to the left and each y-coordinate 3 units down.

| Original Point | Image Point |
|---|---|
| A(2, 3) | A'(−3, 0) |
| B(5, 6) | B'(0, 3) |
| C(7, 1) | C'(2, −2) |

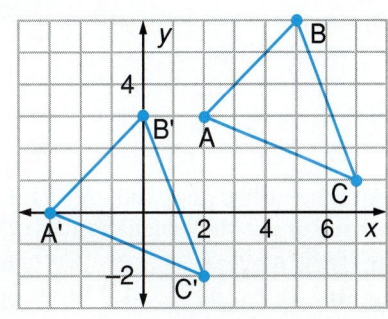

11-8 Translations

**Try This**

2. Graph the points $R(1, 4)$, $S(1, 6)$ $T(3, 4)$, and $V(3, 6)$. Connect the points to form a square. Find and graph the image of square $RSTV$ under this translation:

$$(x, y) \rightarrow (x + 2, y - 8).$$

Two figures are **congruent** if they have the same size and shape. Under a translation, the original figure and its image are always congruent. In Example 2, triangle $ABC$ is congruent to triangle $A'B'C'$.

- In Example 2, which segment is congruent to side $BC$? to side $AB$? to side $AC$?

# EXERCISES

**Objectives:** To identify image points under a translation
To find translations of segments and triangles

## Check Your Understanding

**Skill Check** Write the letter of the correct answer.

1. Which is the image of $A(3, 4)$ under the translation $(x, y) \rightarrow (x + 1, y - 2)$?
    a. $A'(4, 6)$
    b. $A'(3, 2)$
    c. $A'(3, 6)$
    d. $A'(4, 2)$

2. Which describes how the $x$-coordinate moves under the translation $(x, y) \rightarrow (x - 1, y - 3)$?
    a. One unit to the left
    b. One unit to the right
    c. One unit up
    d. One unit down

3. Which statement is **not** true?
    a. Congruent figures have the same shape.
    b. Under a translation, the image of a figure is congruent to the original figure.
    c. Congruent figures do not have the same size.
    d. Under the translation $(x, y) \rightarrow (x + 1, y)$, triangle $ABC$ is congruent to its image.

4. A segment has endpoints $M(-1, 3)$ and $P(2, 4)$. What are the endpoints under the translation $(x, y) \rightarrow (x + 2, y - 1)$?
    a. $M'(2, 5)$; $P'(4, 3)$
    b. $M'(1, 2)$; $P'(4, 3)$
    c. $M'(-3, 2)$; $P'(4, -5)$
    d. $M'(-3, 2)$; $P'(4, 3)$

442  CHAPTER 11

## Practice and Apply

**Describe each translation in words for a given point $P(1, -2)$.**

**5.** $(x, y) \rightarrow (x - 4, y + 5)$  **6.** $(x, y) \rightarrow (x + 1, y - 3)$

**Graph the image of the given points or figures. Use the given points.**

$P(2, 4)$  $Q(-2, -5)$  $R(-1, -3)$  $S(3, 0)$

**7.** $P$, under the translation $(x, y) \rightarrow (x + 1, y - 2)$
**8.** $Q$, under the translation $(x, y) \rightarrow (x, y + 5)$
**9.** $R$, under the translation $(x, y) \rightarrow (x - 1, y - 1)$
**10.** $S$, under the translation $(x, y) \rightarrow (x - 3, y + 3)$
**11.** A segment with endpoints $R$ and $S$, under the translation $(x, y) \rightarrow (x - 4, y + 3)$
**12.** A triangle with vertices at $P$, $Q$, and $S$, under the translation $(x, y) \rightarrow (x - 4, y + 2)$

## Connect and Extend

**13.** In the figure at the right, the letter H was translated from Quadrant I to its new position in Quadrants II and III. Complete the translation rule: $(x, y) \rightarrow (?, ?)$.

**14.** Under a translation, point $A(3, 1)$ was moved to $A'(4, 3)$. Complete the translation rule: $(x, y) \rightarrow (?, ?)$.

**15.** Describe the image point of $R(4, 5)$ under the change $(x, y) \rightarrow (y, x)$.

**16.** Describe the image triangle when the triangle formed by joining the points $A(1, 1)$, $B(2, 5)$, and $C(3, 3)$ is moved under the change $(x, y) \rightarrow (-x, -y)$. Are the triangles congruent?

## Maintain Your Skills

**17.** Is $x(x + 4) + 5(x + 4)$ equivalent to $(x + 5)(x + 4)$? Explain. **(Pages 59–63)**

**Evaluate each expression. (Pages 150–153)**

**18.** $2x^3z^2$, for $x = -2$ and $z = -3$   **19.** $12 + 9a^3b$, for $a = -1$ and $b = -3$

**Solve. (Pages 221–225, 334–337)**

**20.** A train travels 250 miles in 5 hours. How far will it travel in 7 hours?

**21.** There are 4 boys for every 5 girls attending Rosetree High School. If there are 265 girls, how many boys attend the school?

**22.** 12 is what percent of 10?

**23.** $12\frac{1}{2}\%$ of what number is 4?

# SUMMARY

## KEY TERMS

Axes (p. 410)
Closed half plane (p. 433)
Congruent (p. 442)
Coordinate graph (p. 410)
Image point (p. 440)
Open half plane (p. 432)
Origin (p. 410)
Quadrants (p. 411)
Slope (p. 423)
Slope-intercept form
　of a linear equation (p. 430)
Standard form of a
　linear equation (p. 418)
System of equations (p. 436)
Translation (p. 440)
x-axis (p. 410)
x-coordinate (p. 410)
x-intercept (p. 418)
y-axis (p. 410)
y-coordinate (p. 410)
y-intercept (p. 418)

## KEY IDEAS

**A.** When you graph a point with coordinates $(x, y)$, the x-coordinate gives the distance of the point to the left or right of the y-axis. The y-coordinate gives the distance of the point above or below the x-axis.

**B.** Slope of a line: $\frac{\text{change in } y\text{-coordinates}}{\text{change in } x\text{-coordinates}}$

**C.** The slope of any line parallel to the x-axis is 0.

**D.** The slope of any line parallel to the y-axis is undefined.

**E.** The graph of an equation of the form $y = mx + b$ is a straight line with slope $m$ and y-intercept $b$.

**F.** The graph of a linear equation separates the coordinate plane into three sets of points, the points in the two half-planes and the points on the line

**G.** If the graphs of two lines in the same coordinate plane intersect, the coordinates of the point of intersection is a solution of both equations.

# REVIEW

**Name the coordinates of each point. (Pages 410–413)**
1. Point $E$
2. Point $F$
3. Point $G$
4. Point $H$
5. Point $T$
6. Point $J$

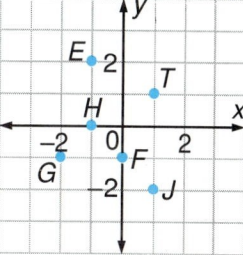

**Name the quadrant or axis where each point is located. (Pages 410–413)**
7. $K(0, -2)$
8. $M(4, -3)$
9. $N(-5, -6)$

**Graph each linear equation in the coordinate plane. (Pages 414–417)**
10. $y = 3x - 2$
11. $y = x + 5$
12. $y = 4 - x$

**Find the $x$- and $y$-intercepts. Use the $x$- and $y$-intercepts to draw each graph. (Pages 418–421)**
13. $x - 3y = -9$
14. $2x + 4y = -12$
15. $-4x - 4y = 8$

**Find the slope of the line passing through the two given points. Then graph the line. (Pages 422–426)**
16. $(5, 7)$ and $(4, 0)$
17. $(3, 5)$ and $(1, -1)$
18. $(2, 3)$ and $(3, 3)$

**Find the slope of the graph of each equation. (Pages 422–426)**
19. $x - 3y = 9$
20. $y + 2 = 0$
21. $x = 1$

**Graph each equation using the slope and $y$-intercept. (Pages 422–426)**
22. $x - 2y = -12$
23. $6x + 4y = 8$
24. $2y - 3x = 24$

**Graph each inequality. (Pages 430–434)**
25. $3x + 6y \geq 6$
26. $2x + y < 1$
27. $-y \leq -4$

**Solve each system by graphing. Check your solutions. (Pages 435–439)**
28. $x - y = 1$
    $x + y = 3$
29. $-y + x = 2$
    $y = 3x$
30. $2y = x - 8$
    $2y - 3x = -16$

**Write two equations in two variables to solve the problem. Then solve. (Pages 435–439)**
31. Shad saved $35 more than Peggy for a skiing vacation. Together they saved $165. How much did each person save?

**Graph the image of the given point under the given translation. (Pages 440–443)**
32. $P(-4, 7)$ under the translation $(x, y) \rightarrow (x - 2, y + 3)$
33. $Q(0, 5)$ under the translation $(x, y) \rightarrow (x + 5, y - 5)$

# CHAPTER TEST

Name the coordinates of each point. Then name the quadrant or axis where each point is located.

1. Point A
2. Point B
3. Point C
4. Point D
5. Point E
6. Point F

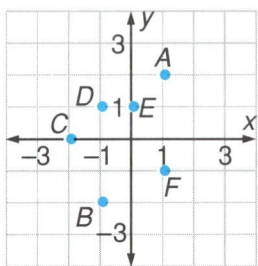

7. In which quadrants is the $x$-coordinate always negative?
8. In which quadrants is the $y$-coordinate always positive?
9. In which quadrants do the $x$- and $y$-coordinates have opposite signs?

**Graph each linear equation in the coordinate plane.**

10. $y = -4x$
11. $y = 2x + 3$
12. $y = 2 - 2x$

**Find the $x$- and $y$-intercepts.**

13. $y - 3x = 1$
14. $5y - 10x = 20$
15. $6x + 12y = 24$

**Find the slope of the line passing through the two given points. Then graph the line.**

16. $(\ 4,\ 5)$ and $(\ 4, 1)$
17. $(5,\ 6)$ and $(3,\ 2)$
18. $(0,\ 4)$ and $(\ 8, 0)$

**Find the slope of the graph of each equation.**

19. $2y = 6x - 8$
20. $-3x + 2y = 9$
21. $7x + 4y = 13$

**Graph each equation using the slope and $y$-intercept.**

22. $-2y - 5x = -10$
23. $2x - 3y = 24$
24. $x - y = 0$

**Graph each inequality.**

25. $x - y \leq -5$
26. $x \geq 2$
27. $-4(x - 1) < y$

28. Solve by graphing. Check the solution.
$$x = 3y$$
$$3y - 2x = 6$$

29. A segment has endpoints $A(-2, 5)$ and $B(3, 4)$. What are the end-points under the translation $(x, y) \rightarrow (x + 2, y - 3)$?

**Write two equations in two variables to represent the situation. Then solve.**

30. Eighty-five more student tickets than adult tickets were sold for the school play. The total number sold was 343. How many of each kind were sold?

**Solve and check each equation. (Pages 101–104)**

**1.** $\frac{c}{-4} - 11 = 12$   **2.** $9.4 = 2.3x - 18.2$   **3.** $-\frac{5}{8}m + 14 = -26$

**Write the greatest prime factor of each number. (Pages 143–146)**

**4.** 150   **5.** 204   **6.** 315   **7.** 261   **8.** 492

**Solve. (Pages 171–174, 221–225, 282–286, 330–333)**

**9.** At Chio's Auto Dealership, $\frac{3}{4}$ of the cars for sale have 4 doors. One third of the 4-door cars are red. What fraction of the cars are red with 4 doors?

**10.** A survey found that 6 out of 10 adults recycle aluminum cans for cash. In a town of 3,245 adults, how many can be expected to recycle aluminum cans?

**11.** Vicki plans to change the oil in her car. She bought an oil filter for $3.99, an air filter for $5.99, and 6 quarts of oil at $0.89 a quart. How much money did she spend?

**12.** Sam wants to buy a new coat. At one store, the list price of the coat is $95 with a discount of 10%. At another store the list price is $100 with a discount of 15%. Which is the better buy?

**13.** Construct a box and whisker plot based on the data in the stem and leaf plot below.(**Pages 370–372**)

```
5 | 1
4 | 2, 3
3 | 0, 3, 4, 7, 9
2 | 3, 5, 6
1 | 2, 8
0 | 3
```

**Write a description of each graph. (Pages 377–381)**

**14.** (number line with open circle at 0 and closed circle at 4)

**15.** (number line with open circle at 0)

**Solve each inequality. (Pages 393–397)**

**16.** $14 < 8 - 2y$   **17.** $5x + 3 \geq -7$   **18.** $6 - y < 4$

Cumulative Review: Chapters 1–11

# UNIT 5

### PROJECT 1
The Parks and Recreation Department has recently acquired five acres of land adjacent to an existing park. On it, they want to include as many areas as possible for field and court sports and other recreational activities that interest the community's adults and children. Work as a team to determine recreational areas that can be included, and determine how they could be positioned on the property.

# Using Real Numbers

## PROJECT 2
Once a teenager is earning money and saving to buy things, how do you think the teenager would like to spend the money? Some of the things you probably thought of were clothes, travel, a car, and education. This project will help you take a look at what it costs to buy and operate a car.

## PROJECT 3
Research has shown that to maintain an acceptable health standard, a person needs to follow a balanced nutritional program combined with consistent exercise. Your assignment will be to make a list of healthy foods to eat, to determine your calorie needs, and to identify activities and exercises that will help you keep fit.

## CONTENTS

### 12 Square Roots and Right Triangles
Problem Solving Exploration:
   Square Roots
Using Square Roots
The Pythagorean Theorem
Problem Solving Strategies:
   Formulating Questions
Similar Triangles
The Tangent Ratio

### 13 Polynomials
Adding Polynomials
Subtracting Polynomials
Problem Solving Exploration:
   Using a Multiplication Model
Multiplying Binomials
Problem Solving Strategies:
   Too Much/Too Little Data
Using the Distributive Property
Special Products
Common Factors
Factoring a Trinomial

### LESSON 12-1

# Problem Solving Exploration
# Square Roots

**D**ave owns a remodeling business. A customer wants him to build a square patio having an area of 10 square meters.

How long should each side be?

Recall that the area of a square with side of length $s$ is $s \cdot s$, or $s^2$. That is, the area of a square with a side 3 meters long, is $3 \cdot 3$, or 9 square meters. If the length of a side is 4 meters, then the area of the square is $4 \cdot 4$, or 16 square meters.

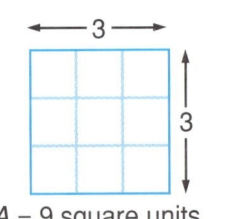

A = 9 square units

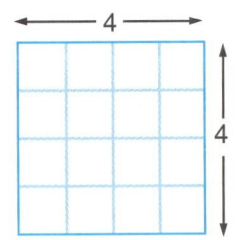

A = 16 square units

So the length of a side of the square having an area of 10 square meters must be a little greater than 3 meters. Dave decides that the length of the patio must be between 3 meters and 4 meters.

**450** CHAPTER 12

Using a calculator, Dave determines that
$$(3.1)^2 = 9.61 \quad \text{and} \quad (3.2)^2 = 10.24.$$

**A.** Use a calculator to estimate the length of a square having an area of 10 square meters. Copy and complete the table.

| | Side | $A = s^2$ | | | Side | $A = s^2$ |
|---|---|---|---|---|---|---|
| 1. | 3.12 | $(3.12)^2 = \underline{?}$ | | 2. | 3.14 | $(3.14)^2 = \underline{?}$ |
| 3. | 3.16 | $(3.16)^2 = \underline{?}$ | | 4. | 3.18 | $(3.18)^2 = \underline{?}$ |

5. What is a good estimate for the length of the patio?
6. Continue to use a calculator to estimate the length to the nearest thousandth (three decimal places).

If you continue to approximate the length to eight decimal places, you will find that it is about 3.16227766. So $(3.16227766)^2 \approx 10$. You can carry the estimate to as many decimal places as you wish, but the decimal will not show a pattern.

To express the exact value of the decimal, you use the square root symbol, $\sqrt{\phantom{x}}$. That is, the length of a side of a square with area 10 is $\sqrt{10}$ (read: square root of 10). This means that

$$(\sqrt{10})^2 = 10 \text{ and } \sqrt{10} \approx 3.162.$$

- Why is $\sqrt{10}$ an irrational number?

You can use the square root key on a calculator to approximate $\sqrt{10}$.

$10\ \boxed{\sqrt{\phantom{x}}}$

$A = (\sqrt{10})^2 = 10$

**B.** Copy and complete this table. First, estimate the square root to one decimal place. Then use a calculator to check whether the tenths place in the number is the same as your estimate.

| | Square Root | Estimate | | | Square Root | Estimate |
|---|---|---|---|---|---|---|
| 7. | $\sqrt{3}$ | ? | | 8. | $\sqrt{4}$ | ? |
| 9. | $\sqrt{8}$ | ? | | 10. | $\sqrt{12}$ | ? |
| 11. | $\sqrt{24}$ | ? | | 12. | $\sqrt{49}$ | ? |
| 13. | $\sqrt{79}$ | ? | | 14. | $\sqrt{95}$ | ? |
| 15. | $\sqrt{100}$ | ? | | 16. | $\sqrt{110}$ | ? |
| 17. | $\sqrt{125}$ | ? | | 18. | $\sqrt{196}$ | ? |

19. What strategies did you use to estimate the square roots?
20. Between which two whole numbers is $\sqrt{29}$? $\sqrt{110}$?

# LESSON 12-2 Using Square Roots

The square with area 16 has a side length of 4. Because $4^2 = 16$, we call 4 a **square root** of 16. Notice that $(-4)^2 = 16$. So $-4$ is also a square root of 16.

We use the **radical symbol**, $\sqrt{\phantom{x}}$, to indicate the positive square root. That is, $\sqrt{16} = 4$. Using a calculator, you can estimate other square roots. For example,

$$\sqrt{12} \approx 3.464.$$

Note that $\sqrt{12}$ is called a **radical**.

The numbers 1, 4, 9, 16, 25, 36, 49, 64, 81, 100, and so on are called **perfect squares** because their positive roots are the whole numbers 1, 2, 3, 4, 5, 6, 7, 8, 9, 10, and so on.

**EXAMPLE 1** Between what two whole numbers is $\sqrt{51}$?

**Solution** Since 51 lies between the perfect squares 49 and 64, $\sqrt{51}$ lies between their square roots, 7 and 8.

**Check** Using a calculator, $\sqrt{51} \approx 7.14.$

**Try This** Between which two whole numbers is each square root?
1. $\sqrt{14}$  2. $\sqrt{76}$  3. $\sqrt{166}$  4. $\sqrt{58}$

Sometimes you can simplify radicals. For example,
$$\sqrt{16} + \sqrt{9} = 4 + 3 = 7$$
$$\sqrt{16 + 9} = \sqrt{25} = 5$$

Notice that $\sqrt{16} + \sqrt{9} = 7$ and that $\sqrt{16 + 9} = 5.$ So $\sqrt{16} + \sqrt{9} \neq \sqrt{16 + 9}.$

How can $\sqrt{2} + \sqrt{5}$ be simplified?

To add expressions involving square roots, the expressions must have the same number under the radical symbol.

Since 2 and 5 are not perfect squares, and they have no factors that are perfect squares, $\sqrt{2} + \sqrt{5}$ cannot be simplified. To check this, use a calculator to show that $\sqrt{2} + \sqrt{5} \neq \sqrt{7}.$

**EXAMPLE 2** Simplify, if possible.
a. $\sqrt{2 \cdot 32}$  b. $\sqrt{75 + 25}$  c. $\sqrt{4} + \sqrt{9}$  d. $\sqrt{3} + \sqrt{7}$

**Solutions**
a. $\sqrt{2 \cdot 32} = \sqrt{64} =$ **8**
b. $\sqrt{75 + 25} = \sqrt{100} =$ **10**

c. **Think:** 4 and 9 are perfect squares.
So $\sqrt{4} + \sqrt{9} = 2 + 3 =$ **5.**

d. The numbers 3 and 7 are not perfect squares, and do not have any factors which are perfect squares.
So $\sqrt{3} + \sqrt{7}$ is **already simplified.**

**Try This**  **Simplify.**
5. $\sqrt{29 + 7}$  6. $\sqrt{24 \cdot 6}$  7. $\sqrt{25} + \sqrt{100}$

The product of two radicals can sometimes be simplified. Since $\sqrt{4} \cdot \sqrt{9} = 2 \cdot 3 = 6$, and $\sqrt{4 \cdot 9} = \sqrt{36} = 6$,
$$\sqrt{4} \cdot \sqrt{9} = \sqrt{4 \cdot 9}.$$

This suggests a rule for multiplying radicals.

> **Rule for Multiplying Radicals**
> If $a \geq 0$ and $b \geq 0$, then
> $\sqrt{a} \cdot \sqrt{b} = \sqrt{ab}$  and  $\sqrt{ab} = \sqrt{a} \cdot \sqrt{b}$.

You can use the Rule for Multiplying Radicals to simplify radicals. You can simplify a radical if the number under the radical symbol is a perfect square or it is has any factors that are perfect squares.

**EXAMPLE 3** Simplify: a. $\sqrt{8}$  b. $\sqrt{75}$

**Solutions**
a. **Think:** $8 = 4 \cdot 2$ and 4 is a perfect square.

| | |
|---|---|
| Write $\sqrt{8}$ as a product. | $\sqrt{8} = \sqrt{4 \cdot 2}$ |
| Apply the rule for multiplying radicals. | $= \sqrt{4} \cdot \sqrt{2}$ |
| Simplify. | $= 2 \cdot \sqrt{2}$, or $2\sqrt{2}$ |

**Check:** Use a calculator.
$\sqrt{8} \approx 2.828$ and $2\sqrt{2} \approx 2(1.4142) = 2.8284$

**b. Think:** 75 = 25 · 3 and 25 is a perfect square.

Write $\sqrt{75}$ as a product. $\quad\quad \sqrt{75} = \sqrt{25 \cdot 3}$
Apply the rule for multiplying radicals. $\quad = \sqrt{25} \cdot \sqrt{3}$
Simplify. $\quad = 5\sqrt{3}$

**Check:** Use a calculator.
$\sqrt{75} \approx 8.66$ and $5\sqrt{3} \approx 5(1.732) = 8.66$

**Try This**

**Simplify.**

**8.** $\sqrt{12}$        **9.** $\sqrt{48}$        **10.** $\sqrt{120}$

Sometimes the denominator of a fraction or rational expression contains a radical. To simplify such an expression, multiply both the numerator and the denominator by the radical. This will result in having a whole number in the denominator.

For example, to simplify $\frac{2}{\sqrt{3}}$, multiply by $\frac{\sqrt{3}}{\sqrt{3}}$.

$$\frac{2}{\sqrt{3}} = \frac{2}{\sqrt{3}} \cdot \frac{\sqrt{3}}{\sqrt{3}}$$
$$= \frac{2\sqrt{3}}{\sqrt{3} \cdot \sqrt{3}}$$
$$= \frac{2\sqrt{3}}{\sqrt{9}} = \frac{2\sqrt{3}}{3}$$

This is called **rationalizing the denominator.**

# EXERCISES

**Objectives:** To calculate square roots with a calculator
To simplify square roots and radical expressions

## Check Your Understanding

**Skill Check** Write the letter of the correct answer.

**1.** Between which two whole numbers is $\sqrt{11}$?
    **a.** 1 and 2     **b.** 2 and 3
    **c.** 3 and 4     **d.** 5 and 6

**2.** Which of these is **not** a perfect square?
    **a.** $\sqrt{100}$     **b.** $\sqrt{49}$
    **c.** $\sqrt{400}$     **d.** $\sqrt{99}$

**3.** Which **cannot** be simplified?
    **a.** $\sqrt{2 \cdot 8}$     **b.** $\sqrt{3} \cdot \sqrt{12}$
    **c.** $\sqrt{3} + \sqrt{2}$     **d.** $\sqrt{4} + \sqrt{16}$

**4.** Which has the same value as $\frac{10}{\sqrt{5}}$?
    **a.** $\sqrt{5}$     **b.** $10\sqrt{5}$
    **c.** 2     **d.** $2\sqrt{5}$

**5.** *Complete:* For $\sqrt{73}$, $\sqrt{\phantom{x}}$ is called a ? and $\sqrt{73}$ is called a ?.

**6.** Is $\sqrt{7}$ a rational or an irrational number? Explain.

**7.** Which is greater, $\sqrt{8}$ or $2\sqrt{3}$? How do you know?

## Practice and Apply

Between which two whole numbers does the given number lie?

**8.** $\sqrt{23}$   **9.** $\sqrt{12}$   **10.** $\sqrt{73}$   **11.** $\sqrt{48}$
**12.** $\sqrt{125}$   **13.** $\sqrt{149}$   **14.** $\sqrt{215}$   **15.** $\sqrt{239}$
**16.** $\sqrt{525}$   **17.** $\sqrt{390}$   **18.** $\sqrt{460}$   **19.** $\sqrt{500}$

Simplify, if possible.

**20.** $\sqrt{3 \cdot 27}$   **21.** $\sqrt{16} + \sqrt{36}$   **22.** $\sqrt{5+20}$   **23.** $\sqrt{7} \cdot \sqrt{7}$
**24.** $\sqrt{4} + \sqrt{5}$   **25.** $\sqrt{32}$   **26.** $\sqrt{39}$   **27.** $\sqrt{2 \cdot 8}$
**28.** $\sqrt{121} + \sqrt{64}$   **29.** $\sqrt{5 \cdot 5}$   **30.** $\sqrt{48}$   **31.** $\sqrt{50}$
**32.** $\sqrt{6} \cdot \sqrt{6}$   **33.** $\sqrt{10} + \sqrt{15}$   **34.** $\sqrt{144}$   **35.** $\sqrt{72}$
**36.** $\sqrt{21} + \sqrt{15}$   **37.** $\sqrt{40}$   **38.** $\sqrt{8} \cdot \sqrt{8}$   **39.** $\sqrt{200}$
**40.** $\dfrac{2}{\sqrt{2}}$   **41.** $\dfrac{1}{\sqrt{3}}$   **42.** $\dfrac{4}{\sqrt{2}}$   **43.** $\dfrac{3}{\sqrt{5}}$
**44.** $\dfrac{6}{\sqrt{3}}$   **45.** $\dfrac{6}{\sqrt{7}}$   **46.** $\dfrac{5}{\sqrt{10}}$   **47.** $\dfrac{2}{\sqrt{50}}$

## Connect and Extend

You can use this graph of $y = x^2$ to estimate the square root of a number.

**EXAMPLE:**

To find $\sqrt{14}$, find 14 on the $x^2$-axis. Move directly right to the curve.

From this point on the curve, move down to the $x$-axis. Estimate the value of $x$.

**Estimate:** $\sqrt{14} \approx$ **3.7**

Use the graph to estimate each square root.

**48.** $\sqrt{26}$   **49.** $\sqrt{8}$
**50.** $\sqrt{11}$   **51.** $\sqrt{30}$

**52.** How can you use the graph to estimate $(4.7)^2$?

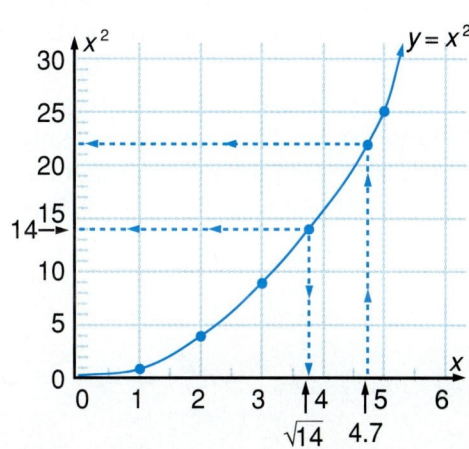

**Use a calculator to determine whether the equation is true.**

**53.** $2\sqrt{3} + 5\sqrt{3} = 7\sqrt{3}$

**54.** $3\sqrt{5} - 2\sqrt{5} = \sqrt{5}$

**55.** What property is demonstrated in Exercises 53 and 54?

**Use a calculator to estimate the values. Then plot the points on a number line.**

**56.** $A: \sqrt{2}$

**57.** $B: \sqrt{5}$

**58.** $C: 2 + \sqrt{3}$

**59.** $D: 2 - \sqrt{3}$

## Maintain Your Skills

**Solve. (Pages 64–68)**

**60.** The perimeter of Beverly's square herb garden is 96 meters. Find the area of the garden.

**61.** The perimeter of a rectangle is 30 feet. If the width of the rectangle is 6 feet, what is the area?

**Solve each equation. (Pages 105–109)**

**62.** $-3(-6 - 4a) = -6$

**63.** $4(e - 10) + 6e = -90$

**64.** $\frac{1}{2}(6t + 8) = 37$

**Evaluate for $a = 4$, $b = -3$, $c = -2$, and $d = -4$. (Pages 150–153)**

**65.** $6b^3$

**66.** $5cd^2$

**67.** $-3a^2b$

**68.** $11 + 6ac^4$

**69.** $68 - bd^3$

# Mathematical Footnote

**Topology** is a branch of mathematics that studies shapes and how they can be stretched, twisted, and bent. Try this topology experiment. Take a piece of thin wire or a pipe cleaner and form it into the letter C. Then bend it to see which of these letters you can form without having the wire cross itself or touch itself.

I    L    M    N    S    V    W    Z

The letter C and the letters above are all equivalent shapes in topology. The letters C and D are not topologically equivalent, because you would have to break the D to form a C. A sugar bowl and a button with two holes are topologically equivalent solid figures. Can you tell why from the drawing below?

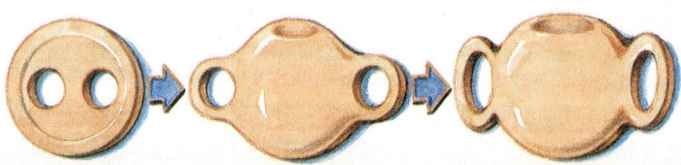

## LESSON 12-3

# The Pythagorean Theorem

After the summer flooding of the Nile River, the ancient Egyptians had to lay out the boundaries of their fields. They found a way to construct square corners by knotting a rope into 12 equal spaces and stretching it around stakes to form a right triangle. (A **right triangle** has a square corner; that is, an angle of 90°.) To do this, the Egyptians learned to place the stakes so the triangle had sides of 3, 4, and 5 units.

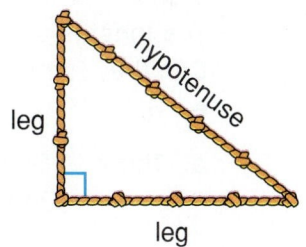

The ancient Greeks adopted this trick from the Egyptians. Later on, a group called the Pythagoreans began to think of the triangle's sides as the sides of three squares. This led them to an amazing fact.

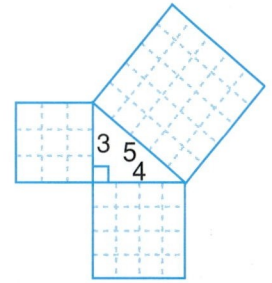

For the 3-4-5 triangle, the area of a square having a side of 3 plus the area of a square having a side of 4 equals the area of a square having a side of 5.

Then the Pythagoreans found that this pattern could be extended to all right triangles. Their generalization came to be known as the *Pythagorean Theorem*.

> **Pythagorean Theorem**
> If a right triangle has legs $a$ and $b$, and hypotenuse $c$, then
> $$a^2 + b^2 = c^2$$
> or $\quad c^2 = a^2 + b^2.$

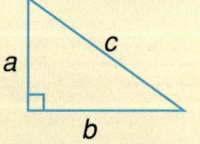

It is also true that if a triangle has sides $a$, $b$, and $c$, such that $a^2 + b^2 = c^2$, then it is a right triangle. Note that the **hypotenuse** of a right triangle is always the longest side, and that the other two sides are called the **legs** of the right triangle.

- If the lengths of the sides of a right triangle are 5, 12, and 13, which is the length of the hypotenuse?

**EXAMPLE 1** The lengths of the sides of a triangle are given. Can they form a right triangle?
   **a.** 6, 8, 10            **b.** 7, 8, 9            **c.** 5, 12, 13

**Solution**   **a.** Think:   Does $6^2 + 8^2 = 10^2$?
                      Does $36 + 64 = 100$?
                            $100 = 100$    Yes ✓

       **b.** Think:   Does $7^2 + 8^2 = 9^2$?
                      Does $49 + 64 = 81$?
                            $113 \neq 81$    No ✓

       **c.** Think:   Does $5^2 + 12^2 = 13^2$?
                      Does $25 + 144 = 169$?
                            $169 = 169$    Yes ✓

**Try This**   For each exercise, the lengths of the sides of a triangle are given. Can they form a right triangle?
   **1.** 12, 16, 20          **2.** 2, 4, 5          **3.** 10, 24, 26

When you know the lengths of two sides of a right triangle, you can find the length of the third side.

**EXAMPLE 2** Solve for x. Round answers to the nearest tenth.

a.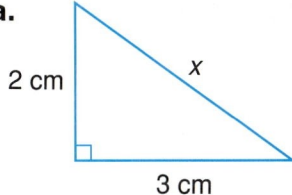

b.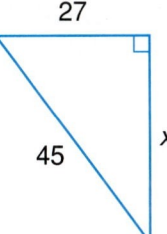

**Solutions**

a. Use the Pythagorean Theorem.

| | |
|---|---|
| Substitute 2 for a, 3 for b, and x for c. | $a^2 + b^2 = c^2$ |
| Square 2 and 3. | $2^2 + 3^2 = x^2$ |
| | $4 + 9 = x^2$ |
| | $13 = x^2$ |
| Think: If $x^2 = 13$, $x = \sqrt{13}$. | $x = \sqrt{13}$ |
| Find $\sqrt{13}$. Use a calculator. | $x \approx 3.6$ |

The length of the hypotenuse is about **3.6 centimeters.**

| | |
|---|---|
| b. Substitute 27 for a, x for b, and 45 for c. | $a^2 + b^2 = c^2$ |
| | $27^2 + x^2 = 45^2$ |
| Square 27 and 45. | $729 + x^2 = 2025$ |
| Subtract 729 from both sides. | $x^2 = 1296$ |
| Think: If $x^2 = 1296$, $x = \sqrt{1296}$. | $x = \sqrt{1296}$ |
| Find $\sqrt{1296}$. Use a calculator. | $x = 36$ |

The unknown length is **36.**

**Check** Looking back at the given lengths in each triangle, you can see that 3.6 and 36 are reasonable answers.

**Try This** Solve for x.

4.

5.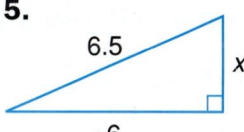

6.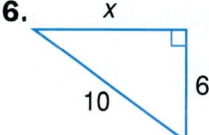

The Pythagorean Theorem can be used to solve many types of problems.

**EXAMPLE 3**  Julio wants to bury a water pipe from one corner of his field to the opposite corner. How many feet of pipe does he need if his rectangular field is 200 feet by 300 feet? Round the answer to the nearest foot.

**Solution**  Draw a diagram. Since the field is a rectangle, look at one of the right triangles formed by the pipe and two sides of the rectangle. The pipe is the hypotenuse.

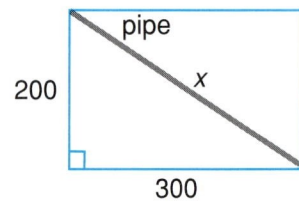

$$x^2 = 200^2 + 300^2$$
$$x^2 = 40{,}000 + 90{,}000$$
$$x^2 = 130{,}000$$
$$x = \sqrt{130{,}000} \approx 360.55$$

Julio will need at least **361 feet** of pipe.

**Check**  From the diagram, it appears that 361 is a reasonable answer.

**Try This**  7. A baseball diamond is a square. Each side is 90 feet long. What is the distance from first base to third base? Round the answer to the nearest foot.

# EXERCISES

**Objectives:** To determine the length of a side of a right triangle
To solve problems involving right triangles

## Check Your Understanding

**Skill Check**  Write the letter of the correct answer.

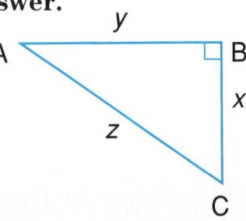

1. Which statement about triangle $ABC$ is **not** true?
   a. It is a right triangle.
   b. The hypotenuse is represented by $y$.
   c. One leg is represented by $x$.
   d. The hypotenuse is represented by $z$.

**2.** Which can form a right triangle?
  **a.** 2, 2, 3   **b.** 30, 40, 50
  **c.** 5, 11, 12   **d.** 15, 15, 20

**3.** The legs of a right triangle are 5 inches and 7 inches long. What is the length of the hypotenuse?
  **a.** $\sqrt{35}$ inches   **b.** $\sqrt{21}$ inches
  **c.** $\sqrt{74}$ inches   **d.** $\sqrt{12}$ inches

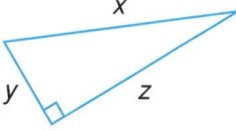

**4.** Which statement about the right triangle is true?
  **a.** $x^2 = y^2 + z^2$   **b.** $x^2 + y^2 = z^2$
  **c.** $y^2 - z^2 = x^2$   **d.** $x + y = z$

**5.** Which side of a right triangle is always the longest?

**6.** You are asked to find the length of the leg of a right triangle. What information do you need?

## Practice and Apply

**Copy and complete the table. Use a calculator. Round answers to the nearest tenth.**

|     | Leg | Leg | Hypotenuse |
|-----|-----|-----|------------|
| 7.  | 8   | 12  | ?          |
| 8.  | 10  | ?   | 26         |
| 9.  | ?   | 8   | 17         |
| 10. | 5   | 9   | ?          |
| 11. | ?   | 6   | 25         |
| 12. | 24  | 45  | ?          |
| 13. | ?   | 15  | 20         |

**Solve each problem. Round answers to the nearest whole number.**

**14.** A ladder 25 feet long leans against a building and reaches the edge of the roof. The base of the ladder is 12 feet from the building. How many feet above the ground does the ladder touch the roof?

**15.** A rectangular swimming pool in a park is 60 feet long and 25 feet wide. Marsha swims from one corner of the pool to the opposite corner and back 10 times. How many feet does she swim?

**16.** A hiker leaves camp and walks 5 miles west. Then he walks 7 miles south. How far is he from the camp?

**17.** Write a problem of your own that involves using the Pythagorean Theorem to find a distance. Solve the problem. Then challenge a friend to find the solution. Do your solutions agree?

## Connect and Extend

For a right triangle with legs $a$ and $b$ and hypotenuse $c$, find the missing length. Simplify radicals, if possible.

**18.** $a = 1, b = 1$     **19.** $a = 2, b = 2$     **20.** $a = 8, b = 8$

**21.** What is similar about the answers to Exercises 18–20?

**22.** If each leg of a right triangle has its length equal to $n$, what is the length of the hypotenuse?

You have seen that you can form right triangles with sides of these lengths.

    3, 4, and 5      9, 12, and 15      5, 12, and 13

These are called **Pythagorean Triples**.

**Copy and complete the table. The first entry is completed for you.**

| | $x$ | $y$ | $a = x^2 - y^2$ | $b = 2xy$ | $c = x^2 + y^2$ | $a, b, c$ |
|---|---|---|---|---|---|---|
| | 5 | 3 | $a = 5^2 - 3^2 = 16$ | $b = 2 \cdot 5 \cdot 3 = 30$ | $c = 5^2 + 3^2 = 34$ | 16, 30, 34 |
| **23.** | 4 | 2 | $a = \underline{?}$ | $b = \underline{?}$ | $c = \underline{?}$ | $\underline{?}$ |
| **24.** | 6 | 4 | $a = \underline{?}$ | $b = \underline{?}$ | $c = \underline{?}$ | $\underline{?}$ |
| **25.** | 7 | 6 | $a = \underline{?}$ | $b = \underline{?}$ | $c = \underline{?}$ | $\underline{?}$ |

**26.** Show that the triples obtained in Exercises 23–25 are Pythagorean triples by showing that $a^2 + b^2 = c^2$.

**27.** Create 5 more sets of Pythagorean triples.

**28.** Explain how you can use a piece of string and the Pythagorean Theorem to locate $\sqrt{2}$ on this number line.

## Maintain Your Skills

**Solve and check each proportion. (Pages 216–220)**

**29.** $\dfrac{c}{40} = \dfrac{3}{5}$     **30.** $\dfrac{15}{e} = \dfrac{10}{8}$     **31.** $\dfrac{12}{30} = \dfrac{f}{25}$     **32.** $\dfrac{7}{8} = \dfrac{h}{4}$

**Solve each inequality. (Pages 382–389)**

**33.** $p + 2 \leq -1$     **34.** $-7s < 28$     **35.** $-36 \geq -18q$     **36.** $r - 2\tfrac{1}{4} > 1\tfrac{1}{4}$

**Solve. (Pages 440–443)**

**37.** A segment has endpoints $Q(-3, -1)$ and $R(-1, 4)$. What are the endpoints under the translation $(x, y) \rightarrow (x + 3, y - 2)$?

**38.** A triangle has vertices at $A(0, -3)$, $B(1, 1)$ and $C(-3, 0)$. What are the vertices under the translation $(x, y) \rightarrow (x - 1, y - 4)$?

**39.** A triangle has vertices at $A(1, -1)$, $B(3, 3)$, and $C(6, -2)$. What are the vertices under the translation $(x, y) \rightarrow (x - 2, y + 4)$?

# LESSON 12-4
## Problem Solving Strategies
## Formulating Questions

The Booster Club is planning a special halftime show at the football game. The club would like to have students march onto the field and form a giant X as shown above. You are on the planning committee that is arranging the halftime show.

Not all mathematical problems come in neat packages. Sometimes, as in the paragraph above, the given information doesn't contain any numbers. How will you go about solving a problem like this?

One strategy that you can use is to ask questions that will help you to formulate the subproblems you will need to solve.

How many students will you need?
Will you have the same number of students in each line?
How many students will there be in each line?
How far apart will the students stand?

**12-4** Problem Solving Strategies **463**

The questions on page 463 will lead to other questions, such as:

How long is the football field between the goal posts?
How wide is the field between the goal posts?
How long is the diagonal between the goal posts?

The Exercises will lead you through one way to formulate the problems you will need to solve and the strategies you can use to solve them.

# EXERCISES

1. Name at least two ways to find out how long the lines of students must be.
2. Name at least two ways to find out if the two lines must have the same length.
3. Write and solve a math problem that will tell you how long the lines of students must be.
4. How can you determine how much space is taken up by the average student standing in the line?
5. Decide how many students you will need for each line.
6. Design a way to keep the lines straight.
7. Plan a way to be sure that the lines cross in the middle.
8. Plan how the students will march onto the field, form the X, and march off again.

**Between which two whole numbers does the given number lie?**
(Pages 452–456)

1. $\sqrt{20}$
2. $\sqrt{55}$
3. $\sqrt{122}$
4. $\sqrt{274}$
5. $\sqrt{444}$

**Simplify, if possible.** (Pages 452–456)

6. $\sqrt{196} + \sqrt{81}$
7. $\sqrt{38 + 11}$
8. $\sqrt{12 \cdot 2}$
9. $\sqrt{12} \cdot \sqrt{12}$
10. $\sqrt{117}$
11. $\sqrt{6} + \sqrt{8}$
12. $\dfrac{1}{\sqrt{5}}$
13. $\dfrac{2}{\sqrt{6}}$

**Solve for $x$. Round answers to the nearest tenth.** (Pages 457–462)

14.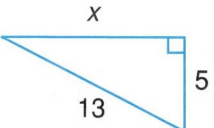

15.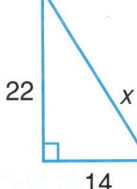

16.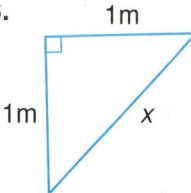

**Solve.** (Pages 457–462)

17. A ramp outside a building is 37 feet long. The beginning of the ramp is 35 feet away from the base of the building. How many feet above the ground does the edge of the ramp connect with the building?

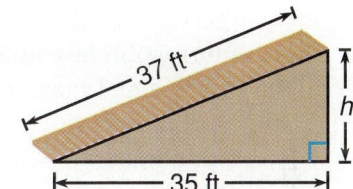

# Mathematical Footnote

When Dr. Fanya S. Montalvo was a seventh grader in Chicago, she built a tiny rocket, fueled by the head of a match. She worked very hard in high school and by her senior year was taking honors courses in mathematics. Later, as a doctoral student in computer science, she held a job as a drummer in a rock band. She also sings folk and country songs and plays the guitar and the piano.

Dr. Montalvo is now a research scientist in artificial intelligence.

# LESSON 12-5
# Similar Triangles

**A**retha wants to estimate the height of the tree in the photograph at the right. She knows that she can use similar triangles to find the estimate indirectly.

**Similar triangles** are triangles that have the same shape. Two triangles are similar if their corresponding angles have equal measures and their corresponding sides have the same ratio.

Look at the two triangles at the right.

1. Corresponding angles have equal measures.
2. Corresponding sides have the same ratio. That is,

$$\frac{8}{16} = \frac{10}{20} = \frac{15}{30} = \frac{1}{2}$$

So the two triangles are similar.

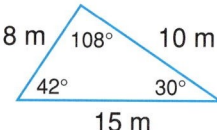

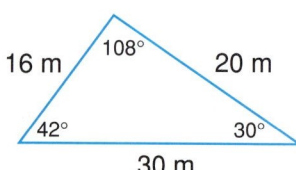

Notice that in similar triangles, sides that correspond are opposite angles having equal measures.

**466** CHAPTER 12

**EXAMPLE 1** Determine whether each pair of triangles is similar. Equal angles are marked by the same number of curved lines.

a.

b.

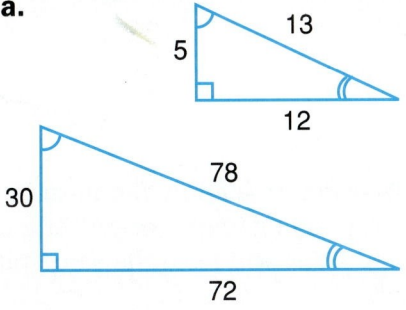
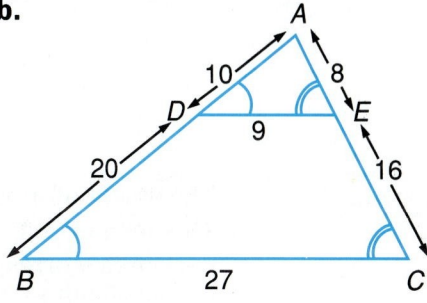

**Solutions**

a. Corresponding angles have equal measures.

Ratio of sides: $\frac{30}{5} = \frac{72}{12} = \frac{78}{13} = \frac{6}{1}$

Since corresponding angles have equal measures, and corresponding sides have the same ratio, the triangles **are similar.**

b. It is easier to solve this problem if you look at each triangle separately. Both triangles contain angle A. Angles D and B have the same measure, and angles E and C have the same measure.

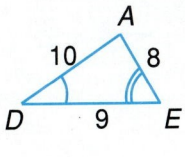
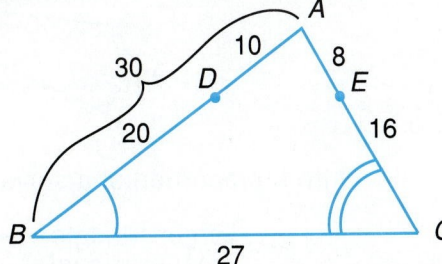

Ratio of sides: $\frac{10}{30} = \frac{8}{24} = \frac{9}{27} = \frac{1}{3}$

Since corresponding angles have equal measures, and corresponding sides have the same ratio, the triangles **are similar.**

• Are all right triangles similar? Explain.

12-5 Similar Triangles **467**

**Try This**   Determine whether each pair of triangles is similar.

1. 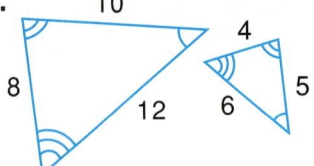   2.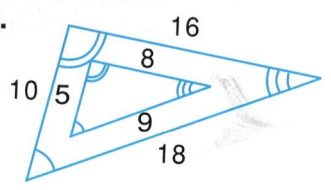

The shape of a triangle is determined by the measures of its angles. This means that if corresponding angles of two triangles have equal measures, corresponding sides will have the same ratio and the triangles will be similar.

> **Angle-Angle-Angle (AAA) Property**
>
> If corresponding angles of two triangles have equal measures, then the ratios of corresponding sides will be equal and the corresponding triangles will be similar.

When two polygons are similar, a proportion can be used to find an unknown length.

**EXAMPLE 2**   Solve for the unknown length, $n$, in the pair of similar triangles. Round the answer to the nearest tenth.

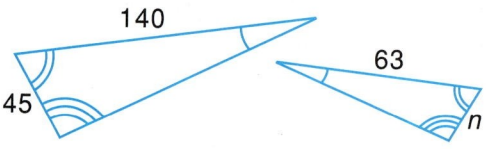

**Solution**   Write a proportion and solve for $n$.

Larger triangle $\longrightarrow$ $\dfrac{140}{63} = \dfrac{45}{n}$ $\longleftarrow$ Larger triangle
Smaller triangle $\longrightarrow$ $\phantom{\dfrac{140}{63} = \dfrac{45}{n}}$ $\longleftarrow$ Smaller triangle

Find the cross products.   $140 \cdot n = 63 \cdot 45$

Solve for $n$.   $\dfrac{140n}{140} = \dfrac{2835}{140}$

$n = 20.25$

The length of the unknown side is about **20.3.**

Since the ratio of the corresponding sides is $\dfrac{140}{63}$, or $\dfrac{20}{9}$, and 45 is a little more than twice 20.25, a length of 20.3 is reasonable.

**Check**

Does $\frac{140}{63} = \frac{45}{20.25}$?

Does $(140)(20.25) = (63)(45)$?

$2835 = 2835$ ✓

**Try This** Find the unknown length in each pair of similar triangles.

3.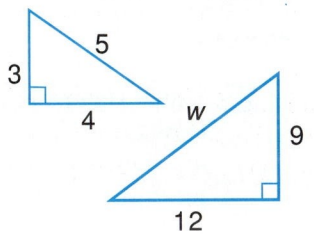

4.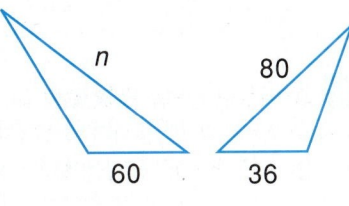

To solve Aretha's problem on page 466, you can use the AAA Property.

**EXAMPLE 3** Assume that the tree and Aretha form right angles with the ground. Now, consider the two similar triangles determined by the tree, Aretha, and the two shadows. The angle of the sun's rays is the same for both triangles so the triangles are similar. If Aretha is five feet tall, she can measure her shadow and the tree's shadow. Then she can write a proportion and solve for the height, $h$, of the tree to the nearest foot.

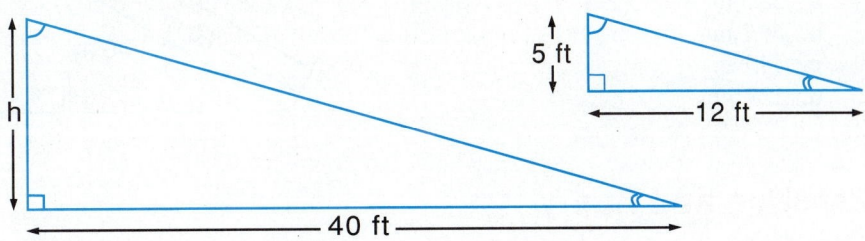

**Solution**

Write a proportion.  Tree ⟶ $\frac{h}{5} = \frac{40}{12}$ ⟵ Tree's Shadow
Aretha ⟶           ⟵ Aretha's Shadow

Find the cross products.   $12 \cdot h = 5 \cdot 40$

Solve.   $h = \frac{200}{12} = 16.7$

The tree is about **17 feet** tall. A height of 17 feet is reasonable.

**Try This**

5. Bill is 6 feet tall. On a sunny day, Bill's shadow is 8 feet long when the shadow of a high diving platform is 52 feet long. Find the height of the diving platform.

# EXERCISES

**Objective:** To solve problems involving similar figures

## Check Your Understanding

**Skill Check** Write the letter of the correct answer.

1. Which statement is true?
   a. Any two triangles are similar.
   b. Any two right triangles are similar.
   c. Corresponding angles of similar triangles have the same measure.
   d. Corresponding sides of similar triangles have the same length.

2. Which statement about similar triangles is **not** true?
   a. They have the same shape.
   b. Corresponding sides have the same ratio.
   c. Corresponding sides are next to corresponding angles.
   d. The ratio of the corresponding sides can be 1:1.

3. Which is the unknown length for the similar triangles?
   a. 8.5 ft
   b. 17 ft
   c. 11.8 ft
   d. 34 ft

4. Which is the unknown length for the similar triangles?
   a. 28.2 m
   b. 56.7 m
   c. 44.7 m
   d. 127.5 m

## Practice and Apply

**Solve for $x$.**

5. $\dfrac{x}{4} = \dfrac{3}{6}$

6. $\dfrac{5}{x} = \dfrac{15}{45}$

7. $\dfrac{7}{4} = \dfrac{3.5}{x}$

8. $\dfrac{2}{x} = \dfrac{1\frac{1}{2}}{8}$

**Show that the corresponding sides of the similar triangles have the same ratio. Corresponding angles having the same measure are marked.**

9.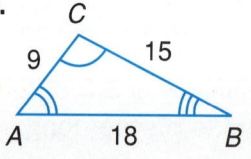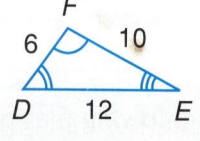

10.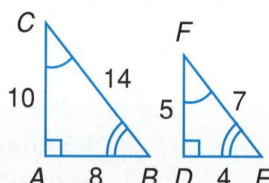

470   CHAPTER 12

**Find the unknown length for each pair of similar triangles.**

11.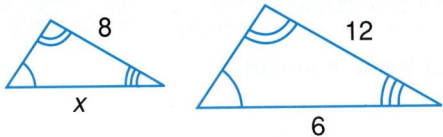
12.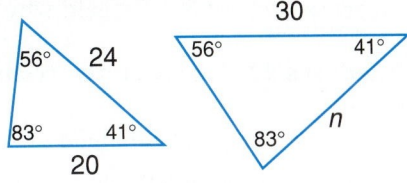

13.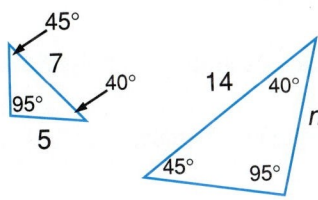
14.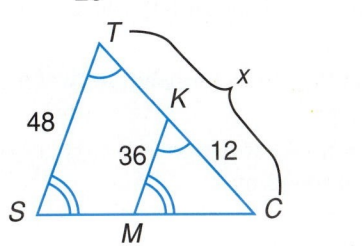

**Solve. Begin by drawing and labeling a diagram.**

15. A child one meter tall casts a shadow 2 meters long. At the same time, a telephone pole casts a shadow 20 meters long. What is the height of the telephone pole?

16. Norman is 1.8 meters tall. On a sunny day, Patty measured Norman's shadow and the shadow of the school. Use the similar triangles shown at the right to find the height of the school. Round your answer to the nearest tenth of a meter.

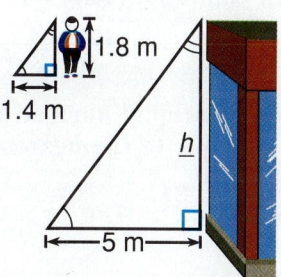

## Connect and Extend

17. What positive values of $x$ will make this proportion true: $\frac{x}{2} = \frac{8}{x}$? Describe how two figures with dimensions to fit the proportion might look.

18. Two rectangles are similar if the ratio, length:width, for one rectangle equals the ratio, length:width, for the second rectangle. Are the rectangles below similar? Explain.

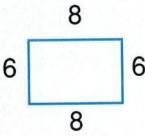 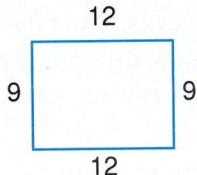

19. Are all squares similar? Explain.
20. Are all non-square rectangles similar? Make a drawing to explain your answers.

12-5  Similar Triangles  **471**

**21.** A right triangle has an acute angle of 35°. Another right triangle has an acute angle of 55°. Explain why the triangles are similar.

**For Exercises 22–25, graph the triangles on the same coordinate plane.**

**22.** Graph the triangle determined by $A(0, 2)$, $B(2, 4)$, and $C(3, 0)$.

**23.** Graph the image of triangle $ABC$ determined by $A'(0, 6)$, $B'(6, 12)$, and $C'(9, 0)$.

**24.** Describe the relationship between the triangles in Exercises 22 and 23.

**25.** Describe what happens if each $x$- and $y$-coordinate of the image triangle is multiplied by 2.

## Maintain Your Skills

**Solve each equation. (Pages 95–99, 183–188)**

**26.** $\frac{a}{26} = 7$   **27.** $-\frac{3}{8} = -\frac{t}{16}$   **28.** $\frac{1}{6}k = -\frac{12}{9}$   **29.** $-6 = \frac{s}{34}$

**Express each rational number as a decimal. If the decimal repeats, round the answer to the nearest thousandth. (Pages 277–280)**

**30.** $\frac{9}{20}$   **31.** $\frac{2}{11}$   **32.** $\frac{1}{6}$   **33.** $\frac{28}{8}$   **34.** $\frac{1}{8}$   **35.** $\frac{1}{3}$

## ★ Math Team Problems

**36.** Find the lengths of the seams in the Trojan helmet at the right. The length of the first seam is $\sqrt{2}$.

**37.** Graph the points $A(4, 2)$ and $B(1, 6)$ in the coordinate plane. Draw a right triangle and use the Pythagorean Theorem to find the distance between the points.

**38.** Explain why this statement is true. Let $A(a, b)$ and $B(c, d)$ be any two points in the coordinate plane. Then the distance, $D$, between points $A$ and $B$ can be found by this formula.
$$D = \sqrt{(a - c)^2 + (b - d)^2}$$

**39.** Use the formula in Exercise 38 to find the distance between the pairs of points.
  **a.** $(4, 4)$ and $(-1, 5)$   **b.** $(-3, 6)$ and $(2, -12)$

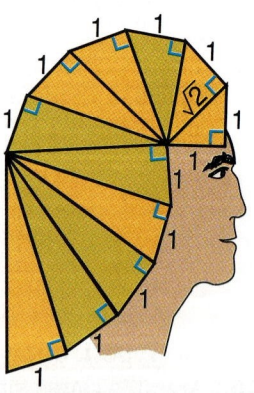

# COMPUTER EXPLORATION

## Similar Figures

**Objective:** To explore the scale factor for similar figures

1. a. With the Holt *Pre-Algebra* disk in drive A (or drive 1), run the program, *Similar Figures*, as follows.
      **IBM:** Type **a:h** and press **RETURN** (or **ENTER**).
      **Apple:** Turn on the computer. When the menu appears on the screen, type **H**.

2. a. Read the opening messages and type **R**. Type **L** to have the computer draw a larger rectangle similar to the original rectangle. The computer prints the **scale factor** 1.2. This means that the ratio of each side of the new rectangle to the corresponding side of the original rectangle is 1.2 : 1.
   b. Type **L** three more times. By how much does the factor increase each time? Are all the rectangles similar to the original rectangle and to each other?
   c. Keep typing **L** until the computer beeps. The beep tells you that the rectangle cannot be made any larger and still fit on the screen.

3. a. Type **S** to make the rectangle smaller. When the rectangle reaches its original size, what is the scale factor?
   b. Predict how many times more you can type **S** and shrink the rectangle before the computer beeps. Explain.

4. a. Type **N** to obtain a new figure and type **T** to obtain a triangle.
   b. Type **S** to obtain a smaller triangle. Are the 2 triangles similar?
   c. Predict how many more times you would type **S** to obtain a triangle with a scale factor of 0.4?
   d. Type **L** until the scale factor is 1. Starting with this triangle, how many times must you press **L** to obtain a triangle whose sides are 1.8 times as long as the corresponding sides of the original triangle?

# LESSON 12-6
# The Tangent Ratio

## EXPLORE

**E**ach of the right triangles below has an acute angle with a measure of 30°. **Acute angles** have measures less than 90°.

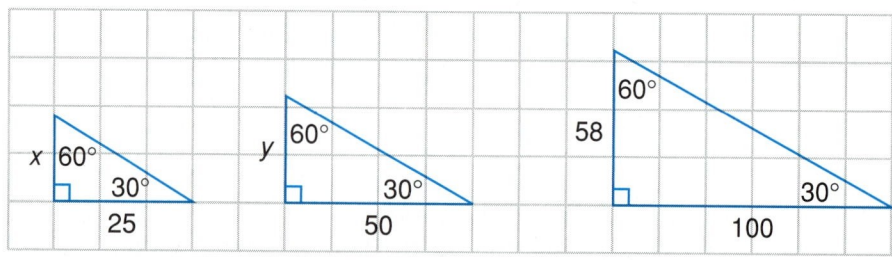

1. Why are the triangles similar?

2. Copy and complete the proportions using $x$ and $y$.

$$\frac{?}{25} = \frac{58}{100} \qquad \frac{?}{50} = \frac{58}{100}$$

3. Solve the proportions for $x$ and $y$.

4. Which decimal represents each ratio in the proportion?
   **a.** 5.8      **b.** 0.58      **c.** 0.058

5. Use right triangle $ABC$ to tell whether the statement at the left below is true or false.

The ratio $\frac{a}{b}$ is 0.58.

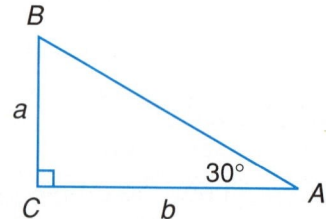

Recall that in a right triangle, the side opposite the right angle is called the *hypotenuse*. The sides opposite the acute angles are called *legs* and are referred to as *opposite* or *adjacent* to an acute angle. This is illustrated at the top of page 475.

474   CHAPTER 12

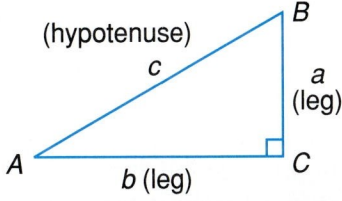 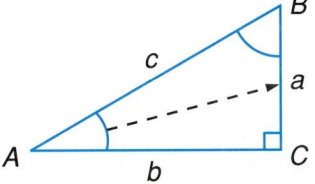 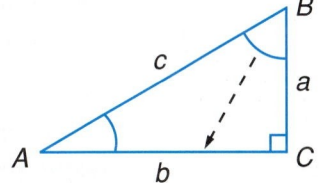

In right triangles, the ratio of the length of the side opposite an acute angle to the length of the side adjacent to the acute angle is called the *tangent ratio*.

The tangent of an acute angle $A$ is usually written tan $A$ (read: tangent of angle $A$).

> **Tangent Ratio**
> In any right triangle $ABC$, where $C$ is a right angle,
> $$\tan A = \frac{\text{length of side opposite angle } A}{\text{length of side adjacent to angle } A}$$
> $$\tan B = \frac{\text{length of side opposite angle } B}{\text{length of side adjacent to angle } B}$$

- In right triangle $ABC$ at the top of this page, does tan $A = \frac{a}{b}$ or $\frac{b}{a}$?

**EXAMPLE 1** Find each ratio. Write a decimal for the ratio. Round non-terminating decimals to four decimal places.

  **a.** tan $A$          **b.** tan $B$

**Solution** Refer to the right triangle $ABC$.

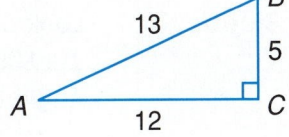

**a.** $\tan A = \dfrac{\text{side opposite angle } A}{\text{side adjacent to angle } A}$

$= \dfrac{5}{12}$

$\tan A =$ **0.4167**

**b.** $\tan B = \dfrac{\text{side opposite angle } B}{\text{side adjacent to angle } B}$

$= \dfrac{12}{5}$

$\tan B =$ **2.4**

12-6 The Tangent Ratio

**Try This**   Find each ratio. Write a decimal, rounded to four decimal places, for the ratio.

1. tan A
2. tan B

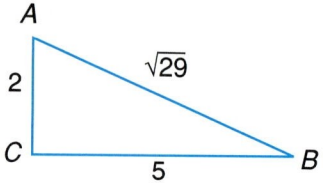

The value of the tangent ratio depends only on the measure of the acute angle of the right triangle. The table on page 601 gives approximate values for some acute angles and actual values for others. In this textbook, we will round tangent values to four decimal places by using a table or a calculator.

**EXAMPLE 2**   Find tan 41°.

**Solution**   Find 41 in the **Angle**-column. Look directly right to read 0.8693 in the **Tangent**-column.

| Angle | Tangent |
|---|---|
| 40° | 0.8391 |
| 41° | 0.8693 |
| 42° | 0.9004 |

**Using a calculator:** 41 [tan] ⟦0.869286737⟧

tan 41° = **0.8693** (rounded to 4 decimal places)

**Try This**   Use the table on page 601 or use a calculator to find each ratio.

3. tan 45°
4. tan 77°

When the value of the tangent of an acute angle is known, you can use the table or a calculator to find its measure.

**EXAMPLE 3**   What is the measure of an angle whose tangent is 1.1106? Round the measure of the angle to the nearest degree.

**Solution**   Look down the **Tangent** column. Stop at 1.1106. Then look directly left to the **Angle** column. Read 48°.

| Angle | Tangent |
|---|---|
| 47° | 1.0724 |
| 48° | 1.1106 |
| 49° | 1.1504 |

**Using a calculator:** 1.1106 [INV][tan] ⟦47.99967895⟧

The measure of the angle is **48°**, to the nearest degree.

**Try This**   Use the table on page 601 or use a calculator.

5. What is the measure of an angle whose tangent is 0.3443?
6. What is the measure of an angle whose tangent is 5.6713

The tangent ratio can also be used to solve problems involving measures of angles and lengths of sides.

**EXAMPLE 4**  A flagpole forming one side of triangle ABC is shown at the right. Point A is 14 meters from point C and the measure of angle A is 47°. How high is the flagpole to the nearest meter?

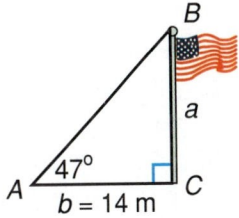

**Solution**  You know the measure of angle A and the length of the side adjacent to angle A. Use the tangent ratio.

$$\tan 47° = \frac{a}{14}$$

**Find tan 47°.**  $1.0724 = \frac{a}{14}$

**Solve for a.**  $(1.0724)14 = a$

$15.0136 = a$

The flagpole is about **15 meters** high.

**Try This**  Find x to the nearest meter.

7.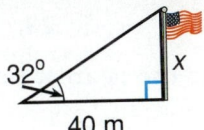

8.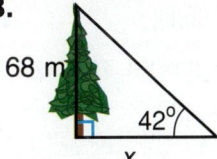

# EXERCISES

**Objective:** To find and use the tangent ratio

## Check Your Understanding

**Skill Check**  Write the letter of the correct answer.

**1.** Find tan A.
   a. 0.6000
   b. 0.8000
   c. 1.3333
   d. 0.7500

**2.** Find tan B.
   a. $\frac{y}{x}$
   b. $\frac{x}{y}$
   c. $\frac{y}{z}$
   d. $\frac{x}{y}$

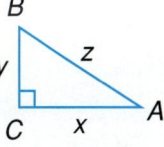

12-6   The Tangent Ratio   **477**

**3.** Which is the value of tan 36°?
  **a.** 0.5870   **b.** 0.8090
  **c.** 0.7265   **d.** 1.3467

**4.** If tan B = 0.4663, what is the measure of angle B to the nearest degree?
  **a.** 70°   **b.** 30°
  **c.** 65°   **d.** 25°

**True or False? When a statement is false, explain why.**

**5.** All right triangles whose acute angles have measures of 30° and 60° are similar.

**6.** All the tangent values in the table on page 601 are exact values.

**7.** The tangent of an acute angle of a right triangle is defined as
$\frac{\text{length of side adjacent to the angle}}{\text{length of side opposite the angle}}$.

# Practice and Apply

**Find tan A for each value of angle A. Use a table or use a calculator. Round values to four decimal places.**

**8.** 16°   **9.** 82°   **10.** 45°   **11.** 37°   **12.** 71°
**13.** 60°   **14.** 58°   **15.** 6°   **16.** 22°   **17.** 80°

**For each value of tan B, find the measure of angle B to the nearest degree.**

**18.** tan B = 0.0699   **19.** tan B = 3.7321   **20.** tan B = 0.6494
**21.** tan B = 1.0000   **22.** tan B = 11.4301   **23.** tan B = 5.1446

**For Exercises 24–29, find the length of side a to the nearest tenth.**

| | Measure of Angle A | b |
|---|---|---|
| **24.** | 30° | 100 yd |
| **25.** | 45° | 80 in |
| **26.** | 70° | 60 ft |
| **27.** | 60° | 100 ft |
| **28.** | 55° | 30 yd |
| **29.** | 15° | 100 mi |

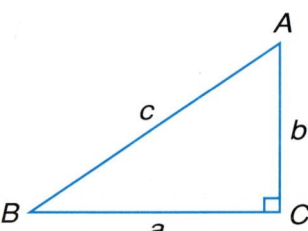

**30.** In the figure below, find the length of side BC, if the measure of angle B is 48° and the length of side AC is 80 meters.

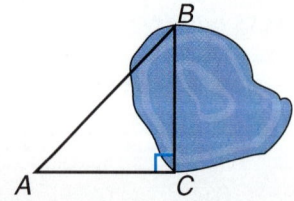

478   CHAPTER 12

**31.** The triangle at the right is a right triangle. Use the tangent ratio to find the distance between Tom's home and John's home, through the park.

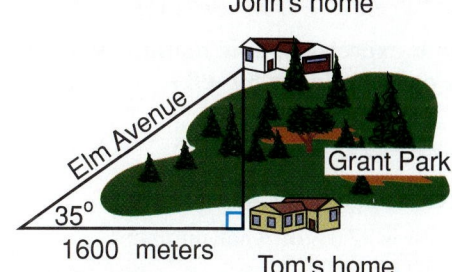

## Connect and Extend

In the figure below, side $AC$ has the same measure in all 8 triangles. As m$\angle A$ increases, the measure of the side opposite $\angle A$ increases.

**For Exercises 32–37, find the tangent ratio for each given angle measure. Round the ratio to the nearest tenth.**

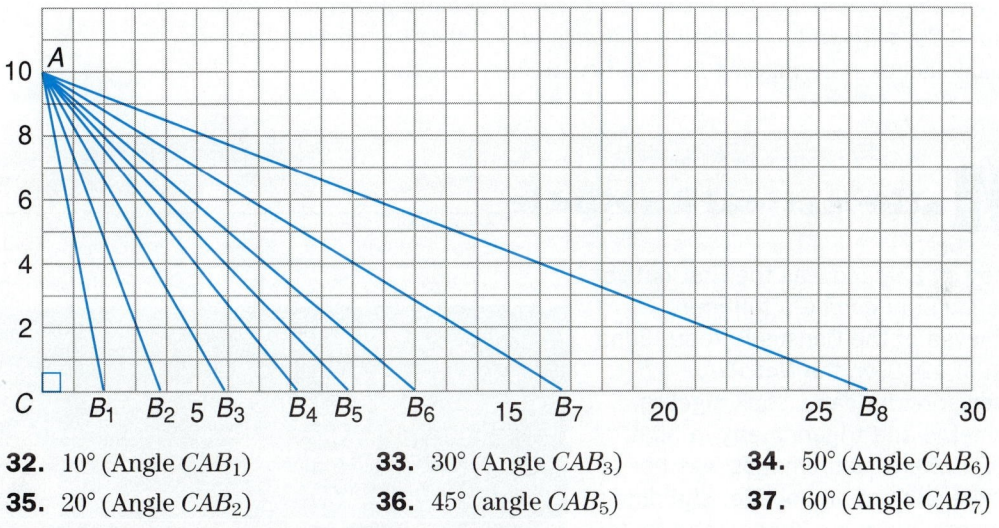

**32.** 10° (Angle $CAB_1$)
**33.** 30° (Angle $CAB_3$)
**34.** 50° (Angle $CAB_6$)
**35.** 20° (Angle $CAB_2$)
**36.** 45° (angle $CAB_5$)
**37.** 60° (Angle $CAB_7$)

**Refer to your answers to Exercises 32–37 to answer these questions.**

**38.** Does the value of the tangent ratio increase as the measure of the angle doubles? (Compare, for example, tan 10° and tan 20°, and tan 60° and tan 30°.)

**39.** For which angles is the tangent ratio less than 1?

**40.** When the measure of angle $A$ is 45°, what can you say about the length of the side opposite angle $A$ and the length of the side adjacent to angle $A$?

**41.** Is it possible to find the length of the side opposite angle $A$ if you know that the measure of angle $A$ is 45°? Explain.

## Maintain Your Skills

**Which exercises show instances of the Distributive Property? Write Yes or No. (Pages 59–63)**

**42.** $2a + 7a = (2 + 7)a$
**43.** $9 \cdot r + s = 9(r + s)$
**44.** $(5 + z)z = 5z + z^2$
**45.** $(r + s) + t = t + (r + s)$

**Solve. (Pages 334–337)**

**46.** 14 is 12.5% of what number?
**47.** 3 is what percent of $1\frac{1}{2}$?
**48.** In baseball practice, Sam got a hit 28% of the times he batted. If he batted 25 times, how many hits did he get?

**Simplify. (Pages 452–456)**

**49.** $\sqrt{2} \cdot \sqrt{32}$
**50.** $\sqrt{49} + \sqrt{9}$
**51.** $\sqrt{89 + 107}$
**52.** $\dfrac{10}{\sqrt{15}}$

## ★ Math Team Problem

**53.** A customer wanted a 25-foot piece of wire. The clerk incorrectly measured the wire with a yardstick that was two inches too short. How many inches were missing from the customer's length of wire?

## Mathematical Footnote

Edna Lee Paisano was the first native American to become a fulltime employee of the Census Bureau. Edna, born in 1948 on the Nez Perce Reservation in Idaho, took algebra, geometry, and trigonometry in high school where mathematics was her favorite subject. In college, she didn't know how math could be useful to her people, so she majored in sociology. Later, in graduate school, she studied statistics. At the Census Bureau, she said that she realized how important it is for native Americans to know demography, computer programming, and statistics . . . because it is very important to have people who can interpret data correctly.

- Do you know what demography is?

# SUMMARY

## KEY TERMS

Acute angle (p. 474)
Hypotenuse (p. 458)
Leg (p. 458)
Perfect square (p. 452)
Pythagorean theorem (p. 458)
Pythagorean triple (p. 462)
Radical (p. 452)

Radical symbol (p. 452)
Rationalizing the denominator (p. 454)
Right triangle (p. 457)
Similar triangles (p. 466)
Square root (p. 452)
Tangent ratio (p. 475)

## KEY IDEAS

**A. Rule for Multiplying Radicals:** If $a \geq 0$ and $b \geq 0$, then $\sqrt{a} \cdot \sqrt{b} = \sqrt{ab}$ and $\sqrt{ab} = \sqrt{a} \cdot \sqrt{b}$.

**B. Pythagorean Theorem:** If a right triangle has legs $a$ and $b$ and hypotenuse $c$, then $a^2 + b^2 = c^2$.

**C.** If a triangle has sides $a$, $b$, and $c$, such that $a^2 + b^2 = c^2$, then the triangle is a right triangle.

**D.** The hypotenuse of a right triangle is always the longest side.

**E. Angle-Angle-Angle (AAA) Property**
If corresponding angles of two triangles have equal measures, then the ratios of corresponding sides will be equal and the corresponding triangles will be similar.

# REVIEW

**Between which two whole numbers does the given number lie? (Pages 452–456)**

1. $\sqrt{34}$
2. $\sqrt{91}$
3. $\sqrt{177}$
4. $\sqrt{425}$

**Simplify, if possible. (Pages 452–456)**

5. $\sqrt{4} \cdot \sqrt{4}$
6. $\sqrt{80}$
7. $\sqrt{42}$
8. $\sqrt{64} + \sqrt{169}$
9. $\sqrt{79 + 42}$
10. $\dfrac{3}{\sqrt{3}}$
11. $\sqrt{108}$
12. $\sqrt{6 \cdot 24}$

**Solve for x. (Pages 457–462)**

13.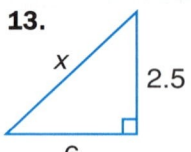

14. 

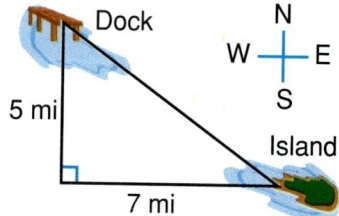

15. A sailboat travels from the dock on the mainland to an island. It sails south a distance of 5 miles. Then it sails east a distance of 7 miles. Find the shortest distance between the mainland and the island. Round the answer to the nearest tenth. **(Pages 457–462)**

**Find the unknown length for the similar figures. (Pages 466–472)**

16.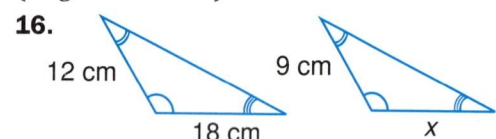

17.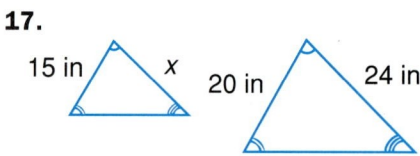

18. Jessica and her father stand next to each other. Jessica is 1.2 meters tall and her father is 1.8 meters tall. When Jessica's shadow is 0.9 meter long, how long is her father's shadow? **(Pages 466–472)**

**For Exercises 19–26, use a table or use a calculator.**

**Find tan A for each value of angle A. Round values to four decimal places. (Pages 474–480)**

19. 29°  20. 89°  21. 7°
22. 60°  23. 73°  24. 48°

**For each value of tan B, find the measure of angle B to the nearest degree. (Pages 474–480)**

25. tan B = 0.8693  26. tan B = 4.7046
27. tan B = 0.0349  28. tan B = 1.4281

**Solve. Round lengths to the nearest tenth. (Pages 474–480)**

29. Find the height, $h$ of the mountain.

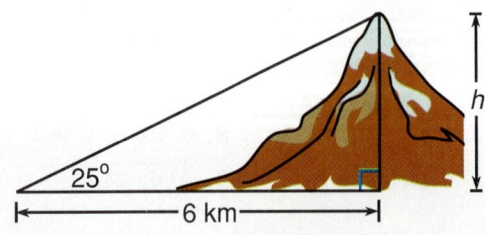

30. Find the distance, $d$, across the pond.

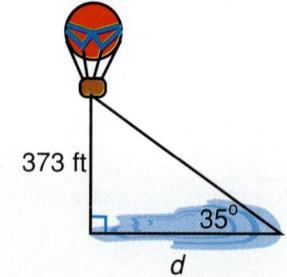

# TEST

**Simplify, if possible.**

1. $\sqrt{54}$
2. $\sqrt{8} \cdot \sqrt{8}$
3. $\sqrt{36} + \sqrt{24}$
4. $\sqrt{98}$
5. $\sqrt{121 + 104}$
6. $\dfrac{10}{\sqrt{5}}$
7. $\sqrt{6 \cdot 54}$
8. $\dfrac{9}{\sqrt{6}}$

9. The lengths of the sides of a triangular jogging path are 20 yards, 21 yards, and 25 yards. Does the path form a right triangle?

10. A ladder leans against a building and reaches the edge of the roof which is 20 feet high. The base of the ladder is 15 feet from the building. How long is the ladder?

**Find the missing length for the similar figures.**

11.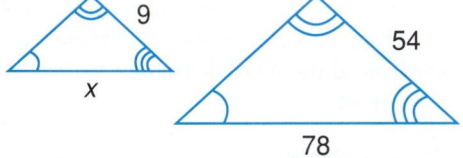

12.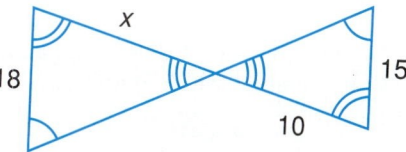

13. A scout used similar triangles to find the width of a river. The scout constructed similar triangles and made the measurements shown. Find the width, $w$, of the river.

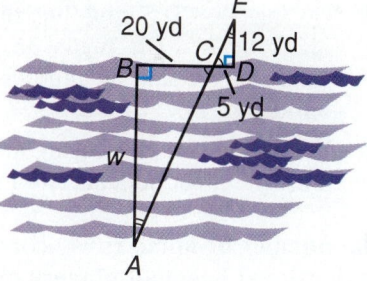

**Find tan $A$ for each value of angle $A$. Round values to four decimal places.**

14. $70°$
15. $13°$
16. $84°$

**For each value of tan $B$, find the measure of angle $B$ to the nearest degree. Use a table or a calculator.**

17. $\tan B = 11.4301$
18. $\tan B = 1.6003$
19. $\tan B = 0.6494$

**Solve. Round lengths to the nearest tenth.**

20. Find the height, $h$, of the building.

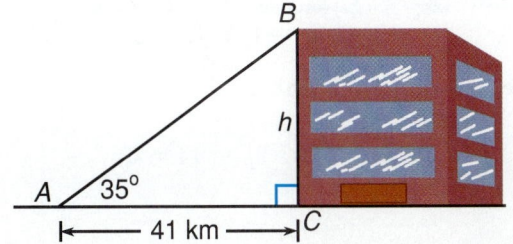

21. Find the distance, $x$, from the ship to the lighthouse.

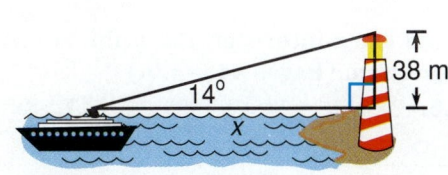

**Solve and check each equation. (Pages 101–104, 267–270)**

1. $c + \frac{1}{5} = -\frac{4}{5}$
2. $f - \frac{5}{6} = -\frac{2}{3}$
3. $\frac{k}{6} - \frac{1}{9} = -\frac{13}{9}$

**Which amount in each pair is greater? (Pages 226–230)**

4. 11,800 feet or $2\frac{1}{2}$ miles.
5. 19 cups or $9\frac{1}{4}$ pints

**Write each number in scientific notation. (Pages 289–294)**

6. The sun is about 3,666,000,000 miles from the planet Pluto.
7. A red blood cell has a diameter of about 0.0000273 inches.

**Solve. (Pages 319–322, 338–341)**

8. In a survey of 1,181 student council delegates, about 81% felt comfortable using a word processor. How many students was this? Round the answer to the nearest whole number.
9. The average price of a new car in 1980 was about $7500. In 1990, the average price of a new car was about $16,300. What was the percent increase? Round the answer to the nearest whole percent.

**Solve. (Pages 330–333, 359–363)**

10. On July 14, CD Sound, Inc. discounted all compact disks that cost $12.80 by 15%. On August 15, the store discounted all the compact disks not yet sold by another 25%. What is the sale price of the disks on August 16?

11. Which measure, the mean, median, or mode would you use to determine the most popular automobile color for the year 1991? Explain.

The number of home runs scored by 30 professional baseball players are listed at the right. (Pages 355–357)

| 15 | 17 | 5  | 3  | 13 | 15 |
|----|----|----|----|----|----|
| 5  | 11 | 25 | 3  | 19 | 14 |
| 12 | 16 | 5  | 22 | 5  | 8  |
| 12 | 4  | 4  | 5  | 3  | 26 |
| 8  | 3  | 5  | 20 | 2  | 27 |

12. Make a frequency table that shows the number of home runs. Use the intervals 1–5, 6–10, 11–15, and so on.

13. Make a histogram from your frequency table.

**Write a conjunction for each graph. (Pages 399–402)**

14.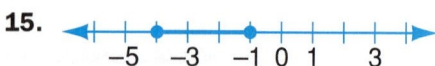
15.

**Find the $x$-intercept, the $y$-intercept, and the slope of each linear equation. (Pages 418–426)**

16. $2x - y = 1$
17. $6x + 3y = 12$
18. $2y + 14 = 6x$

# POLYNOMIALS

## LESSON 13-1  Adding Polynomials

Tama decided to plant a border around her rectangular flower garden. The perimeter of the garden is 32 feet. If the length of one side of the garden is 1 foot longer than twice the other, what are the dimensions of Tama's garden?

Tama drew a diagram to help her solve the problem. She let $a$ represent the length of the shorter side and $2a + 1$ represent the length of the longer side. Since Tama knew the perimeter of the garden, she wrote and solved this equation.

$$l + w + l + w = P$$
$$(2a + 1) + a + (2a + 1) + a = 32$$

Combine like terms.
$$6a + 2 = 32$$
$$6a = 30$$
$$a = 5$$

Replace $a$ with 5 to find $2a + 1$.
$$2a + 1 = 2 \cdot 5 + 1$$
$$= 11$$

The dimensions of the garden are 5 feet by 11 feet.

In the equation Tama wrote, $2a$ is a **monomial** because it contains one term. The expression $(2a + 1)$ is a **binomial** because it contains two terms. A **polynomial** is a monomial or the sum or difference of two or more monomials.

Here are some examples of polynomials.

$$-6x^2 \qquad 5t + 7 \qquad y^2 - 3y + 7 \qquad r^2 - z^2$$

- Which of the polynomials above are monomials?
- Which of the polynomials above are binomials?

Computation with polynomials is similar to computation with whole numbers, integers, and rational numbers. You can model polynomials with tiles that look like this.

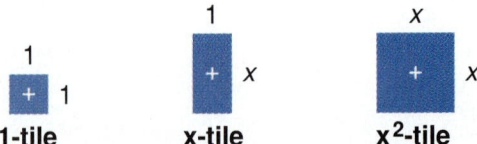

Notice that the 1-tile is a square. Each side is 1 unit long and its area is 1 square unit. An $x$-tile is a rectangle. One side is 1 unit long; the other side is $x$ units long. So the area of an $x$-tile is $x$ square units. An $x^2$-tile is a square. Each side is $x$ units long and its area is $x^2$ square units.

**EXAMPLE 1** Use tiles to model each polynomial.

| Polynomial | Model |
|---|---|
| a. $2x^2 + 5x + 1$ | |
| b. $x^2 + 2$ | |
| c. $2x + 1$ | |

**Try This** Use tiles to model each polynomial.

1. $3x^2 + 2x + 1$
2. $3x^2 + 5$
3. $6x + 3$

For each positive tile, there is a corresponding negative tile.

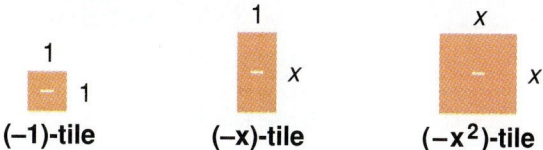

Just as with the positive and negative counters you used to model integers, a (+)-tile and a (−)-tile of the same kind form a neutral pair.

To model a polynomial such as $2x^2 - 3x - 1$, think of it as $2x^2 + (-3x) + (-1)$.

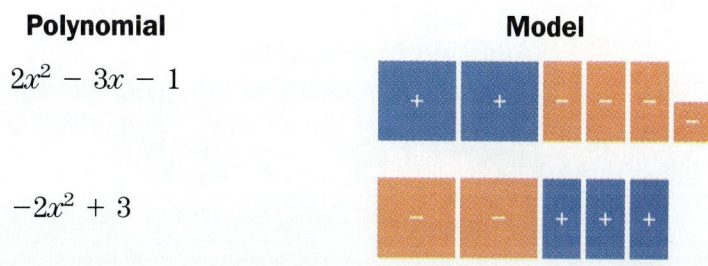

- What tiles will you need to model the polynomial $2x^2 - 3$?
- What tiles will you need to make neutral pairs with the polynomial $-x^2 + x - 1$?

You can use tiles to model addition of polynomials.

**EXAMPLE 2** Use tiles to find each sum.

a. $3x^2 + 2x + 5$
   $2x^2 - 3x - 3$

b. $x^2 - 3x + 2$
   $-2x^2 + 4x - 2$

**Solutions** Model each polynomial. Then remove neutral pairs and count the remaining like tiles.

a.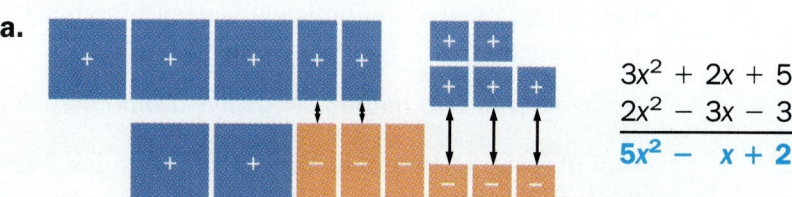

$3x^2 + 2x + 5$
$2x^2 - 3x - 3$
$\overline{5x^2 - x + 2}$

13-1 Adding Polynomials **487**

b.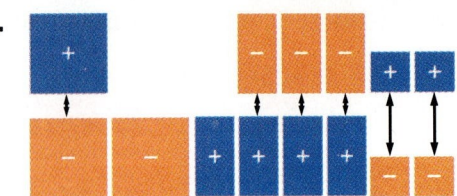

$$x^2 - 3x + 2$$
$$-2x^2 + 4x - 2$$
$$\overline{-x^2 + x}$$

**Try This**  Use tiles to find each sum.

4. $x^2 - 2x + 3$
   $3x^2 + x + 1$

5. $4x^2 + x - 5$
   $-3x^2 - 2x + 1$

The results of Example 2 suggest a rule for adding polynomials.

> **Addition of Polynomials**
> To add polynomials, add the coefficients of the like terms and add the constant (numerical) terms.

**EXAMPLE 3**  Add.
a. $3x^2 + 5x - 4$
   $x^2 - 4x + 5$

b. $(-2x^2 + x - 4) + (4x^2 + 2x - 2)$

**Solutions**  Write like terms directly below each other. Then add the like terms and the constant terms.

a. $3x^2 + 5x - 4$
   $x^2 - 4x + 5$
   $\overline{4x^2 + x + 1}$

b. $-2x^2 + x - 4$
   $4x^2 + 2x - 2$
   $\overline{2x^2 + 3x - 6}$

**Try This**  Add.

6. $4x^2 - 3x - 1$
   $x^2 - 5x + 8$

7. $(-3y^2 + 4y + 9) + (7y^2 + y - 1)$

- Is the sum of two monomials always a binomial? If not, give a counterexample.

- Is the sum of two binomials always a binomial? Explain.

- Can the sum of two polynomials, each having 3 terms, be a monomial? Give an example to explain your answer.

# EXERCISES

**Objective:** To add polynomials

## Check Your Understanding

**Skill Check** Write the letter of the correct answer.

1. Which is a binomial?
   **a.** $2x^2 + 4x + 5$   **b.** $3t^2 + 1$   **c.** $y^2$   **d.** $\dfrac{r}{3}$

2. Which can be used to model the polynomial $2x^2 - 1$?
   
   **a.**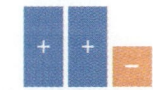
   
   **b.**

   **c.**
   
   **d.**

3. Which is the sum of $(x^2 + 2x + 4)$ and $(x + 5)$?
   **a.** $3x + 9$   **b.** $x^2 + 3x + 9$   **c.** $x^2 + x + 9$   **d.** $x^2 + 3x - 1$

4. Which is the sum of $(2x^2 + 3x - 1)$ and $(x^2 - 2x + 1)$?
   **a.** $3x^2 + 5x + 1$   **b.** $3x^2 + x + 1$   **c.** $3x^2 + 5x$   **d.** $3x^2 + x$

## Practice and Apply

Add.

5. $4x + 3$
   $\underline{2x - 5}$

6. $-3b + 8$
   $\underline{2b + 1}$

7. $5a - 6$
   $\underline{-2a - 4}$

8. $x^2 + 3x + 1$
   $\underline{\phantom{x^2 +\,} 4x - 3}$

9. $2x^2 + 4x - 3$
   $\underline{x^2 - 2x + 5}$

10. $-3y^2 - 2y + 9$
    $\underline{2y^2 + 2y - 5}$

11. $2d^2 - 3d$
    $\underline{2d^2 + 5d - 3}$

12. $5m^2 + 8$
    $\underline{-2m^2 + 4m}$

13. $4t^2 - 2t + 4$
    $\underline{4t^2 + 2t - 5}$

14. $(3r^2 + 5r) + (2r^2 + 2r)$

15. $(6t^2 - 3t + 1) + (2t^2 + 5)$

16. $(5z^2 - 2z + 4) + (-3z^2 + z - 6)$

17. $(8x + 6) + (4x^2 + 3x - 8)$

18. $(k^2 + 4k - 3) + (k^2 - 4k + 4)$

19. $(2t^2 + 3t + 9) + (-5t^2 + 3t)$

20. If $x$ is 3 meters, which is greater, the perimeter of a square whose sides are each $(3x - 1)$ units or the perimeter of a rectangle whose length is $(x + 9)$ and whose width is $(2x - 3)$?

For Exercises 21–22, the perimeter of the figure is given. Find the length of each side.

**21.** Perimeter: 22 cm

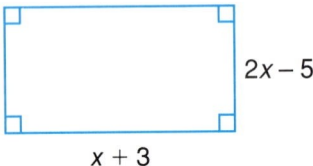

**21.** Perimeter: 60 in

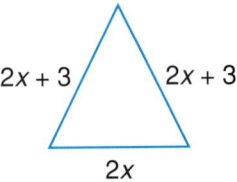

## Connect and Extend

**Find the sum.**

**23.** $(5 - 3x - x^2) + (2x^2 - 5x - 1)$

**24.** $(a^2 - 9) + (12a^2 - 3a)$

**25.** $(4r - 2r^2) + (5r^2 - r + 1)$

**26.** $(3a^2 - 2b) + (2a^2 + b)$

For Exercises 27–28, the perimeter of the figure is given. Find the length of each side.

**27.** Perimeter: 53 meters

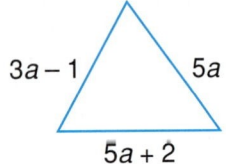

**28.** Perimeter: 137 yards

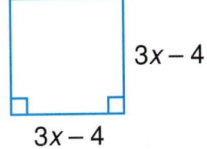

**29.** Study the addition problem at the right.

$$\begin{array}{r} 24 = 2 \cdot 10 + 4 \\ + \; 53 = 5 \cdot 10 + 3 \\ \hline 7 \cdot 10 + 7 \end{array}$$

How is this related to adding $(2x + 4)$ and $(5x + 3)$?

## Maintain Your Skills

**Write the letter of the expression that is not equivalent to each given expression. (Pages 59–63)**

**30.** $-7a + 5a$
   **a.** $(-7 + 5)a$
   **b.** $2a$
   **c.** $-2a$

**31.** $8r - s - 5r - 8s$
   **a.** $(r - s)8 - (s - 5r)$
   **b.** $3r - 9s$
   **c.** $(8 - 5)r - s(1 + 8)$

**32.** $-4t^2 - t^2$
   **a.** $-5t^2$
   **b.** $(4 + 1)t^2$
   **c.** $-t^2(4 + 1)$

**Write each number in expanded form. (Pages 271–274)**

**33.** 4.063
**34.** 0.51
**35.** 6.07
**36.** 0.089

**Simplify, if possible. (Pages 452–456)**

**37.** $\sqrt{3 \cdot 7}$
**38.** $\sqrt{98}$
**39.** $\sqrt{3} + \sqrt{6}$
**40.** $\dfrac{7}{\sqrt{21}}$

# LESSON 13-2  Subtracting Polynomials

You can also use tiles to model subtraction of polynomials.

**EXAMPLE 1**  Use tiles to subtract.

$$\begin{array}{r} 3x^2 - 2x - 3 \\ -(2x^2 - 3x - 2) \end{array}$$

Use tiles to model $3x^2 - 2x - 3$.

Remove 2 positive $x^2$-tiles. There is one $x^2$-tile left.

To remove 3 negative $x$-tiles when there are only 2, you have to add a neutral pair of $x$-tiles.

Neutral Pair

Now remove 3 negative $x$-tiles. There is one positive $x$-tile left.

Finally, remove 2 negative 1-tiles. There is one negative 1-tile left.

$$\begin{array}{r} 3x^2 - 2x - 3 \\ -(2x^2 - 3x - 2) \\ \hline x^2 + x - 1 \end{array}$$

The model shows that the result is $x^2 + x - 1$.

**Try This**  Use tiles to subtract.

1. $\begin{array}{r} 2x^2 - x - 3 \\ -(x^2 - x + 1) \end{array}$

2. $\begin{array}{r} -x^2 + 2x + 1 \\ -(-x^2 + x - 1) \end{array}$

You can also subtract by adding the opposite. For example, to add the opposite of $2x^2 - 3x - 2$ in Example 1, you write the opposite of each term of $2x^2 - 3x - 2$.

**Polynomial:** $2x^2 - 3x - 2$   **Opposite:** $-2x^2 + 3x + 2$

So $\begin{array}{r} 3x^2 - 2x - 3 \\ -(2x^2 - 3x - 2) \end{array}$ is the same as $\begin{array}{r} 3x^2 - 2x - 3 \\ +(-2x^2 + 3x + 2) \\ \hline x^2 + x - 1 \end{array}$

13-2  Subtracting Polynomials   **491**

> **Subtraction of Polynomials**
> To subtract a polynomial, add its opposite.

**EXAMPLE 2** Subtract.
a. $5x^2 + 2x - 4$
   $-(3x^2 - 2x + 5)$

b. $(-b^2 + 5) - (2b^2 - 4b + 1)$

**Solutions** Rewrite as addition by writing the opposite of the polynomial being subtracted. Line up like terms. When terms are missing, write the missing variable with a coefficient of zero.

a. $\phantom{-}5x^2 + 2x - 4$
   $\underline{-3x^2 + 2x - 5}$
   $\phantom{-}2x^2 + 4x - 9$

b. $-1b^2 + 0b + 5$
   $\underline{-2b^2 + 4b - 1}$
   $-3b^2 + 4b + 4$

**Try This** Subtract.
3. $4x^2 + 3x - 6$
   $-(5x^2 - 4x + 6)$

4. $(-5a^2 - 4a - 6) - (7a^2 - 5)$

# EXERCISES

**Objective:** To subtract polynomials

## Check Your Understanding

**Skill Check** Write the letter of the correct answer.

1. Which is the opposite of $2x^2 - 3x + 1$?
   a. $-2x^2 - 3x - 1$
   b. $2x^2 + 3x + 1$
   c. $-2x^2 + 3x - 1$
   d. $-2x^2 - 3x - 1$

2. Find the difference. $x^2 - 1 - (x^2 + 2)$
   a. $2x^2$
   b. $2x^2 - 2$
   c. $-3$
   d. $-2x^2 - 2$

3. Subtract: $2x^2 + 4x + 1$
   $-(x^2 - 3x - 4)$
   a. $x^2 + 7x + 5$
   b. $3x^2 + 7x - 3$
   c. $3x^2 - x + 5$
   d. $x^2 + 7x - 3$

4. Subtract: $(-2a^2 - 7) - (-3a^2 + 2a - 5)$
   a. $5a^2 + 2a - 12$
   b. $a^2 + 2a + 12$
   c. $a^2 - 2a + 2$
   d. $a^2 - 2a - 2$

**Subtract.**
5. $x^2 - (-4x^2)$
6. $0 - (9y^3)$
7. $-4t - (-t)$
8. $s - (-3s)$

## Practice and Apply
**Subtract.**

9. $6t + 4$
   $-(3t - 4)$

10. $-5p - 6$
    $-(2p - 4)$

11. $3x + 5$
    $-(-2x - 3)$

12. $5m^2 + 2m + 4$
    $-(2m^2 - 2m + 4)$

13. $3q^2 - 4q - 2$
    $-(q^2 - 3q + 5)$

14. $4z^2 + 4z - 9$
    $-(-2z^2 - 4z + 3)$

15. $8d^2 - 4d + 6$
    $-(9d^2 + 3d + 4)$

16. $-2b^2 - 5b - 7$
    $-(2b^2 + 5b - 7)$

17. $11n^2 - 2n + 9$
    $-(7n^2 - 2n + 6)$

18. $(4q^2 - 2q + 2) - (3q^2 + 5)$
19. $(10y^2 + 4y) - (6y^2 + 3y + 2)$
20. $(5m - 6) - (3m^2 - 4)$
21. $(3c^2 - 9c - 4) - (2c^2 + 9c - 5)$
22. $(4k^2 - 2k - 5) - (3k^2 + 6k + 8)$
23. $(2s^2 + 3s - 8) - (-4s^2 - 2s)$

## Connect and Extend
**Subtract.**

24. $(6 - 2r + 2r^2) - (2r^2 + 3 - r)$
25. $(p^2 - t^2 - 1) - (t^2 - p^2 - 1)$

**Simplify.**

26. $(3x^2 + 5x + 2) + (2x^2 - 3x + 5) - (x^2 - 2x + 6)$
27. $(2a^2 - 3a + 1) - (4a - a^2 + 1) - (4 - a^2 + 2a)$
28. If the difference of $x^2 + 3x + 1$ and another polynomial is 1, what is the polynomial?
29. Write 347 and 227 in expanded notation and subtract. Explain how the subtraction can be related to subtracting polynomials.

## Maintain Your Skills

**Show how you can use the Distributive Property to evaluate each expression. (Pages 59–63)**

30. $(-9)18 + (-9)2$
31. $(-6)(-28) + (-6)(228)$

**Solve. (Pages 105–109, 267–270)**

32. $20x - 4.5 = 73.5$
33. $23 = \frac{x}{12} + 5$
34. $61.9 = 4.8x - 19.7$

**Name the quadrant or axis where each point is located. (Pages 410–413)**

35. $M(1, -3)$
36. $N(-3, 1)$
37. $P(0, -1)$
38. $Q(-3, -1)$

39. The length of the hypotenuse of a right triangle is 20 meters. The length of one leg is 16 meters. Find the length of the third side. **(Pages 457–462)**

# LESSON 13-3
# Problem Solving Exploration Using a Multiplication Model

You can also use tiles to model multiplication of polynomials. In this model, you find the product by building a rectangle.

How would you model $x(2x)$?

**A.** To model $x(2x)$, place one $x$-tile at the left of the grid as shown. Then place two $x$-tiles on their sides at the top of the grid.

Use these tiles to build a rectangle that will have the same height as one $x$-tile and the same length as two $x$-tiles.

Count the tiles in the rectangle you built and write the product.

**B.** To model $x(2x + 1)$, use the model as in A above, but with a 1-tile added to the top row.

Build a rectangle having the same height as an $x$-tile and the same length as two $x$-tiles plus a 1-tile. Count the tiles in the rectangle you built and write the product.

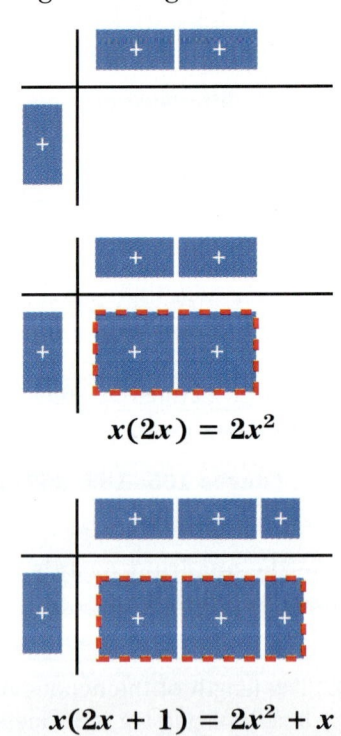

$x(2x) = 2x^2$

$x(2x + 1) = 2x^2 + x$

**C.** To model the polynomial $2x - 1$ with tiles, think of $2x - 1$ as $2x + (-1)$.

To model $(x)(2x - 1)$, start with an $x$-tile in the left side of the grid. Place two $x$-tiles and a negative 1-tile at the top of the grid.

Build a rectangle having the same height as an $x$-tile and a length equal to the length of a two $x$-tiles plus one negative 1-tile. When choosing tiles to build the rectangle, follow the rules for multiplying integers. That is, two positive $x$-tiles give a positive $x^2$-tile. A positive $x$-tile and a negative 1-tile give a negative $x$-tile.

Count the tiles. Write the product.

$x(2x - 1) = 2x^2 - x$

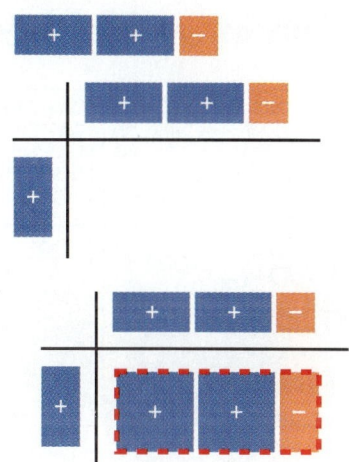

The model at the right shows the product of $(x + 1)$ and $(x - 3)$.

$(x + 1)(x - 3) = x^2 + x - 3x - 3$
$\phantom{(x + 1)(x - 3)} = \mathbf{x^2 - 2x - 3}$

## EXERCISES

Check the result obtained in each lesson model by substituting 10 for $x$.

**1.** $(x)(2x) = 2x^2$
**2.** $(x)(2x + 1) = 2x^2 + x$
**3.** $(x)(2x - 1) = 2x^2 - x$
**4.** $(x + 1)(x - 3) = x^2 - 2x - 3$

Use tiles to model each product. Check the results for each model by substituting 10 for $x$.

**5.** $x(1 - x)$
**6.** $(2x)(-2x)$
**7.** $(x + 1)(x + 2)$
**8.** $(x + 1)(x - 2)$
**9.** $(x - 1)(x + 2)$
**10.** $(x - 1)(x - 2)$
**11.** $(x + 1)(x + 1)$
**12.** $(x + 1)(x - 1)$

Compare the models you used to find the products for each given pair of exercises. Explain how the models are alike and how they are different.

**13.** Exercises 7 and 10
**14.** Exercises 8 and 9

# Calvin and Hobbes                              by Bill Watterson

## LESSON 13-4  Multiplying Binomials

The figure at the left below represents a rectangle with a height of $3x$ units and a width of $(x + 2)$ units. To find the area, you find the product of $3x$ and $(x + 2)$, or

$$3x(x + 2).$$

Notice that $3x$ is a monomial, and $(x + 2)$ is a binomial.

The figure at the right below shows how you can use tiles to find the area of the rectangle.

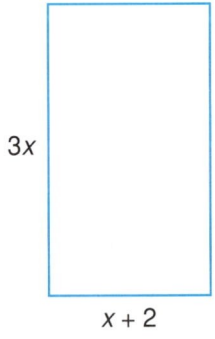 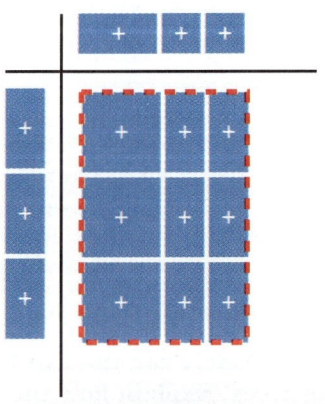

The area of the rectangle is $3x^2 + 6x$. So $3x(x + 2) = 3x^2 + 6x$.

Example 1 uses tiles to find the product of two binomials.

**EXAMPLE 1** Use tiles to find this product: $(x + 3)(x + 2)$

**Solution**

1. Build a rectangle with dimensions of $(x + 3)$ and $(x + 2)$.

2. To complete the rectangle, you need to add tiles having the width of an x-tile. So add six 1-tiles.

3. Count the tiles in the rectangle and write the product.
   $(x + 3)(x + 2) = x^2 + 2x + 3x + 6$, or $\mathbf{x^2 + 5x + 6.}$

**Try This**

1. Use tiles to find the product: $(x + 1)(x + 4)$

**EXAMPLE 2** Use tiles to find $(x + 5)(x - 4)$.

**Solution**

1. Build a rectangle with dimensions of $(x + 5)$ and $(x - 4)$.

2. To complete the rectangle, you need tiles having the same width as an x-tile. Use 20 negative 1-tiles.

3. Count the tiles in the rectangle and write the product.

$(x + 5)(x - 4) = x^2 + (-4x) + 5x + (-20)$
$= \mathbf{x^2 + x - 20}$

13-4  Multiplying Binomials

**Try This**    **Use tiles to find each product.**
     **2.** $(x + 2)(x - 1)$      **3.** $(x + 4)(x - 5)$

**EXAMPLE 3**    Use tiles to find this product:   $(3x + 1)(x + 2)$

**Solution**
1. Build a rectangle with dimensions of $(3x + 1)$ and $(x + 2)$.

2. Use positive tiles to complete the rectangle.

$(3x + 1)(x + 2) = 3x^2 + 6x + x + 2$
$= \mathbf{3x^2 + 7x + 2}$

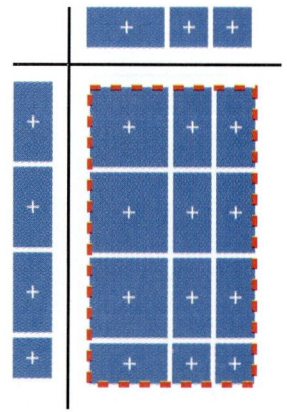

**Try This**    **Use tiles to find each product.**
     **4.** $(2x + 1)(x + 3)$      **5.** $(3x + 1)(x + 1)$

# EXERCISES

**Objectives:** To use models to multiply a binomial by a monomial
To use models to multiply two binomials

## Check Your Understanding

**Skill Check**   Write the letter of the correct answer.

**1.** Which diagram shows that $x(x - 1) = x^2 - x$?

a.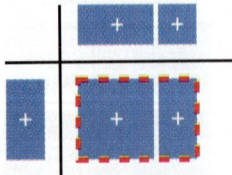

b.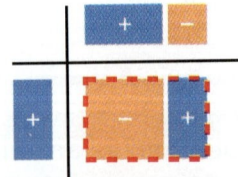

c.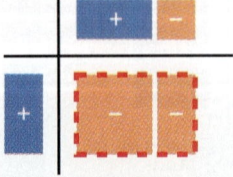

d.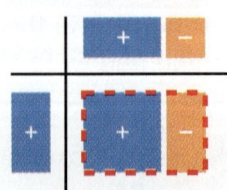

**2.** Which product is shown by the diagram?
  a. $x(2x + 2) = 2x^2 + 2x$
  b. $x(2x + 2) = 2x^2 + 2$
  c. $x(2x + 2) = 2x^2 - 2x$
  d. $x(2x + 2) = 2x^2 - 2$

**3.** Which product is shown by the diagram?
  a. $(x + 1)(x - 2) = x^2 - x - 2$
  b. $(x + 1)(x - 2) = x^2 - 3x - 2$
  c. $(x + 1)(x - 2) = x^2 + x - 2$
  d. $(x + 1)(x - 2) = x^2 + 3x - 2$

**4.** Which product is shown by the diagram?
  a. $(2x + 1)(x + 1) = 2x^2 + 2x + 1$
  b. $(2x + 1)(x + 1) = 2x^2 + 3x - 1$
  c. $(2x + 1)(x + 1) = 2x^2 + 3x + 1$
  d. $(2x + 1)(x + 1) = 2x^2 - 3x + 1$

## Practice and Apply

**Draw a diagram to show each product.**

**5.** $2x(2x + 1)$    **6.** $(x + 1)(x + 1)$    **7.** $(x + 1)(x + 3)$
**8.** $(x - 2)(x - 1)$    **9.** $(x + 3)(x - 1)$    **10.** $(2x + 3)(x + 2)$

**Find the product. Draw a diagram to help.**

**11.** $x(2x - 1)$    **12.** $(2x + 1)(2x)$    **13.** $(x + 1)(x - 1)$
**14.** $(x + 3)(x + 1)$    **15.** $(x + 2)(x - 4)$    **16.** $(x + 4)(x - 1)$
**17.** $(2x + 1)(x - 1)$    **18.** $(x + 1)(2x - 1)$    **19.** $(3x + 1)(x + 1)$
**20.** $(2x - 1)(x - 1)$    **21.** $(3x - 1)(x + 2)$    **22.** $(x + 3)(2x - 1)$

## Connect and Extend

**23.** Explain how the diagram at the right shows that $6 \cdot 23 = 138$.

|   | 20 | 3 |
|---|---|---|
| 6 | 120 | 18 |

**Draw a diagram to show the multiplication. Write the product.**

**24.** $4 \cdot 15$    **25.** $3 \cdot 125$    **26.** $7 \cdot 208$

27. Explain how this diagram shows that 14 · 12 = 168.

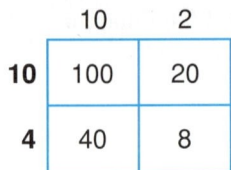

**Draw a diagram to show the multiplication. Write the product.**

28. 25 · 25
29. 12 · 21
30. 15 · 125

31. Explain why this diagram shows that $(x + 1)(x - 1) = x^2 - 1$.

|   | x | −1 |
|---|---|----|
| x | x² | −x |
| 1 | x | −1 |

## Maintain Your Skills

**Use the Distributive Property to complete each statement.**
**(Pages 59–63)**

32. $6(-d + e) = \underline{\phantom{?}}d + \underline{\phantom{?}}e$

33. $-4x - 4y = \underline{\phantom{?}}(x + \underline{\phantom{?}})$

**Solve. (Pages 69–74, 330–333)**

34. A parallelogram has a base of 22 inches and a height of 10 inches. Two inches will be added to either the base or the height. Which increase will give the greater area?

35. Derrick bought a cassette player on sale for 30% off and saved $51. What was the original price of the cassette player?

**Write the letter of the graph at the right that corresponds to each equation.**
**(Pages 430–434)**

36. $y = \frac{1}{3}x$
37. $y = \frac{1}{3}x - 1$
38. $y = \frac{1}{3}x + 1$

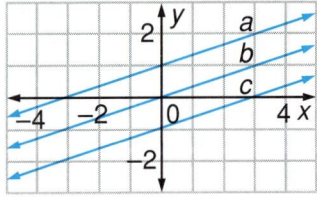

## ★ Math Team Problems

The length of one side of a square is decreased by 3 inches and the length of the base is increased by 6 inches. The area of the resulting rectangle is the same as that of the original square.

39. Write the expressions for the area of the original square and of the resulting rectangle.
40. Write an equation relating these areas.
41. Use the equation you wrote in Exercise 40 to find the area of the rectangle.

# Problem Solving Strategies
## Too Much/Too Little Data

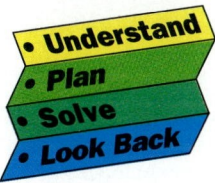

**W**hen solving a problem, it is important to identify what information is useful and whether enough information is given.

If enough information is given to solve the problem, find the solution. Then state whether you used *Clue 1 only, Clue 2 only,* or *both Clue 1 and Clue 2.* If not enough information is given, write "Can't tell."

### EXAMPLE 1

**Clue 1**  $5 = 2 \cdot x\%$
**Clue 2**  $x > 100$    What is the value of $x$?

### SOLUTION

Since $5 = 2 \cdot x\%$, $x\% = \frac{5}{2}$, or 2.5.

Since $x\% = 2.5$,  $\frac{x}{100} = 2.5$. ⟵ $x\% = \frac{x}{100}$
$\qquad\qquad\qquad x = 2.5(100)$
$\qquad\qquad\qquad x = $ **250**    **Clue 1 only** was used.

### EXAMPLE 2

**Clue 1**  Ariel has 13 nickels and dimes.
**Clue 2**  The coins have a total value of 95¢.
How many dimes does Ariel have?

### SOLUTION

Use guess and check.

**Think:** Since the total value is 95¢, Ariel has an odd number of nickels.

**Guess 1:**   5 nickels and 8 dimes
**Check 1:**   25¢ + 80¢ = $1.05   ⟵ **Too high**

**Guess 2:**   11 nickels and 2 dimes
**Check 2:**   55¢ + 20¢ = 75¢    ⟵ **Too low**

**Guess 3:**   7 nickels and 6 dimes
**Check 3:**   35¢ + 60¢ = 95¢ ✓

Ariel has **6 dimes. Both clues** were used.

# EXERCISES

**Objective:** To apply strategies in problem solving

If enough information is given to solve the problem, find the solution. Then state whether you used *Clue 1 only, Clue 2 only,* or *both Clue 1 and Clue 2*. If not enough information is given, write "Can't tell."

1. **Clue 1** Janet and her roommate divide their rent evenly.
   **Clue 2** Janet's rent is 20% more than the $200 she paid last month.
   How much is Janet's rent this month?

2. **Clue 1** Video tapes sales were 30% of Monday's total sales.
   **Clue 2** Video tapes were on sale for $4.99 each.
   What was the total of Monday's video tape sales?

3. **Clue 1** The ratio of two numbers is 2:3.
   **Clue 2** The smaller number is $\frac{2}{3}$ of the larger number.
   What is the smaller number?

4. **Clue 1** The square of an integer is between 40 and 50.
   **Clue 2** The integer is odd.
   What is the integer?

5. **Clue 1** A trophy weighs 20 ounces.
   **Clue 2** The trophy is made of pewter. Pewter is 60% tin and 20% lead.
   How much tin was used to make the trophy?

6. **Clue 1** In a 10-mile race, Beth ran the first 3 miles in 25 minutes.
   **Clue 2** She ran the final 7 miles in 1 hour 35 minutes.
   What was Beth's average speed for the race, in miles per hour?

7. **Clue 1** $s + t = 12$ and $s + w = 5$
   **Clue 2** $r + s = 15$ and $t + w = 5$
   What is the value of $r + s + t + w$?

8. **Clue 1** $10^{3k} \cdot 10^2 = 10^8$
   **Clue 2** $k$ is an integer.
   What is the value of $k$?

9. **Clue 1** One number is three more than a second number.
   **Clue 2** Both numbers are integers.
   What are the numbers?

**Add. (Pages 485–490)**

1. $(-5h - 2) + (3h + 7)$
2. $(j^2 - 3j + 4) + (4j - 3)$
3. $(-5k^2 + 3k - 7) + (-2k^2 - 3k + 6)$
4. $(-5t^2 + 1) + (4t^2 - 3)$
5. $(4a^2 - a + 5) + (-a^2 + 6a - 3)$
6. $(13 - 2r^2) + (r^2 - 4r - 6)$

**Subtract. (Pages 491–493)**

7. $(6f - 1) - (3f - 7)$
8. $(5z^2 + 2z) - (6z^2 - 5z)$
9. $(e^2 + e - 1) - (-7e + 2)$
10. $(c^2 + 5c - 2) - (-3c^2 - 5c - 8)$
11. $(-4w^2 + 2w - 1) - (3w^2 + 3w - 4)$
12. $(-v^2 + v - 1) - (-7v + 1)$

**Find the product. (Pages 496–500)**

13. $(x + 2)(3x)$
14. $2x(2x - 1)$
15. $(x + 1)(x - 2)$
16. $(x - 3)(x - 4)$
17. $(2x - 1)(x + 3)$
18. $(3x - 1)(x - 1)$

## Mathematical Footnote

David Blackwell is a natural-born teacher. He says that his chief interest is not research, but rather, "I'm interested in *understanding*, which is quite a different thing." He often goes to the chalkboard in his office to "share something beautiful with someone else."

David Blackwell grew up in a small town in southern Illinois. He credits his high-school geometry teacher with engendering his first interest in mathematics. He says that he didn't think the rest of mathematics was particularly interesting until his third year in college when he really "fell in love with mathematics." He went to college with the intention of becoming an elementary school teacher, but ended up as a distinguished professor of mathematics at the University of California in Berkeley. Blackwell has received many honors for his work in statistics, probability, game theory, and information theory.

## Calvin and Hobbes
by Bill Watterson

# LESSON 13-6
# Using the Distributive Property

**Y**ou can use tiles to model this problem.

What is the area of a rectangle with width $2x$ and length $(x + 3)$?

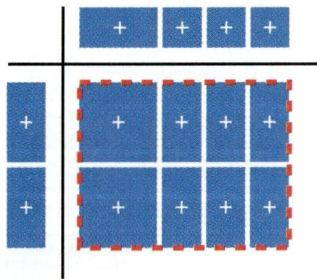

The model shows that the area is $2x^2 + 6x$.

You can also use the Distributive Property to find the product mentally.

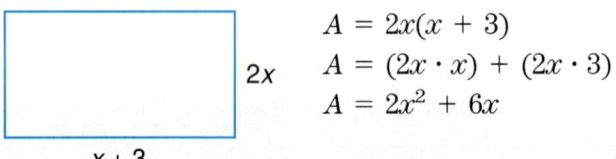

$$A = 2x(x + 3)$$
$$A = (2x \cdot x) + (2x \cdot 3)$$
$$A = 2x^2 + 6x$$

**EXAMPLE 1** Find the product using the Distributive Property.
    **a.** $x(x - 3)$      **b.** $x(2x - 5)$      **c.** $(3x + 2)(-2x)$

**Solutions**
    **a.** $x(x - 3) = x \cdot x - (x \cdot 3)$
                   $= x^2 - 3x$

    **b.** $x(2x - 5) = x \cdot 2x - (x \cdot 5)$
                     $= 2x^2 - 5x$

    **c.** $(3x + 2)(-2x) = 3x \cdot (-2x) + 2 \cdot (-2x)$
                         $= -6x^2 - 4x$

**Try This** Find the product using the Distributive Property.
    **1.** $5x(x + 3)$      **2.** $(x - 5)(-3x)$      **3.** $-x(x - 1)$

Recall that the Distributive Property states that for all numbers $a$, $b$, and $c$,

$$a(b + c) = ab + ac.$$

This generalization can be used to find the product of two binomials. For example,

$$\begin{aligned}(x + 2)(x + 4) &= (x + 2)x + (x + 2)4 \\ &= x(x + 2) + 4(x + 2) \\ &= x^2 + 2x + 4x + 8 \\ &= x^2 + 6x + 8\end{aligned}$$

**EXAMPLE 2** Multiply.
    **a.** $(x + 1)(x + 4)$      **b.** $(x + 3)(2x + 1)$      **c.** $(x - 3)(x - 1)$

**Solutions**
    **a.** $(x + 1)(x + 4) = x(x + 4) + 1(x + 4)$
                          $= x^2 + 4x + x + 4$
                          $= x^2 + 5x + 4$

    **b.** $(x + 3)(2x + 1) = x(2x + 1) + 3(2x + 1)$
                           $= 2x^2 + x + 6x + 3$
                           $= 2x^2 + 7x + 3$

    **c.** $(x - 3)(x - 1) = x(x - 1) - 3(x - 1)$
                         $= x^2 - x - 3x + 3$
                         $= x^2 - 4x + 3$

**Try This**  Multiply.
4. $(x + 2)(x - 1)$
5. $(2x + 3)(x - 2)$
6. $(x - 3)(2x - 4)$

The Distributive Property can be used to multiply with a vertical arrangement.

$$\begin{array}{r} x + 2 \\ \underline{x + 3} \\ 3x + 6 \\ \underline{x^2 + 2x\phantom{00}} \\ x^2 + 5x + 6 \end{array} \begin{array}{l} \leftarrow 3(x + 2) \\ \leftarrow x(x + 2) \\ \leftarrow x^2 + (3x + 2x) + 6 \end{array}$$

- How can this method be used to find the product of 24 and 15?

# EXERCISES

**Objective:** To multiply polynomials using the Distributive Property

## Check Your Understanding

**Skill Check**  Write the letter of the correct answer.

1. Which is the product of $x$ and $(x - 1)$?
   a. $x^2 - 1$
   b. $x^2 + x$
   c. $x^2 + 1$
   d. $x^2 - x$

2. Find the product of $(2x - 3)$ and $2x$.
   a. $4x^2 - 3$
   b. $4x^2 - 6x$
   c. $-4x^2 + 6x$
   d. $-4x^2 - 6x$

3. Multiply: $(x + 1)(x - 2)$
   a. $x^2 - x - 2$
   b. $x^2 + 3x - 2$
   c. $x^2 - 3x + 2$
   d. $x^2 - 3x - 2$

4. Multiply: $(2x - 1)(x + 2)$
   a. $2x^2 - 2$
   b. $2x^2 - 3x - 2$
   c. $2x^2 - x - 2$
   d. $2x^2 + 3x - 2$

## Practice and Apply

Multiply.

5. $(a + 2)(a + 3)$
6. $(d + 4)(d - 3)$
7. $(x - 4)(x - 2)$
8. $(c - 1)(c - 4)$
9. $(2x - 1)(x - 5)$
10. $(d + 3)(2d + 3)$
11. $(x + 9)(x - 9)$
12. $(4m - 3)(2m + 3)$
13. $(5r + 3)(r - 4)$
14. $(q - 6)(q - 6)$
15. $(n - 1)(2n + 3)$
16. $(p + 2)(2p - 1)$
17. $(-2z + 3)(z - 1)$
18. $(4b - 3)(b - 4)$
19. $(2y - 1)(-y + 3)$
20. $(s - 4)(-s + 4)$
21. $(-t - 3)(-2t + 1)$
22. $(-5a - 1)(-a - 1)$

Find the area of each figure.

**23.** Rectangle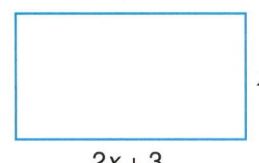

**24.** Square

**25.** Right Triangle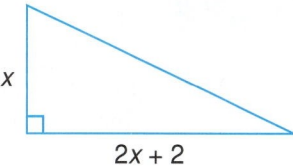

A mental shortcut used to multiply two binomials is called the FOIL Method.

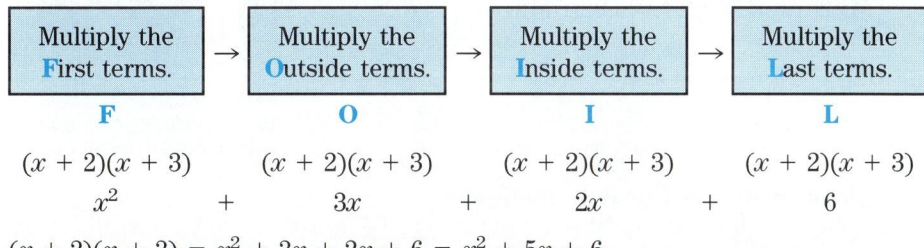

$(x + 2)(x + 3) = x^2 + 3x + 2x + 6 = x^2 + 5x + 6$

**Use the FOIL Method to multiply.**

**26.** $(x + 1)(x + 2)$
**27.** $(a + 2)(a - 3)$
**28.** $(b - 1)(b - 2)$
**29.** $(2d + 1)(d - 3)$
**30.** $(p + 1)(2p - 2)$
**31.** $(2q - 3)(-q - 1)$

## Connect and Extend

**Explain how the FOIL Method can be used to find each product.**

**32.** $(20 + 1)(20 - 1)$
**33.** $13 \cdot 45$
**34.** $2\frac{1}{2} \cdot 1\frac{1}{5}$

**Find the area of the shaded region in each rectangle.**

**35.**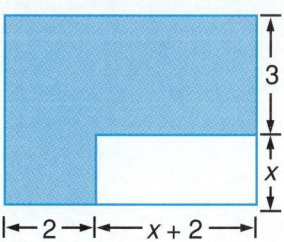

**36.**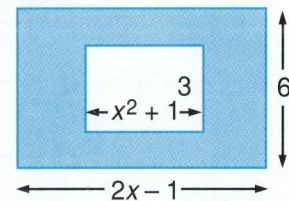

## Maintain Your Skills

**Simplify. (Pages 150–153)**

**37.** $24 + (-5)^3$
**38.** $(10 + 7)^2$
**39.** $(-4)^3 + (1 - 7)^2$
**40.** $(8 - 6)^5 - (9 - 4)^2$

**Simplify. (Pages 267–270)**

**41.** $-\frac{5}{6} \cdot -\frac{4}{5}$
**42.** $\frac{5}{6} \cdot (-9)$
**43.** $-\frac{2}{3} \div \frac{1}{3}$
**44.** $-\frac{10}{3} \div -2$

# LESSON 13-7 Special Products

**W**hat is the area of a square with sides of $x + 2$?

To find the area, you will need to find $(x + 2)(x + 2)$, or $(x + 2)^2$. Here are three ways to solve the problem.

**Method 1:** Use a model (see the figure at the right).

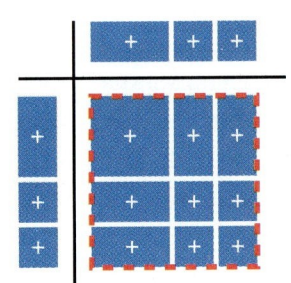

**Method 2:** Use the Distributive Property.
$$(x + 2)(x + 2) = x(x + 2) + 2(x + 2)$$
$$= x^2 + 2x + 2x + 4$$
$$= x^2 + 4x + 4$$

**Method 3:** Use the FOIL method.

$(x + 2)(x + 2) = x^2 + 4x + 2$

Using each method, you have shown that $(x + 2)^2 = x^2 + 4x + 4$.

**EXAMPLE 1** Find each product.
   **a.** $(x + 1)^2$            **b.** $(x + 3)^2$

**Solutions**
**a.** $(x + 1)^2 = (x + 1)(x + 1)$
$\phantom{(x + 1)^2} = x^2 + x + x + 1$
$\phantom{(x + 1)^2} = \mathbf{\mathit{x}^2 + 2\mathit{x} + 1}$

**b.** $(x + 3)^2 = (x + 3)(x + 3)$
$\phantom{(x + 3)^2} = x^2 + 3x + 3x + 9$
$\phantom{(x + 3)^2} = \mathbf{\mathit{x}^2 + 6\mathit{x} + 9}$

**Try This** Multiply.
  **1.** $(x + 5)^2$      **2.** $(x + 7)^2$      **3.** $(x + 11)^2$

## EXPLORE

Look for the patterns in the squares of these binomials.
$$(x + 1)^2 = x^2 + 2x + 1 \qquad (x + 2)^2 = x^2 + 4x + 4$$
$$(x + 3)^2 = x^2 + 6x + 9$$

1. What patterns do you see?
2. Predict the result of $(x + 5)^2$.
3. Check your prediction by using one of the methods on page 508.

Find each product:  4. $(x - 1)^2$   5. $(x - 2)^2$   6. $(x - 3)^2$

7. Describe the pattern in Exercises 4–6.
8. How are the products of $(x + 2)^2$ and $(x - 2)^2$ alike? How are they different?

The square of a binomial is called a **perfect square trinomial.**

To find the square of a binomial:

a. Square the first term.
b. Find twice the product of the two terms. The sign of this product will be the same as the sign that comes before the second term.
c. Square the second term.

The pattern for perfect square trinomials is generalized below.

> **Perfect Square Trinomial Pattern**
> $(a + b)^2 = a^2 + 2ab + b^2$
> $(a - b)^2 = a^2 - 2ab + b^2$

**EXAMPLE 2**  Find $(2x - 1)^2$.

**Solution**  Use the generalization for perfect square trinomials.
$$(2x - 1)^2 = (2x)^2 - (2)(2x)(1) + (1)^2$$
$$4x^2 - 4x + 1$$

**Try This**  Find each product:  4. $(x + 8)^2$   5. $(c - 6)^2$   6. $(2b + 3)^2$

## EXPLORE

**Find the area of the rectangle.**

9.

$x - 2$

$x + 2$

10.

$x - 3$

$x + 3$

11.

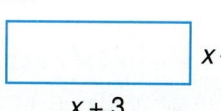

$x - 4$

$x + 4$

13-7  Special Products

12. Describe the pattern that appears in the product when you find the area in Exercises 9–11.

To find the product of sum and difference of the same two terms:

   a. Square the first term.
   b. Subtract the square of the second term.

The generalization for the product of the sum and difference of two terms is shown below.

> **Product of the Sum and Difference of the Same Two Terms**
> $$(a + b)(a - b) = a^2 - b^2$$

**EXAMPLE 3**  Multiply: $(2c + 1)(2c - 1)$

**Solution**  Use the generalization for the product of the sum and difference of the same two terms.

$$(2c + 1)(2c - 1) = (2c)^2 - (1)^2$$
$$= 4c^2 - 1$$

**Try This**  Multiply.

7. $(a - 5)(a + 5)$     8. $(2a + 4)(2a - 4)$     9. $(3a + 2)(3a - 2)$

# EXERCISES

**Objectives:** To square a binomial
To find the product of the sum and difference of the same two terms

## Check Your Understanding

*Skill Check*  Write the letter of the correct answer.

1. Which is the product of $(x + 2)^2$?
   a. $x^2 + 4$
   b. $x^2 + 2$
   c. $x^2 + 2x + 4$
   d. $x^2 + 4x + 4$

2. Which of these is a perfect square trinomial?
   a. $x^2 + 6x + 3$
   b. $x^2 + 6x + 9$
   c. $x^2 + 3x + 9$
   d. $x^2 + 4x + 9$

**3.** Which statement is true?
  **a.** $(a + b)(a - b) = a^2 + b^2$
  **b.** $(c + d)(c - d) = c^2 + d^2$
  **c.** $(2 - a)(2 + a) = 2 - a^2$
  **d.** $(6 + b)(6 - b) = 36 - b^2$

**4.** Which is the product of $(x - 1)^2$?
  **a.** $x^2 - 2x + 1$   **b.** $x^2 - x - 1$
  **c.** $x^2 + 2x - 1$   **d.** $x^2 - 2x - 1$

## Practice and Apply
Find the product.
**5.** $(c + 5)^2$
**6.** $(a + 2)(a - 1)$
**7.** $(m - 6)^2$
**8.** $(2t - 1)^2$
**9.** $(2 + r)^2$
**10.** $(3p + 2)(3p - 2)$
**11.** $(2y - 3)^2$
**12.** $(4x - 1)(4x + 1)$
**13.** $(q + 1)^2$
**14.** $(3 - a)(3 + a)$
**15.** $(s - 4)^2$
**16.** $(3d - 4)^2$
**17.** $(-a + 1)^2$
**18.** $(-2t + 3)^2$
**19.** $(-p + 3)(-p - 3)$
**20.** $\left(\frac{1}{2}a + 1\right)\left(\frac{1}{2}a - 1\right)$
**21.** $\left(b + \frac{1}{2}\right)^2$
**22.** $\left(t - \frac{2}{3}\right)^2$

## Connect and Extend
You can square a number by writing it as a sum or difference of two numbers.

$(22)^2 = (20 + 2)^2 = (20)^2 + 2(20)(2) + (2)^2 = 400 + 80 + 4 = 484$

$(39)^2 = (40 - 1)^2 = (40)^2 - 2(40)(1) + (1)^2 = 1600 - 80 + 1 = 1521$

**Use this pattern to square each number.**
**23.** $(8.1)^2$   **24.** $(9.9)^2$   **25.** $\left(1\frac{1}{4}\right)^2$   **26.** $\left(5\frac{3}{4}\right)^2$

Use the pattern for the product of the sum and difference of the same two numbers to find these products.
**27.** $(10 + 5)(10 - 5)$   **28.** $(31)(29)$   **29.** $(56)(64)$   **30.** $(4.4)(3.6)$

**31.** The area of a square is $x^2 + 4x + 4$. What is the perimeter of the square?

## Maintain Your Skills
**Find the GCF of each pair of numbers. (Pages 154–158)**
**32.** 35 and 84   **33.** 42 and 105   **34.** 50 and 78

**Evaluate each expression. (Pages 262–266)**
**35.** $\dfrac{a^2 - 3ab}{-6(a + b) - 2}$, for $a = 4$ and $b = 7$
**36.** $\dfrac{(e + t)^2}{t^2 - e^2}$, for $e = 10$ and $t = -8$

**Write each number in scientific notation. (Pages 289–294)**
**37.** 48,000   **38.** 0.0031   **39.** 0.0396   **40.** 0.000104

## LESSON 13-8: Common Factors

**To factor** an expression means to write it as a product. You can factor a polynomial if all of its terms contain a common factor.

For example, the greatest common factor of 6x and 12 is 6. To factor $6x + 12$, you "factor out" its greatest common factor, 6.

$$6x + 12 = 6 \cdot x + 6 \cdot 2$$
$$= 6(x + 2)$$

Notice that you have used the Distributive Property in reverse.

**EXAMPLE 1** Factor, if possible.
 a. $5x + 15$  b. $6c - 8$  c. $12m - 16$  d. $3x + 2$

**Solutions** First, identify the whole number that is the greatest common factor of both terms. Then factor.

| Binomial | GCF | Factored Form |
|---|---|---|
| a. $5x + 15$ | 5 | $5 \cdot x + 5 \cdot 3 =$ **5(x + 3)** |
| b. $6c - 8$ | 2 | $2 \cdot 3c - 2 \cdot 4 =$ **2(3c - 4)** |
| c. $12m - 16$ | 4 | $4 \cdot 3m - 4 \cdot 4 =$ **4(3m - 4)** |
| d. $3x + 2$ | 1 | **Not possible** |

**Try This** Factor, if possible.
 1. $4x + 6$  2. $5a + 6$  3. $16m - 24$  4. $30d - 42$

Sometimes the greatest common factor of two terms contains a variable. For example, the GCF of $6x^2$ and $8x$ is $2x$. To factor $6x^2 + 8x$, you "factor out" $2x$.

$$6x^2 + 8x = (2x)(3x) + (2x)(4)$$
$$= 2x(3x + 4)$$

You can use the Distributive Property to check.

$$2x(3x + 4) = (2x)(3x) + (2x)(4)$$
$$= 6x^2 + 8x \checkmark$$

Since the product is the same as the original binomial, the factoring is correct.

**EXAMPLE 2** Factor.

    **a.** $a^2 + 6a$      **b.** $2b^2 + 6b$      **c.** $10n^2 + 5n$

**Solutions** First, identify the GCF of both terms. Then factor.

| | Binomial | GCF | Factored Form |
|---|---|---|---|
| **a.** | $a^2 + 6a$ | $a$ | $a \cdot a + 6 \cdot a =$ **$a(a + 6)$** |
| **b.** | $2b^2 + 6b$ | $2b$ | $(2b)(b) + (2b)3 =$ **$2b(b + 3)$** |
| **c.** | $10n^2 + 5n$ | $5n$ | $(5n)(2n) + (5n)(1) =$ **$5n(2n + 1)$** |

**Try This** Factor.

    **5.** $6t^2 + 3t$      **6.** $4s^2 - 4s$      **7.** $10p^2 - 12p$

An expression can also be factored if each of its terms contains a **common binomial factor.** For example, for the expression,

$$3(x + 1) + x(x + 1),$$

the common binomial factor is $(x + 1)$. So you "factor out" $(x + 1)$.

$$3(x + 1) + x(x + 1) = (3 + x)(x + 1)$$

• How can you check whether you factored correctly?

**EXAMPLE 3** Factor.

    **a.** $x(x + 4) + 2(x + 4)$      **b.** $(x + 5) \cdot x + (x + 5) \cdot 3$
    **c.** $x(x + 5) - 4(x + 5)$      **d.** $3x(x - 2) - 5(x - 2)$

**Solutions** First, identify the common binomial factor. Then factor the expression.

    **a.** GCF: $(x + 4)$      $x(x + 4) + 2(x + 4) =$ **$(x + 4)(x + 2)$**
    **b.** GCF: $(x + 5)$      $(x + 5) \cdot x + (x + 5) \cdot 3 =$ **$(x + 5)(x + 3)$**
    **c.** GCF: $(x + 5)$      $x(x + 5) - 4(x + 5) =$ **$(x + 5)(x - 4)$**
    **d.** GCF: $(x - 2)$      $3x(x - 2) - 5(x - 2) =$ **$(x - 2)(3x - 5)$**

**Try This** Factor.

    **8.** $2(x + 1) + x(x + 1)$      **9.** $y(y - 3) - 1(y - 3)$
    **10.** $2y(y + 1) - 3(y + 1)$      **11.** $(t - 4) \cdot t + (t - 4) \cdot 9$

• How can you use mental math to simplify this expression?

$$3(x + 5) + 7(x + 5) - 2(x + 5)$$

13-8    Common Factors

# EXERCISES

**Objective:** To factor polynomials by finding common monomial or binomial factors

## Check Your Understanding

**Skill Check** Write the letter of the correct answer.

**1.** Which of these statements is **not** true?
  **a.** The GCF of $5x$ and 10 is 2.
  **b.** The GCF of $3x$ and 2 is 1.
  **c.** The GCF of $8x$ and 12 is 4.
  **d.** The GCF of $10x$ and 15 is 5.

**2.** Which is the factored form of $24x - 32$?
  **a.** $2(3x - 4)$
  **b.** $4(6x - 8)$
  **c.** $8(3x + 4)$
  **d.** $8(3x - 4)$

**3.** Which is the factored form of $9b^2 - 3b$?
  **a.** $3(3b - 1)$  **b.** $3(3b - 3)$
  **c.** $3b(b - 3)$  **d.** $3b(3b - 1)$

**4.** Which is the factored form of $4(x - 3) - x(x - 3)$?
  **a.** $(x - 3)(4 - x)$
  **b.** $(x - 3)(4 + x)$
  **c.** $(x - 3)(x + 4)$
  **d.** $(x - 3) + (4 + x)$

## Practice and Apply

**Factor, if possible.**

**5.** $8a + 4$
**6.** $3d - 2$
**7.** $5y - 15$
**8.** $7m + 7$
**9.** $8m + 20$
**10.** $9q + 16$
**11.** $25t - 15$
**12.** $12y - 16$
**13.** $6y - 36$
**14.** $14y + 21$
**15.** $6p - 9$
**16.** $20c + 24$

**Factor.**

**17.** $2d^2 + 4d$
**18.** $6x^2 - 9x$
**19.** $15x^2 + 10x$
**20.** $8y^2 + 8y$
**21.** $12m^2 - 16m$
**22.** $14n^2 - 21n$
**23.** $18p^2 + 27p$
**24.** $15c^2 - 30c$
**25.** $40a^2 - 16a$

**Factor.**

**26.** $y(y - 1) + 2(y - 1)$
**27.** $b(b + 3) - 3(b + 3)$
**28.** $a(a - 5) + 8(a - 5)$
**29.** $2t(t + 4) - 3(t + 4)$
**30.** $5c(c - 3) + 4(c - 3)$
**31.** $(x + 1) \cdot 2 + (x + 1) \cdot 3$
**32.** $5n(n - 1) + 4(n - 1)$
**33.** $-2(p + 1) + 3(p + 1)$
**34.** $8(m - 4) + m(m - 4)$
**35.** $8(t - 1) - 1(t - 1)$

514  CHAPTER 13

## Connect and Extend

36. Explain how factoring can be used to show that $(x^2 + 5x) \div x = x + 5$.
37. Explain why this sentence is true.
    If a binomial has one monomial factor, then the other factor is a binomial.
38. Explain how factoring can be used to simplify $\dfrac{6a^2 + 3a}{3a}$. Write the rational expression in lowest terms.
39. The area of a square is $121c^2$. Find the length of a side.
40. The area of a rectangle is $15b^2 - 10b$. The length is $5b$. What is the width?

## Maintain Your Skills

**Solve. (Pages 236–246)**

41. One postcard will be drawn out of 300 received for a ticket to a playoff game. You mailed 15 postcards. What are your chances of getting the ticket?
42. A bag contains 4 red cubes, 5 black cubes, and one white cube. One cube is drawn at random and not replaced. Then a second cube is drawn. Find P(white, then red).

**Find the values of $x$ and $y$ in the ordered pairs, $(x, 0)$ and $(0, y)$, for each equation. (Pages 435–439)**

43. $x + y = -2$
44. $2y - x = 6$
45. $4x + 2y = 5$

**Solve. (Pages 474–480, 496–500)**

46. A pole is supported by a wire anchored to the ground. How high is the pole, $p$? Round the height to the nearest tenth of a meter.

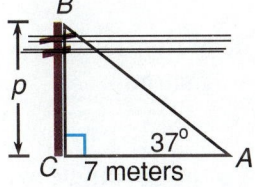

47. The length of a rectangle is $(x + 7)$ and the width is $(x - 6)$. Find the area of the rectangle.
48. The length of a side of a square is $(2x - 5)$. Find the area of the square.

## ★ Math Team Problems

49. In the figure at the right, a belt runs over six wheels in the direction of the arrows. How many wheels are turning clockwise?
50. Six water glasses are placed in a row. Three of the glasses are empty and three are full. The arrangement of the glasses is empty, empty, empty, full, full, full. By moving exactly one glass, change the arrangement to empty, full, empty, full, empty, full.

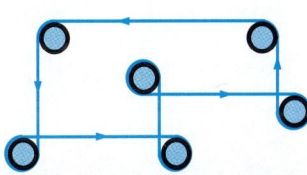

13-8  Common Factors  **515**

# LESSON 13-9

## Factoring a Trinomial

You have used tiles to find the area of a rectangle. If the area of a rectangle is given as $x^2 + 5x + 6$, can you factor to find expressions for the length and the width? Can you use tiles to factor $x^2 + 5x + 6$?

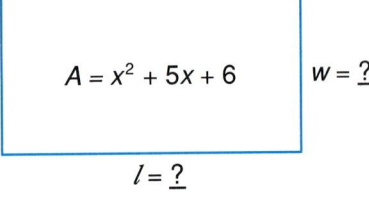

**a.** Use tiles to model $x^2 + 5x + 6$.

**b.** Build a rectangle. Position the $x^2$-tile. Then arrange the $x$-tiles.

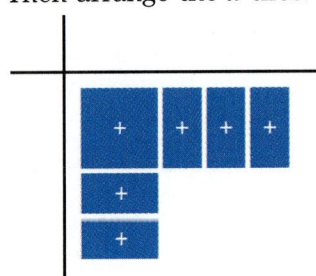

**c.** Complete the rectangle with the six 1-tiles.

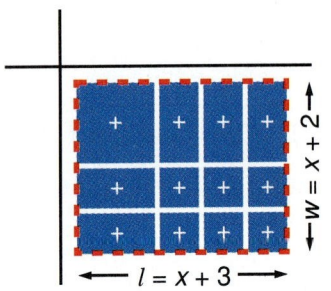

**d.** Write the length and width of the rectangle. These are the factors.

**Factors:** $(x + 3)(x + 2)$

So the rectangle with area $x^2 + 5x + 6$ can have a length of $(x + 3)$ and a width of $(x + 2)$. To check, use the Distributive Property or the FOIL method.

**Check:** Using the Distributive Property:
$$(x + 3)(x + 2) = x(x + 2) + 3(x + 2)$$
$$= x \cdot x + x \cdot 2 + 3 \cdot x + 3 \cdot 2$$
$$= x^2 + 5x + 6$$

**Check:** Using the FOIL method:

$$(x + 3)(x + 2) = x^2 + 2x + 3x + 6$$
$$= x^2 + 5x + 6$$

By examining the product of two binomials, you can find clues that will help you to factor a trinomial.

$$(x + 3)(x + 2) = x^2 + 2x + 3x + 2 \cdot 3$$
$$= x^2 + 5x + 6$$

Notice that the last term of $x^2 + 5x + 6$ is the product of $2 \cdot 3$, and that the numerical coefficient, 5, of the middle term equals the sum, $2 + 3$.

So to factor $x^2 + 5x + 6$, think of completing the pattern,

$$(x + ?)(x + ?).$$

You need to find two numbers which, when added give 5, *and* when multiplied give 6.

First, write pairs of integers whose product is 6.

$$1 \cdot 6 \qquad 2 \cdot 3 \qquad (-1) \cdot (-6) \qquad (-3) \cdot (-2)$$

Identify the pair having a sum of 5.      2 and 3

The factors are $(x + 2)$ and $(x + 3)$.

**EXAMPLE 1**   Factor $x^2 + 6x + 8$.

**Solution**   **Think:** To factor $x^2 + 6x + 8$, find two numbers whose product is 8 and whose sum is 6.

| Integers with Product of 8 | Sum of Integers |
|---|---|
| $1 \cdot 8$ | $1 + 8 = 9$ |
| $-1 \cdot (-8)$ | $-1 + (-8) = -9$ |
| $2 \cdot 4$ | $2 + 4 = 6$ ← This pair works! |

So, $x^2 + 6x + 8 =$ **$(x + 2)(x + 4)$**.

**Try This**   Factor each trinomial.
1. $x^2 + 3x + 2$      2. $x^2 + 7x + 10$      3. $x^2 + 7x + 12$

What are the factors of $x^2 - 2x - 3$? This time, you have to complete the pattern $(x + ?)(x + ?)$ by finding two numbers whose product is $-3$ and whose sum is $-2$.

| Integers with Product of $-3$ | Sum of the Integers |
|---|---|
| $-1 \cdot 3$ | $-1 + 3 = 2$ |
| $-3 \cdot 1$ | $-3 + 1 = -2$ ← This works! |

So $x^2 - 2x - 3 = [x + (-3)][x + 1]$, or **$(x - 3)(x + 1)$**.

**EXAMPLE 2** Factor.
    **a.** $x^2 + x - 6$      **b.** $x^2 - 2x - 15$      **c.** $x^2 - x - 6$

**Solutions**

**a.** Factors of $-6$:    $-1 \cdot 6$    $1 \cdot (-6)$    $-2 \cdot 3$
    Sum of factors:     5           $-5$         1 ✓

So $x^2 + x - 6 = (x - 2)(x + 3)$.

**b.** Factors of $-15$:    $-15 \cdot 1$    $15 \cdot (-1)$    $3 \cdot (-5)$
    Sum of factors:     $-14$        14        $-2$ ✓

So $x^2 - 2x - 15 = (x + 3)(x - 5)$.

**c.** Factors of $-6$:    $-6 \cdot 1$    $6 \cdot (-1)$    $2 \cdot (-3)$
    Sum of factors:     $-5$          5        $-1$ ✓

So $x^2 - x - 6 = (x + 2)(x - 3)$.

**Try This**    Factor.
**4.** $x^2 + x - 2$      **5.** $x^2 - 3x - 10$      **6.** $x^2 - 7x + 12$

# EXERCISES

**Objective:** To factor a quadratic trinomial

## Check Your Understanding

**Skill Check**   Write the letter of the correct answer.

**1.** Which two integers have a sum of $-13$ and a product of 36?
    **a.** 3 and 12
    **b.** $-18$ and 2
    **c.** $-6$ and $-7$
    **d.** $-4$ and $-9$

**2.** Which two integers have a sum of 1 and a product of $-72$?
    **a.** $-9$ and $-8$
    **b.** $-9$ and 8
    **c.** $-8$ and 9
    **d.** 8 and 9

**3.** Which are the factors of $x^2 + x - 20$?
    **a.** $(x - 4)$ and $(x - 5)$
    **b.** $(x + 5)$ and $(x + 4)$
    **c.** $(x + 5)$ and $(x - 4)$
    **d.** $(x + 4)$ and $(x - 5)$

**4.** Which are the factors of $x^2 - x - 20$?
    **a.** $(x + 5)$ and $(x + 4)$
    **b.** $(x - 5)$ and $(x - 4)$
    **c.** $(x + 5)$ and $(x - 4)$
    **d.** $(x - 5)$ and $(x + 4)$

## Practice and Apply

**Factor.**

5. $x^2 + 3x + 2$
6. $y^2 + 7y + 10$
7. $s^2 - s - 6$
8. $p^2 + 4p - 12$
9. $x^2 + 6x + 5$
10. $q^2 - 4q - 5$
11. $t^2 - 6t - 27$
12. $c^2 - 12c - 28$
13. $d^2 + 8d + 15$
14. $x^2 - 16x + 64$
15. $a^2 - 14a - 51$
16. $h^2 + 7h + 6$
17. $m^2 - 14m + 49$
18. $v^2 + 7v - 18$
19. $x^2 - 3x - 10$
20. $a^2 + 10a + 25$
21. $z^2 + 16z - 17$
22. $n^2 - n - 56$

## Connect and Extend

23. The area of a rectangle is $x^2 + x - 2$. If the length is $x + 2$, what is the width?
24. A car travels $(x^2 + 9x + 14)$ miles in $(x + 2)$ hours. Write an expression for the car's speed in miles per hour.

**Graph $y = x^2 + 2x - 3$ on a coordinate plane. Refer to your graph for Exercises 25–28.**

25. Is the graph linear?
26. In how many places does the graph cross the $x$-axis?
27. What are the $x$-coordinates of the points where the graph crosses the $x$-axis?
28. The solution set of $(x - 1)(x + 3) = 0$ is $\{1, -3\}$. How do the $x$-coordinates you found in Exercise 26 relate to these solutions?
29. Graph $y = x^2 + 2x - 8$ in a coordinate plane. Find the $x$-coordinates, that is, where the graph crosses the $x$-axis. Write a paragraph summarizing how you did this.

## Maintain Your Skills

30. Show how you can use prime factorizations to prove that $15 \cdot 77 = 21 \cdot 55$. **(Pages 143–146)**

31. In Pedro's Restaurant, a customer has 6 sandwiches and 4 salads to choose from. How many different sandwich and salad orders are possible? **(Pages 252–257)**

**Write a mixed number or a fraction in lowest terms for each decimal. (Pages 271–274)**

32. $-0.6$
33. $0.75$
34. $7.08$
35. $-2.625$

**Find two solutions for each inequality. (Pages 377–381)**

36. $y > -2x - 4$
37. $3y - 6x \geq -12$
38. $4(x - y) > 20$

# SUMMARY

## KEY TERMS
Binomial (p. 486)  
Factor (p. 512)  
Monomial (p. 486)  
Perfect square trinomial (p. 509)  
Polynomial (p. 486)

## KEY IDEAS

**A.** To add polynomials, add the coefficients of the like terms, and add the constant (numerical) terms.

**B.** To subtract a polynomial, add its opposite.

**C. The FOIL method for Finding the Product of Two Binomials**
Add the product of the first terms, the product of the outside terms, the product of the inside terms, and the product of the last terms.

**D.** To find the square of a binomial:
1. Square the first term.
2. Find twice the product of the two terms. The sign of this product will be the same as the sign that comes before the second term.
3. Add to this the square of the second term.

**E.** To find the product of the sum and difference of the same two terms:
1. Square the first term.
2. Subtract the square of the second term.

# REVIEW

**Add. (Pages 485–490)**

**1.** $(5c - 4) + (c + 3)$
**2.** $(8g - 6) + (-9g + 7)$
**3.** $(h^2 - 6h + 7) + (7h + 10)$
**4.** $(-w^2 + 8w - 15) + (4w^2 + 6w + 9)$
**5.** $(4e^2 - 5e) + (2e^2 - e)$
**6.** $(-3a^2 - 7) + (5a + 10)$
**7.** $(5r^2 - r + 3) + (-r^2 - r - 5)$
**8.** $(7r^2 - 2r - 9) + (3r^2 + 6r + 2)$

**Subtract. (Pages 491–493)**

**9.** $(3b + 4) - (2b + 8)$
**10.** $(-7f - 5) - (10f - 3)$
**11.** $(-6t^2 + 3t - 1) - (4t^2 - 2t + 3)$
**12.** $(q^2 + 5q + 11) - (q^2 - 10q + 25)$
**13.** $(2s^2 - 4) - (7s^2 + 2s)$
**14.** $(5h^2 - 2h - 3) - (-6h - 4)$

**Find each product. (Pages 496–499)**

**15.** $j(7j + 8)$
**16.** $(3g)(g - 9)$
**17.** $(k + 4)(k - 4)$
**18.** $(m - 2)(m + 6)$
**19.** $(2a - 5)(a - 3)$
**20.** $(3x - 2)(x - 6)$

**If enough information is given to solve the problem, find the solution. Then state whether you used *Clue 1 only*, *Clue 2 only*, or *both Clue 1 and Clue 2*. If not enough information is given, write "Can't tell." (Pages 501–502)**

**21.** **Clue 1** Tamara has 12 quarters and nickels.
    **Clue 2** The coins have a total value of $1.20.

How many nickels does Tamara have?

**22.** **Clue 1** Fruit juices are sold in two quantities, 32 ounces and 64 ounces.
    **Clue 2** Mrs. Hawkins spent $12.59 on fruit juices.

How many of each size did she buy?

**Multiply. (Pages 504–507)**

**23.** $(u + 6)(u + 5)$
**24.** $(v + 4)(v - 7)$
**25.** $(3z - 8)(z - 7)$
**26.** $(-4y - 5)(y + 9)$
**27.** $(-6g + 1)(-g - 1)$
**28.** $(-7e - 2)(-e - 2)$

**Find the area of each figure. (Pages 504–507)**

**29.**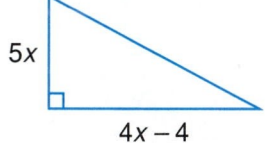
5x, 4x − 4

**30.**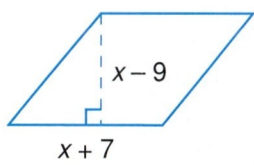
x − 9, x + 7

**Find the product. (Pages 508–511)**

**31.** $(h + 6)(h - 6)$
**32.** $(n + 8)(n - 8)$
**33.** $(2t + 5)^2$
**34.** $(-4t + 7)(-4t - 7)$
**35.** $(-9t - 2)^2$
**36.** $\left(-\frac{1}{2}g - \frac{4}{5}\right)^2$

**Factor, if possible. (Pages 512–515)**

**37.** $12r - 30$
**38.** $17s + 84$
**39.** $16p - 36$
**40.** $4w^2 + 28w$
**41.** $84z^2 - 63z$
**42.** $10v^2 - 10v$

**Factor. (Pages 516–519)**

**43.** $x^2 + 6x + 8$
**44.** $b^2 + 3b - 10$
**45.** $c^2 - 9c + 18$
**46.** $g^2 - g - 72$
**47.** $k^2 - 4k + 3$
**48.** $t^2 + 6t - 72$

# TEST

**Add.**
1. $(-r + 3) + (4r - 7)$
2. $(-7p^2 - 7p) + (6p^2 + 5p)$
3. $(2w^2 - 3w + 5) + (-w^2 + w - 12)$
4. $(-3u^2 + 4u - 1) + (u^2 + u)$
5. $(-6c^2 + 10) + (-9c - 12)$
6. $(-e^2 - 5e + 7) + (-4e^2 + 5e + 9)$

**Subtract.**
7. $(5h - 7) - (3h - 2)$
8. $(-4q^2 + 3q) - (-5q^2 + 2q)$
9. $(-3v^2 + 7v - 5) - (4v^2 - v + 10)$
10. $(-a^2 + 2a - 6) - (7a^2 - 3a - 11)$
11. $(12c^2 - c) - (9c^2 - 6)$
12. $(d^2 + d - 1) - (-7d + 2)$

**Find the product.**
13. $2k(k - 4)$
14. $(-v)(v + 8)$
15. $(g - 6)(g + 7)$
16. $(2x - 3)(x - 1)$
17. $(-2j - 5)(-j - 4)$
18. $(3y + 4)(-y - 6)$
19. $(-k - 12)(-k + 12)$
20. $(4z + 5)^2$
21. $(-6m - 2)^2$
22. $(3v - 7)(3v + 7)$

**Find the area of each figure.**

23.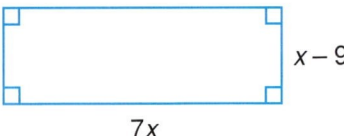

24.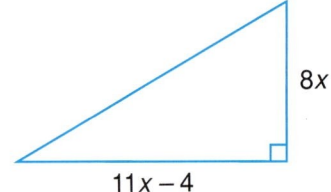

**Factor.**
25. $20m - 25$
26. $14c^2 + 16c$
27. $-y^2 - y$
28. $-7s(s + 2) - 9(s + 2)$
29. $a^2 + 9a + 18$
30. $u^2 + 5u - 24$
31. $z^2 - 5z - 14$
32. $f^2 - 17f + 66$

33. If the area of a square can be represented by $s^2 + 16s + 64$, write a binomial that can represent a side of the square.

34. If the area of a rectangle can be represented by $r^2 - 12r + 35$, write binomials that can represent the length and width of the square.

**Solve. (Pages 183–188, 297–300)**

1. William works 20 hours and completes $\frac{4}{5}$ of a job. How long will it take him to do the entire job?
2. A car rental agency charges $29.95 per day plus $0.24 per mile. What is the charge for renting a car for 5 days and driving it 377 miles?

**Write a percent for each ratio. (Pages 305–308)**

3. $\frac{3}{10}$   4. $\frac{1}{2}$   5. $\frac{3}{4}$   6. $\frac{2}{5}$   7. $\frac{3}{2}$

**The city mileage ratings in miles per gallon for the ten most fuel efficient automobiles of a recent year are listed at the right. (Pages 359–362)**

| Mileage Ratings (miles per gallon) | | | | |
|---|---|---|---|---|
| 39 | 38 | 50 | 46 | 37 |
| 38 | 53 | 37 | 38 | 45 |

8. Find the mean, median, and mode of the data.
9. If you were a car salesman, which of these measures, the mean, median, or mode, would you use to try to sell a car to a prospective buyer?

**Solve each inequality. (Pages 393–397)**

10. $a + 6 \leq 8$
11. $3c \geq -12$
12. $-6f > 15$
13. $d - 9 > -7$

**Solve. (Pages 393–397, 410–413)**

14. The sum of two consecutive even integers is less than 50. Find the pair of numbers with the greatest sum.
15. Identify the figure that is formed when the following points are graphed and the points are connected in order by line segments.
    A(2, 2)   B(3, −3)   C(−5, −3)   D(−2, 2)

**Solve. (Pages 457–462)**

16. Dawna placed a ladder 2 yards from the foot of her house. It reached a ledge 5 yards high. How long is the ladder?

17. Clarence painted a diagonal stripe on his door. The door's dimensions are 3 feet by 7 feet. How long was the stripe?

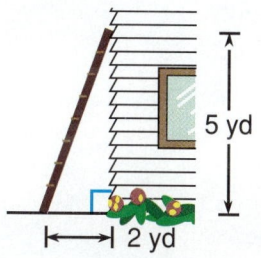
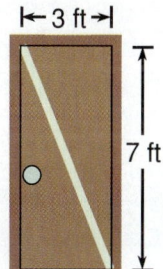

Cumulative Review: Chapters 1–13   **523**

# UNIT 6

### PROJECT 1
Just as there are different styles of music, so there are different styles of buildings. As a team, look at buildings in your neighborhood, city, or county. Then choose one that has a particular style which the group finds interesting. For that building, you will prepare an architectural notebook (or report).

### PROJECT 3
To stop a car traveling at 55 miles per hour takes a minimum of 200 feet. That figure, however, varies depending on road conditions. For this project, your team will investigate reaction, braking, and stopping distances for cars. You will also check around your community for speed limits that you think should be changed because of existing road conditions.

# Using Equations

## CONTENTS

### 14 Equations in Geometry
Angles and Angle Measures
Parallel and Perpendicular Lines
Problem Solving Strategies:
    Extending Patterns
Triangles
Polygons
Circumference and Area
Circle Graphs

### 15 Volume and Surface Area
Problem Solving Exploration:
    Surface Area and Volume
Volume of a Rectangular Prism
Surface Area of a Rectangular
    Prism
Volume of a Cylinder
Volume of a Pyramid
Volume of a Cone
Problem Solving Strategies:
    Making a Model
Volume and Area of a Sphere

## PROJECT 2
Your family has just moved into a new home. For the first time, you will have your own room and will be allowed to decorate it. You will have to stay within a budget, however. How will you plan this project? How will you organize the list of items to be bought and the costs? Work with a team to research needs and costs.

## LESSON 14-1

# Angles and Angle Measures

**T**hink of a plane as the set of points that make up a flat surface. You can think of a notebook page as representing part of a plane. An **angle** is a plane figure that has two rays with a common endpoint. The common endpoint is called the **vertex.** An angle can be named in several ways.

| Name | Symbol |
|---|---|
| Angle $ABC$ | $\angle ABC$ |
| Angle $CBA$ | $\angle CBA$ |
| Angle $B$ | $\angle B$ |
| Angle 1 | $\angle 1$ |

The vertex of the angle above is at $B$. Notice that "B" is the middle letter when you write $\angle ABC$. The rays which form the sides of $\angle ABC$ are called ray $BC$ (also written $\overrightarrow{BC}$) and ray $BA$ (also written $\overrightarrow{BA}$).

• How many angles can you find in the the photo above?

**526**  CHAPTER 14

**EXAMPLE 1** Write four different names for the angle at the right.

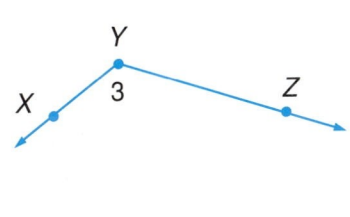

**Solution**
a. ∠PQR   b. ∠RQP
c. ∠Q     d. ∠4

**Try This** Write four names for each angle.

1.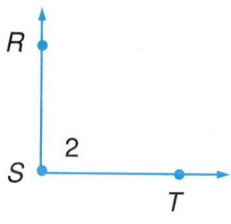
2.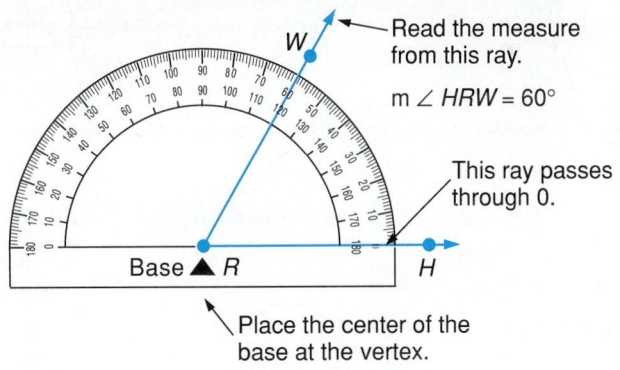

Angles are measured in degrees. You can use a protractor to measure an angle. To find the measure of ∠WRH below, place the center of the base of the protractor at the vertex of the angle. Place the protractor so that one ray of the angle passes through "0" on the protractor. Read the measure of the angle from the second ray. The symbol m∠HRW means the measure of angle HRW.

Angles are classified by their measures.

**Right Angle**         **Acute Angle**         **Obtuse Angle**

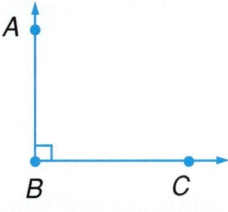 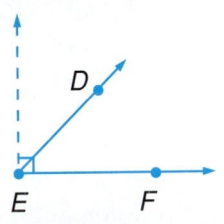 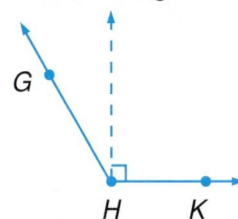

m∠ABC is exactly 90°.   m∠DEF is less than 90°.   m∠GHK is greater than 90° and less than 180°.

14-1 Angles and Angle Measures

**EXAMPLE 2** In each triangle, classify each angle as acute, right, or obtuse.

a.

b.

c.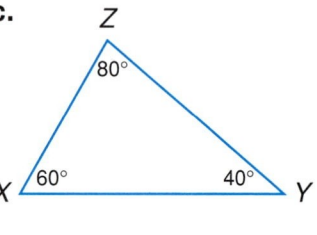

**Solutions**

a. Triangle ABC
∠A: **right** angle
∠B: **acute** angle
∠C: **acute** angle

b. Triangle QPR
∠Q: **acute** angle
∠P: **obtuse** angle
∠R: **acute** angle

c. Triangle XYZ
∠X: **acute** angle
∠Y: **acute** angle
∠Z: **acute** angle

**Try This** In each triangle, classify each angle as acute, right, or obtuse.

3.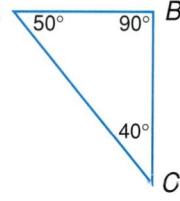

4.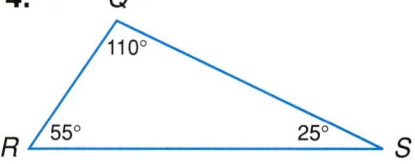

Some pairs of angles can be classified as *complementary* or *supplementary*.

Two angles are **complementary** if the sum of their measures is 90°.

Two angles are **supplementary** if the sum of their measures is 180°.

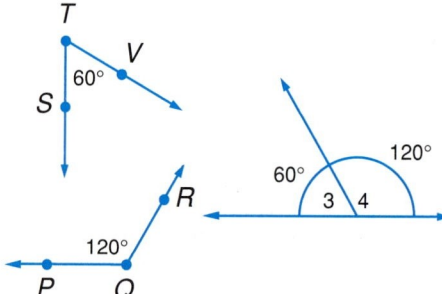

Since m∠J + m∠F = 90°, ∠J and ∠F are complementary.

Since m∠1 + m∠2 = 90°, ∠1 and ∠2 are complementary.

Since m∠T + m∠Q = 180°, ∠T and ∠Q are supplementary.

Since m∠3 + m∠4 = 180°, ∠3 and ∠4 are supplementary.

**EXAMPLE 3**

a. Write and solve an equation to find m∠1 in the figure at the right.

b. Two angles are supplementary and the measure of one angle is 50° more than the other. Find the measures of the angles.

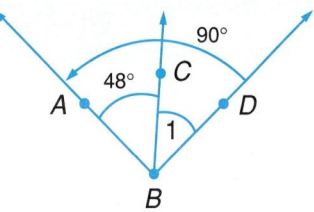

**Solutions**

a. m∠DBC + m∠CBA = 90
    m∠1 + 48 = 90
    m∠1 = 42

So m∠1 is **42°**.

b. **Think:** Since the angles are supplementary, the sum of their measures is 180.

**Plan:** Use variables to represent the measures of the angles. Then write and solve an equation.

Let x = the measure of the smaller angle.
Then x + 50 = the measure of the larger angle.

x + (x + 50) = 180
2x + 50 = 180
2x = 130
x = 65   ← Don't forget to find (x + 50).
x + 50 = 115

The measures of the angles are **65°** and **115°**.

**Try This**  Write and solve an equation to find m∠ABC.

5.

6.

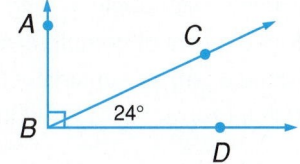

7. Two angles are complementary. The measure of one angle is half the measure of the other. Find the measures of the two angles.

• Can two acute angles be supplementary? Explain.

# EXERCISES

**Objectives:** To measure and classify angles
To determine the measures of complementary and supplementary angles

## Check Your Understanding

**Skill Check** Write the letter of the correct answer.

**1.** Which is **not** a correct way to name this angle?
a. $\angle DEF$
b. $\angle E$
c. $\angle EFD$
d. $\angle FED$

**2.** How many angles are there in this figure?
a. 3
b. 4
c. 5
d. 6

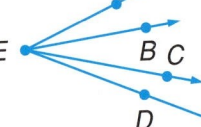

**3.** Two angles are supplementary and the measure of one angle is 50° less than the other. Which equation could you use to find the measures of the angles?
a. $x + x + 50 = 90$
b. $x + x - 50 = 90$
c. $x + x - 50 = 180$
d. $x = x - 50$

**4.** The measure of an angle is $\frac{2}{3}$ the measure of its supplementary angle. Which equation could you use to find the measures of the angles?
a. $\frac{2}{3}x + x = 180$   b. $x + \frac{2}{3}x = 90$
c. $x - \frac{2}{3}x = 180$   d. $x - \frac{2}{3}x = 90$

## Practice and Apply

For Exercises 5–9, refer to the figure at the right.

5. Name an acute angle.
6. Name an obtuse angle.
7. Name a right angle.
8. Name a pair of complementary angles.
9. Name a pair of supplementary angles.
10. Use a protractor to draw angles of 50°, 100°, and 27°.

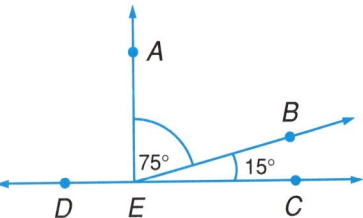

Write and solve an equation to find $m\angle ABC$.

**11.**

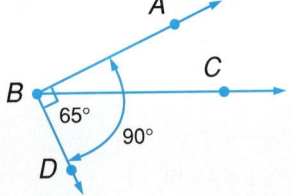

**12.**

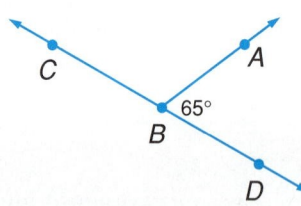

530   CHAPTER 14

For Exercises 13–16, write and solve an equation to find the angle measures.

**13.** The complement of an angle is 32°. What is the measure of the angle?

**14.** The supplement of an angle is 58°. What is the measure of the angle?

**15.** The smaller of two supplementary angles is 40° less than the other. Find the measures of the angles.

**16.** Two complementary angles have equal measures. What is the measure of each angle?

**17.** If m∠ABC = $x$, what is the measure of its supplement?

**18.** Explain why the supplement of a right angle is also a right angle.

**19.** If m∠ABC = 90°, explain why it has no complement.

## Connect and Extend

In Exercises 20–21, find the measure of each angle.

**20.** 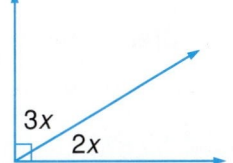   **21.**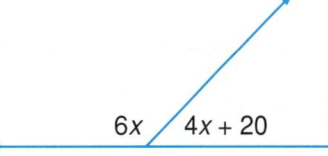

**22.** The measure of an angle is 20° less than the measure of its supplement. Find the measure of the angle, the measure of its supplement, and the measure of its complement.

**23.** Two angles are supplementary. The measure of one angle is 30° more than twice the measure of the other. Find the measure of each angle.

**24.** Find the difference between the measure of the supplement and the measure of the complement of an angle of 60°.

**25.** Find a counterexample for this statement.
If two angles are supplementary, one of the angles must be acute.

## Maintain Your Skills

**Find the slope of the line passing through the two given points. (Pages 422–426)**

**26.** (7, 1) and (5, 9)     **27.** (−8, 2) and (10, −6)     **28.** (−5, −1) and (−6, −4)

**Find the point of intersection of each system of equations. (Pages 435–439)**

**29.** $y = 2x - 3$
    $y = x + 1$

**30.** $x + y = 8$
    $y - x = 2$

**31.** $y = 3x + 5$
    $x + y = -3$

**Add. (Pages 485–490)**

**32.** $(r^2 + 8) + (3r^2 - 11)$

**33.** $(-2x^2 + 2x - 4) + (5x^2 - x + 3)$

# LESSON 14-2
# Parallel and Perpendicular Lines

If you draw two lines in the same plane, they will either intersect at a single point or they will be parallel. If the two lines intersect and form right angles, the lines are **perpendicular**.

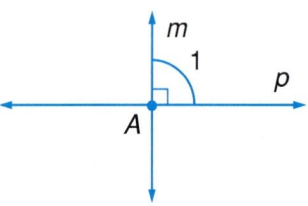

Line $p$ intersects line $m$ at point $A$, such that m∠1 = 90°.

Line $p$ is perpendicular to line $m$.

In symbols: $p \perp m$

Line $q$ and line $r$ do not intersect.

Line $q$ is **parallel** to line $r$.

In symbols: $q \parallel r$

When two lines intersect, two pairs of opposite angles are formed. These opposite angles are called **vertical angles**. Vertical angles are equal. ∠1 and ∠3 are vertical angles. So m∠1 = m∠3.

Since ∠2 and ∠4 are vertical angles, m∠2 = m∠4.

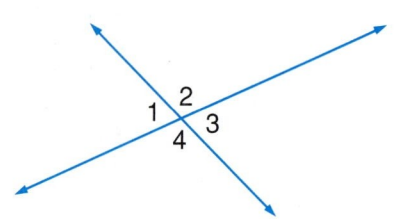

**EXAMPLE 1** In each figure, m∠1 = 60°. Find m∠2 and m∠3.

a.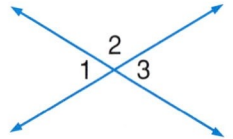

b.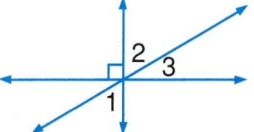

**Solutions**

a. **Think:** ∠1 and ∠3 are vertical angles. So m∠1 = m∠3. Therefore, m∠3 = **60°**.

**Think:** ∠1 and ∠2 are supplementary. Since m∠1 = 60°, m∠2 = 180° − 60° = **120°**.

**b.** Think: ∠1 and ∠2 are vertical angles.
So m∠1 = m∠2 = **60°**.

Think: ∠2 and ∠3 are complementary.
Therefore, m∠3 = 90° − m∠2 = 90° − 60° = **30°**.

**Try This**

In each figure, let m∠1 = 70°. Find m∠2 and m∠3.

1. 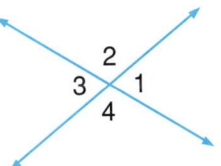   2.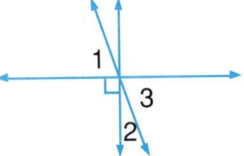

A line that intersects each of two or more lines in different points is called a **transversal**. Transversal *t*, in the figure at the right, intersects lines *a* and *b*.

∠1, ∠2, ∠7, and ∠8 are called **exterior angles**.

∠3, ∠4, ∠5, and ∠6 are called **interior angles**.

Certain pairs of angles are given special names.

| **Corresponding Angles** | **Alternate Interior Angles** | **Alternate Exterior Angles** |
|---|---|---|
| ∠1 and ∠5 | ∠3 and ∠6 | ∠1 and ∠8 |
| ∠3 and ∠7 | ∠4 and ∠5 | ∠2 and ∠7 |
| ∠2 and ∠6 | | |
| ∠4 and ∠8 | | |

## EXPLORE

In the figure at the right, lines *m* and *n* are parallel and are intersected by line *t*.

1. Use a protractor to find the measure of each angle.

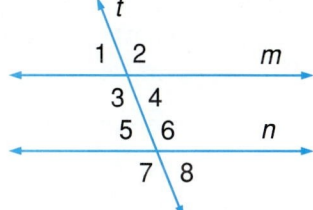

Tell how each of these pairs of angles are related.

2. ∠1 and ∠5    3. ∠2 and ∠6
4. ∠3 and ∠7    5. ∠4 and ∠8

14-2 Parallel and Perpendicular Lines   **533**

6. Use the results of Exercises 1–5 to complete this statement. If two parallel lines are intersected by a transversal, the measures of the corresponding angles are __?__.

7. Repeat the activity for the pairs of alternate interior angles. Write a statement about the relationship between the measures of pairs of alternate interior angles.

8. Repeat the activity for pairs of alternate exterior angles. Write a statement about the relationship between the measures of pairs of alternate exterior angles.

The results of Exercises 1–8 are summarized below.

> **Summary**
> If two parallel lines are intersected by a transversal, then:
> a. the measures of corresponding angles are equal.
> b. the measures of alternate interior angles are equal.
> c. the measures of alternate exterior angles are equal.

**EXAMPLE 2** In the figure below, parallel lines $c$ and $d$ are intersected by transversals $r$ and $t$. Find m∠1, m∠2, m∠3, and m∠4.

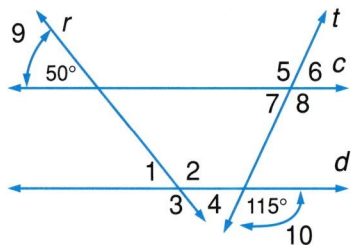

**Solutions**

Since ∠1 and ∠9 are corresponding angles, m∠1 = **50°**.

Since ∠9 and ∠4 are alternate exterior angles, m∠4 = **50°**.

Since ∠4 and ∠2 are supplementary, m∠2 = 180° − 50° = **130°**.

Since ∠2 and ∠3 are vertical angles, m∠3 = **130°**.

**Try This** Refer to the figure in Example 2 to find the measure of each angle.

3. m∠5    4. m∠6    5. m∠7    6. m∠8

# EXERCISES

**Objective:** To find the measures of vertical angles, corresponding angles, alternate interior angles, and alternate exterior angles

## Check Your Understanding

**Skill Check** Write the letter of the correct answer.

In the figure at the right, parallel lines *s* and *r* are intersected by transversal *t*. Refer to this figure for Exercises 1–4.

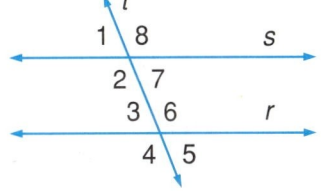

1. Which pair of angles are corresponding angles?
   a. ∠1 and ∠8
   b. ∠7 and ∠3
   c. ∠7 and ∠5
   d. ∠4 and ∠6

2. Which pair of angles are vertical angles?
   a. ∠1 and ∠5
   b. ∠8 and ∠6
   c. ∠3 and ∠4
   d. ∠2 and ∠8

3. Which pair of angles are supplementary?
   a. ∠3 and ∠7
   b. ∠7 and ∠5
   c. ∠1 and ∠7
   d. ∠3 and ∠4

4. Which pair of angles have equal measures?
   a. ∠4 and ∠5
   b. ∠3 and ∠7
   c. ∠4 and ∠1
   d. ∠6 and ∠5

## Practice and Apply

In the figure at the right, line *a* is parallel to line *b*. Refer to this figure for Exercises 5–14.

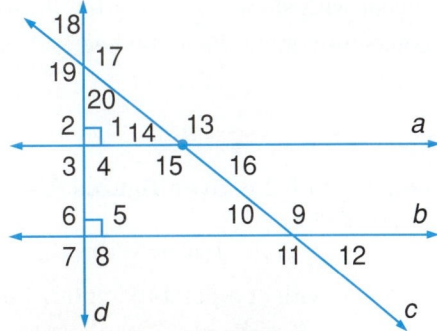

5. Name two pairs of perpendicular lines.
6. Name three pairs of lines that intersect but are not perpendicular.
7. Name two transversals.
8. Name all pairs of vertical angles.
9. Name all pairs of congruent corresponding angles.
10. Name all pairs of congruent alternate interior angles.
11. Name all pairs of congruent alternate exterior angles.
12. Name all right angles.
13. Name 10 pairs of congruent angles.
14. Name 10 pairs of supplementary angles.

14-2  Parallel and Perpendicular Lines   **535**

For Exercises 15–17, find m∠1, m∠2, and m∠3. In Exercises 16 and 17, line *r* is parallel to line *p*.

15.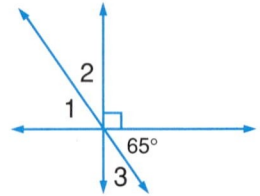
16.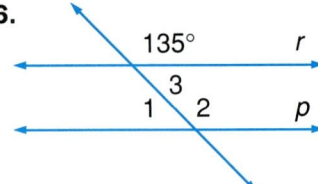
17.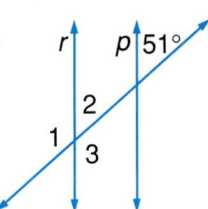

## Connect and Extend

For Exercises 18–19, write an equation and find m∠*PQR*. In Exercise 19, line *a* is parallel to line *b*.

18.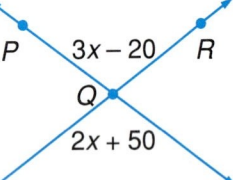
19.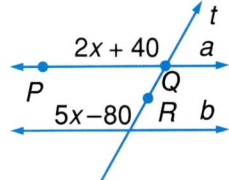

20. Write a logical argument to show that if two parallel lines are intersected by a transversal, the interior angles on the same side of the transversal are supplementary. Refer to the figure at the right.

21. On a coordinate plane, draw a line with slope $\frac{2}{3}$ and a segment with slope $-\frac{3}{2}$. Study the drawing and write a conjecture about what you see.

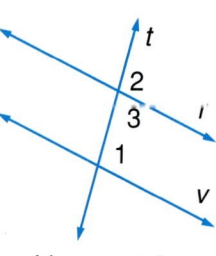

Lines *r* and *v* are parallel.

## Maintain Your Skills

Find the image of the given figures. Use these as the given points. (Pages 440–443)

$E(4, 5) \quad F(0, 2) \quad G(-4, -6) \quad H(-2, -3)$

22. A segment with end points $F'$ and $G'$, under the translation $(x, y) \rightarrow (x, y + 3)$.

23. A triangle with vertices at $E$, $G$, and $H$, under the translation $(x, y) \rightarrow (x - 3, y + 7)$.

For each value of tan *A*, find the measure of angle *A* to the nearest degree. (Pages 474–480)

24. tan *A* = 0.1944
25. tan *A* = 3.0777
26. tan *A* = 2.0503

Find the product. (Pages 496–500)

27. $(x - 11)^2$
28. $(3x - 1)(3x + 1)$
29. $(2x + 3)^2$

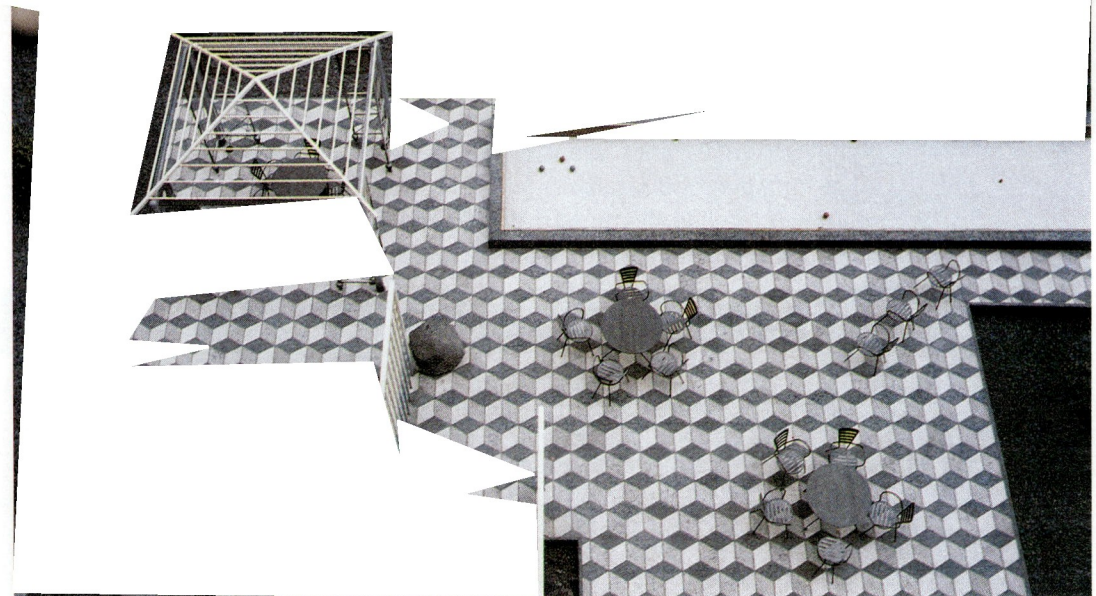

## LESSON 14-3

# Problem Solving Strategies
# Extending Patterns

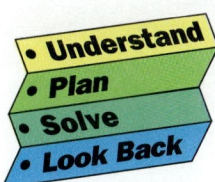

While waiting at a restaurant for the food he had ordered, Elvis noticed the design on three sides of his placemat. His hand was covering the design on the fourth side.

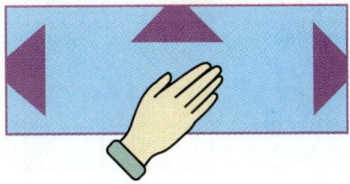

Elvis asked himself these questions.

If you start with the figure at the left, how does the figure change from each position to the next? What will the figure under my hand look like?

To answer these questions, Elvis pictured the design like this. He thought: "The second square is turned 90° to the right. The third square is turned 90° to the right of the second. So, the fourth square will be turned 90° to the right of the third square." Is Elvis' reasoning correct?

14-3 Problem Solving Strategies **537**

Elvis thought of the pattern on the placemat as a sequence of geometric figures. To identify a pattern in such a sequence, look for what changes and what remains the same. Then you can extend the sequence by visualizing and drawing the next figure in the sequence.

# EXERCISES

**Objective:** To apply strategies in problem solving

**Identify the pattern. Then extend the sequence by drawing the next figure.**

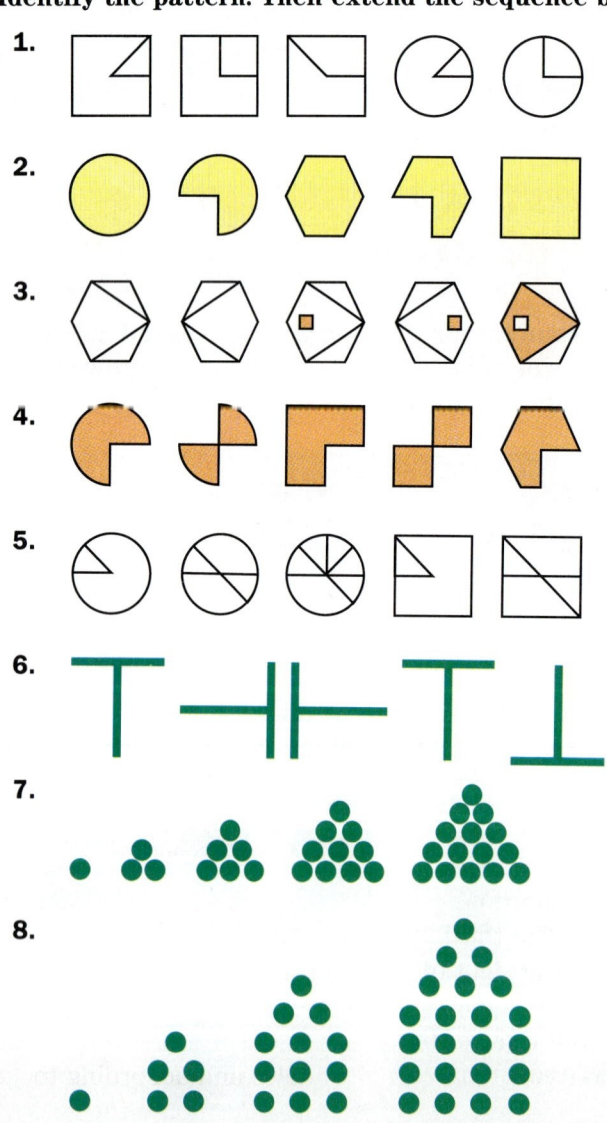

## Triangles

**W**hen three pieces of wood or metal are bolted together at their endpoints to form a triangle, they form a rigid figure that is very strong. For this reason, triangles are used in many kinds of constructions.

• How many triangles do you see in the figure above?

A **triangle** is made up of three line segments that meet at their endpoints. In triangle $PQR$ at the right, line segments $\overline{PQ}$, $\overline{QR}$, and $\overline{RP}$ are the **sides** of the triangle. The **endpoints** of the sides are $P$, $Q$, and $R$.

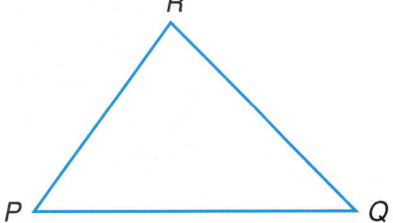

Triangle $PQR$ is written as $\triangle PQR$.

Segment $QR$ is written as $\overline{QR}$.

The measure of $\overline{QR}$ is written as $QR$.

Triangles are classified according to their angles and according to the number of equal sides.

**14-4 Triangles 539**

| Triangle | Type | Property |
|---|---|---|
| △ABC | Equiangular | All angles have the same measure. |
| △AMC | Right | Contains a right angle. Recall that the symbol ⌐ indicates a right angle. |
| △ABC | Acute | All angles are less than 90°. |
| △LMN | Obtuse | One angle is greater than 90°. |
| △POM | Scalene | No two sides have the same length. |
| △LMN | Isosceles | Two sides have the same length. |
| △ABC | Equilateral | All sides have the same length. |

**EXAMPLE 1** Classify each triangle according to its angles and according to its sides.

a. 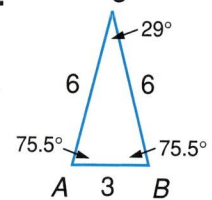  b. 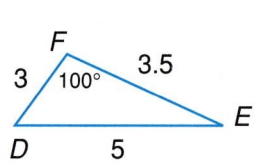  c.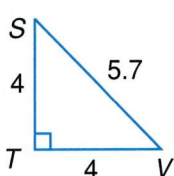

**Solutions**

a. △ABC is **acute** and **isosceles** because the measures of all angles are less than 90° and two sides have the same length.

b. △DEF is **obtuse** and **scalene** because the m∠F is greater than 90° and no two sides have the same lengths.

c. △STV is **right** and **isosceles** because it contains one right angle and two sides have the same length.

**Try This**

1. Draw a scalene triangle, an isosceles triangle, and an equilateral triangle. Use a ruler to measure the sides and label them.
2. Draw an obtuse triangle, an acute triangle, and a right triangle. Use a protractor to measure the angles.
3. Draw a triangle that is isosceles and obtuse.
4. Draw a triangle that is scalene and acute.

# EXPLORE

**A.** Follow these directions.

**a.** Copy △ABC on tracing paper.

**b.** Color the vertices A, B, C as shown.

**c.** Tear off the vertices. Rearrange them like this.

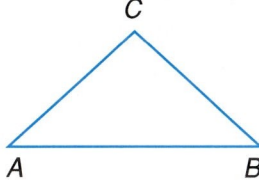 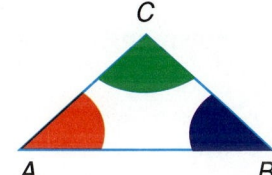 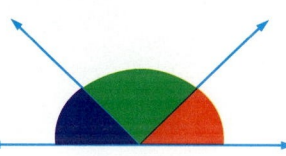

1. Now draw a right triangle, an obtuse triangle, and an isosceles triangle. Repeat the activity in A for each of these triangles.

2. True or False? The sum of the measures of the angles of a triangle equals the sum of the measures of a pair of supplementary angles. Explain your answer.

The Explore activities suggest the following.

> **Triangle-Sum Property**
> The sum of the measures of the angles in any triangle equals 180°.

**EXAMPLE 2** For each triangle, write and solve an equation to find m∠A.

**a.** 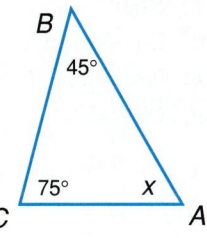     **b.**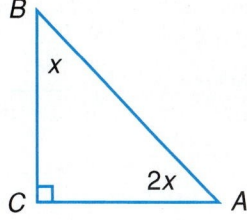

**Solutions** Use the Triangle-Sum Property to write an equation.

**a.** m∠A + m∠B + m∠C = 180

$x + 45 + 75 = 180$

$x + 120 = 180$

$x = 60$    So m∠A = **60°**.

14-4 Triangles    **541**

**b.** m∠A + m∠B + m∠C = 180
2x + x + 90 = 180
3x + 90 = 180
3x = 90
x = 30   ← Don't forget to find 2x.
2x = 60

So m∠A = **60°**.

**Try This**   Find the unknown angle measures in each triangle.

5.

6.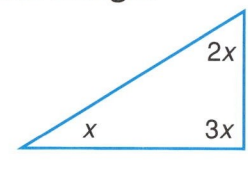

• How can you check your answers to Try This Exercises 5 and 6?

# EXERCISES

**Objectives:** To classify triangles
To find the measures of the angles of a triangle

## Check Your Understanding

**Skill Check**   Write the letter of the correct answer.

**1.** Classify △CDE according to the lengths of its sides.
   a. Right
   b. Scalene
   c. Isosceles
   d. Obtuse

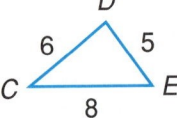

**2.** Classify △ABC according to its angle measures.
   a. Right
   b. Equilateral
   c. Isosceles
   d. Obtuse

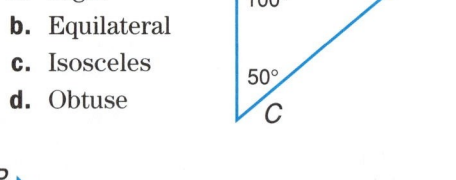

**3.** Classify △PDQ according to the lengths of its sides and according to the measure of its angles.
   a. Obtuse, isosceles triangle
   b. Isosceles right triangle
   c. Scalene right triangle
   d. Scalene obtuse triangle

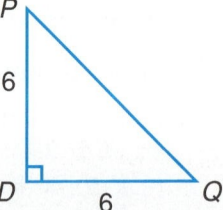

542   CHAPTER 14

**4.** Which statement is false?
  **a.** An obtuse triangle has one obtuse angle and two acute angles.
  **b.** A right triangle has two right angles and one acute angle.
  **c.** In a scalene triangle, no two sides have the same measure.
  **d.** In an equilateral triangle, all three sides have the same measure.

## Practice and Apply

**Classify each triangle according to the measures of its angles and the lengths of its sides.**

**5.** 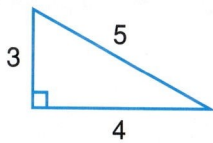    **6.** 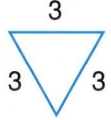    **7.** 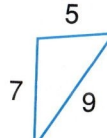    **8.**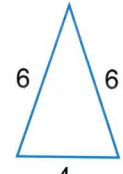

**For Exercises 9–12, write and solve an equation to find the measure of the unknown angles in the triangle.**

**9.** 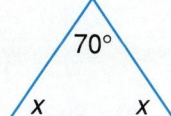    **10.** 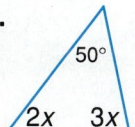    **11.** 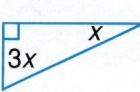    **12.**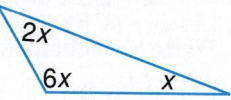

**13.** Can a right triangle be obtuse? Explain.
**14.** Can a right triangle be equilateral? Explain.
**15.** Can an acute triangle be scalene? Explain.
**16.** Can an obtuse triangle be equilateral? Explain.
**17.** Can a right triangle be isosceles? Explain.
**18.** Can an obtuse triangle be isosceles? Explain.

## Connect and Extend

**19.** A right triangle has one acute angle of 38°. What is the measure of the other acute angle? How are the two acute angles related?

**20.** All the angles of a triangle have the same measure. Find the measure of each angle of the triangle.

**21.** In an acute triangle, the measure of the largest angle is 20° more than the measure of the smallest. The remaining angle is 40° less than twice the measure of the smallest angle. Find the measures of all the angles.

22. Use a protractor to draw a triangle that contains two 50° angles. Then use a ruler to measure the number of millimeters in the sides of the triangle which are opposite the two 50° angles. How are the lengths of the sides related? Write a generalization about the sides opposite two angles of a triangle that have equal measures.

23. $\angle 4$ is called an **exterior angle** of $\triangle ABC$. $\angle 1$ and $\angle 2$ are called the **remote interior angles** for $\angle 4$. Write a logical argument to show that the measure of an exterior angle of a triangle is equal to the sum of the measures of its remote interior angles.

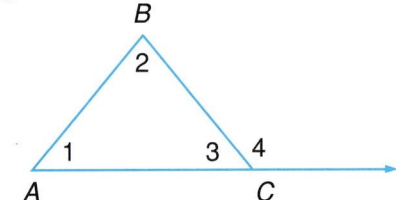

24. Two parallel lines are intersected by a transversal so that all corresponding angles and all alternate interior angles have equal measures. How are the parallel lines related to the transversal?

## Maintain Your Skills

**Evaluate the expression. (Pages 262–266)**

25. $\frac{(n-2)180}{n}$, for $n = 8$

26. $\frac{c^2 + 2cd + d^2}{c + d}$, for $c = 3$ and $d = -7$

**Write a decimal for each percent. (Pages 309–312)**

27. 25%    28. 7%    29. 0.4%    30. 112%    31. 89.3%

**Solve. (Pages 457–462)**

32. How high up on a wall will a 15-foot ladder reach if the foot of the ladder is placed 5 feet from the building?

33. Bill's room is 3 meters wide and 5 meters long. How long is the diagonal connecting two opposite corners of the room?

## ★ Math Team Problems

34. Show that when the measure of a given angle is added to twice the measure of its complement, the sum equals the measure of its supplement.

35. Find half the sum of the measure of a given angle and the measure of its complement.

36. The cost of a watch and a bracelet is $77. The cost of the watch and a necklace is $95. The cost of the necklace and the bracelet is $108. Find the cost of each item.

**Solve. In Exercises 3 and 5, lines $t$ and $k$ are parallel. (Pages 526–531)**

**1.** The supplement of an angle is 71°. What is the measure of the angle?

**2.** The complement of an angle is 43°. What is the measure of the angle?

**For Exercises 3–5, find $m\angle 1$, $m\angle 2$, and $m\angle 3$. (Pages 532–536)**

**3.**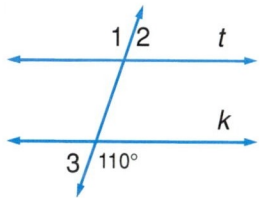

**4.**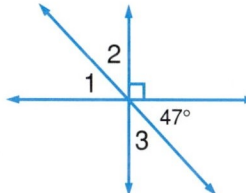

**5.**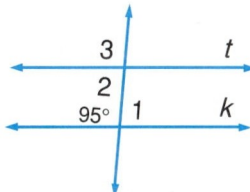

**Classify each triangle according to the measures of its angles and the lengths of its sides. Then write and solve an equation to find the missing measures of the other angles. (Pages 539–544)**

**6.**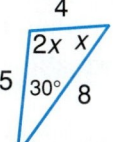

**7.**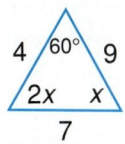

**8.**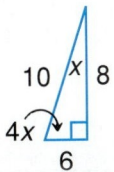

**9.**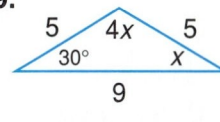

# Mathematical Footnote

William P. Thurston likes to work with abstract mathematics. He points out that it would be silly to have one course in how to count people and another course in learning to count miles. Abstract mathematics, for example, the idea of 3 + 4, can be applied to both people and miles. On the other hand, you can't have half a person, so fractions have applications in some situations and not in others.

Thurston says: "I think that most mathematicians love mathematics for mathematics' sake. . . . But . . . in most cases, the mathematics they generate will ultimately have significant applications."

Thurston is a topologist, someone who studies what happens to shapes when they are twisted, stretched, and shrunk. To a topologist, a coffee cup and a bagel have the same shape!

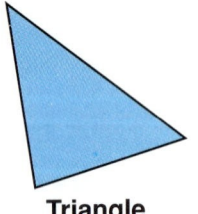
Triangle

Quadrilateral

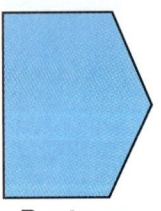
Pentagon

Hexagon

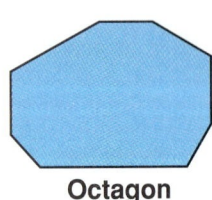
Octagon

## LESSON 14-5 Polygons

A **simple closed curve** begins at a point and returns to that point without crossing itself. A simple closed curve made up of line segments is called a **polygon.** Polygons are named according to the number of sides. The most common polygons are shown above.

Some quadrilaterals have special names.

| Quadrilateral | Name | Description |
|---|---|---|
|  | Parallelogram | A quadrilateral whose opposite sides are parallel and equal |
| | Rectangle | A parallelogram with four right angles |
| | Square | A rectangle with four equal sides |
| | Rhombus | A parallelogram with four equal sides |
| | Trapezoid | A quadrilateral with exactly two parallel sides |

A line segment, other than a side, that joins two vertices of a polygon is called a **diagonal.** All the diagonals from a single vertex are drawn for each polygon in the table on page 547. Try to find the relationship between the number of sides of the polygon and the number of triangles formed.

| Polygon | Number of Sides of Polygon | Number of Triangles Formed | Number of Degrees in Polygon |
|---|---|---|---|
| | 4 | 2 | 2(180°) = 360° |
| | 6 | 4 | 4(180°) = 720° |
| | 8 | 6 | 6(180°) = 1080° |

- If a polygon has $n$ sides and you draw all the diagonals from one vertex, are $n$ triangles or $(n - 2)$ triangles formed?
- If $(n - 2)$ triangles are formed, what is the sum of the measures of the angles of the polygons?

**EXAMPLE 1** Find the sum of the measures of the angles of the polygon at the right.

**Solution** **Think:** The polygon has 7 sides. If I draw all possible triangles from one vertex, 7 − 2, or 5 triangles will be formed.

Sum of angles: (7 − 2)(180) = 5(180)
= 900

The sum of the measure of the angles is **900°**.

**Try This** **Find the sum of the measures of the angles of a polygon with the given number of sides.**
  1. 10 sides
  2. 12 sides

Since all quadrilaterals have four sides, the sum of the measures of its angles is (4 − 2)180, or 360°. That is,
**the sum of the measures of the angles of a quadrilateral is 360°.**

14-5 Polygons   **547**

**EXAMPLE 2** For each quadrilateral, write and solve an equation to find m∠A.

**Solutions**

a.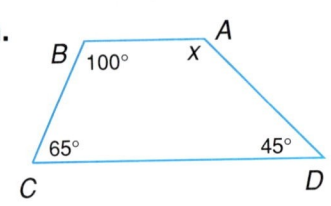

$65 + 45 + 100 + x = 360$
$210 + x = 360$
$x = 150$
m∠A = **150°**

b.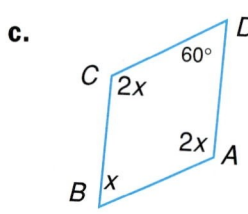

$(x + 8) + 110 + 100 + x = 360$
$218 + 2x = 360$
$2x = 142$
$x = 71$
m∠A = **71°**

c.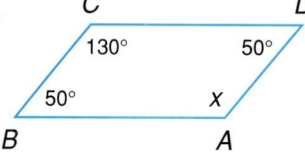

$60 + 2x + 2x + x = 360$
$60 + 5x = 360$
$5x = 300$
$x = 60$
$2x = 120$
m∠A = **120°**

**Try This** For each quadrilateral, write and solve an equation to find m∠A.

3.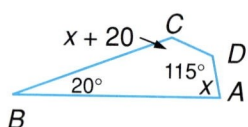

4.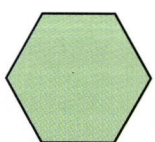

- How can you check your answers to Try This Exercises 3 and 4?

In a **regular polygon,** all the sides have the same length and all angles have the same measure. So a regular hexagon has 6 sides of equal length and 6 angles of equal measure.

How can you find the number of degrees in each angle?

**Think:** There are 6 sides. So the total number of degrees in the hexagon is $(6 - 2)180$.

$$(6 - 2)180 = 4 \cdot 180 = 720°$$

Since there are 6 angles of equal measure, divide 720 by 6.

$$720 \div 6 = 120$$

There are 120° in each angle of the hexagon.

# EXERCISES

**Objective:** To classify polygons
To find the measures of the angles of a quadrilateral

## Check Your Understanding

**Skill Check** Write the letter of the correct answer.

1. Which name describes the polygon?
   a. Trapezoid
   b. Pentagon
   c. Hexagon
   d. Octagon

2. Which segment is a diagonal?
   a. $\overline{AC}$
   b. $\overline{AB}$
   c. $\overline{AE}$
   d. $\overline{DC}$

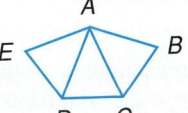

3. How many degrees are there in an octagon?
   a. 1440°   b. 135°   c. 1080°   d. 120°

4. In the figure at the right, $\overline{AB} \parallel \overline{DC}$ and $\overline{AD} \parallel \overline{BC}$. None of the angles are right angles. Which name **cannot** be applied to the figure?
   a. Rhombus
   b. Square
   c. Parallelogram
   d. Quadrilateral

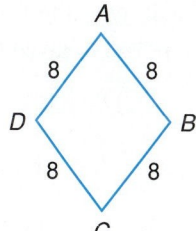

## Practice and Apply

Give the name of each polygon.

5.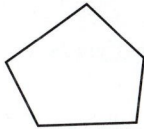

6.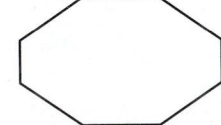

7.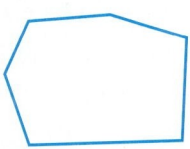

Find the sum of the measures of the angles of a polygon having the given number of sides.

8. 7 sides   9. 10 sides   10. 5 sides   11. 20 sides

14-5  Polygons  **549**

For each quadrilateral, write and solve an equation to find m∠A.

12.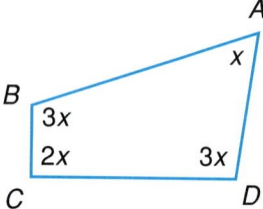
13.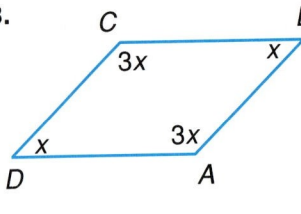
14.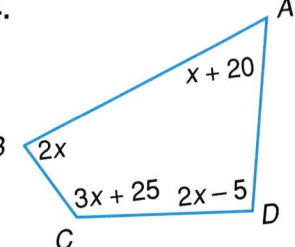

Find the measure of each angle of the regular polygon.

15. Triangle
16. Pentagon
17. Octagon

18. On a baseball field, home plate has 3 right angles and two angles, ∠1 and ∠2, that have the same measure. Find m∠2.

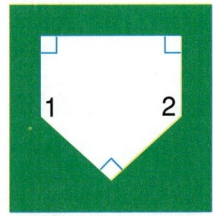

## Connect and Extend

**True or False? When a statement is false, explain why.**

19. Every rectangle is a parallelogram.
20. Every parallelogram is a rectangle.
21. Every parallelogram is a quadrilateral.
22. Every square is a rectangle.
23. Every rectangle is a square.
24. Every square is a parallelogram.
25. A regular polygon contains 1440°. Find the measure of each angle.
26. Write an argument to show that the number of degrees in a polygon can be found by the formula $(n - 2)180$ where $n$ is the number of sides of the polygon. Refer to the table on page 547.
27. Explain how the formula $\dfrac{(n-2)180}{n}$, where $n$ is the number of sides of a polygon, gives the measure of each angle of a regular polygon.

## Maintain Your Skills

28. Write the ratio, in lowest terms, of the number of girls to the number of boys in your Pre-Algebra class. **(Pages 216–220)**

Use the compound interest formula to find the compound amount. The interest is compounded yearly. Round answers down to the nearest cent. **(Pages 324–328)**

29. On $250 at 9% for 4 years
30. On $3,000 at 8.4% for 3 years

**Find the product. (Pages 504–507)**

31. $(w + 3)(-4w - 5)$
32. $(-3v - 2)^2$
33. $\left(\frac{1}{2}z + 2\right)\left(\frac{1}{2}z - 2\right)$

Replace each ? with + or − to make a true statement.
**(Pages 516–519)**

34. $a^2 - 3a - 28 = (a \; ? \; 4)(a \; ? \; 7)$
35. $s^2 - 14s + 45 = (s \; ? \; 5)(s \; ? \; 9)$

# LESSON 14-6

## Circumference and Area

**M**any objects that we see and use every day have the shape of a circle. Look at the stopwatch above. Notice that the hand turns about a central point and that all the minute marks on the watch are the same distance from the central point. This suggests a way of describing a circle.

A **circle** is a set of points in a plane that are all the same distance from a given point called the **center** (point $O$). A segment from the center to any point on a circle is called a **radius** ($\overline{OC}$). A segment between any two points on the circle is called a **chord** ($\overline{AB}$). If the chord passes through the center of the circle, it is called a **diameter** ($\overline{AB}$).

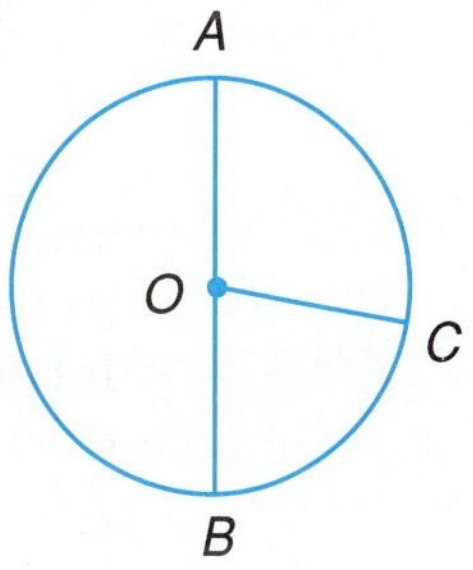

### EXPLORE

The distance around a circle is called its **circumference.** In the figure at the top of the next page, the diameter and circumference of a can have been found by using string.

14-6 Circumference and Area **551**

The ratio of the circumference of the can to its diameter can be computed to the nearest hundredth.

| Object | Circumference | Diameter | Ratio: $\frac{C}{d}$ |
|---|---|---|---|
| | 31.4 cm | 10 cm | $\frac{31.4}{10} = 3.14$ |

1. Find the measure of the circumference and diameter of three circular objects of different sizes by using string.

2. Record the results in a table. Find the ratio, $\frac{C}{d}$, to the nearest hundredth in each case.

3. Compare the ratios, $\frac{c}{d}$. What pattern do you see?

The ratio of the circumference of a circle to its diameter is called **pi** (pronounced pie). The symbol for pi is $\pi$. Many approximations for $\pi$ have been computed since ancient times. In 1988, a computer calculated a value of $\pi$ to 201,326,000 decimal digits.

You learned in Chapter 12 that $\pi$ is an irrational number. It cannot be named exactly by a decimal or a fraction. Rational number approximations for $\pi$ that are often used are 3.14 and $\frac{22}{7}$. We will use 3.14 as the approximate value of $\pi$ in this textbook.

When you know the length of a diameter of a circle, you can use the ratio $\pi = \frac{C}{d}$ to find the circumference.

Multiply each side of $\pi = \frac{C}{d}$ by $d$.

So $\pi d = C$ or $C = \pi d$.

Since a diameter is twice a radius, you can also write $C = 2\pi r$.

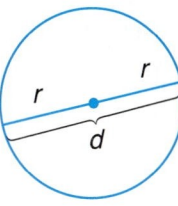

> **Circumference of a Circle**
> The circumference, $C$, of a circle with radius, $r$, and diameter, $d$, is
> $$C = \pi d \text{ or } C = 2\pi r.$$

**EXAMPLE 1** Find the circumference of a circle with the given diameter, $d$, or radius, $r$.

a. $d = 8$          b. $r = 2$ feet

**Solutions**

a. $C = \pi d$
$C = (3.14)(8)$
$C = \mathbf{25.12}$

b. $C = 2\pi r$
$C = 2(3.14)(2)$
$C = \mathbf{12.56\ ft}$

**Try This**    Find the circumference of a circle with the given radius or diameter.

**1.** $d = 4$ cm          **2.** $r = 2.5$ in

If you press the $\boxed{\pi}$ key on your calculator to compute the circumference of a circle with $d = 8$, you get this answer.

$8\ \boxed{\times}\ \boxed{\text{2ndF}}\ \boxed{\pi}\ \boxed{=}\ \boxed{25.132741}$    • Why is it different?

The **area** of a circle is the region enclosed within the circle. This area can be approximated by dividing the circle into pie-shaped equal parts, cutting out the pie-shaped pieces, and placing them side-by-side as shown.

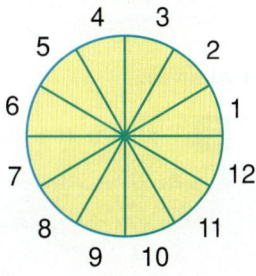
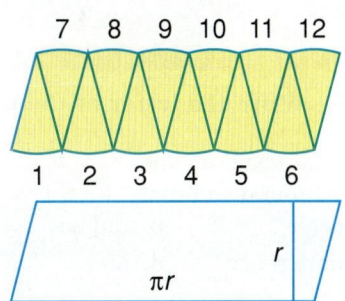

14-6   Circumference and Area   **553**

Notice that the new figure looks like a parallelogram. The base of the new figure has **half** the length of the circumference ($2\pi r$), of the circle; that is, its length is $\frac{1}{2}(2\pi r)$, or $\pi r$. Its height equals a radius, $r$, of the circle. So the area of the circle can be written as follows.

$$A = \pi r \cdot r$$
$$A = \pi r^2$$

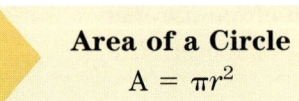

**Area of a Circle**
$A = \pi r^2$

Recall that area is measured in square units, such as square centimeters ($cm^2$), square inches ($in^2$), square feet ($ft^2$), and so on.

**EXAMPLE 2**  Find the area of a circle with the given radius or diameter.
  **a.** $r = 5$  **b.** $d = 12$ in

**Solutions**
**a.** $A = \pi r^2$
$A \approx (3.14)(5)^2$
$A \approx (3.14)(25)$
$A \approx$ **78.5** square units

**b. Think:** $r = \frac{1}{2}(12) = 6$
$A = \pi r^2$
$A \approx (3.14)(6)^2$
$A \approx (3.14)(36)$
$A \approx$ **113.04** $in^2$

**Try This**  Find the area of a circle with the given radius or diameter.
  **3.** $r = 9$ cm  **4.** $d = 7$ yd

# EXERCISES

**Objective:** To find the circumference and area of a circle

## Check Your Understanding

**Skill Check**  Write the letter of the correct answer.

**1.** What is the name for the distance around a circle?
  **a.** diameter  **b.** radius
  **c.** circumference  **d.** area

**2.** Which is the ratio for $\pi$?
  **a.** $\frac{d}{C}$  **b.** $\frac{r}{C}$
  **c.** $\frac{C}{d}$  **d.** $\frac{C}{r}$

3. Which can be used to find the circumference of a circle?
   a. $C = 2\pi d$   b. $C = 2\pi r^2$
   c. $C = 2\pi r$   d. $C = d^2$

4. Which can be used to find the area of a circle?
   a. $A = \pi d^2$   b. $A = \pi r^2$
   c. $A = 2\pi r$   d. $A = 2\pi d$

**True or False? When a statement is false, tell why.**
5. A diameter of a circle equals half a radius of the circle.
6. A diameter of a circle is also a chord of the circle.
7. Every radius of a circle has one endpoint at the center of the circle.
8. A circle has exactly one diameter.

## Practice and Apply

**Find the circumference and area of a circle having the given diameter or radius.**

9. $r = 2$ in
10. $d = 6$ cm
11. $r = 4$ ft
12. $r = 2.5$ m
13. $d = 6.4$ yd
14. $d = 14$ ft
15. $r = 40$ in
16. $d = 20$ mm

17. The radius of a bicycle wheel is 3 meters. What is the circumference of the wheel?

18. The diameter of a circular rose garden is 12 feet. What is the area of the garden?

19. Suppose that a cement path, 1 foot wide, surrounds the garden in Exercise 18. What is the total area of the garden and the walk?

**Find the area of the shaded region.**

20. 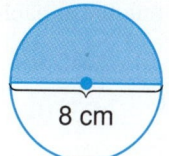 8 cm

21. 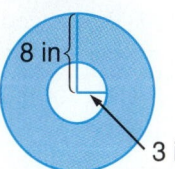 8 in, 3 in

22. 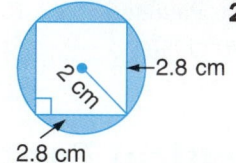 2 cm, 2.8 cm, 2.8 cm

23. 7 m, 5 m

## Connect and Extend

24. Find the area of the shaded region in the figure at the right.

25. If the area of a circle is 12.56 cm², what is the length of a radius of the circle?

26. The circumference of George's circular flower garden is 50 feet. What is its total area? (Hint: Draw a diagram.)

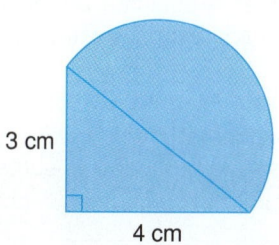
3 cm, 4 cm

**27.** Copy and complete the table. Refer to this table for Exercise 28.

| Radius | 1 | 2 | 4 | 8 |
|---|---|---|---|---|
| Area | 3.14 | ? | ? | ? |

**28.** If the radius of a circle is doubled, what happens to its area?

**29.** If the radius of a circle is 2n, what is its area?

**The value of $\pi$ can be calculated using this expression.**

$$\frac{\pi}{4} = 1 - \frac{1}{3} + \frac{1}{5} - \frac{1}{7} + \frac{1}{9} \ldots$$

**30.** Write the next 3 terms of the expression.

**31.** Use these 8 terms and a calculator to find a value for $\pi$. Round the value to the nearest hundredth.

## Maintain Your Skills

**Write a percent for each ratio. (Pages 305–308)**

**32.** $\frac{2}{5}$     **33.** $\frac{3}{20}$     **34.** $\frac{9}{40}$     **35.** $\frac{8}{10}$     **36.** $\frac{7}{4}$

**Solve. (Pages 334–337)**

**37.** What is 75% of 360?

**38.** 6 is 24% of what number?

**39.** 3 is what percent of 8?

**40.** What is 200% of 0.5?

**Write two equations in two variables to solve each problem. Then solve. (Pages 435–439)**

**41.** Pauline and Ella jogged a total of 12 miles. Ella jogged 3 miles more than twice the distance Pauline jogged. How far did each person jog?

**42.** Marshall saved $10 less than 4 times the amount his brother saved. Together they saved $20. How much did each person save?

## Mathematical Footnote

You already know about the number $\pi$, the number that is the ratio of the circumference of a circle to its diameter. But have you ever heard of the number e? Like $\pi$, e is an irrational number. This means that when e is written as a decimal, the decimal never ends or repeats itself. You can find an approximate value for e by adding this series.

$$1 + \frac{1}{1!} + \frac{1}{2!} + \frac{1}{3!} + \frac{1}{4!} + \ldots$$

The mathematician Leonard Euler calculated the value of e to 23 decimal places. The number e is sometimes called Euler's number.

# COMPUTER EXPLORATION

## Circumference and Area

**Objective:** To explore the circumference and area formulas

1. **a.** With the Holt *Pre-Algebra* disk in drive A (or drive 1), run the program, *Circumference and Area*, as follows.
   **IBM:** Type **a:i** and press **RETURN**.
   **Apple:** Turn on the computer. When the menu appears on the screen, type **I**.

2. **a.** Type 1 and run the program for the circumference formula. Refer to the program to answer these questions.
   **b.** About how many times is the radius marked off on the circumference?
   **c.** What is the ratio of the circumference to the radius?
   **d.** What is the formula for the circumference C of a circle in terms of r, where $\pi \approx 3$.
   **e.** About how many times can the diameter be marked off around a circle?
   **f.** What is the ratio of the circumference of a circle to its diameter, d?
   **g.** What is the formula for the circumference C of a circle in terms of the diameter, d, where $\pi \approx 3$?

3. **a.** Type 2 and run the program for the area formula.
   **b.** Copy this table. As you run the program, complete the entries in the table. There will be 7 entries in all. The first one is done for you. NOTE: There will always be one rectangle in the figure on the screen that is difficult to see because it is so small. It is the rectangle at the extreme right.

   | Number of Rectangles | Area: $\frac{1}{4}$ Circle | Area: Entire Circle |
   |---|---|---|
   | 12 | 0.74 | 2.95 |

   **c.** Study the table and look for patterns. As the number of rectangles increases, what happens to the area of the entire circle?
   **d.** What is the area of a circle with radius 1, where $\pi = 3.14$?
   **e.** How close is the approximate area shown in the last entry in your table to the area in 3d?

# Circle Graphs

**G**inny is the manager of a small Health Food Bar in the mall. She made a table to show the percent of each kind of sandwich sold on Saturdays.

| Sandwich | % Sold |
|---|---|
| Lettuce and Tomato | 40% |
| Chicken Salad | 30% |
| Tuna Salad | 20% |
| Vegetable Crunch | 10% |

For her monthly report, Ginny decided to draw a circle graph to illustrate the results of her survey. A **circle graph** is a circle divided into non-overlapping sections. Each section represents a percent of the whole circle. A circle graph shows how the parts are related to the whole.

Since a circle contains 360°, Ginny began by finding the number of degrees in 40%, 30%, 20%, and 10% of the circle.

$$40\% \text{ of } 360° = \tfrac{2}{5} \cdot 360 = 144° \qquad 20\% \text{ of } 360° = \tfrac{1}{5} \cdot 360 = 72°$$

$$30\% \text{ of } 360° = \tfrac{3}{10} \cdot 360 = 108\% \qquad 10\% \text{ of } 360° = \tfrac{1}{10} \cdot 360 = 36°$$

Then Ginny checked that the sum of the percents was 100 and that the sum of the degrees was 360.

$$40\% + 30\% + 20\% + 10\% = 100\% \qquad 144° + 108° + 72° + 36° = 360°$$

Here is how Ginny made a circle graph.

1. Using a compass, she drew a circle and marked the center, $O$.

2. Then she used a protractor to draw angles of 144°, 108°, 72°, and 36° in the circle. She used the center of the circle as the vertex of each angle.

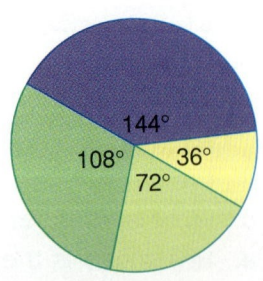

558 CHAPTER 14

3. Finally, she labeled the graph and chose a title for it.

**Saturday Sandwich Choices**

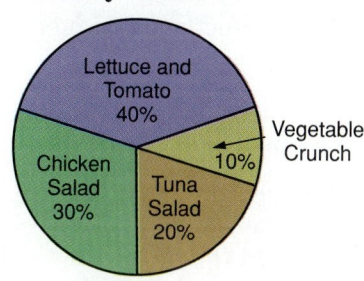

**EXAMPLE** Brenda receives an allowance of $10.00 per week. She spends $5.00 on movies, $3.00 on snacks, and saves $2.00. Draw a circle graph to illustrate how she uses her weekly allowance.

**Solution**

1. Find what percent of the total she spends on each item.

   Movies: $\frac{5}{10} = \frac{1}{2} = 50\%$    Snacks: $\frac{3}{10} = 30\%$

   Savings: $\frac{2}{10} = \frac{1}{5} = 20\%$

2. Use the percents in Step 1 to find the number of degrees for each part of the circle graph.

   Movies:  50% of 360° = $\frac{1}{2} \cdot 360 = 180°$

   Snacks:  30% of 360° = (360)(0.3) = 108°

   Savings: 20% of 360° = (360)(0.2) = 72°

3. Draw and label the graph.

**How Brenda Spends Her Weekly Allowance**

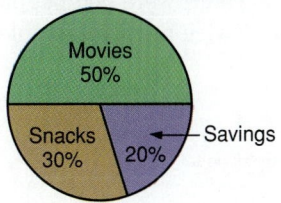

**Check**   50% + 30% + 20% = 100%
180° + 108° + 72° = 360°

**Try This** Miss Jackson told her Pre-Algebra class that their grades would be determined in the following way.

Tests: $\frac{1}{2}$    Quizzes: $\frac{1}{8}$    Homework: $\frac{1}{4}$    Class Participation: $\frac{1}{8}$

Make a circle graph to show this information.

# EXERCISES

**Objective:** To construct and interpret a circle graph

## Check Your Understanding

**Skill Check** Write the letter of the correct answer.

Ted sells tickets at a movie theater. One night, he drew a circle graph to show the attendance at the four movies being shown. He labeled the movies A, B, C, and D. Refer to the graph for Exercises 1–4.

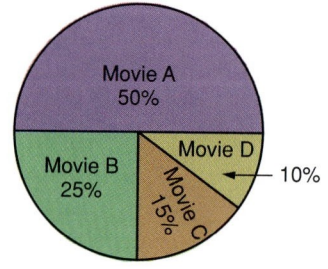

1. Which statement is **not** true?
   a. The least popular movie is D.
   b. More people saw movie B than movie D.
   c. Fewer people attended movie B than movie C.
   d. The most popular movie was A.

2. If 400 people in all attended movies that night, which statement is true?
   a. 180 people attended film A.
   b. 80 people attended film D.
   c. 60 people attended film C.
   d. 90 people attended film B.

3. Which statement about the circle graph is **not** true?
   a. The number of degrees in the pie-shaped region for film A is 180.
   b. The number of degrees in the region for film B is 90.
   c. The number of degrees in the region for film C is 54.
   d. The number of degrees in the region for film D is 40.

4. Suppose that 300 people attended movie A that night. Which statement is true?
   a. 50 people attended movie B.
   b. 20 people attended movie D.
   c. There were 600 people in all who attended the movies.
   d. Twice as many people saw movie B as saw movie C.

## Practice and Apply

Use the given data to make a circle graph.

5. **Vehicles Passing Corner**

| Autos | 65% |
| Motorcycles | 2% |
| Trucks | 18% |
| Vans | 10% |
| Bicycles | 5% |

6. **Amount Spent in 1 Month**

| Food | $50.00 |
| Clothes | $12.00 |
| Entertainment | $16.00 |
| Personal | $10.00 |
| Savings | $12.00 |

For Exercises 7–10, refer to the circle graph at the right. The budget is for a monthly budget of $1600.

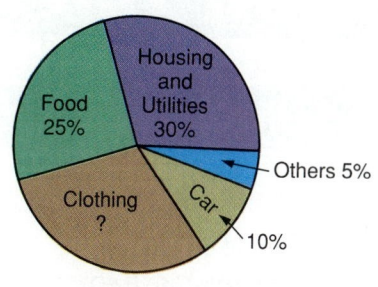

7. What amount is budgeted for food?
8. What amount is budgeted for car expenses?
9. How much is budgeted for clothing?
10. If monthly income is increased 25%, what is the effect on the amount available to spend on food?

These are the estimated populations in millions of 6 major world regions.

| North America | 260 | Europe | 500 | Asia | 2,600 |
| Latin America | 360 | USSR | 280 | Africa | 500 |

11. Use a calculator to find what percent of the whole population lives in each region. Round the percent to the nearest whole number.
12. Draw a circle graph to show the percent of the population living in each region.

## Connect and Extend

13. Toss a number cube 50 times. Make a tally sheet recording how many times each of the numbers, 1, 2, 3, 4, 5, and 6 appears on the top face. Then make a circle graph to show the results.
14. Take a survey of your class to find each student's favorite soft drink. Make a circle graph to show the results.
15. Find a circle graph in a newspaper or magazine and bring it to class. Write an explanation of what the graph shows.
16. Write a problem to go with the circle graph at the right. Include some questions for your classmates to answer.

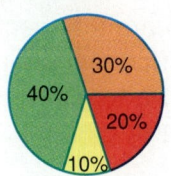

## Maintain Your Skills

**Solve.** (Pages 105–109, 393–397)

17. $4(m - 3) - 3 = 25$
18. $-\frac{1}{5}e + 1 \leq \frac{2}{5}$
19. $22 < -2(4x - 3)$

**Use the Pythagorean Theorem to find the length of the unknown side.** (Pages 457–462)

20. $a = 6, c = 10, b = \underline{\ ?\ }$
21. $a = 15, b = 8, c = \underline{\ ?\ }$

**Find each product.** (Pages 496–500)

22. $(t + 5)(t - 5)$
23. $(r - 3)(r + 3)$
24. $(q - 3)(q + 1)$

**Factor.** (Pages 512–515)

25. $c(c + 4) - 5(c + 4)$
26. $-7n(n + 9) + 2(n + 9)$

14-7 Circle Graphs **561**

# SUMMARY

## KEY TERMS

acute angle (p. 527)
acute triangle (p. 540)
angle (p. 526)
area of a circle (p. 553)
center of a circle (p. 551)
circle (p. 551)
circle graph (p. 558)
circumference (p. 553)
complementary angles (p. 528)
diagonal (p. 546)
diameter (p. 551)
equilateral triangle (p. 540)
isosceles triangle (p. 540)
obtuse angle (p. 527)

obtuse triangle (p. 540)
parallel lines (p. 532)
perpendicular lines (p. 532)
polygon (p. 546)
radius (p. 551)
rhombus (p. 546)
right angle (p. 527)
scalene triangle (p. 540)
supplementary angles (p. 528)
triangle (p. 539)
transversal (p. 533)
trapezoid (p. 546)
vertex (p. 526)
vertical angles (p. 532)

## KEY IDEAS

**A.** Vertical angles are equal.

**B.** If two parallel lines are intersected by a transversal, then:
   **a.** the measures of corresponding angles are equal.
   **b.** the measures of alternate interior angles are equal.
   **c.** the measures of alternate exterior angles are equal.

**C.** Triangle-Sum Property
   The sum of the measures of the angles in any triangle equals 180°.

**D.** Circumference of a circle: $C = \pi d$ or $C = 2\pi r$

**E.** Area of a circle: $A = \pi r^2$

# REVIEW

**Classify each angle measure as acute, obtuse, or right. (Pages 526–531)**

**1.** 93°  **2.** 8°  **3.** 90°  **4.** 60°  **5.** 179°  **6.** 45°

**Solve. (Pages 526–531)**

**7.** The supplement of an angle is 81°. What is the measure of the angle?

**8.** The complement of an angle is 28°. What is the measure of the angle?

In the figure at the right, line $a$ is parallel to line $b$. Lines $c$ and $d$ are transversals. Refer to this figure to find the measure of each angle. (Pages 532–536)

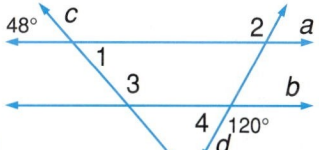

**9.** $m\angle 1$  **10.** $m\angle 2$  **11.** $m\angle 3$  **12.** $m\angle 4$

**Classify each triangle. Then write and solve an equation to find the unknown angle measures. (Pages 539–544)**

**13.**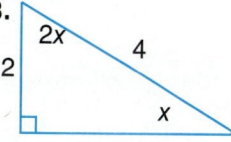

**14.**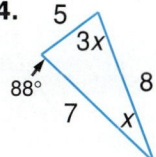

**15.**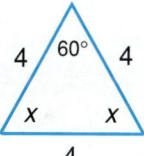

**For each quadrilateral, write and solve an equation to find $m\angle A$. (Pages 546–550)**

**16.**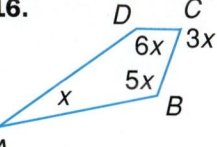

**17.**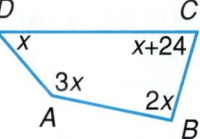

**18.**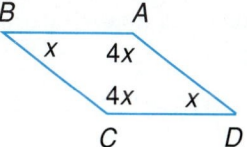

**Find the circumference and area of a circle having the given diameter or radius. (Pages 551–556)**

**19.** $r = 4$ ft  **20.** $d = 10$ m  **21.** $d = 12$ in  **22.** $r = 3.1$ cm

**Use the given data to make a circle graph. (Pages 558–561)**

**23.** Boys Favorite Sport to Play

| | |
|---|---|
| Basketball | 31% |
| Volleyball | 15% |
| Baseball | 24% |
| Football | 21% |
| Other | 9% |

**24.** Selected Characteristics of Women Readers

| | |
|---|---|
| Book Readers | 812 |
| Newspaper and Magazine Readers | 504 |
| Non-Reader | 84 |

Chapter 14 Review **563**

# TEST

**1.** The complement of an angle is 52°. What is the measure of the angle?

**2.** The supplement of an angle is 41°. What is the measure of the angle?

In the figure at the right, line *a* is parallel to line *b*. Lines *c* and *d* are transversals. Refer to this figure to find the measure of each angle.

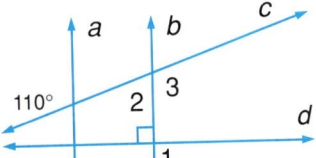

**3.** m∠1  **4.** m∠2  **5.** m∠3

Classify each triangle. Then write and solve an equation to find the unknown angle measures.

**6.** 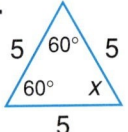  **7.** 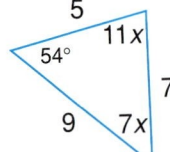  **8.**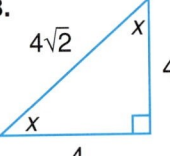

Give the name of each polygon.

**9.**   **10.**   **11.** 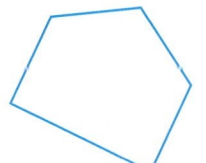  **12.**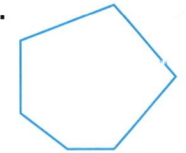

For each quadrilateral, write and solve an equation to find m∠A.

**13.** 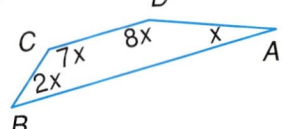  **14.** 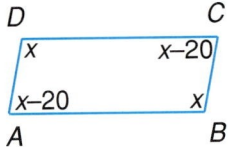  **15.**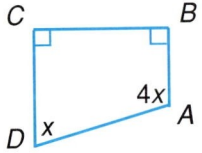

Find the circumference and area of a circle having the given diameter or radius.

**16.** $d = 6$ ft  **17.** $r = 7$ cm  **18.** $r = 4.1$ in  **19.** $d = 4.8$ m

**20.** Use the given data to make a circle graph. Give the number of degrees in each pie-shaped region of the graph.

| Favorite Bicycle Colors | |
|---|---|
| Black | 24% |
| Blue | 23% |
| Red | 22% |
| White | 8% |
| Other | 23% |

# CUMULATIVE REVIEW: CHAPTERS 1–14

The table at the right shows the weight in pounds of 19 different cats and dogs. (Pages 366–369)

1. Make a stem and leaf plot to show the data.
2. What percent of those animals weigh less than 20 pounds? Round the percent to the nearest tenth.
3. What percent of these animals weigh more than 23 pounds? Round the percent to the nearest tenth.

| Weight | | | | | | |
|---|---|---|---|---|---|---|
| 36 | 40 | 18 | 21 | 13 | 22 | 45 |
| 22 | 18 | 22 | 9 | 38 | 21 | |
| 27 | 41 | 16 | 24 | 34 | 37 | |

**Solve and graph. (Pages 403–405)**
4. $-8x > -56$ or $5 \geq -25x$
5. $-3 + 2y > 3$ or $5 + 2y \leq -1$

**Graph each equation or inequality. (Pages 430–434)**
6. $5y + 10x = -5$
7. $4x - y \geq -2$
8. $3y + 9x > -12$

9. The difference between two positive numbers is 5. One number is 1 less than 4 times the other number. What are the numbers? (Pages 435–439)

**Simplify if possible. (Pages 452–456)**
10. $\sqrt{80}$
11. $\sqrt{6} + \sqrt{10}$
12. $\dfrac{9}{\sqrt{3}}$
13. $\sqrt{24} \cdot \sqrt{6}$

14. Catalina is 4 feet tall. On a sunny day, her shadow is 10 feet long. At the same time, a flagpole casts a shadow 25 feet long. How tall is the flagpole? (Pages 466–472)

**Subtract. (Pages 496–500)**
15. $(2g^2 - 1) - (5g^2 + 4g)$
16. $(7v^2 - 3v - 6) - (v^2 - 2v - 8)$

**Find the product. (Pages 496–500)**
17. $(-4z + 3)(z - 9)$
18. $(-h + 7)(-h - 7)$
19. $\left(\dfrac{2}{3}c - 3\right)^2$

Refer to the figure at the right for Exercises 20–25. Lines $c$ and $d$ are transversals, and lines $a$ and $b$ are parallel. (Pages 532–536)

20. Name two acute angles.
21. Name one obtuse angle.
22. Name one pair of supplementary angles.
23. Name one pair of vertical angles.
24. Name one pair of alternate exterior angles.
25. Name one pair of complementary angles.

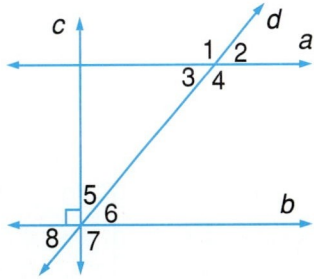

Cumulative Review: Chapters 1–14   **565**

## LESSON 15-1
## Problem Solving Exploration
## Surface Area and Volume

A. For this activity, you will need four rectangles, each 10 centimeters long and 8 centimeters wide, made from centimeter graph paper.

1. Make a 1-square box.

a. Start with the rectangle.

b. Remove a square, 1 cm × 1 cm, from each corner.

c. Turn up the sides to complete the 1-square box.

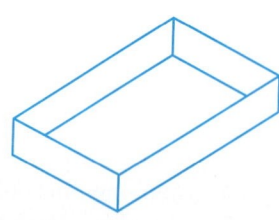

2. Make a 2-square box by removing a square, 2 cm × 2 cm, from each corner.
3. Make a 3-square box by removing a square, 3 cm × 3 cm, from each corner.
4. What happens when you try to make a 4-square box?

Copy and complete this table.

| Box | Area of a Bottom | Area of a Larger Side | Area of a Smaller Side |
|---|---|---|---|
| 5. 1-square | 8 · 6 = 48 cm² | 8 · 1 = 8 cm² | 6 · 1 = 6 cm² |
| 6. 2-square | ? | ? | ? |
| 7. 3-square | ? | ? | ? |

B. Suppose that each box has a top. Then the **surface area** of the box equals the total of its top, bottom, and side areas.

8. Explain why the surface area of the 1-square box can be found by either of these computations.

   a. 48 + 8 + 6 = 62
      2 · 62 = 124            Surface area: 124 cm²
   b. 2 · 48 = 96    2 · 8 = 16    2 · 6 = 12
      96 + 16 + 12 = 124    Surface Area: 124 cm²

9. What is the surface area of a 2-square box?
10. What is the surface area of a 3-square box?
11. Explain how you can find the surface area of a box $x$ units long, $y$ units wide, and 3 units high in two different ways.

C. 12. How many 1-centimeter cubes can be stacked in a 1-square box?
13. How many 1-centimeter cubes can be stacked in a 2-square box?
14. How many 1-centimeter cubes can be stacked in a 3-square box?
15. If the dimensions of a box are given as $l$, $w$, and $h$, write a formula to tell how many 1-centimeter cubes can be stacked in the box.

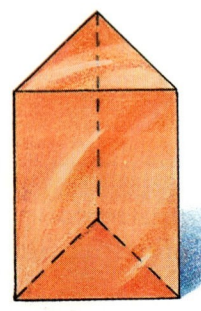  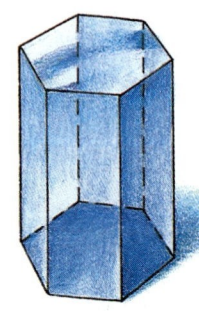 

Triangular Prism    Pentagonal Prism    Hexagonal Prism    Octagonal Prism

# Volume of a Rectangular Prism

**P**risms have many shapes. The two bases of a **prism** are polygons. A prism is named according to the polygon that forms its bases. In this chapter, you will study **rectangular prisms,** such as the one shown at the right.

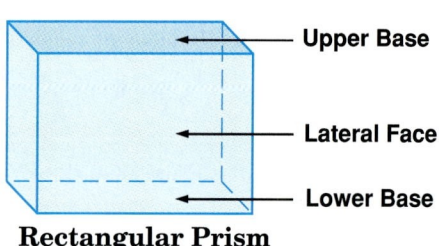

The **volume** of a prism is the amount of space it occupies. The volume is determined by finding how many cubic units can be fitted into the prism.

To find the volume of a rectangular prism, first find the number of cubic units in the base.

Then multiply by the number of layers of cubic units.

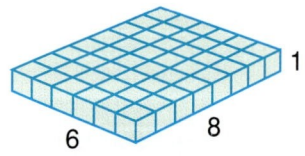    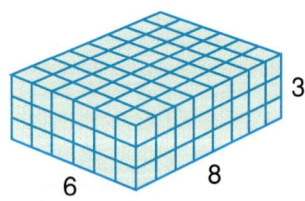

$6 \cdot 8 = 48$ cubic units     $6 \cdot 8 \cdot 3 = 144$ cubic units

**568**    CHAPTER 15

So the number of cubic units in a rectangular prism with dimensions 6 ft × 8 ft × 3 ft is

$$6 \cdot 8 \cdot 3 = 144 \text{ cubic feet, or } 144 \text{ ft}^3.$$

Then the number of cubic units in a rectangular prism with dimensions $l \times w \times h$, equals the number of cubic units in its base times its height. So the volume of the rectangular prism is

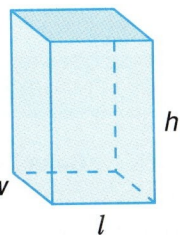

$$V = lwh.$$

Since $l \cdot w$ is the area of the base, the formula for the volume of a rectangular prism can also be written as

$$V = Bh, \text{ where } B \text{ is the area of the base.}$$

**EXAMPLE 1** What is the volume of a rectangular prism with dimensions 12.5 m × 6 m × 2 m?

**Solution**

Write the formula.  $V = lwh$
Substitute the values for  $V = (12.5)(6)(2)$
$l$, $w$, and $h$, and solve for V.  $V = 150$

The volume is **150 m³**.

**Try This** **Find the volume of each rectangular prism.**

1. $l = 16$ cm, $w = 8$ cm, $h = 8$ cm
2. $l = 2\frac{2}{3}$ yd, $w = 1\frac{1}{2}$ yd, $h = \frac{1}{2}$ yd

A **cube** is a special kind of rectangular prism. Its length, width, and height are equal. If $s$ is the length of a side of the cube, the formula for its volume can be written as

$$V = s \cdot s \cdot s, \text{ or } V = s^3.$$

**EXAMPLE 2** Find the volume of a cube that is $1\frac{1}{2}$ inches on a side.

**Solution**

Write the formula.  $V = s^3$

Substitute $1\frac{1}{2}$ for s and  $V = \left(1\frac{1}{2}\right)^3$
solve for V.
$V = \left(\frac{3}{2}\right)^3 = \frac{27}{8}, \text{ or } 3\frac{3}{8}$

The volume of the cube is $3\frac{3}{8}$ **in³**

15-2 Volume of a Rectangular Prism

**Try This**  **Find the volume of the cube with the given side, s.**

**3.** s = 3.6 cm    **4.** s = $\frac{2}{3}$ ft    **5.** s = 0.5 in

# EXERCISES

**Objective:** To find the volume of a rectangular prism

## Check Your Understanding

*Skill Check*  Write the letter of the correct answer.

**1.** What is the volume of a rectangular prism with $l = 2$ ft, $w = 4$ ft, and $h = 2$ ft?
  **a.** 16 ft$^3$     **b.** 40 ft$^2$
  **c.** 40 ft$^3$    **d.** 16 ft$^2$

**2.** What is the volume in cubic units of a cube with side $t$?
  **a.** $t^2$     **b.** $6t^2$
  **c.** $4t^3$   **d.** $t^3$

**3.** How many cubic units are there in a rectangular prism with dimensions $p \times q \times r$?
  **a.** $p \cdot q \cdot r$   **b.** $p + q + r$   **c.** $p^3 + q^3 + r^3$   **d.** $p^3 \cdot q^3 \cdot r^3$

**4.** Which statement is **not** true?
  **a.** Every cube is a rectangular prism.
  **b.** A rectangular prism has 4 lateral faces.
  **c.** The area of the base of a cube with side $s$ is $s^2$.
  **d.** The bases of a rectangular prism can be octagons.

**Find the volume of each rectangular prism.**

**5.** 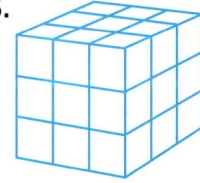    **6.** 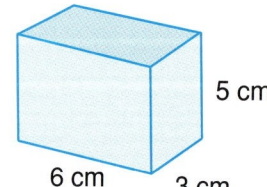    **7.**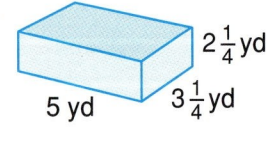

## Practice and Apply

**Find the volume of the rectangular prism.**

| | Length | Width | Height | | Length | Width | Height |
|---|---|---|---|---|---|---|---|
| **8.** | 4 m | 3 m | 5 m | **9.** | 2.5 m | 4 m | 1 m |
| **10.** | $2\frac{2}{3}$ yd | $1\frac{1}{2}$ yd | $1\frac{1}{4}$ yd | **11.** | 5 m | 10 m | 6 m |
| **12.** | 7.6 m | 3 m | 4.5 m | **13.** | $6\frac{2}{3}$ in | 8 in | $10\frac{1}{2}$ in |

14. How many cubic feet are there in 1 cubic yard?
15. How can you change 5 cubic yards to cubic feet?
16. How can you change 270 cubic feet to cubic yards?
17. A contractor excavates a basement for a new home. The excavation pit is 45 feet long, 27 feet wide, and 9 feet deep. How many cubic yards did the contractor excavate?

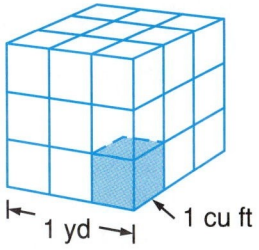

18. Concrete is usually measured in cubic yards. Mr. Ellis wants to build a concrete patio that is 36 feet wide, 18 feet long and 6 inches deep. If concrete costs $10 per cubic yard, how much will the patio cost?
19. A rectangular drainage ditch is to be filled with gravel. The dimensions of the ditch are 6 m × 4.5 m × 5.5 m. A truck carries 3.5 m³ of gravel each trip. How many truckloads are needed to fill the ditch?

## Connect and Extend

**Two similar rectangular prisms have dimensions $4 \times 3 \times 2$ and $8 \times 6 \times 4$.**

20. Compute the volume of each prism.
21. What is the ratio of their volumes?
22. What is the ratio of the lengths of their sides?

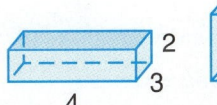
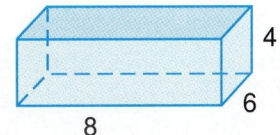

23. The length of a side of a cube is doubled. Compare the ratio of the sides to the ratio of the volumes.

## Maintain Your Skills

**Evaluate each pair of expressions for $e = \frac{1}{2}, f = -6, g = 8$, and $h = -2.2$. Then write the expression having the smaller value. (Pages 262-266)**

24. $eh$ or $e + h$
25. $-(e + f)^2$ or $-(e - f)^2$
26. $\dfrac{(f + g)^2}{-(f + g)}$ or $\dfrac{(f - g)^2}{f - g}$

**Write an equation of the line having the given slope and y-intercept. (Pages 430–434)**

27. slope: 2
    y-intercept: (0, 4)
28. slope: −1
    y-intercept: (0, 3)
29. slope: $\frac{1}{5}$
    y-intercept: (0, −2)

**Replace each ? with the missing factor. (Pages 516–519)**

30. $y^2 - 5y + 4 = (y - 4)(\underline{?})$
31. $e^2 + 7e + 10 = (e + 5)(\underline{?})$
32. $a^2 - 6a - 16 = (a + 2)(\underline{?})$
33. $z^2 - 10z - 11 = (z - 11)(\underline{?})$
34. $t^2 + 12t + 35 = (t + 5)(\underline{?})$
35. $r^2 - 8r + 12 = (r - 2)(\underline{?})$

# LESSON 15-3 Surface Area of a Rectangular Prism

The box Diego is wrapping has the shape of a rectangular prism. Each side of the prism is called a **face**. The top and bottom faces are also called **bases**. To find the amount of wrapping paper needed to cover the box, Diego must find its surface area. The **surface area** is the total area of its faces.

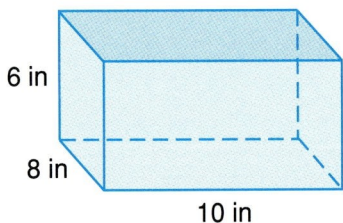

To find the surface area, Diego completed these computations.

Area of top: 10 · 8 = 80      Area of bottom: 10 · 8 = 80
Area of left side: 8 · 6 = 48     Area of right side: 8 · 6 = 48
Area of facing side: 10 · 6 = 60     Area of back side: 10 · 6 = 60

Total area: 2 · 80 + 2 · 48 + 2 · 60 = 160 + 96 + 120
                                                 = 376

The surface area of the box is 376 in$^2$.

- Why will Diego need more than 376 in$^2$ of paper to wrap the present?

Diego's computation suggests a formula for finding the surface area of a rectangular prism.

| Surface Area | = | Area of Top and Bottom | + | Area of Front and Back Sides | + | Area of Left and Right Sides |
|---|---|---|---|---|---|---|
| $S$ | = | $2lw$ | + | $2lw$ | + | $2wh$ |

> **Surface Area of a Rectangular Prism**
>
> For a rectangular prism of dimensions $l \times w \times h$,
> $$S = 2lw + 2lh + 2wh$$
> or
> $$S = 2(lw + lh + wh)$$

**EXAMPLE 1** Find the surface area of the rectangular prism below.

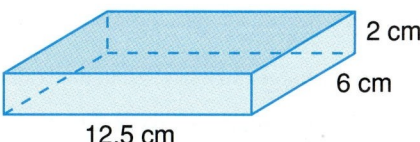

**Solution**  Write the formula.  $S = 2lw + 2lh + 2wh$
Substitute the values for  $S = 2(12.5)(6) + 2(12.5)(2) + 2(6)(2)$
$l$, $w$, and $h$.  $S = 150 + 50 + 24$
$S = 224$

The surface area is **224** cm².

**Try This**  **Find the surface area of the rectangular prism having the given dimensions.**
  1. $l = 16$ cm, $w = 8$ cm, $h = 8$ cm
  2. $l = 2\frac{1}{2}$ yd, $w = 1\frac{1}{2}$ yd, $h = \frac{1}{2}$ yd

Recall that a **cube** has 6 faces. Each face has the same area. So the surface area of a cube with side $s$ is $6s^2$.

**EXAMPLE 2**  Find the surface area of the cube below.

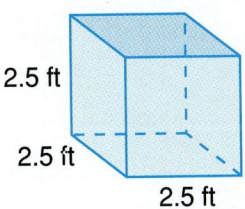

**Solution**  Write the formula.  $S = 6s^2$
Substitute 2.5 for $s$.  $S = 6(2.5)^2$
$S = 6(6.25)$
$S = 37.5$

The surface area is **37.5** ft².

**Try This**  **Find the surface area of the cube with side $s$.**
  3. $s = 1\frac{1}{2}$ yd    4. $s = 8$ m    5. $s = 1$ cm

• Write a formula for the surface area of a cubical box that has no top.

# EXERCISES

**Objective:** To find the surface area of a rectangular prism

## Check Your Understanding

**Skill Check**  Write the letter of the correct answer.

1. Which statement about a rectangular prism is **not** true?
   a. Surface area is measured in cubic units.
   b. The surface area of a rectangular prism is $S = 2lw + 2lh + 2wh$.
   c. Surface area is measured in square units.
   d. A cube is a rectangular prism in which $l = w = h$.

2. Which is **not** a formula for surface area?
   a. $S = 2lw + 2lh + 2wh$
   b. $S = 2(lw + lh + wh)$
   c. $S = lw + lw + lh + lh + wh + wh$
   d. $S = 2l + 2w + 2h$

3. Which is the surface area of a rectangular prism with $l = 3$ cm, $w = 2$ cm, and $h = 5$ cm?
   a. 30 cm$^3$
   b. 62 cm$^2$
   c. 62 cm$^3$
   d. 30 cm$^2$

4. Which is the surface area of a cube with side $r$?
   a. $6r^2$
   b. $5r^3$
   c. $4r^2$
   d. $4r^3$

## Practice and Apply

**Find the surface area of each rectangular prism.**

|    | Length | Width | Height |    | Length | Width | Height |
|----|--------|-------|--------|----|--------|-------|--------|
| 5. | 4 m    | 3 m   | 5 m    | 6. | 2.5 m  | 4 cm  | 1 cm   |
| 7. | 10 cm  | 15 cm | 8 cm   | 8. | 1 m    | 0.05 m| 2.5 m  |
| 9. | 5 ft   | $2\frac{1}{2}$ ft | $2\frac{1}{2}$ ft | 10. | 1.5 cm | 5 cm | 8 cm |

11. What is the minimum amount of cardboard needed to make a box that is $4\frac{1}{2}$ inches long, 2 inches wide, and 1 inch deep?

12. Rhonda is painting a toy chest that has the shape of a rectangular prism. The chest is 4 feet long, $2\frac{1}{2}$ feet wide, and 2 feet tall. She has enough paint to cover 45 square feet. Will she need to buy more paint? Explain why or why not.

13. Plywood costs 18¢ per square foot. Todd is going to use plywood to make the four walls of his dog house. If the dog house is to be 6 feet long, 4 feet wide, and 3 feet high, how much will the plywood cost?

## Connect and Extend

**Copy the pattern at the right.**

14. Cut out the pattern and fold along the lines to make a cube. Find the surface area and the volume of the cube.

15. What is the ratio of the surface area of the cube to its volume?

16. Make a sketch to show how the box at the right will look if it is folded flat. If you use one rectangular piece of cardboard to cut the box from, what are the smallest dimensions you can use? How much cardboard will be wasted?

17. The surface area of a cube is 54 square units. What is its volume?

18. The surface area of a rectangular prism has the same number of square units as the volume has cubic units. Describe the prism.

19. Explain why the surface area of a cube is $6s^2$.

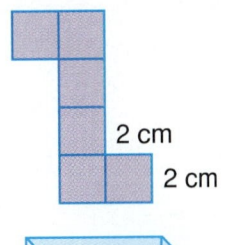

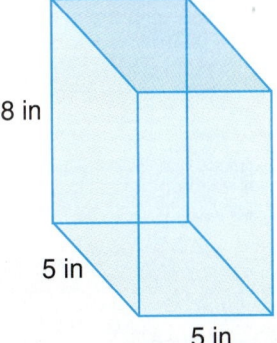

## Maintain your Skills

**Solve. (Pages 221–225, 551–556)**

20. If 4 pounds of fish cost $27.96, how much does 1 pound of fish cost?
21. What is the circumference of a circle with a diameter of 12.5 inches?

**Replace each ? with <, >, or = to make a true statement. (Pages 226–230)**

22. 1,800 lb ? 0.8T
23. 17,000 ft ? $3\frac{1}{4}$ miles
24. $9\frac{1}{4}$ pt ? 148 fl oz

**Solve. (Pages 334–337)**

25. Michael cut 40 inches from a board that was 180 inches long. What percent of the length of the board is left? Round your answer to the nearest whole percent.

26. Mrs. McBean bought a used car for $5,790. She made a 15% down payment. How much was her down payment?

## ★ Math Team Problem

27. In the book *Gulliver's Travels* by Jonathan Swift, Gulliver's first trip brought him to the land of Lilliput. The Lilliputians discovered that Gulliver's body was similar to theirs and that the ratio of Gulliver's height to theirs was 12:1. They therefore concluded that he would require the same amount of food needed for 1,728 Lilliputians. Explain how they arrived at that number.

## LESSON 15-4

# Volume of a Cylinder

A right **cylinder** has circles for bases. The **radius,** $r$, of a cylinder is the radius of either base. The **height** $h$, of a cylinder is the perpendicular distance between the bases.

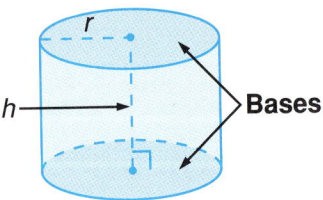

Cylinders are useful for many purposes because of two properties. First, the form of a cylinder is rigid, and secondly, for a given volume, it requires less material in its construction than a comparable prism. For this reason, cylindrical columns are used to support heavy loads and cylindrical containers are used to hold liquids.

- What are some uses for cylinders in the home?

- What are some uses for cylinders in industry?

You used a parallelogram to approximate the area of a circle. In somewhat the same way, you can use a prism to approximate the volume of a cylinder.

**576**  CHAPTER 15

1. Think of dividing the cylinder into pie-shaped wedges. The radius of a base of the cylinder is $r$ and its height is $h$.

2. Reassemble the cylinder to form a prism as shown. The length of the prism is half the circumference, $C$, of a base of the cylinder, or $\pi r$. The height of the prism is $h$, and its width is $r$.

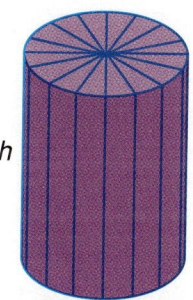

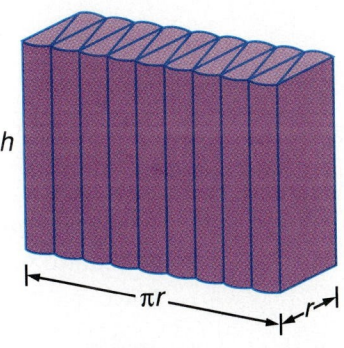

3. Find the volume of the prism.    $V = (\pi r)(r)(h) = \pi r^2 h$

Notice that the area of the base of a cylinder with radius, $r$, is $\pi r^2$. This suggests the formula for the volume of a cylinder.

> **Volume of a Cylinder**
> For any cylinder with radius $r$, height $h$, and $B$ = area of base,
> $$V = Bh \text{ or } V = \pi r^2 h.$$

**EXAMPLE 1** What is the volume of a cylinder with a radius of 3 centimeters and a height of 5 centimeters?

**Solution**

Write the formula.        $V = \pi r^2 h$
Substitute for $r$ and $h$.    $V \approx (3.14)(3)^2(5)$
                $V \approx 141.3$

The volume is about **141.3 cm³**.

**Try This**   Find the volume of each cylinder. Use $\pi = 3.14$.

1. $r = 8$ cm, $h = 8$ cm
2. $r = \frac{1}{3}$ yd, $h = \frac{1}{2}$ yd

15-4  Volume of a Cylinder  **577**

**EXAMPLE 2**  How many yards of concrete will it take to form a cylindrical pillar for a highway support that is 65 feet high and whose base has a diameter of 11 feet if 15% of the volume must be steel rods?

**Solution**  Since $d = 11$ feet, $r = \frac{1}{2}(11) = 5.5$ feet.

Since you are asked to find the volume in cubic yards, change the radius and height to yards.

$5.5 \text{ ft} = \frac{5.5}{3} \approx 1.83 \text{ yd}$       $65 \text{ ft} = \frac{65}{3} \approx 21.67 \text{ yd}$

Write the formula for volume.   $V = \pi r^2 h$
Substitute for $r$ and $h$.            $V \approx (3.14)(1.83)^2(21.67)$
Solve for V. Use a calculator.    $V \approx 227.87$

The volume of the pillar is about 228 yd³.

**Think:** 15% is steel. So 85% is concrete.

Find 85% of 228.        $(0.85)(228) = 193.8$

The amount of concrete needed is about **194 yd³**.

**Try This**  3. a. A cylindrical water tank has a radius of 5 feet and is 20 feet high. How much water can it hold?

b. There are 7.5 gallons in 1 cubic foot. How many gallons of water will the water tank hold?

# EXERCISES

**Objective:** To find the volume of a cylinder

## Check Your Understanding

**Skill Check**  Write the letter of the correct answer.

1. Which statement about a cylinder is **not** true?
    a. The bases are circles.
    b. The formula for the volume is $\pi r h^2$.
    c. The area of a base is $\pi r^2$.
    d. The volume is the product of the area of its base and its height.

**2.** If the height of a cylinder is 8 centimeters and its radius is 5 centimeters, what is its volume?
   **a.** 1005 cm³   **b.** 125.6 cm³
   **c.** 628 cm³   **d.** 200 cm³

**3.** The area of the base of a cylinder is 36 in² and the height of the cylinder is 4 in. What is the volume?
   **a.** 1,808.64 in³   **b.** 144 in³
   **c.** 576 in³   **d.** 452.16 in³

**4.** The height of a cylinder equals the length of its radius. Which is the formula for its volume?
   **a.** $V = \pi r^3$   **b.** $V = 2\pi r^2$   **c.** $V = 2\pi r^2 h$   **d.** $V = 3\pi r^3$

## Practice and Apply

**Find the volume of each cylinder. Use 3.14 for $\pi$.**

**5.**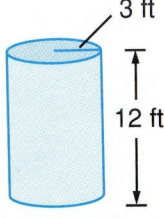

**6.**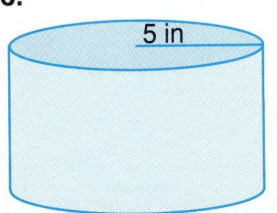

**7.**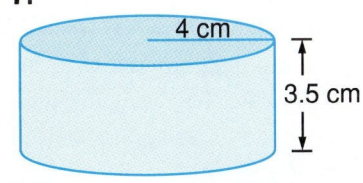

**Solve. Use 3.14 for $\pi$. Round answers to the nearest tenth.**

**8.** The radius of a hot water tank is 0.4 meters and its height is 1.5 meters. Find the volume of the tank.

**9.** A box of salt has the shape of a cylinder. The radius of the box is $1\frac{1}{2}$ inches and its height is 5 inches. How much salt does it hold?

**The water tank at the right is full of water. Its radius is 5 feet and its height is 20 feet. Use this information for Exercises 10–13.**

**10.** If 1 cubic foot is about 7.5 gallons, how many gallons of water are in the tank?

**11.** If 30% of the water is used each day, how many gallons is this?

**12.** Water costs a user $3 for every 150 gallons used. How much is paid for the water used each day?

**13.** If 1 cubic foot of water weighs about 62.5 pounds, what is the weight of the water in tons when the tank is full?

**14.** The base of a square prism is 4 inches on a side and has a height of 7 inches. The base of a cylinder has a diameter of 4 inches and the height of the cylinder is the same as that of the prism. Which has the greater volume, the prism or the cylinder? Why?

## Connect and Extend

**A can of juice has a radius of 3 inches and a height of 8 inches.**

**15.** What is the volume of the can?

**16.** Suppose that the juice company wants to design a new can that has the same height but 25% less volume. What will be the volume of the new can?

**17.** To the nearest hundredth, what will be the radius of the new can. Use a calculator and the guess-and-check strategy. Compute the volume with a radius of 2.9, of 2.8, 2.7, and so on.

**18.** Solve this equation for $r$: $\frac{3}{4}\pi(9)(8) = \pi(r^2)(8)$

**19.** How does your answer to Exercise 17 compare with your answer to Exercise 18? Explain.

**20.** Use the diagram below to find the surface area of the soup can.

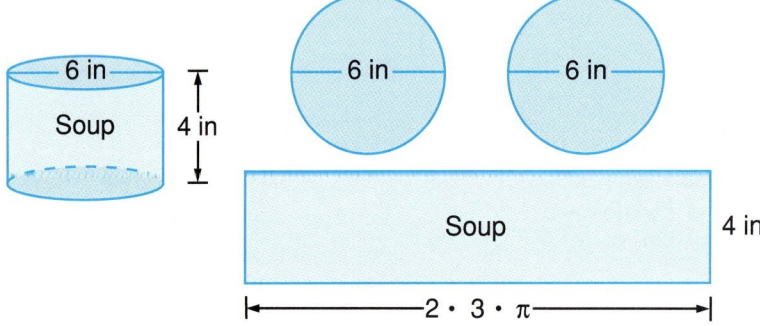

## Maintain Your Skills

**Simplify each expression. (Pages 46–50)**

**21.** $5 + 6 \div 3 \cdot 4$    **22.** $9 - (6 + 3) \cdot 4$    **23.** $(5 + 8 \cdot 2) \div 7$

**Solve. (Pages 457–462)**

**24.** A 26-foot wire is staked to the ground 10 feet from the base of the pole. How tall is the pole?

**25.** A ladder leans against a building and reaches a ledge 25 feet from the ground. The foot of the ladder is 20 feet from the building. How long is the ladder?

**Give the name of the polygon having the given number of sides. (Pages 546–550)**

**26.** 4 sides    **27.** 8 sides    **28.** 6 sides    **29.** 5 sides

**Find the volume of a rectangular prism having the given dimensions.
(Pages 568–571)**

1. $l = 10$ m, $w = 4$ m, $h = 5$ m
2. $l = 6$ yd, $w = 4$ yd, $h = 4\frac{1}{4}$ yd
3. $l = 4.6$ mm, $w = 12$ mm, $h = 0.5$ mm
4. $l = 12$ cm, $w = 8$ cm, $h = 4$ cm

**Find the surface area of each rectangular prism.
(Pages 572–575)**

5. $l = 3$ ft, $w = 4$ ft, $h = 6$ ft
6. $l = 1.1$ yd, $w = 2$ yd, $h = 6$ yd

7. Mark wants to make a wooden jewelry box that has the shape of a cube. Each side of the box will be 14 inches long. How much wood will he need? **(Pages 572–575)**

**Find the volume of each cylinder. Use $\pi = 3.14$ (Pages 576–579)**

8. $r = 10\frac{1}{2}$ ft, $h = 4$ ft
9. $r = 14$ m, $h = 8$ m

10. The radius of a drinking glass is $1\frac{1}{2}$ inches and its height is 6 inches. How much juice is in the glass if it is filled to 1 inch from the top? **(Pages 576–579)**

# **M**athematical Footnote

Florence Nightingale is famous for establishing nursing as a profession. She also pioneered the keeping of accurate medical statistics which she used to convince others of needless deaths in military hospitals.

Although women did not attend universities at the time, Florence was taught mathematics by her father. She experimented with various graphical methods to represent facts in a way everyone could understand. She also struggled to get the study of statistics introduced into higher education. She spoke of statistics as "the most important science in the whole world," because "upon it depends the practical application of every other [science]."

# Volume of a Pyramid

The Egyptian pyramids were built in about 2600-2500 B.C. A **pyramid** has these properties.

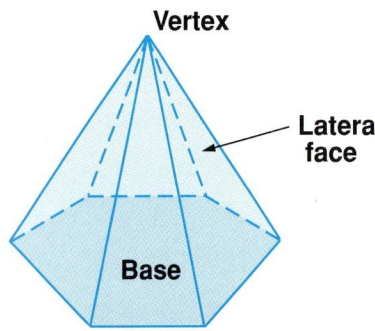

1. All the faces, except one, intersect at a point called the **vertex.**

2. The faces that intersect at the vertex are called **lateral** (side) **faces** and form triangles.

3. The face that does not intersect at the vertex is called the **base** and forms a polygon.

A pyramid is named by the shape of its base. The TransAmerica Building in San Francisco (see the photo) is a square pyramid.

To find the volume of a pyramid, you can perform an experiment that compares its volume to the volume of a prism having a congruent base and equal altitude.

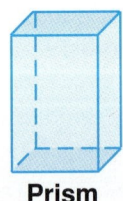

Prism

Fill the pyramid with rice. Then pour the contents into the prism as shown at the right. Repeat until the prism is full.

If you complete the activity, you will find that it takes 3 pyramids of rice to fill the prism. So the volume of the pyramid is one-third the volume of the prism.

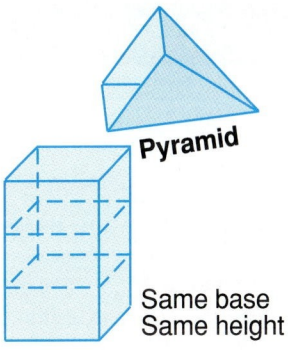

Same base
Same height

Volume of prism: $V = Bh$    Volume of pyramid: $V = \frac{1}{3}Bh$

> **Volume of a Pyramid**
> To find the volume of a pyramid, find the product of $\frac{1}{3}$ the area of the base, $B$, and the height, $h$.
> $$V = \frac{1}{3}Bh$$

**EXAMPLE**  Find the volume of a pyramid with a height of 20 yards and a rectangular base of dimensions 5 yards × 4 yards.

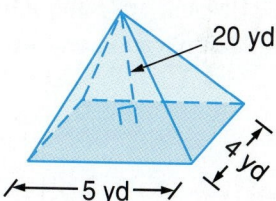

**Solution**  **Think:** Since the base is a rectangle, the area of the base is 5 · 4, or 20 yd².

Write the formula.         $V = \frac{1}{3}Bh$

Substitute for $B$ and $h$.    $V = \frac{1}{3}(20)(20)$

$V = \frac{400}{3}$

$V = 133\frac{1}{3}$

The volume is $133\frac{1}{3}$ **yd³**.

**Try This**  Find the volume of a pyramid with the given dimensions.
  **1.** $B = 256$ in², $h = 8$ in
  **2.** Rectangular base, 3 ft × 4 ft; $h = 12$ ft

15-5  Volume of a Pyramid  **583**

# EXERCISES

**Objective:** To find the volume of a pyramid

## Check Your Understanding

**Skill Check** Write the letter of the correct answer.

1. Which statement is true?
   a. The base of a pyramid is a lateral face.
   b. All pyramids have rectangular bases.
   c. All the faces of a pyramid meet at the vertex.
   d. All the lateral faces of a pyramid are triangles.

2. The volume of a pyramid is 150 cm$^3$. What is the volume of a prism having the same base and height?
   a. 50 cm$^3$    b. 50 cm$^2$
   c. 450 cm$^2$   d. 450 cm$^3$

3. A rectangular pyramid has a base 6 feet long and 4 feet wide. Its height is 8 feet. What is its volume?
   a. 192 ft$^3$    b. 72 ft$^3$
   c. 576 ft$^3$    d. 64 ft$^3$

4. The volume of a pyramid is 8 mm$^3$. The area of the base is 4 mm$^2$. What operations could you use to find the height?
   a. Division and subtraction
   b. Division only
   c. Multiplication and subtraction
   d. Division and multiplication

## Practice and Apply

**Find the volume of each rectangular pyramid.**

5.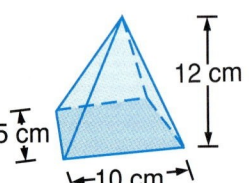

6.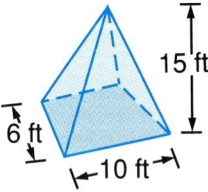

7.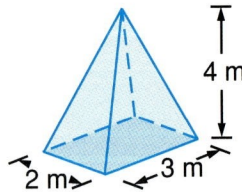

**Solve.**

8. The length of a side of a square pyramid is 2 units. The height of the pyramid is 4 units. Draw a prism having 3 times the volume of the pyramid. Label the dimensions of the prism.

9. Martha and Shari pitched a tent that has the shape of a pyramid. The base of the tent is a rectangle that is 2.1 m long and 2.4 meters wide. The tent is 2 meters high. What is the volume?

10. A city is building a new museum in the shape of a pyramid having a square base. The base is 100 feet long and the building is 60 feet tall. The engineer who is designing the air conditioning for the building designs the system so that it provides one change of fresh air per hour. This means that an amount of fresh air equal to the volume of the building must be put into the building each hour. How many cubic feet of fresh air per minute must be put into the building to meet this requirement?

## Connect and Extend

11. If you know the volume and height of a square pyramid, explain how you would find the area of the base.
12. A pyramid has a square base with $s = 60$ cm. If the volume is 16,000 cm$^3$, what is the height?
13. A pyramid has a square base and a height of $16\frac{1}{2}$ inches. If the volume is $103\frac{1}{8}$ in$^3$, what is the length of a side of the base?

## Maintain Your Skills

**Solve. (Pages 226–230)**

14. How many square inches are there in 1 square foot?
15. How many cubic feet are there in 6 cubic yards?

**Estimate the quotient by rounding. (Pages 297–300)**

16. $55.81 \div 8.43$
17. $68.22 \div 3.89$
18. $126.01 \div 6.77$

**Find the circumference and area of a circle with the given radius or diameter. Use the given value for pi. (Pages 551–556)**

19. $r = 3$ ft; $\pi = 3.14$
20. $d = 11.4$ m; $\pi = 3.14$

## ★ Math Team Problems

21. The diameter of the earth at the equator is about 7800 miles. Suppose that you place a metal band around the earth at the equator at a distance of 1 inch from the surface of the earth. How much greater will be the circumference of the circle formed by the metal band be than the circumference of the earth?
22. Four friends, Luisa, Maria, Carol, and Glenn are sitting in a row. Neither Carol nor Maria is sittin next to Glenn. Carol is sitting just to the right of Maria. Write the seating arrangement from left to right.
23. Two positive integers are each less than 10. The sum of their squares, added to their product, equals a perfect square. Find the integers.

## LESSON 15-6

# Volume of a Cone

You see cones as drinking cups, as yogurt containers, and as nose cones on space vehicles. A **cone** has these properties.

1. Its **base** is a circle.
2. It has a **vertex** as shown.
3. Its **height** is a line segment drawn from the vertex perpendicular to the base.

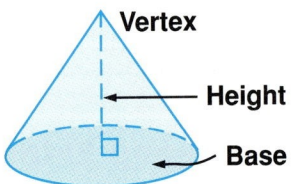

To find the volume of a cone, you can perform an experiment that compares its volume to the volume of a cylinder having the same radius and height.

Fill the cone with rice and empty it into the cylinder. Continue filling and emptying the cone until the cylinder is full.

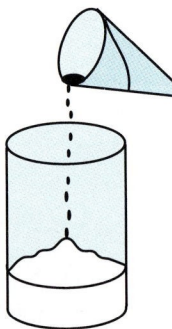

Since it takes 3 cones of rice to fill the cylinder, the volume of the cone is $\frac{1}{3}$ the volume of a cylinder of the same radius and height.

Volume of a cylinder: $V = Bh = \pi r^2 h$

Volume of a cone: $V = \frac{1}{3}Bh = \frac{1}{3}\pi r^2 h$

> **Volume of a Cone**
> The volume of a cone with radius, $r$, and height, $h$, is
> $V = \frac{1}{3}\pi r^2 h.$

**EXAMPLE** What is the volume of a cone with height 8 cm and radius 3 cm?

**Solution**  Write the formula.   $V = \frac{1}{3}\pi r^2 h$

Substitute for $r$ and $h$.   $V = \frac{1}{3}(3.14)(3)^2(8)$

$V = 75.36$

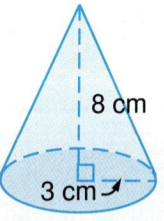

The volume of the cone is about **75 cm³**.

586   CHAPTER 15

# EXERCISES

**Objective:** To find the volume of a cone

## Check Your Understanding

**Skill Check** Write the letter of the correct answer.

1. Which statement is true?
   a. A cone and a cylinder have the same number of bases.
   b. A cone can have a square base.
   c. The height of a cone is always greater than the length of the radius.
   d. The area of the base of a cone with radius, $r$, is $\pi r^2$.

2. Which is the volume of a cone having a radius of 3 centimeters and a height of 10 centimeters?
   a. $30\pi$ cm$^3$   b. $180\pi$ cm$^3$
   c. $90\pi$ cm$^3$   d. $360\pi$ cm$^3$

3. Which is the best estimate for the volume of a cone with a radius of 1 inch and a height of 27 inches?
   a. 81 in$^3$   b. 27 in$^3$
   c. 9 in$^3$   d. 254 in$^3$

4. The volume of a cylinder is 108 ft$^3$. What is the volume of a cone having the same radius and height as the cylinder?
   a. 324 ft$^3$   b. 36 ft$^3$   c. $36\pi$ ft$^3$   d. $324\pi$ ft$^3$

## Practice and Apply

Find the volume of each cone. Use 3.14 for $\pi$. Round answers to the nearest whole number.

5.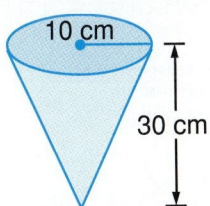

6.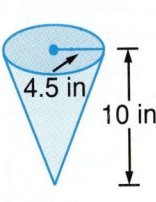

7.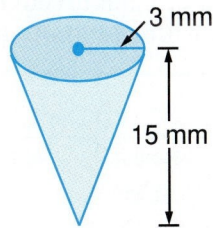

Solve. Use 3.14 for $\pi$. Round answers to the nearest tenth.

8. A cone-shaped hanging planter has a radius of 7 centimeters and a height of 15 centimeters. Find the volume.

9. A cone shaped paper cup has a diameter of 8 inches and a height of 8 inches. Find the volume.

10. A cone-shaped storage tank has a diameter of 5 feet and a height of 16 feet. The tank is filled with water. If there are 7.5 gallons in 1 cubic foot, how many gallons of water are there in the tank?

15-6   Volume of a Cone

## Connect and Extend

**11.** A cone sits inside a cylinder having the same base and height. Write a formula for the space between the cylinder and the cone.

The graph below shows how the volume of a cone changes as the radius changes. Notice that in all cases the height of the cone remains constant at 3 inches.

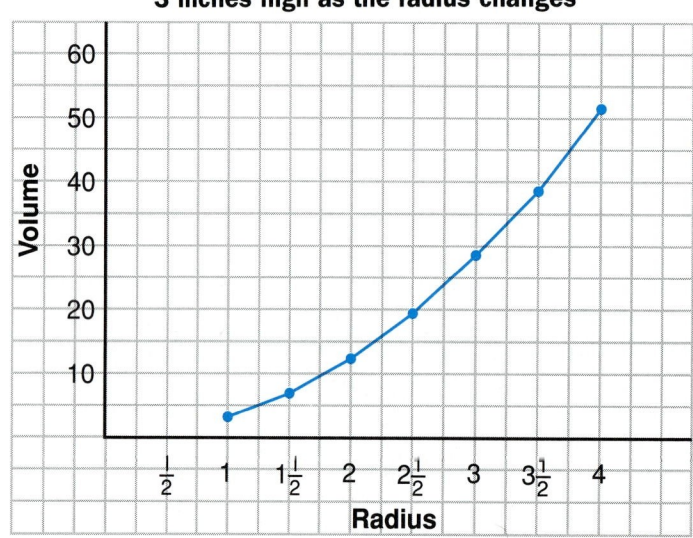

Copy the table below. Use the graph to find the volumes in Columns 1 and 2. Then divide the volume in Column 2 by the volume in Column 1. Round quotients to the nearest tenth.

|     | Column 1 | | Column 2 | | Volume in Column 2 ÷ Volume in Column 1 |
| --- | --- | --- | --- | --- | --- |
|     | Radius | Volume | Radius | Volume | |
| 12. | 1 in | ? | 2 in | ? | ? |
| 13. | $1\frac{1}{4}$ in | ? | $2\frac{1}{2}$ in | ? | ? |
| 14. | $1\frac{1}{2}$ in | ? | 3 in | ? | ? |
| 15. | 2 in | ? | 4 in | ? | ? |

**16.** Write a summary telling how the volume of a cone changes as the radius of the cone doubles in length while the height remains constant.

**17.** Write a paragraph telling how a pyramid and a cone are alike and how they are different.

Many containers are made up of several different figures. To find the volume of the container, you find the volume of each figure separately, then you add to find the total volume.

**Find the volume of each container.**

18.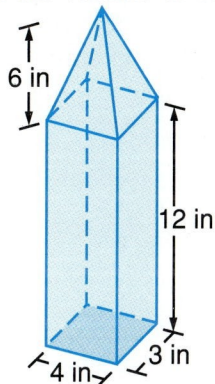

19.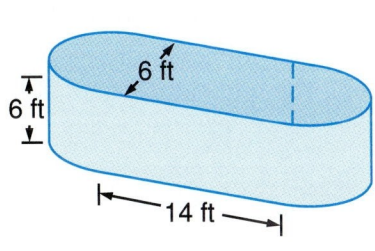

20.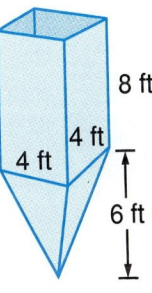

## Maintain Your Skills

**Use fractions to estimate the percent of each number. (Pages 313–316)**

**21.** 72% of 150  **22.** $112\frac{1}{2}$% of 24  **23.** $\frac{1}{2}$% of 60

**Solve. (Pages 393–397)**

**24.** $5x - 9 \leq 36$  **25.** $2.4c + 1.8 > -3.6$

**Factor. (Pages 512–515)**

**26.** $63e - 56$  **27.** $104c^2 + 39c$  **28.** $-4(p + 2) - 5p(p + 2)$

## Mathematical Footnote

You can find a rough approximation for $\pi$ by performing an experiment called "Buffon's Needle." You will need a needle or short piece of wire. On a large piece of paper, draw as many parallel lines as you can so that the distance between the lines equals the length of the needle. From some comfortable height, drop the needle at random on the paper. Keep count of the number of drops and the number of times the dropped needle touches a line. Then use this formula to approximate the value of $\pi$.

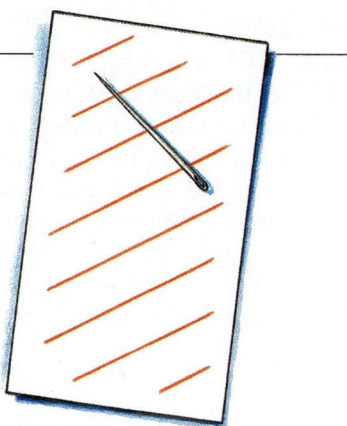

$\pi \approx 2$(number of needle drops) ÷ (number of needle touches)

You will need to drop the needle many times to get a good approximation for $\pi$.

15-6  Volume of a Cone

# LESSON 15-7 Problem Solving Strategies: Making a Model

Here are some patterns you can use to make models of some of the three-dimensional figures you have studied in this chapter. Study each pattern. Think of folding the patterns along the dashed lines and taping the edges to form a solid figure.

A.    B.    C.

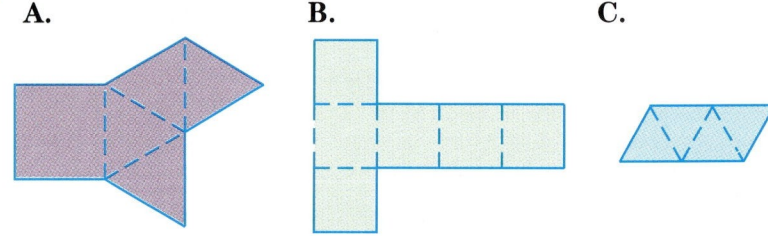

Match each pattern with the name of the solid figure you can make from the pattern.

**a.** cube    **b.** triangular pyramid    **c.** square pyramid

Copy and enlarge each pattern. Cut it out and fold along the dashed line. Tape the cut edges together.

# EXERCISES

**Objective:** To apply strategies in problem solving

**1.** Use the following diagram and these steps to help you draw the pattern for a cone. Begin by drawing on your paper a straight line $l$. Label a point on $l$ as $Q$. At point $Q$, draw line $p$ perpendicular to line $l$. Use $Q$ as a center and draw a circle that has a radius of 2 inches. Label the point where your circle crosses $l$ as point $R$.

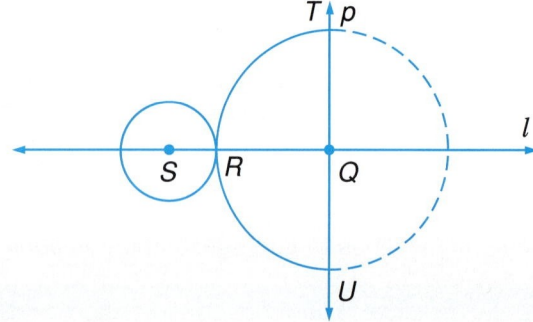

**590** CHAPTER 15

2. In the figure for Exercise 1, if the distance from $Q$ to $R$ is $x$ inches, what is the circumference of circle $Q$? Express your answer in terms of $\pi$.

3. In the figure for Exercise 1, the circumference of the semicircle $TRU$ is the same as the circumference of the small circle $S$. What is the circumference of circle $S$ if $QR$ is $x$ inches? Express your answer in terms of $\pi$.

4. In the figure for Exercise 1, what is the radius of circle $S$ if $QR$ is $x$ inches?

5. Now on your paper, measure 1 inch from $R$ along line $l$ to find point $S$. Use $S$ as the center and draw a circle of radius 1 inch. As you cut out the pattern, leave the two circles connected at point $R$. Cut along line $UQT$, along the arc $TR$, around circle $S$, and along arc $RU$. Curl the semicircle and bend it so that you can tape $\overline{QT}$ next to $\overline{QU}$. Fold the small circle to form the base of the cone and tape it in place.

6. To make a cone from a pattern in this way, what must be the relationship between the radius of the larger circle and the radius of the smaller circle?

7. Which of these patterns can be used to make a cube? Test your answer by making models.

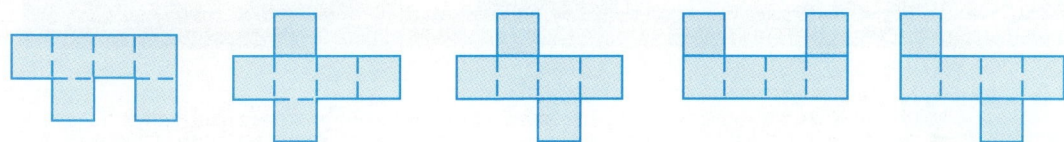

8. Which of these patterns can be used to make a cube? Test your answer by making models.

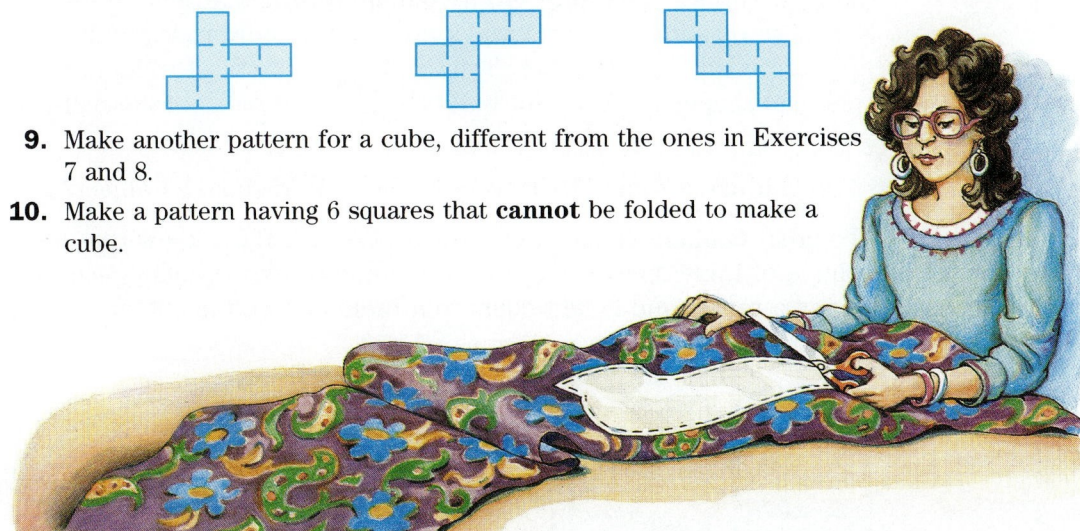

9. Make another pattern for a cube, different from the ones in Exercises 7 and 8.

10. Make a pattern having 6 squares that **cannot** be folded to make a cube.

## LESSON 15-8
# Volume and Area of a Sphere

A **sphere** has many important properties. A spherical shape has the least surface area for a given volume. It is also a rigid figure. For these reasons, spherical tanks are used to store gases under pressure. Also, since a sphere rolls with a minimum of friction, spherical shapes are used for ball bearings, golf balls, basketballs, and so on.

• Can you name some other everyday uses of a spherical object?

The great mathematician Archimedes (287–212 B.C.) knew that the volume of three spheres equals the volume of two cylinders, each having a radius and height equal to a radius of, and diameter (height) of, the sphere.

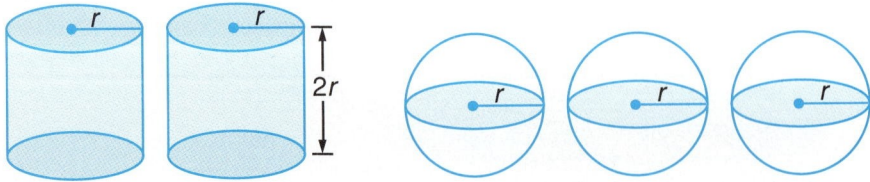

Since the volume of a cylinder is $Bh$ and the height of each cylinder is $2r$, the volume of one cylinder can be written as
$$V = \pi r^2 (2r) = 2\pi r^3.$$
Therefore the volume of the two cylinders is $2(2\pi r^3) = 4\pi r^3$.

But this means that the volume of the three spheres is also $4\pi r^3$. So the volume of one sphere can be written as
$$V = \tfrac{1}{3}(4\pi r^3) = \tfrac{4}{3}\pi r^3.$$

> **Volume of a Sphere**
>
> The formula for the volume of a sphere with radius, $r$, is
> $$V = \tfrac{4}{3}\pi r^3.$$

**EXAMPLE 1** What is the volume of a sphere with a radius of 5 feet? Round the answer to the nearest cubic foot.

**Solution**

Write the formula.    $V = \tfrac{4}{3}\pi r^3$

Substitute for $r$.    $V \approx \tfrac{4}{3}(3.14)(5)^3$

$V \approx \tfrac{4}{3}(3.14)(125)$

$V \approx 523.3$

The volume is about **523 ft³**.

**Try This**

Find the volume of the sphere having the given radius or diameter. Round answers to the nearest cubic unit.

1. radius: 4 in
2. diameter: 5 cm

Archimedes also knew and used the fact that the surface area of a sphere is $\tfrac{2}{3}$ the surface area of a cylinder having a radius and height equal to a radius and diameter (height) of the sphere.

> **Surface Area of a Sphere**
>
> The formula for the surface area of a sphere with radius, $r$, is
> $$S = 4\pi r^2.$$

15-8   Volume and Area of a Sphere

**EXAMPLE 2** What is the surface area of the earth if its radius is approximately 3900 miles?

**Solution**   Write the formula.   $S \approx 4\pi r^2$
Substitute for r.   $S \approx 4(3.14)(3900)^2$
$S \approx 191,037,600$

The surface of the earth is approximately **191,000,000 mi²**.

**Try This**   Find the surface area of the sphere having the given radius or diameter. Round answers to the nearest square unit.

**3.** radius: 8 cm   **4.** diameter: 7 in

# EXERCISES

**Objectives:** To find the volume of a sphere
To find the surface area of a sphere

## Check Your Understanding

**Skill Check**   Write the letter of the correct answer.

**1.** Which statement is **not** true?
   **a.** If the radius of a sphere is 7.5, then its height is 15.
   **b.** The formula for the surface area of a sphere is $S = 4\pi r^2$.
   **c.** The formula for the volume of a sphere is $V = \frac{4}{3}\pi r^2$.
   **d.** The formulas for the surface area and volume of a sphere were known by Archimedes.

**2.** Which is the volume, to the nearest cubic centimeter, of a sphere having a radius of 6 centimeters?
   **a.** 904 cm³   **b.** 2,713 cm³
   **c.** 678 cm³   **d.** 151 cm³

**3.** Which is the surface area, to the nearest square inch, of a sphere having a diameter of 2 inches?
   **a.** 50 in²   **b.** 13 in²
   **c.** 13 in³   **d.** 50 in³

**4.** Which is a formula for finding the volume of a **hemisphere** (half a sphere)?
   **a.** $V = \frac{8}{3}\pi r^3$   **b.** $V = \frac{2}{3}\pi r^3$   **c.** $V = 2\pi r^3$   **d.** $V = 8\pi r^3$

## Practice and Apply

Find the volume and surface area of the sphere having the given radius. Use 3.14 for π. Round answers to the nearest unit.

**5.** 2 in   **6.** 4 in   **7.** 3 cm   **8.** 6 cm   **9.** 12 cm

**For Exercises 10–11, refer to the results of Exercises 5–9.**

10. If the radius of a sphere is doubled, what happens to its surface area?
11. If the radius of a sphere is doubled, what happens to its volume?
12. Mercury is the smallest planet with a diameter of approximately 3000 miles. What is the circumference of Mercury?
13. What is a good estimate for the surface area of Mercury?
14. What calculator sequence gives the volume of a sphere with radius 10?
15. A tennis ball has a diameter of 2.4 inches. A golf ball has a diameter of 1.7 inches. How much larger is the volume of the tennis ball?

## Connect and Extend

**This program computes and prints the volume and surface area of a sphere.**

```
100 PRINT "INPUT THE RADIUS OF A SPHERE."
110 INPUT R
120 LET V = 4 * 3.14159 * R ^ 3 / 3
130 LET A = 4 * 3.14159 * R ^ 2
140 PRINT "THE VOLUME OF THE SPHERE WITH RADIUS ";R;" IS ";V;"."
150 PRINT "THE SURFACE AREA OF THE SPHERE IS ";A;"."
160 PRINT "DO YOU WISH TO CONTINUE (Y/N)";
170 INPUT A$
180 IF A$ = "Y" THEN 100
190 END
```

**Run the program for each value of R. Round answers to the nearest whole number.**

16. R = 16
17. R = 32
18. R = 1.8
19. R = 3.6

20. Use the results of Exercises 16–19 to write an argument supporting your answers to Exercises 10–11.

## Maintain Your Skills

**Solve each equation. (Pages 101–109)**

21. $3v - 8 = -50$
22. $6f - (5f + 18) = 0$
23. $\frac{a}{7} + 4 = -12$

24. Which is greater, 250% of 10 or 350% of 7? **(Pages 319–322)**

**Solve each inequality. (Pages 393–397)**

25. $7y + 4 \geq -10$
26. $4 \leq 16 - 3c$
27. $-9 - 2w < -11$

**Find each product. (Pages 504–511)**

28. $3x(x + 7)$
29. $(t - 2)(-t)$
30. $(x - 3)(x + 2)$
31. $(y + 10)(y - 10)$
32. $(z + 5)(z + 5)$
33. $(q - 8)(q - 8)$

# SUMMARY

### KEY TERMS

cone (p. 586)
cube (p. 573)
face (p. 572)
hemisphere (p. 594)
lateral face (p. 582)
prism (p. 568)
pyramid (p. 582)

rectangular prism (p. 568)
right cylinder (p. 576)
sphere (p. 592)
surface area (p. 572)
vertex (p. 582)
volume (p. 568)

### KEY IDEAS

|    | Solid | Volume | Surface Area |
|----|-------|--------|--------------|
| A. | Rectangular Prism | $V = lwh$ or $V = Bh$ | $S = 2lw + 2lh + 2wh$ |
| B. | Cube | $V = s^3$ | $S = 6s^2$ |
| C. | Cylinder | $V = Bh$ or $V = \pi r^2 h$ | $S = 2\pi r^2 + 2\pi rh = 2\pi r(r + h)$ |
| D. | Pyramid | $V = \frac{1}{3}Bh$ | |
| E. | Cone | $V = \frac{1}{3}\pi r^2 h$ | |
| F. | Sphere | $V = \frac{4}{3}\pi r^3$ | $S = 4\pi r^2$ |

# REVIEW

**Find the volume of the rectangular prism having the given dimensions. (Pages 568–571)**

1. $l = 20$ mm, $w = 10$ mm, $h = 9$ mm
2. $l = 6.4$ m, $w = 5$ m, $h = 4.1$ m
3. $l = 8$ in, $w = 3\frac{1}{2}$ in, $h = 1\frac{2}{3}$ in
4. $l = 5.1$ ft, $w = 5.1$ ft, $h = 5.1$ ft

**Solve. (Pages 568–571)**

5. A fish tank is $2\frac{1}{2}$ feet long, $1\frac{1}{4}$ feet wide, and $1\frac{2}{3}$ feet high. Find the volume.

6. How many cubic inches are there in 1 cubic foot?

**Find the surface area of each rectangular prism. (Pages 572–575)**

**7.** $l = 6$ m, $w = 4$ m, $h = 3$ m
**8.** $l = 4\frac{1}{2}$ ft, $w = 6$ ft, $h = 10\frac{1}{4}$ ft
**9.** $l = 1.5$ yd, $w = 8$ yd, $h = 5.2$ yd
**10.** $l = 1\frac{2}{3}$ in, $w = 1\frac{2}{3}$ in, $h = 1\frac{2}{3}$ in

**Find the volume of each cylinder. Use 3.14 for $\pi$. (Pages 576–580)**

**11.**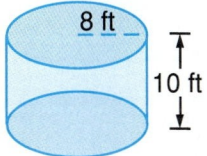
**12.**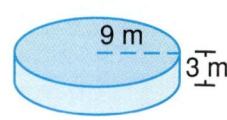
**13.**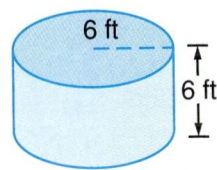

**14.** A cylindrical water tank is $\frac{4}{5}$ full of water. Its radius is 5 feet and its height is 30 feet. If 1 cubic foot is about 7.5 gallons, how many gallons of water are there in the tank? **(Pages 576–580)**

**Find the volume of each rectangular pyramid. (Pages 582–585)**

**15.** $l = 4$ m, $w = 11$ m, $h = 9$ m
**16.** $l = 3.6$ yd, $w = 2.5$ yd, $h = 7$ yd

**17.** Larry has a crystal paperweight that is shaped like a square pyramid. The length of each side of the base is 2 inches. The height of the paperweight is 3 inches. What is the volume? **(Pages 582–585)**

**Find the volume of each cone. Use 3.14 for $\pi$. Round answers to the nearest whole number. (Pages 586–589)**

**18.** **19.** 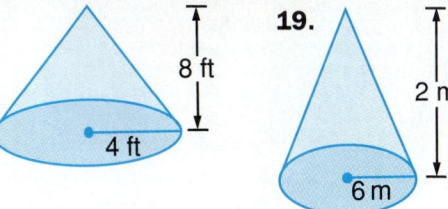 **20.**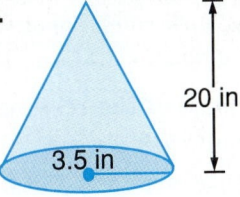

**21.** Which of these patterns can be used to make a closed box? Test your answers by making models. **(Pages 590–591)**

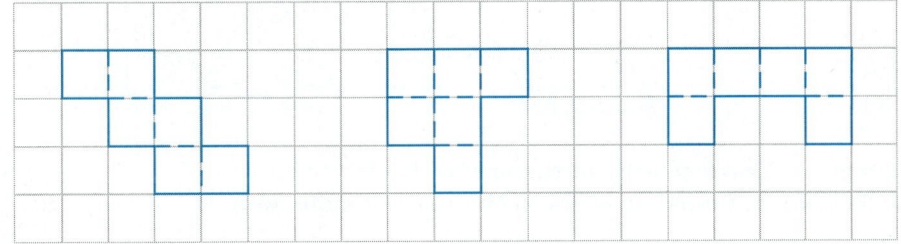

**Find the volume and surface area of the sphere having the given radius. Use 3.14 for $\pi$. Round answers to the nearest unit. (Pages 592–595)**

**22.** $r = 5$ ft
**23.** $r = 7$ in
**24.** $r = 10$ m
**25.** $r = 11$ cm

# TEST

Find the volume of each prism or cylinder. Use 3.14 for π.

1.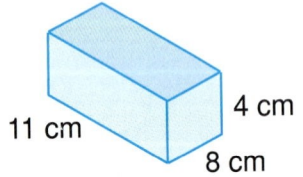
2.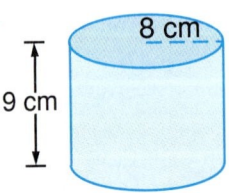
3.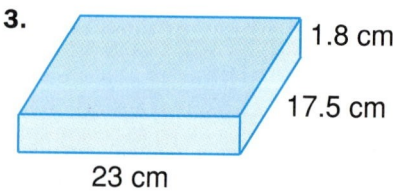

4. A cereal box is 7 inches long, $2\frac{1}{2}$ inches wide and, 10 inches high. How much cereal does the box hold?

5. How many cubic feet are there in 4 cubic yards?

6. The radius of a cylindrical water tank is 2 feet and its height is 15 feet. If 1 cubic foot of water weighs about 62.5 pounds, what is the weight of the water in the tank when the tank is full? Use 3.14 for π.

**Find the surface area of each rectangular prism.**

7. $l = 40$ cm, $w = 50$ cm, $h = 30$ cm
8. $l = 4.4$ m, $w = 2.5$ m, $h = 6$ m
9. The box Nick is wrapping has the shape of a square prism. Each side of the prism is 3 inches. How much paper will it take to cover the box?

**Find the volume of the rectangular pyramid.**

10. $l = 6$ m, $w = 4$ m, $h = 8$ m
11. $l = 1\frac{1}{2}$ ft, $w = 1\frac{1}{2}$ ft, $h = 8$ ft

Find the volume of each cone. Use 3.14 for π. Round answers to the nearest whole number.

12.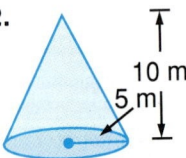
13.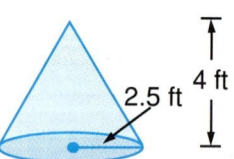
14.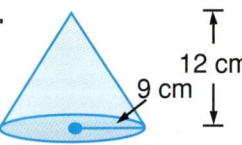

Find the volume and surface area of the sphere having the given radius. Use 3.14 for π. Round answers to the nearest cubic unit or square unit.

15. $r = 4$ yd
16. $r = 8$ m
17. $r = 7$ ft
18. $r = 13$ in

The box plot below shows the mileage ratings in miles per gallon for twenty different cars. (Pages 370–372)

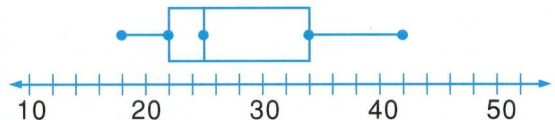

1. About how many gallons of gas does the car with the highest mileage rating use per mile?

2. How many of the cars have a mileage rating of less than 24 miles per gallon?

3. Given the translation $(x, y) \rightarrow (x - 4.5, y + 7)$ and the points $A(-2.5, -8)$ and $B(3, -7)$, find the image points, $A'$ and $B'$. (Pages 440–443)

Solve. Round lengths to the nearest whole meter. (Pages 474–480)

4. Find the height, $h$, of the smokestack.

5. Find $x$.

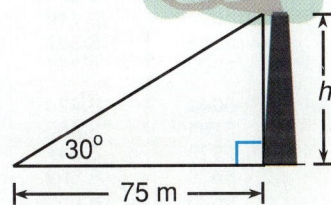
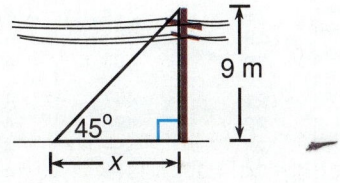

Factor. (Pages 512–519)

6. $f(f - 6) - 7(f - 6)$
7. $c^2 - 4c - 32$
8. $g^2 + g - 72$

Solve. (Pages 539–550)

9. The angles of a triangle are $x$, $7x$, and $2x$. Find the measure of each angle.

10. The angles of a quadrilateral are $4x$, $7x$, $8x$, and $5x$. Find the measure of each angle.

Find the circumference and area of a circle having the given diameter or radius. Use 3.14 for $\pi$. (Pages 551–556)

11. $r = 5$ m
12. $d = 16$ cm
13. $d = 7.6$ ft
14. $r = 60$ in

For Exercises 15–16, refer to the circle graph at the right. Each football practice is 3 hours long. (Pages 558–561)

15. How many minutes of each practice are spent doing drills?

16. How many minutes of each practice are spent scrimmaging?

**How Time is Spent in Football Practice**

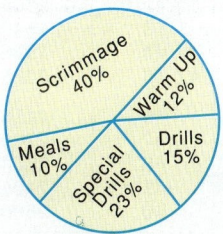

Cumulative Review: Chapters 1–15 **599**

# SQUARES AND SQUARE ROOTS

| n | n² | √n | n | n² | √n |
|---|---|---|---|---|---|
| 1 | 1 | 1.000 | 51 | 2601 | 7.141 |
| 2 | 4 | 1.414 | 52 | 2704 | 7.211 |
| 3 | 9 | 1.732 | 53 | 2809 | 7.280 |
| 4 | 16 | 2.000 | 54 | 2916 | 7.348 |
| 5 | 25 | 2.236 | 55 | 3025 | 7.416 |
| 6 | 36 | 2.449 | 56 | 3136 | 7.483 |
| 7 | 49 | 2.646 | 57 | 3249 | 7.550 |
| 8 | 64 | 2.828 | 58 | 3364 | 7.616 |
| 9 | 81 | 3.000 | 59 | 3481 | 7.681 |
| 10 | 100 | 3.162 | 60 | 3600 | 7.746 |
| 11 | 121 | 3.317 | 61 | 3721 | 7.810 |
| 12 | 144 | 3.464 | 62 | 3844 | 7.874 |
| 13 | 169 | 3.606 | 63 | 3969 | 7.937 |
| 14 | 196 | 3.742 | 64 | 4096 | 8.000 |
| 15 | 225 | 3.873 | 65 | 4225 | 8.062 |
| 16 | 256 | 4.000 | 66 | 4356 | 8.124 |
| 17 | 289 | 4.123 | 67 | 4489 | 8.185 |
| 18 | 324 | 4.243 | 68 | 4624 | 8.246 |
| 19 | 361 | 4.359 | 69 | 4761 | 8.307 |
| 20 | 400 | 4.472 | 70 | 4900 | 8.367 |
| 21 | 441 | 4.583 | 71 | 5041 | 8.426 |
| 22 | 484 | 4.690 | 72 | 5184 | 8.485 |
| 23 | 529 | 4.796 | 73 | 5329 | 8.544 |
| 24 | 576 | 4.899 | 74 | 5476 | 8.602 |
| 25 | 625 | 5.000 | 75 | 5625 | 8.660 |
| 26 | 676 | 5.099 | 76 | 5776 | 8.718 |
| 27 | 729 | 5.196 | 77 | 5929 | 8.775 |
| 28 | 784 | 5.292 | 78 | 6084 | 8.832 |
| 29 | 841 | 5.385 | 79 | 6241 | 8.888 |
| 30 | 900 | 5.477 | 80 | 6400 | 8.944 |
| 31 | 961 | 5.568 | 81 | 6561 | 9.000 |
| 32 | 1024 | 5.657 | 82 | 6724 | 9.055 |
| 33 | 1089 | 5.745 | 83 | 6889 | 9.110 |
| 34 | 1156 | 5.831 | 84 | 7056 | 9.165 |
| 35 | 1225 | 5.916 | 85 | 7225 | 9.220 |
| 36 | 1296 | 6.000 | 86 | 7396 | 9.274 |
| 37 | 1369 | 6.083 | 87 | 7569 | 9.327 |
| 38 | 1444 | 6.164 | 88 | 7744 | 9.381 |
| 39 | 1521 | 6.245 | 89 | 7921 | 9.434 |
| 40 | 1600 | 6.325 | 90 | 8100 | 9.487 |
| 41 | 1681 | 6.403 | 91 | 8281 | 9.539 |
| 42 | 1764 | 6.481 | 92 | 8464 | 9.592 |
| 43 | 1849 | 6.557 | 93 | 8649 | 9.644 |
| 44 | 1936 | 6.633 | 94 | 8836 | 9.695 |
| 45 | 2025 | 6.708 | 95 | 9025 | 9.747 |
| 46 | 2116 | 6.782 | 96 | 9216 | 9.798 |
| 47 | 2209 | 6.856 | 97 | 9409 | 9.849 |
| 48 | 2304 | 6.928 | 98 | 9604 | 9.899 |
| 49 | 2401 | 7.000 | 99 | 9801 | 9.950 |
| 50 | 2500 | 7.071 | 100 | 10000 | 10.000 |

# TRIGONOMETRIC RATIOS

| Angle | sin | cos | tan | Angle | sin | cos | tan |
|---|---|---|---|---|---|---|---|
| 0° | 0.0000 | 1.0000 | 0.0000 | 45° | 0.7071 | 0.7071 | 1.0000 |
| 1° | 0.0175 | 0.9998 | 0.0175 | 46° | 0.7193 | 0.6947 | 1.0355 |
| 2° | 0.0349 | 0.9994 | 0.0349 | 47° | 0.7314 | 0.6820 | 1.0724 |
| 3° | 0.0523 | 0.9986 | 0.0524 | 48° | 0.7431 | 0.6691 | 1.1106 |
| 4° | 0.0698 | 0.9976 | 0.0699 | 49° | 0.7547 | 0.6561 | 1.1504 |
| 5° | 0.0872 | 0.9962 | 0.0875 | 50° | 0.7660 | 0.6428 | 1.1918 |
| 6° | 0.1045 | 0.9945 | 0.1051 | 51° | 0.7771 | 0.6293 | 1.2349 |
| 7° | 0.1219 | 0.9925 | 0.1228 | 52° | 0.7880 | 0.6157 | 1.2799 |
| 8° | 0.1392 | 0.9903 | 0.1405 | 53° | 0.7986 | 0.6018 | 1.3270 |
| 9° | 0.1564 | 0.9877 | 0.1584 | 54° | 0.8090 | 0.5878 | 1.3764 |
| 10° | 0.1736 | 0.9848 | 0.1763 | 55° | 0.8192 | 0.5736 | 1.4281 |
| 11° | 0.1908 | 0.9816 | 0.1944 | 56° | 0.8290 | 0.5592 | 1.4826 |
| 12° | 0.2079 | 0.9781 | 0.2126 | 57° | 0.8387 | 0.5446 | 1.5399 |
| 13° | 0.2250 | 0.9744 | 0.2309 | 58° | 0.8480 | 0.5299 | 1.6003 |
| 14° | 0.2419 | 0.9703 | 0.2493 | 59° | 0.8572 | 0.5150 | 1.6643 |
| 15° | 0.2588 | 0.9659 | 0.2679 | 60° | 0.8660 | 0.5000 | 1.7321 |
| 16° | 0.2756 | 0.9613 | 0.2867 | 61° | 0.8746 | 0.4848 | 1.8040 |
| 17° | 0.2924 | 0.9563 | 0.3057 | 62° | 0.8829 | 0.4695 | 1.8807 |
| 18° | 0.3090 | 0.9511 | 0.3249 | 63° | 0.8910 | 0.4540 | 1.9626 |
| 19° | 0.3256 | 0.9455 | 0.3443 | 64° | 0.8988 | 0.4384 | 2.0503 |
| 20° | 0.3420 | 0.9397 | 0.3640 | 65° | 0.9063 | 0.4226 | 2.1445 |
| 21° | 0.3584 | 0.9336 | 0.3839 | 66° | 0.9135 | 0.4067 | 2.2460 |
| 22° | 0.3746 | 0.9272 | 0.4040 | 67° | 0.9205 | 0.3907 | 2.3559 |
| 23° | 0.3907 | 0.9205 | 0.4245 | 68° | 0.9272 | 0.3746 | 2.4751 |
| 24° | 0.4067 | 0.9135 | 0.4452 | 69° | 0.9336 | 0.3584 | 2.6051 |
| 25° | 0.4226 | 0.9063 | 0.4663 | 70° | 0.9397 | 0.3420 | 2.7475 |
| 26° | 0.4384 | 0.8988 | 0.4877 | 71° | 0.9455 | 0.3256 | 2.9042 |
| 27° | 0.4540 | 0.8910 | 0.5095 | 72° | 0.9511 | 0.3090 | 3.0777 |
| 28° | 0.4695 | 0.8829 | 0.5317 | 73° | 0.9563 | 0.2924 | 3.2709 |
| 29° | 0.4848 | 0.8746 | 0.5543 | 74° | 0.9613 | 0.2756 | 3.4874 |
| 30° | 0.5000 | 0.8660 | 0.5774 | 75° | 0.9659 | 0.2588 | 3.7321 |
| 31° | 0.5150 | 0.8572 | 0.6009 | 76° | 0.9703 | 0.2419 | 4.0108 |
| 32° | 0.5299 | 0.8480 | 0.6249 | 77° | 0.9744 | 0.2250 | 4.3315 |
| 33° | 0.5446 | 0.8387 | 0.6494 | 78° | 0.9781 | 0.2079 | 4.7046 |
| 34° | 0.5592 | 0.8290 | 0.6745 | 79° | 0.9816 | 0.1908 | 5.1446 |
| 35° | 0.5736 | 0.8192 | 0.7002 | 80° | 0.9848 | 0.1736 | 5.6713 |
| 36° | 0.5878 | 0.8090 | 0.7265 | 81° | 0.9877 | 0.1564 | 6.3138 |
| 37° | 0.6018 | 0.7986 | 0.7536 | 82° | 0.9903 | 0.1392 | 7.1154 |
| 38° | 0.6157 | 0.7880 | 0.7813 | 83° | 0.9925 | 0.1219 | 8.1443 |
| 39° | 0.6293 | 0.7771 | 0.8098 | 84° | 0.9945 | 0.1045 | 9.5144 |
| 40° | 0.6428 | 0.7660 | 0.8391 | 85° | 0.9962 | 0.0872 | 11.4301 |
| 41° | 0.6561 | 0.7547 | 0.8693 | 86° | 0.9976 | 0.0698 | 14.3007 |
| 42° | 0.6691 | 0.7431 | 0.9004 | 87° | 0.9986 | 0.0523 | 19.0811 |
| 43° | 0.6820 | 0.7314 | 0.9325 | 88° | 0.9994 | 0.0349 | 28.6363 |
| 44° | 0.6947 | 0.7193 | 0.9657 | 89° | 0.9998 | 0.0175 | 57.2900 |
| 45° | 0.7071 | 0.7071 | 1.0000 | 90° | 1.0000 | 0.0000 | ∞ |

Table of Trigonometric Ratios

# GLOSSARY

**Absolute value:** The distance between an integer and 0 on a number line (Page 4)

**Acute angle:** An angle whose measure is less than 90° (Page 474)

**Algebraic expression:** An expression that combines numbers and variables (Page 39)

**Angle:** A plane figure that has two rays with a common endpoint (Page 526)

**Area:** The number of square units needed to cover a surface (Page 64)

**Average:** The sum of the measures or numbers divided by the number of measures (Page 66)

**Binomial:** An expression containing two terms (Page 486)

**Box and whisker plot:** A graph which shows data grouped and spread (Page 370)

**Chord:** A segment connecting any two points on a circle (Page 551)

**Circle:** The set of all points in a plane that are a given distance from a fixed point in that plane (Page 551)

**Circumference:** The distance around a circle (Page 551)

**Complementary angles:** Two angles whose measures have a sum of 90° (Page 528)

**Composite numbers:** A number greater than 1 that has more than two factors (Page 139)

**Cone:** A three-dimensional figure with one circular base and a curved lateral surface that comes to a point (Page 586)

**Congruent:** Two figures are congruent if they have the same size and shape. (Page 442)

**Conjunction:** A statement formed by joining two sentences with the word *and* (Page 399)

**Coordinate plane:** A plane that has a vertical axis and a horizontal axis that intersect in a point called the origin (Page 410)

**Cross products:** In the equation, $\frac{a}{b} = \frac{c}{d}$, the products $ad$ and $cb$ are cross products. (Page 217)

**Cube:** A three-dimensional solid all of whose faces are squares (Page 573)

**Cylinder:** A three-dimensional figure with two congruent circular bases in parallel planes and having a curved lateral surface (Page 576)

**Dependent event:** An event whose outcome is affected by the outcome of a previous event (Page 242)

**Diagonal:** A line segment, other than a side, that joins two vertices of a polygon (Page 546)

**Diameter:** A chord that contains the center of a circle (Page 551)

**Discount:** An amount that is subtracted from the regular price (Page 330)

**Disjunction:** A statement formed by joining two sentences with the word *or* (Page 403)

**Equation:** A mathematical sentence that shows equality (Page 80)

**Equilateral triangle:** A triangle whose sides all have the same length (Page 540)

**Equivalent expressions:** Expressions that name the same number (Page 40)

**Equivalent fractions:** Fractions that name the same number (Page 164)

**Even number:** A whole number that is divisible by 2 (Page 132)

**Event:** One or more possible outcomes (Page 236)

**Exponent:** An exponent shows how many times a number or base is used as a factor. (Page 150)

**Extremes:** In a proportion, the first and fourth terms are the extremes. (Page 217)

**Factorial:** The symbol $n!$ is read "$n$ factorial." $n! = n(n-1)(n-2) \cdots 3 \cdot 2 \cdot 1$. (Page 253)

**Factoring:** Writing an expression as a product (Page 512)

**Factors:** Numbers or groups of numbers that are multiplied (Page 127)

**Fraction:** A number written in the form $\frac{a}{b}$, where $a$ is a whole number and $b$ is a natural number (Page 163)

**Greatest Common Factor:** The largest common factor of two or more numbers (Page 156)

**Histogram:** A type of bar graph that shows the frequency of data (Page 355)

**Hypotenuse:** The side opposite the right angle in a right triangle (Page 458)

**Improper fraction:** A fraction whose numerator is greater than or equal to the denominator (Page 164)

**Independent events:** Separate events whose outcomes do not affect each other (Page 241)

**Inequality:** A number sentence that uses a symbol such as $>$, $<$, $\geq$, or $\leq$, to compare two expressions (Page 8)

**Integers:** The set of whole numbers and their opposites (Page 3)

**Interest:** Money paid for the use of money (Page 324)

**Inverse operations:** Operations that undo each other (Page 101)

**Irrational number:** A number that cannot be written as the quotient of two integers (Page 264)

**Isosceles triangle:** A triangle with at least two congruent sides (Page 540)

**Least Common Multiple:** The smallest nonzero common multiple of two or more numbers (Page 154)

**Mean:** *See* average. (Page 359)

**Means:** In a proportion, the second and third terms are the means. (Page 217)

**Median:** The middle number in a set of data or the average of the two middle numbers (Page 360)

**Mode:** The number that occurs most often in a set of data (Page 359)

**Monomial:** An expression containing one term (Page 486)

**Multiple:** A product that has a given number as one of its factors (Page 127)

**Net price:** The price of an item after a discount is deducted (Page 331)

**Numerical expression:** An expression that contains numbers only (Page 39)

**Obtuse angle:** An angle whose measure is greater than 90° and less than 180° (Page 527)

**Odd number:** A whole number that is not divisible by 2 (Page 132)

**Opposites:** Two numbers that are the same distance from 0 and are in opposite directions from 0 on the number line (Page 4)

**Ordered pairs:** A pair of numbers in a particular order that correspond to a point in the coordinate plane (Page 137)

**Origin:** The point at which the axes intersect in the coordinate plane (Page 410)

**Parallelogram:** A quadrilateral in which each pair of opposite sides is parallel and equal in length (Page 69)

**Percent:** A ratio with a denominator of 100 (Page 305)

**Perimeter:** The distance around a figure (Page 64)

**Permutation:** An ordered arrangement of some or all of the elements in a set (Page 252)

**Perpendicular lines:** Two intersecting lines that form a right angle (Page 532)

**Pi ($\pi$):** The ratio of the circumference of a circle to its diameter (Page 552)

**Polygon:** A closed plane figure formed by line segments (Page 546)

**Polynomial:** A monomial or the sum or difference of monomials (Page 486)

**Prime factorization:** A factorization in which all factors are prime numbers (Page 143)

**Prime number:** A natural number with exactly two different factors, itself and one (Page 139)

**Principal:** An amount of money on which interest is paid (Page 324)

**Probability:** The ratio of the number of successful outcomes to the number of possible outcomes (Page 236)

**Proper fraction:** A fraction whose numerator is less than its denominator (Page 164)

**Proportion:** A mathematical statement that two ratios are equal (Page 217)

**Pyramid:** A three-dimensional figure whose base is a polygon, and which has triangular lateral faces that meet at a common vertex (Page 582)

**Pythagorean Theorem:** In a right triangle, the square of the length of the hypotenuse is equal to the sum of the squares of the lengths of the other two sides. (Page 458)

**Quadrilateral:** A polygon with four sides (Page 546)

**Radical:** The expression $\sqrt{49}$ is called a *radical* and the $\sqrt{\phantom{x}}$ is a *radical sign*. (Page 452)

**Radius:** A segment from the center to any point on a circle (Page 551)

**Ratio:** A comparison of two numbers by division (Page 216)

**Rational expression:** An algebraic expression written as a ratio (Page 262)

**Rational number:** A number that can be written in the form $\frac{a}{b}$, $b \neq 0$ (Page 263)

**Real numbers:** The set of all rational and irrational numbers (Page 377)

**Reciprocals:** Two numbers whose product is 1 (Page 183)

**Rectangle:** A parallelogram with four right angles (Page 546)

**Regular polygon:** A polygon whose sides all have the same length and whose angles all have the same measure (Page 548)

**Repeating decimal:** A decimal that forms a pattern that repeats without end (Page 277)

**Rhombus:** A parallelogram with four equal sides (Page 546)

**Right angle:** An angle whose measure is 90° (Page 527)

**Right triangle:** A triangle with a right angle (Page 457)

**Sample space:** In probability, a set of possible outcomes (Page 236)

**Scalene triangle:** A triangle with no sides equal (Page 540)

**Scientific notation:** Expressing a number as a product of two factors, when one factor is a power of 10 and the other factor is greater than or equal to 1 and less than 10 (Page 289)

**Similar triangles:** Triangles having the same shape (Page 466)

**Slope:** The steepness of a line measured by a ratio of the vertical change to the horizontal change between any two points on the line (Page 422)

**Sphere:** The set of points in space at a given distance from a fixed point in space (Page 592)

**Square:** A rectangle with four equal sides (Page 546)

**Square root:** One of two equal factors of a number (Page 452)

**Stem-leaf plot:** A way of organizing data in a shape resembling a bar graph (Page 366)

**Supplementary angles:** Two angles whose measures have a sum of 180° (Page 528)

**Surface area:** The total area of the surface of a solid figure (Page 572)

**Tangent ratio:** In a right triangle, the ratio of the length of the side that is opposite an acute angle to the length of the side adjacent to that angle (Page 475)

**Terminating decimal:** A decimal with a finite number of digits (Page 277)

**Translation:** Sliding a graph by adding the same number to one or both coordinates of each point on the graph (Page 440)

**Transversal:** A line that intersects each of two or more lines in different points (Page 533)

**Trapezoid:** A quadrilateral with one pair of parallel sides (Page 71)

**Tree diagram:** A diagram used to find the total number of outcomes in a sample space. (Page 247)

**Triangle:** A polygon with three sides (Page 539)

**Variable:** A letter or other symbol that may represent a number (Page 39)

**Vertex:** The common endpoint of two rays of an angle (Page 526)

**Vertical angles:** The nonadjacent angles formed by two intersecting lines (Page 532)

**Volume:** The measure of the amount of space occupied by a solid figure (Page 568)

***x*-intercept:** The $x$ value of the point where a graph crosses the $x$ axis (Page 418)

***y*-intercept:** The $y$ value of the point where a graph crosses the $y$ axis (Page 418)

# SELECTED ANSWERS

## CHAPTER 1

**Exercises, pages 5–6**
**1.** c   **5.** Negative 8 or the opposite of 8
**9.** Yes   **13.** Yes   **17.** +3   **21.** −6
**25.** 6   **29.** 3   **33.** 23   **37.** 11   **41.** 36
**45.** +3   **49.** −8   **53.** No integer   **57.** 472
**61.** $58.66

**Exercises, pages 9–10**
**1.** a   **5.** <   **9.** >   **13.** >   **17.** >
**21.** −13, −4, 8   **25.** −2, −12, −16
**29.** Answers will vary.   **33.** −80 < 12
**37.** Negative 17 is less than 28.   **41.** 9
**45.** 2   **49.** 5   **53.** The car travels at a constant speed.

**Exercises, pages 17–18**
**1.** d   **5.** −5   **9.**

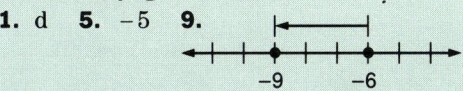

**13.** −5 + 2 = −3   **17.** P   **21.** N
**25.** P   **29.** 0   **33.** 11   **37.** −3   **41.** 10
**45.** Sample answer: Add up all the positive numbers and all the negative numbers. Take the absolute value of both and subtract the smaller from the larger.   **49.** 12   **53.** The sum of the series −100 + (−99) + . . . + 99 + 100 = 0. So the sum of this series plus 101 is 101.   **57.** 527   **61.** 0

**Exercises, pages 24–25**
**1.** c   **5.** N   **9.** 7 + 2   **13.** −11   **17.** 11
**21.** 4   **25.** −1   **29.** 10   **33.** −3 + 3 = 0
**37.** 8   **41.** (−20)   **45.** 945   **49.** >
**53.** −19

**Exercises, pages 30–31**
**1.** c   **5.** False; both are negative.   **9.** 7, 0
**13.** N   **17.** N   **21.** −56   **25.** 45
**29.** −24   **33.** 21   **37.** 108   **41.** Positive
**47.** Yes; changing the sign of only one factor changes the sign of the product.
**51.** 9875   **55.** 180   **59.** −4   **63.** 30

**Exercises, pages 34–35**
**1.** b   **5.** positive   **9.** 33 ÷ (−11) = −3; 33 ÷ (−3) = −11   **13.** N   **17.** N   **21.** 2
**25.** 8   **29.** 7   **33.** 9   **37.** 4   **41.** −252
**45.** −15   **49.** −5   **53.** =   **57.** 0
**61.** −40   **65.** 64

## CHAPTER 2

**Exercises, pages 42–43**
**1.** c   **5.** $m + 5$   **9.** $s \cdot 25$   **13.** $180 - a$
**17.** 2   **21.** 180   **25.** 3   **29.** 6   **33.** 10
**37.** $(w + 3)$ pounds   **41.** $a$ added to 4, the sum of 4 and $a$, 4 increased by $a$, $a$ more than 4, 4 plus $a$   **45.** $n - 5$, $n - 10$, $n - 15$
**49.** 1   **53.** −6   **57.** −48

**Exercises, pages 49–50**
**1.** b   **5.** Sample answers: $7c$, $7(c)$, $(7)c$, $(7)(c)$, $(7) \cdot (c)$, $(7 \cdot c)$, $(7c)$   **9.** −4   **13.** 25   **17.** 19
**21.** 10   **25.** −48   **29.** 2   **33.** No; she did not use the multiplication key after the 9.
**37.** −9   **41.** −20, −12, −10, 10   **45.** $|10|$, $|-10|$, $|0|$, $-|10|$

**Exercises, pages 56–58**
**1.** d   **5.** $a \cdot (b \cdot c)$   **9.** 8; Commutative Property of Multiplication   **13.** −3; Associative Property of Addition   **17.** −16; Commutative Property of Addition   **21.** −9; Commutative Property of Multiplication
**25.** 156   **29.** 93   **33.** −492   **37.** Answers will vary.   **41.** No; no; examples will vary.
**45.** No, as demonstrated by Exercises 43–44.
**49.** ÷   **53.** Yes

**Exercises, pages 61–63**
**1.** b   **5.** 7(13 + 7); 7 · 20 is easier to multiply.   **9.** Yes   **13.** $x$, $y$   **17.** $f$, (−3)
**21.** $6a$   **25.** $-e$   **29.** $5d$   **33.** −6(29) + (−6)(1) = (−6)(29 + 1) = (−6)(30) = −180
**37. a.** 8(40 + 3) = 320 + 24   **b.** (3 + 40)8 = 24 + 320 = 344   **41.** Yes. 7(2 + 3) + 5(2 + 3) = 35 + 25 = 60 and (7 + 5)(2 + 3) = 12 · 5 = 60.   **45.** Answers will vary.
**49.** ÷   **53.** 91.46 + 73 = 164.46

**Exercises, pages 67–68**
**1.** d   **5.** Area   **9.** 36, 81   **13.** 66, 90
**17.** 159   **21.** Arizona and Texas
**25.** Exercise 24: $P = 4s = 4 \cdot 8 = 32$; $A = s^2 = 8 \cdot 8 = 64$   **29.** $P = 21$, $s = 79$, $c = 58$
**33.** $\frac{4}{c}$   **37.** $22.69

**Exercises, pages 72–74**
**1.** b   **5.** True   **9.** 31.5   **13.** 176 square kilometers   **17.** rectangle   **21.** 44   **25.** 64 square feet   **29.** 36 square inches
**33.** Yes; its opposite sides have the same length and are parallel.   **37.** −98
**41.** $-21f$   **45.** 8 people

# CHAPTER 3

**Exercises, pages 82–84**
**1.** b  **5.** c  **9.** Any whole number $< 5$.
**13.** 2  **17.** $\{5\}$  **21.** $\{-5\}$  **25.** $\{-5, -3, -1\}$  **29.** $\{-5, -3\}$  **33.** $\{-5, -3\}$
**37.** $a > 8$  **41.** $2(x + y) = 56$  **45.** $x$ must be negative to obtain a positive product with $-64$.  **49.** 360  **53.** 46, 4, $-8$, $-400$

**Exercises, pages 86–88**
**1.** b  **5.** 4  **9.** $-10$  **13.** 3  **17.** Too high; $-8$  **21.** 261; first guess too low
**25.** 79  **29.** 110  **33.** a  **37.** $4c = 20$
**41.** Answers will vary.  **45.** $-12$  **49.** $4p - pq$ and $p(4 - q)$  **53.** 90 square units

**Exercises, pages 93–94**
**1.** c  **5.** 2  **9.** $-11$  **13.** 12  **17.** $-13$
**21.** 42  **25.** $-121$  **29.** 30  **33.** d
**37.** 42  **41.** 4, because it is the only value in the table that makes $2x - 4 = 8 - x$.
**45.** P

**Exercises, pages 98–99**
**1.** b  **5.** P  **9.** Divide by 10.  **13.** E
**17.** 10  **21.** $-36$  **25.** 32  **29.** 13
**33.** Equation A  **37.** $a$  **41.** $l = \frac{w}{p}$, $p = \frac{w}{l}$
**45.** $-17$  **49.** 9 square meters

**Exercises, pages 103–104**
**1.** b  **5.** 7  **9. a.** Addition Property of Equality  **b.** Additive Inverse Property
**c.** Identity Property of Addition  **d.** Division Property of Equality  **e.** Multiplication Property of Inverses  **f.** Identity Property of Multiplication  **13.** 72  **17.** $-8$
**21.** 144  **25.** 8  **29.** 12  **33.** 1
**37.** $[12 \div (3 \cdot 2)] - [10 \cdot (3 + 1)] = -38$
**41.**

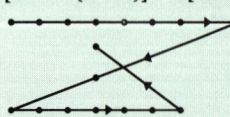

**Exercises, pages 108–109**
**1.** b  **5.** $3x - 6$  **9.** 5  **13.** $-7$  **17.** $-3$
**21.** $-3$  **25.** $-4$  **29.** $-3$  **33.** $-3$
**37.** 12  **41.** Answers will vary.  **45.** 65
**49.** $-80$  **53.** 9  **57.** 68 square units
**61.** 16, 32, 64  **65.** Ex. 62: $1 \times 1 \times 1$, $2 \times 2 \times 2$, $3 \times 3 \times 3$, $4 \times 4 \times 4 \ldots$
Ex. 63: Each term after the first is the sum of the previous two terms.

**Exercises, pages 112–113**
**1.** c  **5.** Sample answer: $c + 4$  **9.** Walt
**13.** Divided by  **17.** Marc  **21.** Answers will vary.  **25.** $j + 3k + 5$  **29.** $(s + t) - 6$
**33.** $w + 5$  **37.** $\frac{n}{9} - 500$  **41.** Exercises 38, 40, and 41

**Exercises, pages 117–118**
**1.** d  **5.** This shows that the number of local runners is 448 more than the number of visiting runners.  **9.** $x + (x + 128) = 1550$; Males: 771; Females: 839  **13.** 45, 46, 47
**17.** Answers will vary.  **21.** 162 miles
**25.** $\div$  **29.** $-4c$

# CHAPTER 4

**Exercises, pages 128–130**
**1.** c  **5.** 1, 2, 3, 6, 7, 14, 21, 42  **9.** True; $9 \cdot 6 = 54$  **13.** False; 16 is a multiple of 4.
**17.** 1, 3, 15  **21.** 1, 2, 4, 7, 14, 28
**25.** 1, 3, 17, 51  **29.** 10, 20, 30, 40, 50
**33.** 16 is divisible by 8.  **37.** 3
**41.** March-June, September-December
**45.** No, a larger number could be a prime.
**49.** a  **53.** 7  **57.** $5a = 30$

**Exercises, pages 134–136**
**1.** a  **5.** 0, 4, 8  **9.** 5  **13.** 2  **17.** 4 and 8  **21.** 3 and 6  **25.** No  **29.** Yes
**33.** No; $26 \div 4 = 6 R2$  **37.** $\{2, 3, 4, 5, 6, 9\}$
**41.** $\{2, 3, 4, 6, 8\}$  **45.** Sample answer: 16
**49.** 24  **53.** 28  **57.** $3x = 12$; 4 goals
**61.** 119

**Exercises, pages 140–141**
**1.** c  **5.** False; it has more than two factors.
**9.** No; prime numbers have exactly two factors and composite numbers have more than two factors.  **13.** Neither; it has only one factor.  **17.** 11, 13  **21.** 2
**25.** Sample answer: $2 + 3$  **29.** They are all odd.  **33.** Sample answer: 31 and 29, 43 and 17, 19 and 41, 23 and 37, 29 and 31, 13 and 47  **37.** $\{-3, -2\}$  **41.** Fred
**45.** $\{2, 4\}$  **49.** $\{2, 5, 10\}$

**Exercises, pages 144–146**
**1.** d  **5.** 3, 15; 3, 3, 5
**9.**      42              **13.**      150
        /  \                        /    \
       2    21                     2      75
            / \                           /  \
           3   7                    2    3    25
                                                / \
                                               5   5

**606** Selected Answers

**17.** 2 · 43   **21.** False; 4 is not prime.
**25.** 5, 11   **29.** 35   **33.** 434,343
**37.** 2 · 2 · 2 · 17   **41.** 31   **45.** 9   **49.** 2, 4, 8   **53.** No

**Exercises, pages 151–153**

**1.** a   **5.** 2401   **9.** −1125   **13.** Base: −6, exponent: 2   **17.** −125   **21.** 169
**25.** −192   **29.** 450   **33.** 216   **37.** 125
**41.** False; $-5^2 = -25$; $(-5)^2 = 25$
**45.** $(-1)^n = 1$ for $n$ = even number   **49.** 12
**53.** 7

**Exercises, pages 156–158**

**1.** d   **5.** 2 · 3; 2 · 5; LCM: 30   **9.** 8
**13.** $2 \cdot 3^2 \cdot 5$   **17.** 20   **21.** 2   **25.** Sample answer: 2, 3   **29.** 48; 96   **33.** 21   **37.** 9
**41.** Sample answer: 15, 30   **45.** $y = -2$
**49.** 10   **53.** 12; 36

## CHAPTER 5

**Exercises, pages 166–167**

**1.** c   **5.** Proper   **13.** $\frac{4 \cdot 4}{9 \cdot 4} = \frac{16}{36}$
**17.** $\frac{6 \cdot 2}{9 \cdot 2} = \frac{12}{18}$   **21.** 18   **25.** $\frac{1}{3}$   **29.** $\frac{21}{25}$
**33.** $\frac{3}{1}$ or 3   **37.** $\frac{4}{5}$ and $\frac{3}{5}$   **41.** Sample answer: $\frac{5}{12}, \frac{6}{13}, \frac{7}{10}, \frac{10}{21}, \frac{14}{27}$   **45.** −4   **49.** 6

**Exercises, pages 173–174**

**1.** c   **5.** Multiply the numerators and multiply the denominators.   **9.** $\frac{1}{2}$   **13.** $\frac{1}{3}$
**17.** $\frac{14}{3}$   **21.** $\frac{3}{8}$ cup   **25.** Less than, because both are proper fractions.   **29.** $2\frac{2}{3}$ pieces
**33.** −26,000,000,000   **37.** $\{-5, -3\}$
**41.** $x + (x+1) + (x+2) = 87$; 28, 29, 30

**Exercises, pages 178–179**

**1.** b   **5.** $6\frac{1}{4}$   **9.** $\frac{97}{10}$   **13.** $1\frac{3}{10}$   **17.** $18\frac{1}{5}$
**21.** $77\frac{1}{4}$   **25.** $61\frac{1}{2}$   **29.** $128\frac{7}{10}$   **33.** It is more difficult than Method 2.   **37.** 48
**41.** 14   **45.** 23

**Exercises, pages 186–188**

**1.** c   **5.** reciprocals   **9.** $\frac{1}{25}$   **13.** 4
**17.** $\frac{17}{18}$   **21.** 40   **25.** $\frac{33}{70}$   **29.** $\frac{3}{2}x =$ 160; $106\frac{2}{3}$ servings   **33.** $\frac{15}{8}x = \frac{45}{8}$; 3 minutes
**37.** 60   **41.** 144   **45.** $\frac{y}{-2}$   **49.** $\frac{4a}{15}$
**53.** 180; 5

**Exercises, pages 191–193**

**1.** d   **5.** 6   **9.** $\frac{2}{5} \cdot \frac{10}{1}$   **13.** $\frac{1}{9}$   **17.** $\frac{8}{9}$

**21.** $1\frac{3}{11}$   **25.** $\frac{1}{3}$   **29.** $\frac{5}{1} \div \frac{5}{6}$ = 6 models
**33.** True   **37.** −2   **41.** Commutative Property of Multiplication   **45.** Composite
**49.** Sample answer: $\frac{4}{10}, \frac{6}{15}$
**53.** The ant will climb 5 − 4, or 1 foot per day. After 25 days, it will have climbed 25 feet. On the 26th day, it will climb 5 feet and reach the top. So the ant will reach the top in 26 days.

**Exercises, pages 197–199**

**1.** c   **5.** $9\frac{7}{8} - 2\frac{1}{8}$   **9.** $3\frac{1}{4}$   **13.** $5\frac{1}{5}$
**17.** 14   **21.** Kale   **25.** $2\frac{1}{12}$ feet
**29.** Sample answer: $4\frac{3}{9} + 1\frac{1}{9}$
**33.** $\frac{3}{4}$   **37.** $10\frac{58}{100} - 8\frac{22}{100} = 2\frac{36}{100}$ = $2.36
**41.** 120   **45.** 2, 5, 5   **49.** 252

**Exercises, pages 203–205**

**1.** c   **5.** Sample answer: $\frac{5}{6}, \frac{9}{10}, \frac{11}{12}$   **9.** $\frac{15}{24}$, $\frac{16}{24}$   **13.** 14   **17.** $5\frac{3}{10}$   **21.** $16\frac{14}{15}$   **25.** $10\frac{4}{15}$
**29.** Class 3   **33.** $\frac{1}{12}$ pound   **37.** A common denominator   **45.** Sample answer: $-\frac{7}{20} + 1\frac{1}{5}$
**49.** 1   **53.** −13   **57.** 15

**Exercises, pages 208–209**

**1.** d   **5.** $3\frac{5}{12}$   **9.** $\frac{15}{16}$   **13.** $7\frac{3}{5}$
**17.** Greater; $\frac{2}{3}$   **21.** $10\frac{15}{16}$ inches   **25.** $-2\frac{1}{6}$
**29.** An infinite number   **33.** $\{9\}$   **37.** 37

## CHAPTER 6

**Exercises, pages 219–220**

**1.** b   **3.** b   **5.** False; 1:3 only   **7.** $\frac{3}{8}$
**9.** $\frac{8}{5}$   **11.** No   **13.** Yes   **15.** 60   **17.** 10
**19.** 8   **21.** 5   **23.** 8   **25.** $\frac{3}{2} = \frac{24}{s}$; 16
**27.** $\frac{5}{2} = \frac{c}{22}$; 55   **29.** Yes; same cross-products.
**31.** Yes; the cross-products are equal.   **33.** 3
**35.** −9   **37.** 11   **39.** No; $\frac{2}{3}$   **41.** Yes
**43.** No; $\frac{4}{7}$

**Exercises, pages 223–225**

**1.** a   **3.** b   **5.** $\frac{135}{3} = \frac{d}{5}$   **7.** 45 mph
**9.** $7 per hour   **11.** $22\frac{1}{2}$ minutes   **13.** 14 inches   **15.** 75 gallons   **17.** Clayton's
**19.** 2,160 times   **21.** $127,500   **23.** $\frac{6}{5}$
**25.** $3\frac{3}{8}$   **27.** $12\frac{8}{9}$   **29.** $\frac{2}{9}$   **31.** $\frac{18}{25}$
**33.** 25   **35.** Draw a diagram. Mary is

ahead, followed by Bob, Tim, and Kathy in that order. So Tim is 25 blocks behind Bob.

**Exercises, pages 228–230**
**1.** a  **3.** d  **5.** Equal to  **7.** $\frac{12}{1} = \frac{i}{5}$; 60 inches  **9.** $\frac{5280}{1} = \frac{f}{4}$; 21,120 ft  **11.** $\frac{8}{1} = \frac{78}{1}$; $9\frac{3}{4}$ c  **13.** $\frac{2000}{1} = \frac{p}{3\frac{1}{2}}$; 7,000 lb
**15.** $\frac{36}{1} = \frac{153}{y}$; $4\frac{1}{4}$ yd  **17.** $1\frac{1}{6}$ yards  **19.** $\frac{3}{1}$
**21.** $3\frac{3}{4}$ T  **23.** Students' drawings should show a square with sides of $1\frac{1}{2}$ inches.
**25.** 39 ft  **27.** $\frac{15}{5} = \frac{12}{4} = \frac{3}{1}$  **29.** The ratio of the areas is the square of the ratios of length:length and width:width.  **31.** $\frac{1620}{20} = \frac{81}{1}$  **33.** No  **35.** True  **37.** False
**Ex. 39, 41:** Sample answers are given.
**39.** 3, 5, 7  **41.** 6, 9, 15  **43.** 1176

**Exercises, pages 238–240**
**1.** c  **3.** d  **5.** T, S, D, O, S, A, E, S
**7.** $\frac{1}{12}$  **9.** $\frac{1}{3}$  **11.** $\frac{2}{3}$  **13.** $\frac{1}{12}$  **15.** 1:1
**17.** Yes  **19.** $\frac{2}{3}$  **21.** $\frac{6}{7}$  **23.** 1; examples will vary.  **Ex. 25, 27:** Examples will vary.
**29.** $\frac{5}{8}$  **31.** Blue area, because it covers more area.  **33.** $\frac{3}{7}$  **35.** $\frac{6}{13}$  **37.** $1\frac{1}{2}$
**39.** 360 calories

**Exercises, pages 244–246**
**1.** d  **3.** d  **5.** Answers will vary.  **7.** $\frac{1}{2}$
**9.** $\frac{1}{4}$  **11.** The event [(H, T) or (T, H)] involves two favorable outcomes out of 4 possible outcomes in the sample space. The event, (H, H), represents only 1 outcome out of 4 possible outcomes in the sample space. So the event [(T, H) or (H, T)] is twice as likely as the event (H, H).  **13.** $\frac{10}{49}$  **15.** $\frac{1}{21}$
**17.** $\frac{5}{21}$  **19.** $\frac{1}{15}$  **21.** $\frac{1}{30}$  **23.** Answers will vary.  **25.** 0  **27.** $\frac{1}{6}$  **29.** $P$(tossing a 5 or a 6) = $P$(tossing a 5) + $P$(tossing a 6)  **31.** +
**33.** −72  **35.** −8  **37.** $2\frac{1}{9}$  **39.** $\frac{2}{5}$
**41.** 44  **43.**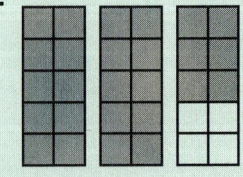
**45.** At 30 miles per hour, she can travel 1 mile in 2 minutes. She will be 1 mile from the bus station. She will not get to the station before the bus leaves.

**Exercises, pages 249–251**
**1.** b  **3.** d  **5.** Red Jacket / Red Skirt / Blue Jacket / Red Jacket / White Skirt / Blue Jacket / Red Jacket / Blue Skirt / Blue Jacket
**7.** 6  **9.** No; she may have a preference for certain color combinations.
**11.** $\frac{1}{64}$  **13.** LLW, LLLW, LLLL  **15.** 4
**17.** Team A
**19.** 60  **21.** 50
**23.** $\frac{3}{5}$  **25.** $\frac{9}{10}$
**27.** 720  **29.** 104
**31.** $\frac{5}{6}$  **33.** $\frac{5}{7}$  **35.** $\frac{5}{51}$

**Exercises, pages 255–257**
**1.** c  **3.** b  **5.** 4!  **7.** 720  **9.** 42  **11.** 8
**13.** 380  **15.** 720  **17.** $_7P_6$  **19.** $_{15}P_{12}$
**21.** 362,880  **23.** 3,628,800  **25.** 120
**27.**
| PERM | EMPR | MEPR | REMP |
| PEMR | EMRP | MERP | REPM |
| PMER | EPMR | MPER | RMEP |
| PMRE | EPRM | MPRE | RMPE |
| PRME | ERMP | MREP | RPEM |
| PREM | ERPM | MRPE | RPME |

There are 4! permutations of PERM. Therefore the list is complete when it contains 24 permutations.  **29.** b  **31.** 30,240
**33.** 9024  **35.** $\frac{1}{30,240}$  **37.** −3  **39.** 6
**41.** Lyn: 34 hours; Stef: 17 hours  **43.** $x = 36$  **45.** $e = 26$  **47.** 30

# CHAPTER 7

**Exercises, pages 264–266**
**1.** d  **3.** b  **5.** True  **7.** False; it is expressed in the form $\frac{a}{b}$, $b \neq 0$.  **9.** $\frac{2}{3}$
**11.** $\frac{1}{2}$  **13.** $\frac{2}{1}$, or 2  **15.** $\frac{-1}{2}$, or $-\frac{1}{2}$
**17.** $\frac{3}{3}$, or 1  **19.** $\frac{12}{33}$  **21.** −1  **23.** −1
**25.** 5  **27.** $\frac{3}{22}$  **29.** $\frac{23}{25}$
**31.**   **33.** The Identity Property for Multiplication states that the product of any number and 1 is the number. Any fraction of the form $\frac{a}{a}$, where $a \neq 0$, equals 1. So if you multiply a whole number by $\frac{a}{a}$, $a \neq 0$, you will get an equivalent expression which, expressed in lowest terms, equals the original

number. Example: $2 = 2 \cdot \frac{1}{1} = 2 \cdot \frac{3}{3} = 2 \cdot \frac{5}{5} \ldots$  **35.** 3  **37.** 0  **39.** 7, 6, −5, −6, −7  **41.** 6, 1, 0, −3, −5  **43.** $\frac{7}{20}$  **45.** $\frac{4}{33}$  **47.** $11\frac{1}{4}$ hours  **49.** Compare the thickness of a nickel to the diameter of a quarter. If the thickness of a nickel is less than $\frac{1}{5}$ the diameter of a quarter, then nickels are preferable; if not, then quarters are preferable.

### Exercises, pages 269–270
**1.** b  **3.** d  **5.** <  **7.** >  **9.** <  **11.** >  **13.** $\frac{3}{-2}, \frac{-3}{4}, \frac{-2}{3}, \frac{-1}{3}, \frac{-1}{5}$  **15.** $\frac{2}{9}$  **17.** $-1\frac{1}{24}$  **19.** $3\frac{3}{4}$  **21.** $-\frac{1}{6}$  **23.** $97\frac{1}{2}$ bags  **25.** $-\frac{1}{108}$  **27.** $1\frac{3}{4}$  **29.** Sample answer: $-\frac{1}{2}$  **31.** $9y$  **33.** $\frac{75}{100}$  **35.** $\frac{37\frac{1}{2}}{100}$  **37.** $\frac{80}{100}$

### Exercises, pages 273–274
**1.** c  **3.** d  **5.** 0.5  **7.** 0.3  **9.** $\frac{0}{10} + \frac{1}{100} + \frac{3}{1000}$  **11.** $100 + 0 + 8 + \frac{6}{10} + \frac{3}{100} + \frac{7}{1000} + \frac{4}{10,000}$  **13.** 479.061  **15.** 50.008  **17.** $47\frac{4}{125}$  **19.** $-12\frac{3}{5}$  **21.** $-93\frac{17}{20}$  **23.** $\frac{1}{250}$  **25.** <  **27.** <  **29.** One-tenth of a meter  **31.** One-tenth of a liter  **33.** $8 + \frac{7}{10^1} + \frac{0}{10^2} + \frac{2}{10^3} + \frac{5}{10^4}$  **35.** $\frac{6}{10^1} + \frac{2}{10^2} + \frac{0}{10^3} + \frac{3}{10^4}$  **37.** $72\frac{13}{50}$; 72.26  **39.** $-\frac{3}{50,000}$; −0.00006  **41.** −5  **43.** 77  **45.** −180  **47.** $2l - 12 = v$  **49.** $\frac{1}{64}$

### Exercises, pages 279–280
**1.** a  **3.** a  **5.** False; $\frac{1}{3} = 0.\overline{3}$  **7.** False; $\frac{3}{5} = 0.6$  **9.** Repeats  **11.** Terminates  **13.** Terminates  **15.** −38.0, −38.03; −38.026  **17.** −1.9, −1.94, −1.940  **19.** $-0.\overline{8}$, −0.89  **21.** $0.0\overline{74}$, 0.07  **23.** $-0.\overline{09}$, −0.09  **25.** Sample answer: $0.\overline{123456}$  **27. a.** Mult Prop of Eq  **b.** Add Prop of Eq  **c.** Distrib Prop  **d.** Div Prop of Eq  **29.** $n = \frac{23}{9}$  **31.** 4, −15  **Ex. 33–37:** Even: 218, 128, 812, 20; Divisible by 4: 128, 812, 20; Divisible by 8: 128  **39.** 360

### Exercises, pages 284–286
**1.** b  **3.** c  **5.** <; 60¢  **7.** <; 30¢  **9.** $3.30; $5.80; $2.40  **Ex. 11, 13, 15, 17:** Estimates will vary.  **11.** 37.324  **13.** −13.763  **15.** 0.896  **17.** 18.228  **19.** −4.195  **21.** 73¢  **23.** Answers will vary.  **25.** 20¢; the store will probably round up.  **27.** 1, 4.3, 100  **29.** 2.25; $\frac{17}{8} = 2\frac{1}{8}$ which is less than $2\frac{1}{4}$  **31.** $15 + (-31) = -16$  **33.** $44 + 17 = 61$  **35.** 75,178; not an odd number  **37.** 30,459  **39.** $3\frac{1}{3}$ yd  **41.** 14 qt

### Exercises, pages 292–294
**1.** d  **3.** b  **5.** 0.00042  **7.** False; the exponent is negative.  **9.** $8.7 \times 10^2$  **11.** $5.5009 \times 10^1$  **13.** $9 \times 10^{-2}$  **15.** $5.7 \times 10^{-4}$  **17.** $5.2 \times 10^7$  **19.** $4 \times 10^6$; $6.3 \times 10^7$  **21.** $1 \times 10^{-8}$  **23.** $10^9$  **25.** $10^0$ or 1  **27.** $10^{11}$  **29.** $10^{19}$  **31.** $5.6 \times 10^4$  **33.** $6 \times 10^{-2}$  **35.** $5.4 \times 10^2$  **37.** $3 \times 10^{-1}$  **39.** $2^{94}$  **41. a.** $(3.2 \times 10^3) \times (4.5 \times 10^5)$  **b.** $1.44 \times 10^9$  **43.** 6000  **45.** 0.016  **47.** −11  **49.** −6  **51.** $\frac{1}{4}$ yd  **53.** $6\frac{2}{5}$  **55.** $-12\frac{1}{20}$

### Exercises, pages 299–300
**1.** b  **3.** b  **5.** 1.334  **7.** 36.6  **Ex. 9, 11, 13:** Estimates will vary.  **9.** 510; U  **11.** 630; U  **13.** 1; O  **15.** 9  **17.** 9  **19.** 15  **Ex. 21, 23:** Estimates will vary.  **21.** 30.21  **23.** $372.53  **25.** Answers will vary.  **27.** Negative; $(-3.5)^5$ is negative because the exponent is odd. $(-2.8)^{32}$ is positive because the exponent is even. The product of a negative number and a positive number is a negative number.  **29.** 57,600 square feet  **31.** $c = 39$  **33.** $f = 8$  **35.** $\frac{1}{8}$  **37.** $\frac{4}{7}$  **39.** $\frac{19}{30}$

## CHAPTER 8

### Exercises, pages 307–308
**1.** b  **3.** c  **5.** $\frac{3}{10} = \frac{n}{100}$  **7.** $\frac{2}{5} = \frac{n}{100}$  **9.** $\frac{20}{19} = \frac{n}{100}$  **11.** 5%  **13.** 6%  **15.** $17\frac{1}{2}$%  **17.** $131\frac{1}{4}$%  **19.** 180%  **21.** $63\frac{7}{11}$%  **23.** 220%  **25.** 38%  **27.** 62%  **29.** 150%  **31.** $4g$%  **33.** $\frac{75}{100}$  **35.** $\frac{24}{100}$  **37.** $\frac{5}{100}$  **39.** $\frac{5}{9}$  **41.** $\frac{1}{6}$  **43.** $\frac{1}{6}$ quart

### Exercises, pages 311–312
**1.** d  **3.** b  **5.** 3; 30; 30  **7.** 60; 60  **9.** $\frac{41}{100}$  **11.** $\frac{98}{100}$  **13.** $\frac{22.3}{100}$  **15.** 5.07  **17.** 0.661  **19.** 0.045  **21.** 0.91  **23.** 0.0025  **25.** 360%  **27.** 4%  **29.** 26%

**31.** 0.3%   **33.** 940%   **35.** $\frac{7}{1000}$   **37.** $\frac{1}{150}$
**39.** $\frac{1}{2}$; 0.5; 50%   **41.** $\frac{3}{8}$; 0.375; 37.5%
**43.** $\frac{1}{2}$; 0.5; 50%   **45.** 1   **47.** 5   **49.** $31.40
**51.** 1   **53.** −1; not positive   **55.** $13\frac{1}{3}$
**57.** $7\frac{1}{2}$   **59.** 210 ways

### Exercises, pages 315–316
**1.** c   **3.** a   **5.** 30   **7.** 152   **9.** 3
**11.** 42   **13.** 12   **15.** 54   **17.** 812
**19.** $3170   **21.** 81% of $9000   **23.** 79% of $1500   **25.** Both equal $1.00.   **27.** 103.6
**29.** 150.4   **31.** Correct   **33.** 2.5   **35.** $\frac{9}{10}$
**37.** $\frac{1}{80}$

### Exercises, pages 321–322
**1.** a   **3.** b   **5.** 32   **7.** 63   **9.** 124.5
**11.** 1392   **13.** 0.312   **15.** 347.9   **17.** 2
**19.** 0.108   **21.** $3.00   **23.** a. 14;   b. 14
**25.** a. 24;   b. 24   **27.** 4   **29.** 10
**31.** 40 square inches   **33.** $\frac{7}{10}$   **35.** −1
**37.** −100   **39.** 0.044   **41.** 0.003

### Exercises, pages 326–328
**1.** c   **3.** a   **5.** True   **7.** False; yearly interest rate   **9.** $60   **11.** $75   **13.** $34.50
**15.** $1030.41   **17.** $957.96   **19.** $18,547.30
**21.** $575   **23.** $845   **25.** Carrie: $5; Bob: $10   **27.** Carrie: $105; Bob: $110
**29.** Carrie: $10.25; Bob: $21   **31.** 2 years: $2,376.20; 4 years: $2,823.16; 6 years: $3,354.20; 8 years: $3,985.12; a little more than 8 years   **33.** Adult: 169; student: 557
**Ex. 35, 37:** Estimates will vary.   **35.** 375; O
**37.** 268; U   **39.** 1.5   **41.** About $19

### Exercises, pages 331–333
**1.** c   **3.** a   **5.** Regular price, or list price
**7.** Regular price, or list price   **9.** $5
**11.** $34.50   **13.** $34.65   **15.** $20.80
**17.** $33\frac{1}{3}$% discount on $69   **19.** 5% discount on $58   **21.** 55%   **23.** 60%   **25.** $20.40
**27.** False. Suppose the original price was $100. Then the first discount gives a first sale price of $100 − $\frac{1}{5}$($100), or $80. The second discount gives a sale price of $80 − $\frac{1}{10}$($80), or $72. A single discount of 30% on $100 gives a sale price of $100 − $\frac{3}{10}$($100), or $70.
**29.** Kimiko: 38 hours; Roscoe: 19 hours
**31.** 12   **33.** 15   **35.** $11.70

### Exercises, pages 336–337
**1.** b   **3.** a   **5.** $\frac{2}{25} \times 21 = n$   **7.** $17 = \frac{p}{100} \times 85$   **9.** 75%   **11.** 84.63   **13.** $43\frac{3}{4}$
**15.** $\frac{8}{100} = \frac{n}{90}$   **17.** $\frac{30}{100} = \frac{102}{n}$   **19.** 90%
**21.** 140%   **23.** 50   **25.** 75%   **27.** 24%
**29.** 24   **31.** 16%   **33.** −8   **35.** No
**37.** 58%   **39.** $8\frac{1}{3}$%   **41.** 150%

### Exercises, pages 339–341
**1.** c   **3.** a   **5.** 80%   **7.** 40%   **9.** $\frac{930}{4650}$; 20%   **11.** $\frac{1.21}{24.20}$; 5%   **13.** $\frac{22.80}{76}$; 30%
**15.** 87%   **17.** 12%   **19.** Debating Team
**21.** Quiz Team   **23.** 25%   Assume the original price is $100. After a 20% discount, the sale price is $80. After the markup, the percent increase is $\frac{20}{80} = \frac{1}{4} = 25\%$. The percent increase is greater than the percent decrease because it is based on a smaller number.
**25.** 33 students   **27.** 12 students   **29.** 44, 48, 52   **31.** $40   **33.** $1.3 \times 10$   **35.** $3 \times 10^{-2}$   **37.** $48.93

## CHAPTER 9

### Exercises, pages 349–350
**1.** It is $1\frac{1}{2}$ times as tall as the Chux box.
**3.** The Super Flakes box is $1\frac{1}{2}$ times as wide as the Chux box.   **5.** No   **7.** The Reader's Store; The Reader's Store   **9.** 2600
**11.** Yes. The bars on the graph of the Reader's Store Book Sales are taller.
**13.** Answers will vary.
**15.**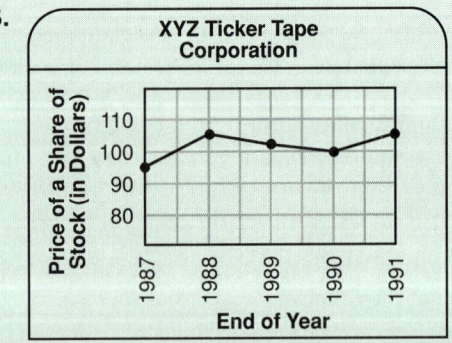
**17.** $\frac{1}{6}$   **19.** $\frac{9}{10}$   **21.** 17   **23.** 92
**25.** Less than

### Exercises, pages 352–354

**1.** c  **3.** b  **5.** 97 feet  **7.** 16 feet
**9.** 44 mph  **11.** 16 mph  **13.** True. The curve of the graph of the braking distance on wet asphalt is steeper than the curve of the graph for dry asphalt.  **15.** 5:00–6:00
**17.** These are the hours when the shoe department has the most customers.
**19.** Answers will vary.  **21.** 10; even
**23.** 40; even  **25.** $11\frac{1}{4}$  **27.** $2\frac{3}{5}$  **29.** 38
**31.** The man can eat 4 apples in 3 days, or $1\frac{1}{3}$ apples per day for $x$ days. The total is $1\frac{1}{3}x$ apples. The man can eat one apple per day for $x + 14$ days. The total is $x + 14$ apples. $1\frac{1}{3}x = x + 14$; $x = 42$ and $x + 14 = 56$. So the man has 56 apples.

### Exercises, pages 356–357

**1.** c  **3.** d  **5.** Add the heights of the first three bars in the histogram.
**7.**

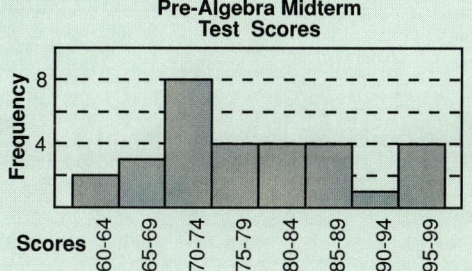

**9.** 57%
**11.**

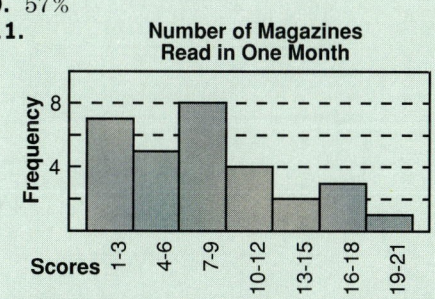

**13.** About 8  **15.** Sample answer: Advantage: A frequency table provides an organized summary of the set of data. Disadvantage: The individual data items are not represented.
**17.** $-13$  **19.** 12  **21.** $\frac{2}{7}$  **23.** $\frac{7}{12}$

### Exercises, pages 360–362

**1.** c  **3.** b  **5.** Mean: 35; Median: 33; Mode: 30  **7.** No; all but 3 of the allowances are between $25 and $35.  **9.** Median: $13,100; Mode: $13,100  **11.** Median or mode

**13.** Median; it is the middle number in the set of data.  **15.** 44  **17.** 130.4  **19.** $\frac{6}{11}$
**21.** $3\frac{3}{5}$  **23.** $16.80  **25.** 75%  **27.** 90%
**29.** 225%

### Exercises, pages 368–369

**1.** b  **3.** c  **5.** 48%  **9.** 14.3%
**11.**  4 | 2,3,6,7,8,9,9
       5 | 0,0,1,1,1,1,2,2,4,4,4,4,5,5,5,5,6,6,7,7,7,7,8
       6 | 0,1,1,1,2,4,4,5,6,8,9

**13.** 23  **15.** 55  **17.** 40  **19.** 46.5
**21.** 20%  **23.** The stem and leaf plot can be used to make a frequency table. Then a histogram can be drawn from the frequency table.  **25.** $2 > -12$; $-12 < 2$
**27.** $-5 > -7$; $-7 < -5$  **29.** 14  **31.** $-1$
**33.** $-5\frac{1}{2}$  **35.** $-2$  **37.** 21  **39.** 30

### Exercises, pages 371–372

**1.** d  **3.** b  **5.** 5 and 27  **7.** Median: 23; Upper quartile: 25.5; Lower quartile: 21
**9.**

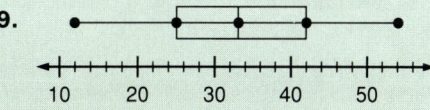

**11.** $68  **13.** $41  **15.** 10  **17.** 10
**19.**

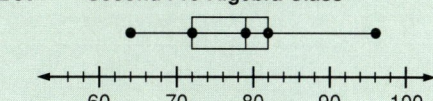

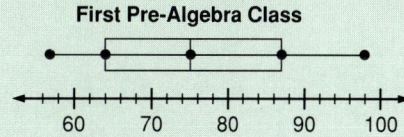

**21.** Mean: 77; Median: 76.5; Modes: 87, 81, 75, 73, and 62  **23.** $\{-1\}$  **25.** $\{0\}$
**27.** 215.43  **29.** 396.88  **31.** $237.15

## CHAPTER 10

### Exercises, pages 380–381

**1.** b  **3.** d  **5.** {All real numbers less than 3}  **7.** {All real numbers greater than or equal to $-\frac{1}{2}$}  **9.** {All real numbers less than $-1$}  **11.** {All real numbers less than $\pi$}
**13.**

**15.** (number line, open circle at 7, shading left; marks 2, 4, 6, 8)

**17.** (number line, closed circle at 1, shading left; marks −2, 0, 2, 4)

**19.** (number line, open circle at 1, shading right; marks −2, 0, 2, 4)

**21.** (number line, closed circle at 4, shading left; marks 0, 2, 4, 6)

**23.** (number line, open circle at 7, shading left; marks 2, 4, 6, 8)

**25.** (number line, closed circle at 8, shading right; marks 4, 6, 8, 10)

**27.** (number line, closed circle at −7, shading left; marks −8, −6, −4, −2)

**29.** {All real numbers greater than or equal to 3}  **31.** {All real numbers greater than or equal to 9}  **33.** $a > b$, $a = b$, or $a < b$.
**35.** $\frac{2}{5}$  **37.** True  **39.** False; the resulting inequality, $x - 1 \leq a - 1$ has the solution $x \leq a$, so the endpoint of the graph does not move.  **41.** {5, 7}  **43.** {−7, −5, 0, 5}
**45.** 4 inches tall; $p - \frac{2}{3}p = 2$  **47.** 16

**Exercises, pages 384–385**
**1.** c  **3.** d
**5.** $x \leq 4$ (number line, closed circle at 4, shading left; marks 2, 4, 6, 8)
**7.** $s \geq 3$ (number line, closed circle at 3, shading right; marks −2, 0, 2, 4)
**9.** $r > -6$ (number line, open circle at −6, shading right; marks −8, −6, −4, −2)
**11.** $p < 0$ (number line, open circle at 0, shading left; marks −2, 0, 2, 4)
**13.** $k < -4$ (number line, open circle at −4, shading left; marks −6, −4, −2, 0)
**15.** $t \geq \frac{3}{4}$ (number line, closed circle at 3/4, shading right; marks −1, 0, 1, 2)
**17.** b  **19.** d  **21.** $130.01  **23.** $1 \leq 0$; false  **25.** $1 \leq 0$; false  **27.** $a + 1$
**29.** Descending; $5 > 1 > 0 > -1 > -2$
**31.** Ascending; $-7 < -3 < -2\frac{1}{2} < -1\frac{1}{2} < -\frac{3}{4}$  **33.** Yes; he should have divided both sides by 3, the coefficient of the variable, $x$.  **35.** 45  **37.** 76  **39.** The area of the football field is 5300 square yards. The area between the two 40-yard lines is $(20 \times 53)$ or 1060 square yards. The probability that a balloon will land between the two 40-yard lines is $\frac{1060}{5300} = \frac{53}{265} = \frac{1}{5}$ and $\frac{1}{5} = 0.2 = 20\%$.

**Exercises, pages 388–389**
**1.** d  **3.** c  **5.** >  **7.** >
**9.** $a \leq 8$ (number line, closed circle at 8, shading left; marks 4, 6, 8, 10)
**11.** $c \leq -3\frac{1}{2}$ (number line, closed circle at −3½, shading left; marks −6, −4, −2, 0)
**13.** $c \leq 2\frac{1}{2}$ (number line, closed circle at 2½, shading left; marks 0, 2, 4, 6)
**15.** $r \leq -6$ (number line, closed circle at −6, shading left; marks −8, −6, −4, −2)
**17.** $r \geq 6$ (number line, closed circle at 6, shading right; marks 2, 4, 6, 8)
**19.** $s \leq 9\frac{1}{2}$ (number line, closed circle at 9½, shading left; marks 8, 10, 12, 14)
**21.** $45.50  **23.** $149.97  **25.** { }
**27.** {0, 1, 2, 3}  **29.** {2, 3}  **31.** 87.5
**33.** −6  **35.** −117  **37.** $6.3 \times 10^3$
**39.** 40%

**Exercises, pages 395–397**
**1.** a  **3.** a  **5.** Add (−5) to each side.
**7.** $x \leq -5$ (number line, closed circle at −5, shading left; marks −6, −4, −2, 0)
**9.** $x \leq -3$ (number line, closed circle at −3, shading left; marks −6, −4, −2, 0)
**11.** $x \geq -1$ (number line, closed circle at −1, shading right; marks −2, 0, 2, 4)
**13.** $x \geq 0$ (number line, closed circle at 0, shading right; marks −2, 0, 2, 4)
**15.** $x > \frac{1}{2}$ (number line, open circle at ½, shading right; marks −2, 0, 2, 4)
**17.** $x \leq 8$ (number line, closed circle at 8, shading left; marks 6, 8, 10, 12)
**19.** $x \leq 1$ (number line, closed circle at 1, shading left; marks −2, 0, 2, 4)
**21.** $x \geq -2$ (number line, closed circle at −2, shading right; marks −8, −6, −4, −2)
**23.** 14, 15, 16, and 17  **25.** $x < 11\frac{1}{4}$

**27.** $x < 2$  **29.** $x < 3$  **31.** All real numbers
**33.** 12.5 hours  **35.** 11 books  **37.** 11 books
**Ex. 39, 41:** LL: Lower left; LR: Lower right; UL: Upper left; UR: Upper right  **39.** (UL + UR) · LL; 20  **41.** UL · UR + LL; 6

### Exercises, pages 401–402
**1.** b  **3.** d  **5.** $-1 < t < 4$
**7.** $-3 < x < 0$
**9.**
**11.**
**13.**
**15.**
**17.**
**19.**
**21.** $167 < w < 185$
**23.** $-1 \le x < 2$

**25.**
Set A: 1, 3, 5, 7  Set B: 2, 4, 6, 8

**27.**
Set A: 1, 15  Set A ∩ Set B: 5, 10  Set B: 0

**29.** 5  **31.** 1, 2, 3, 4, 6, 8, 12, 16, 24, 32, 48, 96  **33.** $\frac{3}{10}$

### Exercises, pages 404–405
**1.** d  **3.** a  **5.** $x \ge 5$  **7.** $x \le -3$
**9.** $x > -1$

**11.** $x \le -3 \text{ or } x > 4$
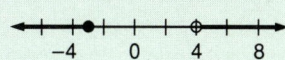

**13.** {All real numbers}
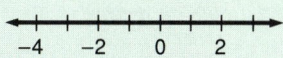

**15.** $x \le -\frac{4}{3} \text{ or } x > 3$
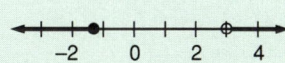

**17.** $\frac{3}{7}$  **19.** 40%  **21.** $y = -14$  **23.** $-192$
**25.** $-1,889,568$  **27.** $\frac{18}{35}$ of a tank

## CHAPTER 11

### Exercises, pages 412–413
**1.** b  **3.** b  **5.** False; the $x$- and $y$-coordinates are reversed.  **7.** False; points on the $x$-axis have coordinates $(x, 0)$.
**9.** $A(-2, 4)$, $B(0, 3)$, $C(-1, -3)$, $D(3, 2)$
**11.** $y$-axis  **13.** $x$-axis  **15.** Quadrant III
**17.** Quadrant III
**31. c.** $P = 10$, $A = 4$
**33.** $(-1, 1)$

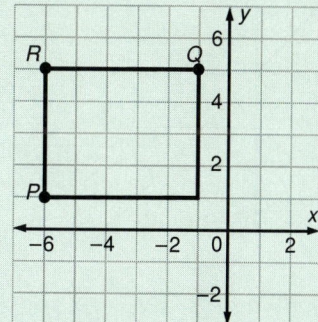

**35.** {3}  **37.** {0}  **39.** $-36$  **41.** 12
**43.** Length: 7 inches; Width: 4 inches

### Exercises, pages 416–417
**1.** c  **3.** c  **5.** No  **7.** No  **9.** Yes
**11.** No  **13.** Sample answer: $(0, -5)$, $(1, -3)$, $(2, -1)$  **15.** Sample answer: $(0, 4)$, $(1, 7)$, $(2, 10)$  **17.** Sample answer: $(0, 3)$, $(1, 7)$, $(2, 11)$

19.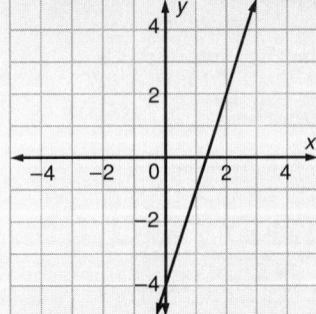

21.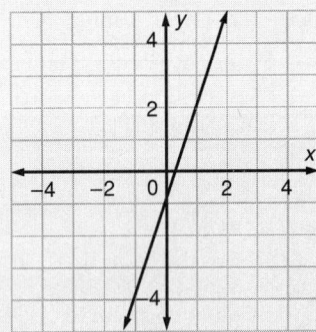

23.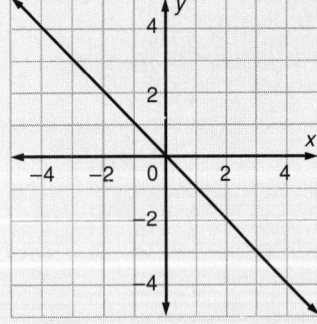

25.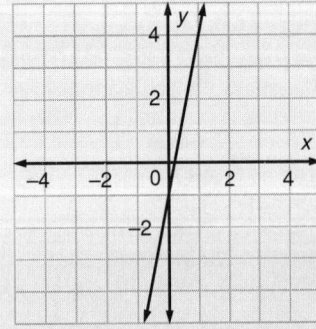

27.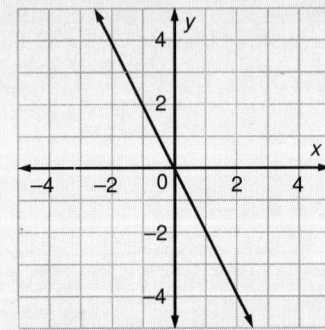

29. $C = 12 + 10w$  31. $C = 5 + 3h$
33. When $x = y$  35. $(5, 0); (0, 5)$
37. Sample answer:

| $x$ | 0 | 1 | 2 |
|---|---|---|---|
| $y$ | 0 | $-1$ | $-2$ |

; $y = -x$

39. $-4$  41. $3\frac{1}{2}$  43. $4\frac{3}{4}$ gallons

45. $z \leq 3$

47. $k \geq -1$

49. The point she passes at the same time each day is the point halfway between her house and LaShana's house.

**Exercises, pages 420–421**
1. b  3. d  5. 0  7. $4x - y = -8$; $x$-intercept: $-2$, $y$-intercept: 8
9. $3x - y = 9$; $x$-intercept: 3, $y$-intercept: $-9$
11. $x - 3y = 9$; $x$-intercept: 9, $y$-intercept: $-3$  13. Sample answers: $3y = -4$, $y = -7, 5y = 2$
15. $x$-intercept: 4; $y$-intercept: 6
17. $x$-intercept: 2; $y$-intercept: $-4$
19. $x$-intercept: $-4$; $y$-intercept: 2
21. $x$-intercept: 5; $y$-intercept: $-1$
23. $x$-intercept: $-2$; $y$-intercept: $-2$
25. $x$-intercept: $-3$; $y$-intercept: $-5$
27. $x$-intercept: $-3$; $y$-intercept: $-6$
29. $x$-intercept: $-2$; $y$-intercept: 3
31. $x = 3$

**33.**
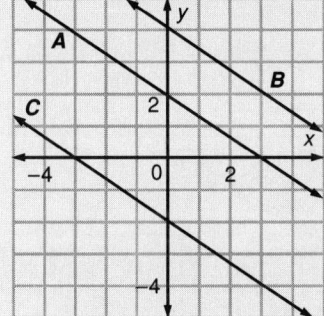

**35.** $\frac{7}{2}$  **37.** 6720  **39.** 3024  **41.** 1.25
**43.** Quadrant IV  **45.** Quadrant III

### Exercises, pages 425–426
**1.** d  **3.** a  **5.** True  **7.** False; the slope is 0.  **9.** True
**11.**
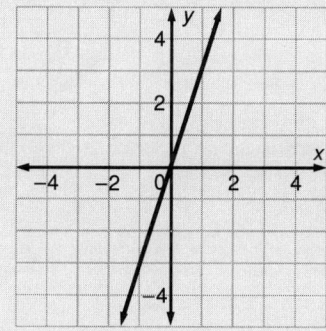

**13.**
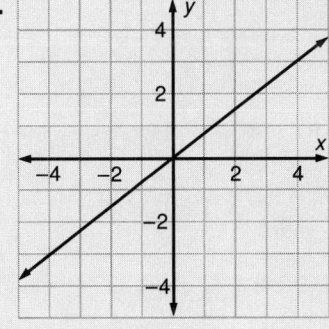

**15.** $-3$
**17.** 1
**19.** 0
**21.** 1
**23.** $-3$
**25.** $\frac{2}{3}$  **27.** $\frac{1}{2}$  **29.** Undefined  **31.** 1

**33.**
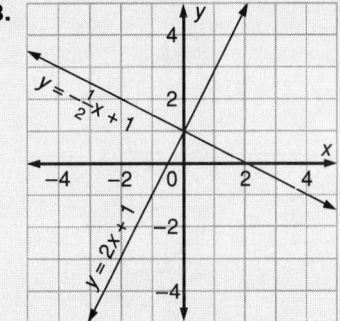

**35.** 2; $-\frac{1}{2}$  **37.** Yes  **39.** $-5, -6, -7$
**41.** 104, 112, 120, 128  **43.** $7.50x > 250$; $x > 33\frac{1}{3}$; at least 34

### Exercises, pages 433–434
**1.** a  **3.** c  **5.** An infinite number  **7.** Yes
**9.** 4; 1  **11.** 0; $-4$  **13.** 2; $-6$

**15.**
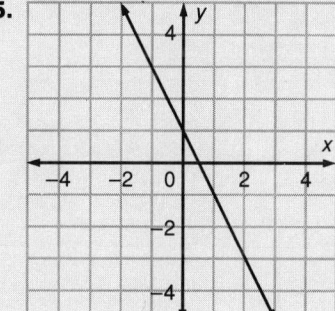

**17.**
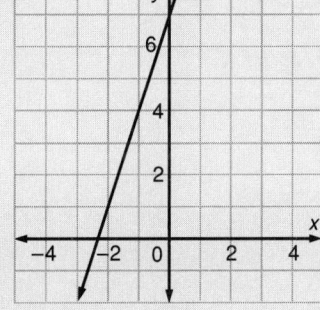

**19.**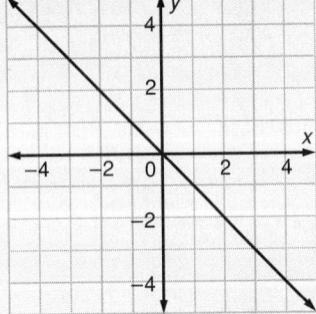

**21.**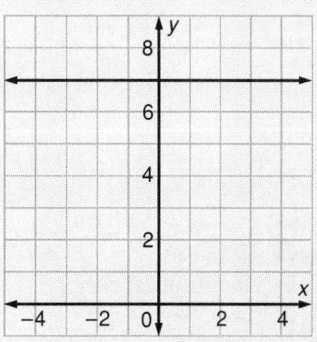

**23.**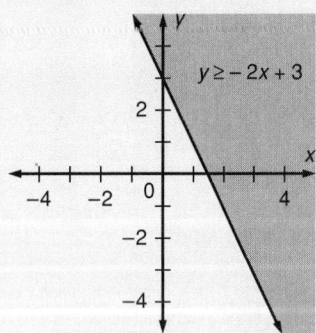

**25.**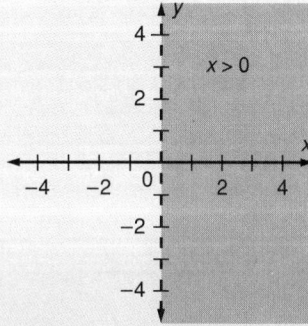

**27.**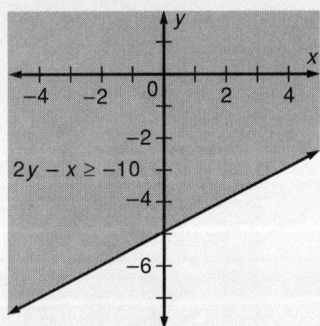

**29.**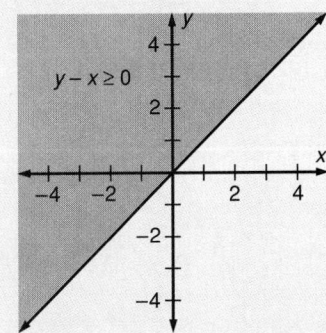

**31.**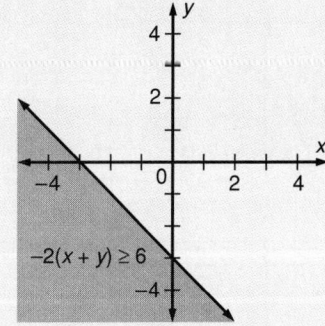

**33.** A closed half-plane    **35.** A straight line
**37.** An open half-plane

**39.**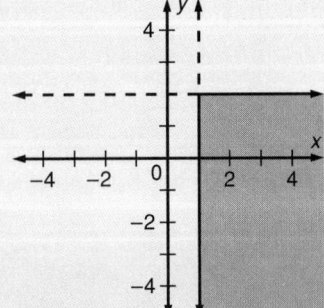

**41.** $-187$  **43.** 23  **45.** 8  **47.** 32
**49.** $y + (y + 1) > 15$  **51.** $y + 5y \geq 72$
**53.** The amount in the jar of dimes. Since a dime is worth twice the value of a nickel, if nickels and dimes were the same size, then half a jar of dimes would have the same value as a full jar of nickels. However, since dimes are smaller than nickels, there will be more than half the number of nickels in half a jar of dimes, and their total value will be greater.

**Exercises, pages 437–439**
**1.** c  **3.** b  **5.** $(4, 6)$  **7.** No solution
**9.** $(7, 3)$  **11.** $(4, 0)$
**13.** $(1, -1)$
**15.** $(1, 3)$
**17.** $(1, -1)$
**19.** $(-2, 4)$
**21.** $-2, 1)$
**23.** $(2, 0)$;
**25.** $y = x + 26, x + y = 68$; 47 tickets
**27.** $x - y = 10, x = 3y$; 15 and 5

**29.**

**31.** For any time less than 2 hours
**33.** About $-18$; $-17.779$  **35.** About 15; 14.991  **37.** 40% increase

**Exercises, pages 442–443**
**1.** d  **3.** c  **5.** The image of $P(1, -2)$ under the translation $(x, y) \to (x - 4, y + 5)$ is $P'(-3, 3)$.
**7.** $P(2, 4) \to (2 + 1, 4 - 2) = P'(3, 2)$
**9.** $R(-1, -3) \to (-1 - 1, -3 - 1) = R'(-2, -4)$

**11.**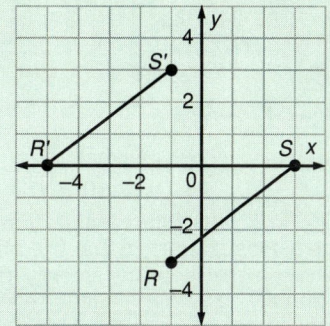

**13.** $(x, y) \to (x - 6, y - 4)$  **15.** The image of $R(4, 5)$ under the change $(x, y) \to (y, x)$ is $R'(5, 4)$.  **17.** Yes. Let $x = 2$ and substitute for the variable in each expression.
**19.** 39  **21.** 212 boys  **23.** 32

## CHAPTER 12

**Exercises, pages 454–456**
**1.** c  **3.** c  **5.** radical symbol, radical
**7.** $2\sqrt{3}$ Since $2\sqrt{3} = \sqrt{4 \cdot 3} = \sqrt{12}$ and $12 > 8$, $\sqrt{12} > \sqrt{8}$ or $2\sqrt{3} > 8$.
**9.** 3 and 4  **11.** 6 and 7  **13.** 12 and 13
**15.** 15 and 16  **17.** 19 and 20  **19.** 22 and 23  **21.** 10  **23.** 7  **25.** $4\sqrt{2}$
**27.** 4  **29.** 5  **31.** $5\sqrt{2}$  **33.** Cannot be simplified  **35.** $6\sqrt{2}$  **37.** $2\sqrt{10}$
**39.** $10\sqrt{2}$  **41.** $\frac{\sqrt{3}}{3}$  **43.** $\frac{3\sqrt{5}}{5}$
**45.** $\frac{6\sqrt{7}}{7}$  **47.** $\frac{\sqrt{2}}{5}$  **49.** 2.8  **51.** 5.5
**53.** Yes  **55.** Distributive Property
**57.** About 2.2

**59.** About 0.3

**61.** 54 square feet  **63.** $-5$  **65.** $-162$
**67.** 144  **69.** 260

**Exercises, pages 460–462**
**1.** b  **3.** c  **5.** Hypotenuse  **7.** 14.4
**9.** 15  **11.** 24.3  **13.** 13.2  **15.** 1300 feet
**17.** Answers will vary.  **19.** $2\sqrt{2}$  **21.** For right triangles with legs of equal length, the length of the hypotenuse equals the length of a leg times $\sqrt{2}$.  **23.** 12; 16; 20; 12, 16, 20
**25.** 13; 84; 85; 13, 84, 85  **27.** Sample answers: 3, 4, 5; 5, 12, 13; 8, 15, 17; 24, 70,

74; 27, 36, 45   **29.** 24   **31.** 10   **33.** $p \leq -3$   **35.** $q \geq 2$   **37.** $Q'(0, -3); R'(2, 2)$
**39.** $A'(-1, 3); B'(1, 7); C'(4, 2)$

**Exercises, pages 470–472**

**1.** c   **3.** a   **5.** 2   **7.** 2   **9.** $\frac{9}{6} = \frac{15}{10} = \frac{18}{12} = \frac{3}{2}$   **11.** $x = 4$   **13.** $n = 10$   **15.** 10 meters   **17.** $x = 4$; One figure would have sides with dimensions 2 and 4, while the other figure would have corresponding sides with dimensions 4 and 8.   **19.** Yes; since the sides of a square are equal, ratios of corresponding sides of two different squares will be equal.
**21.** AAA Property

**Ex. 22, 23:**

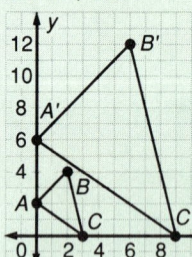

**25.** The new triangle would have vertices (0, 12), (12, 24), and (18, 0), and would be similar to each of the smaller triangles. The ratio of corresponding sides for the new triangle and triangle $ABC$ will be $\frac{6}{1}$, and the ratio of the corresponding sides for the new triangle and triangle $A'B'C'$ will be $\frac{2}{1}$.

**27.** 6   **29.** $-204$   **31.** 0.182   **33.** 3.5
**35.** 0.333   **37.** 5
**39. a.** $\sqrt{26} \approx 5.1$
**b.** $\sqrt{349} \approx 18.7$

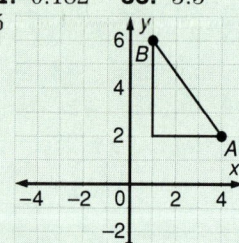

**Exercises, pages 477–480**

**1.** d   **3.** c   **5.** True   **7.** False; the ratio should be $\frac{\text{length of side opposite the angle}}{\text{length of side adjacent to the angle}}$.
**9.** 7.1154   **11.** 0.7536   **13.** 1.7321
**15.** 0.1051   **17.** 5.6713   **19.** 75°
**21.** 45°   **23.** 79°   **25.** 80 inches
**27.** 173.2 feet   **29.** 26.8 yards   **31.** About 1,120 meters   **Ex. 33, 35, 37:** Sample answers are given.   **33.** $\frac{6}{10} = 0.6$
**35.** $\frac{3.8}{10} \approx 0.4$   **37.** $\frac{16.5}{10} \approx 1.65$   **39.** Angles whose measures are less than 45°   **41.** No; you only know that the length of the side opposite angle $A$ and the length of the side adjacent to angle $A$ are the same.   **43.** No
**45.** No   **47.** 200%   **49.** 8   **51.** 14

**53.** Since $25 \div 3 = 8$ R 1, the clerk measured 8 times using the short yardstick. The wire was missing $8 \times 2$, or 16 inches from the 25-foot length the customer wanted.

## CHAPTER 13

**Exercises, pages 489–490**

**1.** a   **3.** b   **5.** $6x - 2$   **7.** $3a - 10$
**9.** $3x^2 + 2x + 2$   **11.** $4d^2 + 2d - 3$
**13.** $8t^2 - 1$   **15.** $8t^2 - 3t + 6$   **17.** $4x^2 + 11x - 2$   **19.** $-3t^2 + 6t + 9$   **21.** $l: 7\frac{1}{3}$ cm; $w: 3\frac{2}{3}$ cm   **23.** $x^2 - 8x + 4$   **25.** $3r^2 + 3r + 1$   **27.** 20 meters, 22 meters, 11 meters
**29.** If 10 is replaced by $x$, the addition is done in the same way.   **31.** a   **33.** $4 + \frac{0}{10} + \frac{6}{100} + \frac{3}{1000}$   **35.** $6 + \frac{0}{10} + \frac{7}{100}$
**37.** Cannot be simplified   **39.** Cannot be simplified

**Exercises, pages 492–493**

**1.** c   **3.** a   **5.** $5x^2$   **7.** $-3t$   **9.** $3t + 8$
**11.** $5x + 8$   **13.** $2q^2 - q - 7$   **15.** $-d^2 - 7d + 2$   **17.** $4n^2 - 3$   **19.** $4y^2 + y - 2$
**21.** $c^2 - 18c + 1$   **23.** $6s^2 + 5s - 8$
**25.** $2p^2 - 2t^2$   **27.** $4a^2 - 9a - 4$
**29.**
$$\begin{array}{r}(3 \times 100) + (4 \times 10) + (7 \times 1) \\ -\ (2 \times 100) + (2 \times 10) + (7 \times 1) \\ \hline (1 \times 100) + (2 \times 10) + 0 \end{array} = 120$$
The subtraction is similar to subtracting $(2x^2 + 2x + 7)$ from $(3x^2 + 4x + 7)$, with $x^2$ replaced by 100 and $x$ replaced by 10.
**31.** $(-6)(-28) + (-6)(228) = (-6)(-28 + 228) = (-6)(200) = -1,200$   **33.** 216
**35.** Quadrant IV   **37.** $y$-axis
**39.** 12 meters

**Exercises, pages 498–500**

**1.** d   **3.** a   **5.** $4x^2 + 2x$   **7.** $x^2 + 4x + 3$
**9.** $x^2 + 2x - 3$   **11.** $2x^2 - x$   **13.** $x^2 - 1$
**15.** $x^2 - 2x - 8$   **17.** $2x^2 - x - 1$
**19.** $3x^2 + 4x + 1$   **21.** $3x^2 + 5x - 2$
**23.** $6 \cdot 20 + 6 \cdot 3 = 120 + 18 = 138$
**25.** 375

| | 100 | 20 | 5 |
|---|---|---|---|
| 3 | 300 | 60 | 15 |

**27.** $10(10 + 2) + 4(10 + 2) = 100 + 20 + 40 + 8 = 168$

618   Selected Answers

**29.**

| | 20 | 1 |
|---|---|---|
| 10 | 200 | 10 |
| 2 | 40 | 2 |

**31.** $(x + 1)(x - 1) = x(x - 1) + 1(x - 1) = x \cdot x + x(-1) + 1(x) + 1(-1) = x^2 - x + x - 1 = x^2 - 1$
**33.** $-4, y$  **35.** $170
**37.** c  **39.** $x^2, (x - 3)(x + 6) = x^2 + 3x - 18$  **41.** 36 square inches

### Exercises, pages 506–507

**1.** d  **3.** a  **5.** $a^2 + 5a + 6$  **7.** $x^2 - 6x + 8$  **9.** $2x^2 - 11x + 5$  **11.** $x^2 - 81$
**13.** $5r^2 - 17r - 12$  **15.** $2n^2 + n - 3$
**17.** $-2z^2 + 5z - 3$  **19.** $-2y^2 + 7y - 3$
**21.** $2t^2 + 5t - 3$  **23.** $2x^2 - 5x - 12$
**25.** $x^2 + x$  **27.** $a^2 - a - 6$  **29.** $2d^2 - 5d - 3$  **31.** $-2q^2 + q + 3$  **33.** Rewrite $13 \cdot 45$ as $(10 + 3)(40 + 5)$. F: $10 \times 40 = 400$; O: $10 \times 5 = 50$; I: $3 \times 40 = 120$; L: $3 \times 5 = 15$; The product is $400 + 50 + 120 + 15 = 585$.  **35.** $5x + 12$  **37.** $-101$  **39.** $-28$
**41.** $\frac{2}{3}$  **43.** $-2$

### Exercises, pages 510–511

**1.** d  **3.** d  **5.** $c^2 + 10c + 25$  **7.** $m^2 - 12m + 36$  **9.** $4 + 4r + r^2$  **11.** $4y^2 - 12y + 9$  **13.** $q^2 + 2q + 1$  **15.** $s^2 - 8s + 16$  **17.** $a^2 - 2a + 1$  **19.** $p^2 - 9$
**21.** $b^2 + b + \frac{1}{4}$  **23.** $65.61$  **25.** $1\frac{9}{16}$
**27.** $75$  **29.** $3,584$  **31.** $4x + 8$  **33.** $21$
**35.** $1$  **37.** $4.8 \times 10^4$  **39.** $3.96 \times 10^{-2}$

### Exercises, pages 514–515

**1.** a  **3.** d  **5.** $4(2a + 1)$  **7.** $5(y - 3)$
**9.** $4(2m + 5)$  **11.** $5(5t - 3)$  **13.** $6(y - 6)$
**15.** $3(2p - 3)$  **17.** $2d(d + 2)$  **19.** $5x(3x + 2)$  **21.** $4m(3m - 4)$  **23.** $9p(2p + 3)$
**25.** $8a(5a - 2)$  **27.** $(b + 3)(b - 3)$
**29.** $(t + 4)(2t - 3)$  **31.** $(x + 1)5$
**33.** $(p + 1)$  **35.** $7(t - 1)$  **37.** If a binomial has one factor, it must be common to each of the two terms of the binomial. The resulting expression when each term of the original binomial is divided by the common monomial factor is a binomial.  **39.** $11c$  **41.** $\frac{1}{20}$
**43.** $x = -2, y = -2$  **45.** $x = \frac{5}{4}, y = \frac{5}{2}$
**47.** $x^2 + x - 42$  **49.** 3 wheels

### Exercises, pages 518–519

**1.** d  **3.** c  **5.** $(x + 1)(x + 2)$
**7.** $(s - 3)(s + 2)$  **9.** $(x + 1)(x + 5)$
**11.** $(t - 9)(t + 3)$  **13.** $(d + 3)(d + 5)$
**15.** $(a - 17)(a + 3)$  **17.** $(m - 7)(m - 7)$ or $(m - 7)^2$  **19.** $(x - 5)(x + 2)$
**21.** $(z - 1)(z + 17)$  **23.** $x - 1$

Ex. 25–28.

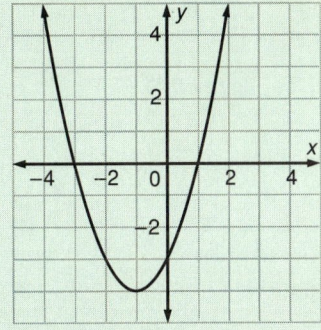

**25.** No  **27.** $x = -3$ and $x = 1$
**29.** $x = -4, x = 2$
Sample answer: Set $y = 0$ in the given equation and then factor the expression $x^2 + 2x - 8$ into $(x + 4)(x - 2)$. The solution set for the equation $0 = (x + 4)(x - 2)$ is $\{-4, 2\}$.  **31.** 24 possibilities
**33.** $\frac{3}{4}$  **35.** $-2\frac{5}{8}$
**37.** Sample answer: $(0, 0), (3, 2)$

## CHAPTER 14

### Exercises, pages 530–531

**1.** c  **3.** c  **5.** $\angle AEB$ or $\angle BEC$  **7.** $\angle DEA$ or $\angle AEC$  **9.** $\angle DEB$ and $\angle BEC$, or $\angle DEA$ and $\angle AEC$  **11.** $m\angle ABC + 65 = 90; 25°$
**13.** $58°$  **15.** $70°; 110°$  **17.** $180 - x$
**19.** Since the measure of the complement is $(90 - m\angle ABC) = 90 - 90 = 0$, there is no angle which is a complement to $\angle ABC$.
**21.** $96°; 84°$  **23.** $50°, 130°$  **25.** Two right angles are supplementary.  **27.** $-\frac{4}{9}$
**29.** $(4, 5)$  **31.** $(-2, -1)$  **33.** $3x^2 + x - 1$

### Exercises, pages 535–536

**1.** c  **3.** d  **5.** $a$ and $d$, $b$ and $d$  **7.** $c, d$
**9.** $\angle 1$ and $\angle 5$, $\angle 2$ and $\angle 6$, $\angle 3$ and $\angle 7$, $\angle 4$ and $\angle 8$, $\angle 9$ and $\angle 13$, $\angle 10$ and $\angle 14$, $\angle 11$ and $\angle 15$, $\angle 12$ and $\angle 16$  **11.** $\angle 1$ and $\angle 7$, $\angle 2$ and $\angle 8$, $\angle 11$ and $\angle 13$, $\angle 12$ and $\angle 14$
**13.** See the pairs of angles given in the answers to Ex. 8–11. Also, any pair of the right angles is congruent.  **15.** $m\angle 1 = 65°$, $m\angle 2 = 25°$, $m\angle 3 = 25°$  **17.** $m\angle 1 = 129°$, $m\angle 2 = 51°$, $m\angle 3 = 129°$  **19.** $2x + 40 = 5x - 80$; $m\angle PQR = 60°$  **21.** The line and line segment are perpendicular.
**23.** $\triangle E'G'H'$, with $E'(1, 12)$, $G'(-7, 1)$, and $H'(-5, 4)$

**25.** 72°  **27.** $x^2 - 22x + 121$
**29.** $4x^2 + 12x + 9$

### Exercises, pages 542–544
**1.** b  **3.** b  **5.** Scalene right  **7.** Scalene obtuse  **9.** $x + x + 70 = 180$; 55°, 55°
**11.** $x + 3x + 90 = 180$; $22\frac{1}{2}°, 67\frac{1}{2}°$
**13.** No; the sum of the other two angles cannot be greater than 90°.  **15.** Yes; see Try This Exercise 4.  **17.** Yes; the two legs can be of equal length.  **19.** 52°; they are complementary angles.  **21.** 50°, 60°, 70°
**23.** Since $\angle 3$ and $\angle 4$ are supplementary angles, $m\angle 3 + m\angle 4 = 180$ or $m\angle 3 = 180 - m\angle 4$. By the Triangle-Sum Property, $m\angle 1 + m\angle 2 + m\angle 3 = 180$. Substituting for $m\angle 3$, $m\angle 1 + m\angle 2 + 180 - m\angle 4 = 180$. Subtracting 180 and adding $m\angle 4$ to each side gives $m\angle 1 + m\angle 2 = m\angle 4$.  **25.** 135  **27.** 0.25
**29.** 0.004  **31.** 0.893  **33.** About 5.8 meters
**35.** Let $x =$ the measure of the angle and $90 - x =$ the measure of its complement. Then $\frac{1}{2}(x + 90 - x) = \frac{1}{2} \cdot 90 = 45°$.

### Exercises, pages 549–550
**1.** c  **3.** c  **5.** Pentagon  **7.** Hexagon
**9.** 1,440°  **11.** 3,240°  **13.** $x + 3x + x + 3x = 360$; $x = 45$; $m\angle A = 135°$
**15.** 60°  **17.** 135°  **19.** True  **21.** True
**23.** False; many rectangles do not have 4 sides of equal length.  **25.** 144°  **27.** There are as many angles in a polygon as there are sides, therefore $n$ is also the number of angles. Since the sum of the angles of a polygon with $n$ sides (and angles) is $(n - 2)180°$ and since all angles of a regular polygon have equal measures, the measure of each of the angles is $\frac{(n - 2)180°}{n}$.  **29.** $352.89
**31.** $-12w^2 - 17w - 15$  **33.** $\frac{1}{4}z^2 - 4$
**35.** $(s - 5)(s - 9)$

### Exercises, pages 554–556
**1.** c  **3.** c  **5.** False; $d = 2r$  **7.** True
**9.** 12.56 in, 12.56 in²  **11.** 25.12 ft, 50.24 ft²
**13.** 20.096 yd, 32.1536 yd²  **15.** 251.2 in, 5,024 in²  **17.** 18.84 m  **19.** 153.86 ft²
**21.** 172.7 in²  **23.** 54.625 m²  **25.** 2 cm
**27.** 12.56; 50.24; 200.96  **29.** $4\pi n^2$, or about $12.56n^2$  **31.** 3.02  **33.** 15%  **35.** 80%
**37.** 270  **39.** 37.5%  **41.** $x + y = 12$, $y = 2x + 3$; Pauline: 3 miles; Ella: 9 miles

### Exercises, pages 560–561
**1.** c  **3.** d  **5.** autos: 234°; motorcycles: 7.2°; trucks: 64.8°; bicycles: 18°; vans: 36°

**7.** $400  **9.** $480 (30%)  **11.** North America: 6%; Latin America: 8%; Europe: 11%; USSR: 6%; Asia: 58%; Africa: 11%
**13.** Answers will vary. Each number should appear about 8 times.  **15.** Answers will vary.  **17.** $m = 10$  **19.** $x < -2$  **21.** 17
**23.** $r^2 - 9$  **25.** $(c - 5)(c + 4)$

## CHAPTER 15

### Exercises, pages 570–571
**1.** a  **3.** a  **5.** 27 cubic units  **7.** $36\frac{9}{16}$ yd³
**9.** 10 m³  **11.** 300 m³  **13.** 560 in³
**15.** Multiply 5 by 27.  **17.** 405 yd³
**19.** 43 truckloads  **21.** 1:8 or 8:1
**23.** Ratio of sides: 1:2 or 2:1; ratio of volumes = 1:8 or 8:1  **25.** $-(e - f)^2$
**27.** $y = 2x + 4$ or $2x - y = -4$  **29.** $y = \frac{1}{5}x - 2$ or $x - 5y = 10$  **31.** $e + 2$
**33.** $z + 1$  **35.** $r - 6$

### Exercises, pages 574–575
**1.** a  **3.** b  **5.** 94 m²  **7.** 700 cm²
**9.** 62.5 ft²  **11.** 31 in²  **13.** $10.80
**15.** 3:1  **17.** 27 cubic units  **19.** Since each edge of the cube is the same length, $s$, the area of each of the six square sides is $s^2$. Thus, the total surface area of the cube is $6s^2$.
**21.** 39.25 in  **23.** <  **25.** 78%  **27.** Since the ratio of their heights is 12:1 and their bodies are similar, visualize Gulliver's body and a Lilliputian's body as similar rectangular prisms in which the ratio of length and width is also 12:1. Thus, Gulliver's body has 12 × 12 × 12, or 1,728, times the volume of a Lilliputian's body.

### Exercises, pages 578–580
**1.** b  **3.** b  **5.** 339.12 ft³  **7.** 175.84 cm³
**9.** 35.3 in³  **11.** 3,532.5 gal  **13.** 49.0625 T
**15.** 226.08 in³  **17.** 2.60 in  **19.** They are the same. $\frac{3}{4}\pi(9)(8)$ represents 75% of the volume of the old can. The new volume can also be represented as $\pi(r^2)8$ where $r$ is the new radius. Since the two expressions are equal, solving the equation in Exercise 18 will give the same value for $r$ as in Exercise 17.
**21.** 13  **23.** 3  **25.** About 32 ft
**27.** Octagon  **29.** Pentagon

**Exercises, pages 584–585**
**1.** d  **3.** d  **5.** 200 cm³  **7.** 8 m³
**9.** 3.36 m³  **11.** Using the formula $V = \frac{1}{3}Bh$, to find the area of the base, $B$, multiply the known volume, $V$, by 3 and then divide this result by the known height, $h$.  **13.** $\frac{5\sqrt{3}}{2}$ in, or about 4.33 in  **15.** 162 yd³  **17.** 17
**19.** 18.84 ft, 28.26 ft²  **21.** Since the metal band is 1 inch from the surface of the earth, the radius of the circle formed by the metal band is 1 inch more than the radius of the earth. Let $r$ = the radius of the earth in miles. Then the circumference of the new circle will be $2\pi(r + 1)$, or $2\pi r + 2\pi$ inches. The difference between the circumference of the band and the circumference of the earth is $(2\pi r + 2\pi - 2\pi r)$ inches = $2\pi$ inches
**23.** Use guess and check. The numbers are 3, 5 and 7, 8.

**Exercises, pages 587–589**
**1.** d  **3.** b  **5.** 3,140 cm³  **7.** 141 mm³
**9.** 134 in³  **11.** $V = \frac{2}{3}Bh$ or $V = \frac{2}{3}\pi r^2 h$
**13.** 5, 19, 3.8  **15.** 13, 52, 4  **17.** Sample answer: Alike: They both have a base and a vertex. The formulas for the volumes are similar. Different: The base of a pyramid is a polygon. The sides are triangles. The base of a cone is a circle; its lateral (side) surface is a curved surface.  **19.** 673.56 ft³
**21.** About 105  **23.** 0.3  **25.** $x > -2.25$
**27.** $13c(8c + 3)$

**Exercises, pages 594–595**
**1.** c  **3.** b  **5.** 33 in³, 50 in²  **7.** 113 cm³, 113 cm²  **9.** 7,235 cm³, 1,809 cm²
**11.** Multiplied by 8.  **13.** Approximately 28,260,000 mi²  **15.** About 4.7 in³
**17.** $V$: 137,258 cubic units; $A$: 12,868 square units  **19.** $V$: 195 cubic units; $A$: 163 square units  **21.** $-14$  **23.** $-112$  **25.** $y \geq -2$
**27.** $w > 1$  **29.** $-t^2 + 2t$  **31.** $y^2 - 100$
**33.** $q^2 - 16q + 64$

# CREDITS

## PHOTO CREDITS

iii, Will & Demi McIntyre / Photo Researchers; iv(tl), Vic Huber / Westlight; iv(tr), Grafton Marshall Smith / Image Bank; iv(bl), James Newberry; v(tl), HRW photo by Richard Haynes; v(tr), Michel Tcherevkoff / Image Bank; v(bl), Vandystadt / Allsport; vi(tl), HRW photo by Eric Beggs; vi(tr), Superstock; vi(bl), HRW photo by Eric Beggs; vii(tl), James Newberry; vii(tr), HRW photo by Eric Beggs; vii(bl), HRW photo by Eric Beggs; vii(br), Antonio Rosario / The Image Bank; viii(tl), Park Street; viii(tr), HRW photo by Eric Beggs; viii(bl), HRW photo by Eric Beggs; viii(br), HRW photo by Eric Beggs; ix(tl), James Newberry; ix(bl), HRW photo by Eric Beggs; ix(br), Martha Cooper / Peter Arnold; x(tl), Park Street; x(tr), FourByFive / Superstock; x(bl), Mark Darlev / Esto; x(br), NASA; xi(tl), FourByFive / Superstock; xi(tr), HRW photo by Eric Beggs; xi(b), James Newberry; xii(l), Bill Helms / Superstock; xii(r), Superstock; 1, FourByFive / Superstock; 2, Bill Ross / Westlight; 5, HRW photo by Richard Haynes; 6, Superstock; 8, David Madison; 10(b), Vic Huber / Westlight; 10(t), Superstock; 13, Superstock; 18, Scott Cunningham / Allsport; 19, Allen Birnbach / Westlight; 20, Bettmann Archive; 25, FourByFive / Superstock; 26, James Newberry; 27, HRW photo by Yoav Levy; 28, Eric Beggs; 32, Shostal / Superstock; 38, Photo by Peter Auster, Courtesy of National Undersea Research Center, University of Connecticut; 39, Peter Aaron / Esto; 42, Michael Sullivan / TexaStock; 45, Superstock; 46, Shostal / Superstock; 50, Robert Harding Associates; 53, FourByFive / Superstock; 58, Park Street; 59, James Newberry; 62, Julie Habel / Westlight; 64, Franklin D. Roosevelt Naval Collection, FDR Library; 67, Boroff / TexaStock; 68, Laurence Parent; 74, Fotex / Drewa / PSI, Chicago; 75, Scott Van Osdol; 77, HRW photo by Russell Dian; 78, David Madison; 80, D.J. McCarthy / Panoramic Stock Images; 84, HRW photo by Eric Beggs; 85, David Madison; 88, Jim Brandenburg / Westlight; 89, Richard Hutchings / Photo Researchers; 90, HRW photo by Eric Beggs; 95, FourByFive / Superstock; 99(t), Jim Strawser / Grant Heilman; 99(b), Ralph Dearden / Ace / PSI, Chicago; 104, Scott Van Osdol; 105, Peter Aaron / Esto; 111, Jeff Person / Stock Boston; 112, James Newberry; 113(t), Richard Hutchings / Photo Researchers; 113(b), Shostal / Superstock; 114, HRW photo by Richard Haynes; 115, David Madison; 116, Gene Stein / Westlight; 117, Jeanne White / Photo Researchers; 118, FourByFive / Superstock; 121, FourByFive / Superstock; 123, David Madison; 124(r), HRW photo by Eric Beggs; 124(l), HRW photo by Eric Beggs; 125, Aldo Mastrocola / Lightwave; 129, Spencer Grant / Photo Researchers; 130, FourByFive / Superstock; 131, HRW photo by Eric Beggs; 135, Lawrence Migdale / Stock Boston; 136, James Newberry; 137, Park Street; 141, Carlos Vergara / Nawrocki Stock Photo; 142, James Newberry; 146, Barrera / TexaStock; 147, Index Stock International; 151, Vandystadt / Allsport; 154, Candee / Nawrocki Stock Photo; 157, Superstock; 158, Fred Bavendam / Peter Arnold; 163, David Frazier; 165, HRW photo by Eric Beggs; 167, M. Beebe / The Image Bank; 168, HRW photo by Dennis Fagan; 170, Wolfgang Kaehler; 171, Mitchell Bleier / Peter Arnold; 173, James Newberry; 174, HRW photo by Eric Beggs; 175, Gilbert Garcia / The Image Bank; 179, Shostal / Superstock; 181, Gerard Lacz / Peter Arnold; 182, FourByFive / Superstock; 186, Park Street; 187, Hans Pfletschinger / Peter Arnold; 192, HRW photo by Eric Beggs; 193, Hans Pfletschinger / Peter Arnold; 194, Park Street; 195, Park Street; 198, Robert Tringali / Sportschrome; 199, Michael Brohm / Nawrocki Stock Photo; 200, FourByFive / Superstock; 204, Park Street; 208, Faith A. Uridel / Nawrocki Stock Photo; 209, HRW photo by Eric Beggs; 211, Eric Beggs; 212, Park Street; 214(r), Superstock; 214(l), FourByFive / Superstock; 215, HRW photo by Eric Beggs; 216, David Kennedy / TexaStock; 219, HRW photo by Eric Beggs; 220, Peter Arnold; 221, HRW photo by Eric Beggs; 222, Park Street; 223, HRW photo by Eric Beggs; 224, Park Street; 226, HRW photo by Eric Beggs; 227, HRW photo by Eric Beggs; 231, HRW photo by Eric Beggs; 232, James Newberry; 223, NASA; 236, HRW photo by Ken Lax; 239, HRW photo by Ken Lax; 240, HRW photo by Eric Beggs; 241, HRW photo by Ken Lax; 242, HRW photo by Eric Beggs; 245(b), HRW photo by Eric Beggs; 245(t), HRW photo by Eric Beggs; 246, HRW photo by Eric Beggs; 247, James Newberry; 249, HRW photo by Eric Beggs; 251, HRW photo by Eric Beggs; 252, James Newberry; 255, HRW photo by Eric Beggs; 256, FourByFive / Superstock; 257, HRW photo by Eric Beggs; 259, HRW photo by Eric Beggs; 260, Park Street; 262, Park Street; 265, Michael Candee / Nawrocki Stock Photo; 266, HRW photo by Eric Beggs; 270(t), J. Ensthaler / Peter Arnold; 270(b), HRW photo by Eric Beggs; 271, Reed Kaestner / Nawrocki Stock Photo; 274, HRW photo by Eric Beggs; 275, James Newberry; 276, HRW photo by Eric Beggs; 278, HRW photo by Eric Beggs; 280, HRW photo by Eric Beggs; 282, Clyde Smith / Peter Arnold; 283, James Newberry; 284, HRW photo by Eric Beggs; 286, Martha Cooper / Peter Arnold; 288, James Newberry; 289, NASA; 291, HRW photo by Eric Beggs; 293, Ted Mahieu / The Stock Market; 294, HRW photo by Eric Beggs; 296, David Madison; 297, Mark Walker / Nawrocki Stock Photo; 298, Park Street; 300, HRW photo by Eric Beggs; 304, Ruben Guzman; 305, Focus on Sports; 308, Park Street; 309, Kent Wood / Peter Arnold; 312, HRW photo by Eric Beggs; 313, Craig Lovell / Viesti Associates; 315, HRW photo by Eric Beggs; 317, David Lissy / Nawrocki Stock Photo; 318, HRW photo by David Phillips; 319, Peter Arnold; 320, J.R. Ferro / Nawrocki Stock Photo; 321, Focus on Sports; 324, T.J. Florian / Nawrocki Stock Photo; 327, Superstock; 329, HRW photo by Eric Beggs; 330, HRW photo by Eric Beggs; 332, Park Street; 333, HRW photo by Eric Beggs; 335, George Garrison / Grant Heilman; 337, Park Street; 338, HRW photo by Eric Beggs; 340, James Newberry; 341(t), Henley & Savage / The Stock Market; 341(b), HRW photo by Eric Beggs; 343, T. J. Florian / Nawrocki Stock Photo; 345, Kennedy / TexaStock; 346(b), David Frazier; 346(t), HRW photo by Eric Beggs; 347, Superstock; 348, Park Street; 350, Superstock; 351, HRW photo by Eric Beggs; 352, Sal Maimone / Superstock; 354, FourByFive / Superstock; 355, HRW photo by Eric Beggs; 356, Superstock; 358, HRW photo by Larry Kolvoord; 362, HRW photo by Eric Beggs; 363, Tuskegee Institute; 364, Bob & Ira Spring; 365, HRW photo by Eric Beggs; 366, Focus on Sports; 368, HRW photo by Eric Beggs; 370, Superstock; 372, HRW photo by Eric Beggs; 376, HRW photo by Eric Beggs; 377, James Newberry; 381, HRW photo by Eric Beggs; 382, Superstock; 385, Beryl Goldberg; 388, Melanie Carr / Nawrocki Stock Photo; 389, Hildegard Adler; 390, James Newberry; 391, HRW photo by Eric Beggs; 392, Fred Bavendam / Peter Arnold; 393, Superstock; 396, James Newberry; 397, Superstock; 398, Comstock; 399, James Newberry; 402, Superstock; 403, James Newberry; 405, HRW photo by Eric Beggs; 407, Superstock; 408, Superstock; 409, Superstock; 413, HRW photo by Eric Beggs; 417, Park Street; 421, Superstock; 426, Park Street; 427, Phillip Harrington / Peter Arnold; 428, HRW photo by Louis Fernandez; 429, Candee Productions / Nawrocki Stock Photo; 434, HRW photo by Eric Beggs; 435, Superstock; 438, Superstock; 440, Superstock; 442, HRW photo / Superstock; 445, Clyde Smith / Peter Arnold; 446, James Newberry; 447, Superstock; 448(t), Ken Levine / Allsport; 448(b), FourByFive / Superstock; 449, HRW photo by Eric Beggs; 450, Sally Cassidy / Nawrocki Stock Photos; 456, Grant Heilman; 457, Superstock; 460, Superstock; 461, Superstock; 464, FourByFive / Superstock; 465, Aldo Mastrocola / Lightwave; 466, Grant Heilman; 473, Garry Gay / The Image Bank; 480, Paul Conklin; 485, Superstock; 494, Co Rentmeester / Image Bank; 502, Scott Halloran / Allsport; 503, HRW photo by Ruben Guzman; 519, FourByFive / Superstock; 521, HRW photo by Eric Beggs; 523, Superstock; 524(l), FourByFive / Superstock; 524(r), Tony Freeman / PhotoEdit; 525, HRW photo by Eric Beggs; 526, Superstock; 535, Don Klumpp / The Image Bank; 537, Mark Darlev / Esto; 539, Peter Aaron / Esto; 545, HRW photo by Eric Beggs; 551(tl), Park Street; 551(tr), HRW photo by James Gilmour; 551(c), Park Street; 551(bl), HRW photo by Eric Beggs; 551(br), Park Street; 555, Park Street; 557, Michel Tcherevkoff / The Image Bank; 558, James Newberry; 560, Park Street; 561, HRW photo by Eric Beggs; 563, FourByFive / Superstock; 565, Walter Chandoha; 566, HRW photo by Eric Beggs; 574, FourByFive / Superstock; 576, FourByFive / Superstock; 579, FourByFive / Superstock; 580, HRW photo by Eric Beggs; 581, Bettmann Archives; 582, Superstock; 584, HRW photo by Eric Beggs; 585, FourByFive / Superstock; 592, Superstock; 595, HRW photo by Eric Beggs; 596, HRW photo by Martha Cooper;

Special effects by Stokes Imaging Services.

## ILLUSTRATIONS

Kathi J. Branson
  Pgs. 66, 132, 149, 152, 230, 285, 422, 544

Chuck Joseph
  Pgs. 70, 161, 338, 349, 422, 461, 465, 471, 472, 477, 478, 479, 482, 483, 515, 523, 599

Michael Krone
  Pgs. 37, 100, 153, 189, 206, 216, 229, 233, 242, 281, 323, 344, 359, 552, 568, 589, 591, 598

John A. Wilson
  Pg. 427

For permission to reprint copyrighted material, grateful acknowledgment is made to the following sources:

*Universal Press Syndicate: Calvin & Hobbes* cartoons. Copyright © 1990 by Universal Press Syndicate. All rights Reserved.
  Pages 496 and 504

# INDEX

## A

Absolute value, 4
Acute angle, 474, 527
Addition Property of Equality, 91
Addition Property of Inequality, 382–383
Additive Identity, *See* Identity Property.
Additive Inverse, *See* Opposites, Property of.
Algebraic expression, 39
Angle(s), 526–529
   acute, 474, 527
   complementary, 528
   corresponding, 466, 533
   exterior, 533
   interior, 533
   obtuse, 527
   right, 527
   supplementary, 528
   vertical, 532
Area
   of a circle, 553–554, 557
   of a parallelogram, 69
   of a rectangle, 64–65
   of a square, 65
   of a trapezoid, 71
   of a triangle, 70
   *See also* Surface area.
Associative Property
   of Addition, 55, 194
   of Multiplication, 55
Average, 66, 317 *See also* Mean.
Axes, 410

## B

Bar graphs, 2, 282, 317, 348, 351, 355
Base, of a power, 150
Binomial(s), 486
   as a common factor, 513
   multiplication of, 496–498, 504–506, 508–510
Box and whisker plots, 370–371

## C

Calculator applications
   and activities, 6, 25, 48, 50, 66, 151, 221–222, 253, 286, 291–292, 294, 297–298, 319–320, 325, 450–451, 454, 476, 553
Circle, 551
   area of, 553–554
   circumference of, 551–553
Circle graphs, 305, 558–559
Circumference, 551–553, 557
Closed interval, 400
Common factors, 512–513
Commutative Property
   of Addition, 54
   of Multiplication, 54
Complementary angles, 528
Complex fractions, 188
Composite numbers, 139
Compound interest, 325–326, 329
Computer activities, 104, 109, 205, 396, 595
Computer Explorations
   adding integers, 19
   area formulas, 75
   circumference and area, 557
   equivalent fractions, 168
   graphing inequalities, 398
   interest, 329
   organizing and presenting data, 358
   prime factorization, 147
   similar figures, 473
Cone, 586
   volume of, 586
Congruence, 442
Conjecture, 202
Conjunction, 399–401
Consumer applications
   cost, 39–41
   discounts, 330–331
   fuel economy, 298
   inflation, 351
   interest, 324–326, 352
   prices, 105, 330–331
   profit, 68, 95
   sales commission, 320
   tips, 321
   wages, 59
Coordinate graphs, 410
Coordinates, 410
Corresponding angles, 466
Corresponding sides, 466
Counterexample, 51
Cross-product, 217
Cube, volume of, 151, 569
Cylinder, 576
   volume of, 576–578

## D

Decimals, 271
   estimating products and quotients of, 297–298
   estimating sums and differences of, 282–284
   in expanded form, 272
   and fractions, 271–272
   and percents, 309
   as real numbers, 377
   repeating, 277–278
   and scientific notation, 289–292
   terminating, 277
Denominator, 163
Density Property of Fractions, 209
Dependent events, 242–243
Digit, 271
Disjunction, 403–404
Distributive Property, 59–61, 105–107, 175–178, 504, 512
Divisibility, 127
   tests for, 131–134
Division Property of Equality, 96
Division Rule for Fractions, 190

## E

Element, of a set, 81
Empty set, 401

Equation(s), 80–81
  equivalent, 92
  graphing equations and
    inequalities, 430–432
  graphing linear
    equations, 414–415
  and inequalities, 79–82
  solving addition and
    subtraction, 91–93
  solving by inverse
    operations, 101–102
  solving by mental math and
    guess-and-check, 85–86
  solving multiplication and
    division, 95–97
  solving multi-step equations,
    105–107
  solving percent equations
    and proportions, 334–335
  solving with reciprocals,
    183–186
  standard form of a linear,
    418–420
  systéms of, 435–437
  writing, 85, 91–93, 95, 101,
    105, 114–116, 186,
    334–335, 435, 529
Estimating
  by analyzing sample data,
    364–365
  to check multiplication of
    mixed numbers, 177
  deciding on estimates, 317
  the percent of a number,
    313–315
  products and quotients of
    decimals, 297–299
  square roots with a
    calculator, 450–451
  sums and differences, 282–284
  sums of mixed numbers,
    200–201
  a tangent ratio from a graph,
    479
Even number, 132
Event(s), 236
  dependent, 242–243
  independent, 241–243
  mutually exclusive, 245
Exponent(s)
  in expanding numbers in
    base ten, 274

  in powers of ten, 287–288
  in prime factorization, 150
  in scientific notation,
    289–292
Expressions, 39–41
  algebraic, 39
  equivalent, 40, 48
  numerical, 39
  order of operations and, 41,
    46–48
  properties of, 54–56, 59–61
  translating word to
    algebraic, 40, 110–111, 270
Exterior angle, 533
Extremes
  in proportions, 217
  in statistical data, 370


───── F ─────

Factoring
  factor tree, 143
  polynomials, 512–513
  prime factorization, 143–144,
    150
  trinomials, 516–518
Factors, 126–128, 137–138
  See also Divisibility, Greatest
    common factor.
Factorial notation, 253
Fibonacci numbers, 235
FOIL method, 507–508
Formulas
  area of a circle, 554
  area of a parallelogram, 69
  area of a rectangle, 65
  area of a square, 65
  area of a trapezoid, 71
  area of a triangle, 70
  average, 66
  circumference, 553
  compound interest, 325
  perimeter of a rectangle, 65
  perimeter of a square, 65
  Pythagorean Theorem, 458
  simple interest, 324
  surface area of a cube, 573
  surface area of a rectangular
    prism, 569
  surface area of a sphere, 593

  volume of a cone, 586
  volume of a cube, 151, 573
  volume of a cylinder, 577
  volume of a pyramid, 583
  volume of a rectangular
    prism, 569
  volume of a sphere, 593
Fractions, 163
  addition and subtraction of,
    194–197, 200–203
  complex, 188
  decimals and, 271–272
  density property of, 209
  division of, 189–191
  equivalent, 164, 168
  improper, 164
  in lowest terms, 165
  and mixed numbers,
    175–178, 189–191
  multiplication of, 169–170,
    171–172
  as a percent, 305–307,
    313–315
  proper, 164
Frequency table, 355
Function, 414
Fundamental Counting
  Principle, 247, 254


───── G ─────

Graphs
  bar graphs, 2, 282, 317, 348,
    351, 355
  box and whisker plots,
    370–371
  circle graphs, 305, 558–559
  coordinate graphs, 410–411
  of equations and inequalities,
    430–432
  frequency tables, 355
  histograms, 355
  of inequalities, 378–379, 398,
    400–401, 403–404, 430–432
  line, 353, 373
  of linear equations, 414–415,
    418–420
  misleading, 348
  pictographs, 129, 158
  stem and leaf plots, 366–367

of systems of equations, 436–437
translations of, 440–442
tree diagrams, 247–248, 252
using data from graphs and tables, 351–352
Greatest common factor, 155–156, 512–513

--- H ---

Half plane, 432
Histogram, 355
Hypotenuse, 458, 474

--- I ---

Identity Property
  of Addition, 55
  of Multiplication, 55
Image point, 440
Improper fractions, 164
Independent events, 241–243
Inequalities, 82
  addition properties of, 382–383
  conjunctions of, 399–401
  disjunctions of, 403–404
  equations and, 79–82
  graphs of, 378–379, 398, 430–432
  multiplication properties of, 386–387
  solving, 393–395
  writing, 393–395
Inequality sign, 8
Infinite set, 128
Integers, 2–3
  addition of, 11–16, 19
  comparing and ordering, 7–8
  division of, 32–34
  multiplication of, 28–30
  subtraction of, 21–24
Intercepts, 48
Interest
  compound, 325–326, 329
  simple, 324–325
Interior angle, 533, 544
Intersection, 400
Inverse operations, 101–102
Irrational numbers, 264, 278, 377, 450–456

--- L ---

Least common denominator, 201
Least common multiple, 154–155
Legs, of a right triangle, 458, 474
Like terms, 60–61, 488
Line(s)
  graphs of, 414–415
  parallel, 532
  perpendicular, 532
  slope of, 422–425
  *See also* Linear equations.
Linear equations, 414–415
  slope of, 424
  slope–intercept form of, 430
  standard form of, 418–420
Line graphs, 353, 373
Lowest terms, 165

--- M ---

Mathematical Footnotes, 20, 50, 53, 88, 108, 142, 153, 170, 182, 233, 281, 294, 323, 363, 389, 392, 427, 439, 456, 465, 480, 503, 545, 556, 581, 589
Math Team Problems, 10, 31, 43, 63, 74, 99, 104, 109, 136, 158, 193, 225, 246, 266, 288, 354, 385, 397, 417, 421, 434, 472, 480, 500, 515, 544, 561, 575, 585
Mean, 359–360
  "assumed", 362
Means, 217
Median, 360
Member, of a set, 81
Mental computation, 18, 54–56, 62, 85–86, 95, 127, 175–176, 195, 218, 228, 267, 295–296, 320–321, 504, 507, 513
Metric system, 295–296
Mixed numbers, 175
  addition of, 200–203
  and decimals, 272
  division of, 189–191
  multiplication of, 175–178
  subtraction of, 206–207
Mode, 359–360
Monomials, 486

Multiple, 126–128
  *See also* Least Common Multiple.
Multiplication Property
  of Equality, 95, 185
Multiplication Property
  of Inequality, 386–387
Multiplication Property
  of Reciprocals, 183
Multiplication Rule for Fractions, 171
Multiplicative Identity, *See* Identity Property.
Multiplicative Inverse *See* Reciprocal.
Mutually exclusive events, 245

--- N ---

Numerator, 163
Numerical coefficient, 60

--- O ---

Obtuse angle, 527
Odd numbers, 132
Odds, 238
Open interval, 400
Opposites, 4
  and multiplication, 107
  Property of, 15
Ordering of integers, 7–8
Order of operations, 41, 46–48
Ordered pairs, 137, 234, 410
Origin, 410
Outcome, 236

--- P ---

Parallel lines, 532–534
Parallelogram, 69, 546
  area of, 69
Patterns, 28–30, 51–52, 60, 109, 131, 137–138, 145–146, 220, 254–255, 257, 287–288, 295, 310, 390, 439, 508–509, 537
Pentagon, 546
Percent, 305–307
  and circle graphs, 558–559
  decimals and, 309–311
  decrease, 338–339
  discount, 330–331
  estimating, 313–315

of systems of equations, 436–437
translations of, 440–442
tree diagrams, 247–248, 252
using data from graphs and tables, 351–352
Greatest common factor, 155–156, 512–513

### H

Half plane, 432
Histogram, 355
Hypotenuse, 458, 474

### I

Identity Property
 of Addition, 55
 of Multiplication, 55
Image point, 440
Improper fractions, 164
Independent events, 241–243
Inequalities, 82
 addition properties of, 382–383
 conjunctions of, 399–401
 disjunctions of, 403–404
 equations and, 79–82
 graphs of, 378–379, 398, 430–432
 multiplication properties of, 386–387
 solving, 393–395
 writing, 393–395
Inequality sign, 8
Infinite set, 128
Integers, 2–3
 addition of, 11–16, 19
 comparing and ordering, 7–8
 division of, 32–34
 multiplication of, 28–30
 subtraction of, 21–24
Intercepts, 48
Interest
 compound, 325–326, 329
 simple, 324–325
Interior angle, 533, 544
Intersection, 400
Inverse operations, 101–102
Irrational numbers, 264, 278, 377, 450–456

### L

Least common denominator, 201
Least common multiple, 154–155
Legs, of a right triangle, 458, 474
Like terms, 60–61, 488
Line(s)
 graphs of, 414–415
 parallel, 532
 perpendicular, 532
 slope of, 422–425
 See also Linear equations.
Linear equations, 414–415
 slope of, 424
 slope–intercept form of, 430
 standard form of, 418–420
Line graphs, 353, 373
Lowest terms, 165

### M

Mathematical Footnotes, 20, 50, 53, 88, 108, 142, 153, 170, 182, 233, 281, 294, 323, 363, 389, 392, 427, 439, 456, 465, 480, 503, 545, 556, 581, 589
Math Team Problems, 10, 31, 43, 63, 74, 99, 104, 109, 136, 158, 193, 225, 246, 266, 288, 354, 385, 397, 417, 421, 434, 472, 480, 500, 515, 544, 561, 575, 585
Mean, 359–360
 "assumed", 362
Means, 217
Median, 360
Member, of a set, 81
Mental computation, 18, 54–56, 62, 85–86, 95, 127, 175–176, 195, 218, 228, 267, 295–296, 320–321, 504, 507, 513
Metric system, 295–296
Mixed numbers, 175
 addition of, 200–203
 and decimals, 272
 division of, 189–191
 multiplication of, 175–178
 subtraction of, 206–207
Mode, 359–360
Monomials, 486

Multiple, 126–128
 See also Least Common Multiple.
Multiplication Property
 of Equality, 95, 185
Multiplication Property
 of Inequality, 386–387
Multiplication Property
 of Reciprocals, 183
Multiplication Rule for Fractions, 171
Multiplicative Identity, See Identity Property.
Multiplicative Inverse See Reciprocal.
Mutually exclusive events, 245

### N

Numerator, 163
Numerical coefficient, 60

### O

Obtuse angle, 527
Odd numbers, 132
Odds, 238
Open interval, 400
Opposites, 4
 and multiplication, 107
 Property of, 15
Ordering of integers, 7–8
Order of operations, 41, 46–48
Ordered pairs, 137, 234, 410
Origin, 410
Outcome, 236

### P

Parallel lines, 532–534
Parallelogram, 69, 546
 area of, 69
Patterns, 28–30, 51–52, 60, 109, 131, 137–138, 145–146, 220, 254–255, 257, 287–288, 295, 310, 390, 439, 508–509, 537
Pentagon, 546
Percent, 305–307
 and circle graphs, 558–559
 decimals and, 309–311
 decrease, 338–339
 discount, 330–331
 estimating, 313–315

Equation(s), 80–81
 equivalent, 92
 graphing equations and
  inequalities, 430–432
 graphing linear
  equations, 414–415
 and inequalities, 79–82
 solving addition and
  subtraction, 91–93
 solving by inverse
  operations, 101–102
 solving by mental math and
  guess-and-check, 85–86
 solving multiplication and
  division, 95–97
 solving multi-step equations,
  105–107
 solving percent equations
  and proportions, 334–335
 solving with reciprocals,
  183–186
 standard form of a linear,
  418–420
 systéms of, 435–437
 writing, 85, 91–93, 95, 101,
  105, 114–116, 186,
  334–335, 435, 529
Estimating
 by analyzing sample data,
  364–365
 to check multiplication of
  mixed numbers, 177
 deciding on estimates, 317
 the percent of a number,
  313–315
 products and quotients of
  decimals, 297–299
 square roots with a
  calculator, 450–451
 sums and differences, 282–284
 sums of mixed numbers,
  200–201
 a tangent ratio from a graph,
  479
Even number, 132
Event(s), 236
 dependent, 242–243
 independent, 241–243
 mutually exclusive, 245
Exponent(s)
 in expanding numbers in
  base ten, 274
 in powers of ten, 287–288
 in prime factorization, 150
 in scientific notation,
  289–292
Expressions, 39–41
 algebraic, 39
 equivalent, 40, 48
 numerical, 39
 order of operations and, 41,
  46–48
 properties of, 54–56, 59–61
 translating word to
  algebraic, 40, 110–111, 270
Exterior angle, 533
Extremes
 in proportions, 217
 in statistical data, 370

## F

Factoring
 factor tree, 143
 polynomials, 512–513
 prime factorization, 143–144,
  150
 trinomials, 516–518
Factors, 126–128, 137–138
 See also Divisibility, Greatest
  common factor.
Factorial notation, 253
Fibonacci numbers, 235
FOIL method, 507–508
Formulas
 area of a circle, 554
 area of a parallelogram, 69
 area of a rectangle, 65
 area of a square, 65
 area of a trapezoid, 71
 area of a triangle, 70
 average, 66
 circumference, 553
 compound interest, 325
 perimeter of a rectangle, 65
 perimeter of a square, 65
 Pythagorean Theorem, 458
 simple interest, 324
 surface area of a cube, 573
 surface area of a rectangular
  prism, 569
 surface area of a sphere, 593
 volume of a cone, 586
 volume of a cube, 151, 573
 volume of a cylinder, 577
 volume of a pyramid, 583
 volume of a rectangular
  prism, 569
 volume of a sphere, 593
Fractions, 163
 addition and subtraction of,
  194–197, 200–203
 complex, 188
 decimals and, 271–272
 density property of, 209
 division of, 189–191
 equivalent, 164, 168
 improper, 164
 in lowest terms, 165
 and mixed numbers,
  175–178, 189–191
 multiplication of, 169–170,
  171–172
 as a percent, 305–307,
  313–315
 proper, 164
Frequency table, 355
Function, 414
Fundamental Counting
 Principle, 247, 254

## G

Graphs
 bar graphs, 2, 282, 317, 348,
  351, 355
 box and whisker plots,
  370–371
 circle graphs, 305, 558–559
 coordinate graphs, 410–411
 of equations and inequalities,
  430–432
 frequency tables, 355
 histograms, 355
 of inequalities, 378–379, 398,
  400–401, 403–404, 430–432
 line, 353, 373
 of linear equations, 414–415,
  418–420
 misleading, 348
 pictographs, 129, 158
 stem and leaf plots, 366–367

finding the percent of a number, 319–321
increase, 338–339
interest, 324–326
solving percent equations and proportions, 334–335
Perfect square, 452
Perfect square trinomial, 509
Perimeter, 64–65
Permutations, 252–254
Perpendicular lines, 532
Pi, 552, 556, 589
Pictographs, 129, 158
Place value, 271
Polygon(s), 546–548
Polynomials, 486
  addition of, 485–488
  common factors, 512–513
  factoring, 516–518
  multiplication of, 494–495
  special products, 508–510
  subtraction of, 491–492
Power(s), 150
  and scientific notation, 289–292
  of ten, 287–288
Prime factorization, 143–144, 147, 150
Prime numbers, 139
  reversal primes, 141
  twin primes, 141
Principal, 324
Prism, 568
  rectangular, 568
Probability, 236–237
  of dependent events, 242–243
  of independent events, 241–243
  of mutually exclusive events, 245
  recording chances, 234–235
Problem solving applications
  braking distance, 353
  falling objects, 151
  measurement and ratio, 226–228
  percent equations and proportions, 334–335
  percent increase and decrease, 338–339

proportions, 221–223
Pythagorean Theorem, 460
similar triangles, 469
supplementary and complementary angles, 529
sum of the measures of the angles of a polygon, 547
tangent ratio, 477
two variables, 435–437
See also Consumer applications.
Problem Solving Exploration
  factors and patterns, 137–138
  making generalizations, 51
  the metric system, 295–296
  modeling multiplication, 169–170
  powers of ten, 287–288
  recording chances, 234–235
  square roots, 450–451
  surface area and volume, 566–567
  using a multiplication model, 494–495
  using an addition model, 11
  using a subtraction model, 21
Problem Solving Strategies
  analyzing sample data, 364–365
  choosing a method of computation, 26–27
  choosing strategies, 89
  collecting data, 231–232
  deciding on estimates, 317
  developing a plan, 114–116
  drawing a diagram, 275–276
  extending patterns, 537
  formulating questions, 463–464
  making a model, 590
  organizing information, 180
  revising the solution, 424
  solving a simpler problem, 148
  too much/too little data, 501
  using a step-by-step process, 44–45
  using generalizations, 390
  writing an equation, 114–116

Proper fraction, 164
Properties
  Angle-Angle-Angle Property, 468
  Associative Property of Addition, 55
  Associative Property of Multiplication, 55
  Commutative Property of Addition, 54
  Commutative Property of Multiplication, 54
  Density Property of Fractions, 209
  Distributive Property of Multiplication Over Addition, 59–61
  Identity Property of Addition, 55
  Identity Property of Multiplication, 55
  Multiplication Property of Reciprocals, 183
  of Opposites, 15
  Triangle-Sum Property, 541
Proportion(s), 217
  cross-products, 217
  for finding a percent ratio, 305–307
  in problem solving, 221–223
  Property of, 217
  in sampling, 365
  in similar triangles, 468
  solving percent equations and, 334–335
Protractor, 527
Pyramid, 582–583
  volume of, 583
Pythagorean Theorem, 457–460
Pythagorean triples, 462

 **Q**

Quadrant, 411
Quadrilateral(s), 546–548
Quartile(s), 370

 **R**

Radical(s), 452–454
Random survey, 251
Range, 367
Rate, 222

discount, 330
interest, 324
unit, 222
Ratio(s), 216
   and measurement, 226–228
   as percents, 305–307
   tangent, 475
Rational expressions, 262–264
Rationalizing the
   denominator, 454
Rational numbers, 263
   addition and subtraction of, 268
   comparing, 267–268
   as decimals, 278
   multiplication and division of, 269
   *See also* Fractions.
Ray, 378
Real numbers, 377–379
Reciprocal(s), 183
   in dividing fractions and mixed numbers, 190
   Multiplication Property of, 183
   solving equations with, 183–186
Rectangle(s), 64, 546
   area of, 64–65
   perimeter of, 64–65
Rectangular coordinate system, 410–411
Rectangular prism, 572
   surface area of, 572–573
   volume of, 568–569
Relatively prime numbers, 158, 165
Repeating decimals, 277–278
Replacement set, 81
Rhombus, 546
Right angle, 527
Right triangles, 457–460, 540

——— S ———

Sampling, 364–365
Sample space, 236
Scientific notation, 289–292
Sequence, 109

Set, infinite, 128
Set notation, 81
Similar figures, 473
Similar triangles, 466–469
Simple interest, 324–325
Slope, of a linear
   equation, 422–425
Slope-intercept form, 430
Solution set
   of an equation, 81
   of an inequality, 82, 379
   of a system of equations, 436
Sphere, 592
   surface area of, 592–594
   volume of, 592–593
Square, 546
   area of, 65
   perimeter of, 65
Square root(s), 452–454
Statistics
   analyzing sample data, 364–365
   collecting data, 231–232
   graphical presentation of data, 348, 351–352, 355–356, 366–367, 369, 370–371
   mean, 359
   median, 359
   mode, 359
   quartile, 370
   random survey, 251
   range, 367
Stem and leaf plots, 366–367, 369
Supplementary angles, 528
Surface area
   of a cube, 573
   of a rectangular prism, 572–573
   of a sphere, 592–594
Systems of equations, 435–437

——— T ———

Tangent ratio, 474–477
Technology *See* Calculator applications, Computer activities, Computer Explorations
Terminating decimal, 277

Translations, 440
Transversal, 533
Trapezoid, 546
   area of, 71
Tree diagram, 247–248, 252
Triangle(s), 539
   area of, 70
   classification of, 540
   right, 457–460, 540
   similar, 466–469
Triangle–Sum Property, 541
Trinomials
   factoring, 516–518
   perfect square, 509

——— U ———

Union, 403
Unit rate, 222

——— V ———

Variable, 39
Vertex
   of an angle, 526
   of a cone, 586
   of a polygon, 546–547
   of a pyramid, 582
Vertical angle, 532
Volume
   of a cone, 586
   of a cube, 151
   of a cylinder, 576–578
   of a pyramid, 582–583
   of a rectangular prism, 568–569
   of a sphere, 592–593

——— W ———

Whole numbers, 3

——— X ———

x-axis, 410
x-coordinate, 410
x-intercept, 418

——— Y ———

y-axis, 410
y-coordinate, 410
y-intercept, 418